Viral Drug Delivery Systems

Ranjita Shegokar • Yashwant Pathak
Editors

Viral Drug Delivery Systems

Advances in Treatment of Infectious Diseases

Springer

Editors
Ranjita Shegokar
CapnoPharm GmbH
Tübingen, Germany

Yashwant Pathak
Taneja College of Pharmacy
University of South Florida
Tampa, FL, USA

ISBN 978-3-031-20539-2 ISBN 978-3-031-20537-8 (eBook)
https://doi.org/10.1007/978-3-031-20537-8

This Springer imprint is published by the registered company Springer Nature Switzerland AG
The registered company address is: Gewerbestrasse 11, 6330 Cham, Switzerland

Preface

The Washington Post of 2016 stirred an intense discussion between the public, scientific communities, and health authorities. How many diseases are precisely known to humankind? At the moment, scientist estimates the presence of more than 10,000 human diseases and only fewer available treatments that too for major diseases[1].

In 2022, the scenario is not far different considering the deliberate speed of academic/industry research, economic up-downs, tougher regulatory policies, complex clinical trial setups, the impact of the Covid-19 pandemic in slowing processes, businesses, and changing world political dynamics and policies. The same question on "availability of effective treatment" is valid now and maybe even for the next 2–3 decades.

Diseases can be genetic or caused by environmental factors (mainly known as infectious diseases). Human infectious diseases are typically classified according to the source of infection as anthroponoses (human–human transmission), zoonoses (animal–human transmission), and sapronoses (abiotic decaying substrate—human). These infectious diseases contribute to the enormous financial burden on the country's economy. By 2001, around 1415 species of organisms had been recorded known to be pathogenic to humans, mainly comprised of bacteria, viruses/prions, fungi, protozoa, and helminths.

This book is a trivial attempt to compile all possible and available information on etiology, pathology, current therapy options available for a wide spectrum of diseases, the role of drug delivery sciences, advances in new techniques, diagnostic tools, and new drug research of various infectious diseases.

Total *four volumes* are compiled to accommodate vast available information.

Volume 1—Malarial drug delivery systems (MDDS)
Volume 2—Tubercular drug delivery systems (TDDS)
Volume 3—Viral drug delivery systems (VDDS)
Volume 4—Infectious disease drug delivery systems (IDDDS)

[1]Are there really 10,000 diseases and just 500 'cures'? – *The Washington Post.* https://www.orpha.net/

Volume 1: MDDS

Malaria is a disease caused by the parasite Plasmodium. The parasite spread to humans through the bites of infected mosquitoes causing high fever, nausea, vomiting, diarrhea, body pain, rapid heart rate, and shaking chills. Each year millions of people get infected by malaria, and many hundred-thousand people die. Some of the most significant risk areas include Sub-Saharan Africa, South and Southeast Asia, Pacific Islands, Central America, and Northern South America. The treatment of malaria mainly comprises the most common antimalarial drugs like chloroquine, primaquine, etc. In the case of drug resistance, artemisinin-based combination therapies (ACTs) are preferred. ACT is an amalgamation of two or more drugs that work against the malaria parasite using a different mechanism of action.

Volume 2: TDDS

Tuberculosis (TB) is a potentially severe infectious disease that affects the lungs and, in some cases, the kidney, spine, and brain. *Mycobacterium* causes tuberculosis via air route. As a result, two TB-related scenarios are possible: latent TB infection (LTBI) and TB disease. If not treated properly, TB disease can be fatal. TB bacteria usually grow in the lungs (pulmonary TB). The typical test used to diagnose TB is the Mantoux tuberculin skin test (TST). The medications used to treat latent TB infection include Isoniazid, Rifapentine, and Rifampin. Classically, the patient may undergo several treatment regimens (1st/2nd /3rd line) recommended as per disease condition and health policy of that specific country. TB treatment can take 4, 6, or 9 months depending on the regimen.

Volume 3: VDDS

Viruses are very tiny infectious germs, which cause infectious diseases such as the common cold, flu, and wart to severe illnesses such as HIV/AIDS, Ebola, and Covid-19 (which caused the recent pandemic where millions of people lost life). They invade living, normal cells and use those cells as host. Depending upon the type of virus, the target body cells are different. Virus infections and diseases are categorized under ten other groups, i.e., contagious, respiratory, gastrointestinal, exanthematous, hepatic, transmission, cutaneous, hemorrhagic, neurologic, and rest of the viruses not in these categories. All viruses have a protein coat and a core of genetic material, either RNA or DNA; unlike bacteria, viruses can't survive without a host. The diagnosis of viral diseases/infections can be performed by viral culture, serological tests, virus antigen detection, and viral nucleic acid or antibody detection. The treatment of viral diseases/infections depends on the type of viral infection. Antibiotics do not work for viral infections. FDA has already approved several antiviral medicines for the treatment of certain illnesses.

Volume 4: IDDDS

Each infectious disease has its specific signs and symptoms. Diagnosis of infectious diseases needs lab testing. Samples of body fluids, e.g., blood, urine, saliva, etc., can reveal evidence of the particular microbe that is causing the illness. While imaging, scans using X-rays, computerized tomography, and magnetic resonance imaging can help pinpoint disease states. Often, local tissue biopsies provide helpful information on the state of infection and adverse observations of disease (if any). This volume is focused on diagnosis, detection, disease models, the link between two or multiple infectious diseases, and vaccine development for the treatment of infectious diseases

This book series compiles all the new treatment avenues that have been explored to treat malaria, tuberculosis, viral infections, and other infectious diseases like Ebola and hepatitis. This series covers various aspects of drug delivery advances for disease targeting, new drug molecules, analysis of currently ongoing clinical trials, vaccine development, and availability of disease models to evaluate drug performance. Dedicated chapters are included on herbal treatment opportunities for each disease. In addition, readers can refer to information on global disease health scenarios, cellular pathophysiology, and drug resistance, full coverage on polymeric nanoparticles, solid lipid nanoparticles, dendrimers, liposome, and micro/nanoemulsions as drug delivery carriers.

Experts from all over the world have shared their knowledge to generate this one-stop resource. This book series is destined to fill the knowledge gap through information sharing and organized research compilation between the diverse area of pharma, medicine, clinical, chemist, and academics to fulfill following specific objectives:

- To discuss opportunities and challenges in the treatment of infectious diseases
- To enlist current efforts by researchers and experts
- To facilitate the insight and knowledge sharing
- To highlight innovative, cutting-edge micro and nanotechnology research
- To establish collaborations between academic scientists, industrial, and clinical researchers

In summary, we are sure this book series will provide you great insights into drug delivery sciences (conventional, micro-nanomedicines, upcoming drug delivery trends) along with updates on clinical and chemical drug research for the treatment of infectious diseases.

Tübingen, Germany — Ranjita Shegokar
Tampa, FL, USA — Yashwant Pathak

Contents

Global Health and Viral Diseases: Past, Present, and Future

Sarika Chauhan, Surya Sankineni, Ranjita Shegokar, and Yashwant Pathak

Abstract Viral diseases are one of the most common and rapidly contagious ways to impair one's health. Hence, finding new ways to combat the spread and treat the viruses can be beneficial to global health overall. Some of these common and dangerous viral diseases include influenza, HIV/AIDS, HPV, and measles, mumps, and rubella (MMR). Prior to modern medicine, thousands faced death at the hand of these diseases; however, today, we can develop interventions and technology that could save most from their devastating effects. Although most of these diseases are highly preventable either via vaccines or preemptive safe practices, their access globally is highly limited by a nation's finances, resources, and initiatives. In order to accomplish the task of eliminating and reducing the spread of viral diseases, the HPV and MMR vaccines must be widely distributed to target populations, perhaps even made a requirement for school admission. Supplementary immunization activities (SAIs) are also recommended to maintain and enforce immunization strategies. For viruses that result in sexually transmitted diseases (STIs), educational resources as well as screening opportunities are needed to aid in the accurate recording of transmission and to execute preventative measures. Viral diseases prove fatal for many all around the world. By initiating plans for intervention and treatment of viral diseases that affect those around the world, we can significantly improve global health access and quality.

Keywords Viral disease · Influenza · HPV · HIV · Mumps · Rubella · Measles · Global

---

S. Chauhan · S. Sankineni · Y. Pathak (✉)
Nova Southeastern University, University of South Florida, Tampa, FL, USA

R. Shegokar
CapnoPharm GmbH, Tübingen, Germany

R. Shegokar, Y. Pathak (eds.), *Viral Drug Delivery Systems*,
https://doi.org/10.1007/978-3-031-20537-8_1

1 Introduction

Understanding viral diseases, their risk factors, and available treatment options is essential to improving health outcomes at the global level. These infectious agents pose such a threat to human health because they are inert and as such can use the biology of a host cell to replicate nearly indefinitely [1]. Through initiatives like the World Health Organization's (WHO) Global Influenza Programme, Cervical Cancer Elimination Initiative, and UNAIDS, public health officials around the world have come together to address the spread of some of the most common—and most fatal—viral diseases.

2 Types of Viral Diseases

The first of these viral diseases is the influenza virus, which has the ability to infest many vertebrates. In humans, the virus invades the respiratory epithelium. There are four variants of the virus that differ depending on their source and effects: *Influenza A virus* (IAV), *Influenza B virus* (IBV), *Influenza C virus* (ICV), and *Influenza D virus* (IDV). The first three types of the influenza virus are extremely transmissible human respiratory pathogens, which eventually allow the reproduction of the viral genome within the host's nucleus. One of the most severe viruses is HIV, or human immunodeficiency virus, which results in acquired immunodeficiency syndrome (AIDS). It is a sexually transmitted infection (STI), meaning it is transferred perinatally and via bodily fluids. The HIV is a *Lentivirus* of the retrovirus family, meaning that the virus is characterized by long incubation periods and leads to long-term expression. Since the syndrome decreases an individual's immunity drastically, it may render them unable or not as fit to fight other illnesses [2].

Another viral disease that is categorized as a STI is the human papillomavirus, HPV. Like HIV, HPV can be spread via sexual contact even though there may be no symptoms present. HPV is most notable for causing cervical cancer as well as several others [3]. Measles, mumps, and rubella (MMR) can be both prevented via the MMR vaccine; it is often given within the first few years of birth to prevent the spread of the viral diseases. Measles is common throughout the world but highly contained in some countries like the United States with the use of the vaccine. However, global control of the diseases is insufficient as vaccine coverage needs to be more widespread [4]. Rubella and mumps were also a very common viral disease before the introduction of the vaccine. One thing to note for rubella is prevention of contracting congenital rubella during pregnancy as it can result in severe defects. Mumps can seldom but notably result in sterility in older males due to the inflammation it causes on the testes; it also results in the inflammation of the ovaries, pancreas, and spinal cord [5].

3 Viral Diseases Across the World

3.1 *Influenza*

H1N1, more commonly known as influenza, is classified as an acute respiratory illness and has again and again proven to be in a class of its own in terms of genetic adaptability. Influenza has been a world health concern since the first identifiable description of an outbreak recorded between 1173 and 1174 in Europe [6]. The official discovery of swine influenza virus occurred in 1933 by Dr. Richard Shope of the Rockefeller Institute in Iowa who had been investigating hog cholera [7]. Following his discovery scientists, Wilson Smith, Christopher Andrews, and Patrick Laidlaw followed Shope's isolation methods to isolate this elusive virus in humans. Three of the most impactful influenza pandemics presented themselves in the twentieth century alone, beginning with the Spanish flu (H1N1) in 1918, followed by the Asian flu (H1N2) in 1957 and Hong Kong flu (H3N2) in 1968. The Spanish flu took approximately 675,000 lives in the United States alone and approximately 50 million worldwide. Influenza has yet to halt genetic evolution, exemplified with the emergence of the new extremely transmissible strain of avian flu (H7N9) in 2019 [8]. Nearly every year, new genetic variations of influenza continue to present themselves. Influenza B is generally a less virulent form of virus, while influenza C is even less harmful, resulting in mild illness at its worst [9].

3.1.1 Prevalence Worldwide/Rate of Infection

The 2020–2021 influenza season resulted in approximately 1675 tested cases of influenza recorded in the United States by the Centers for Disease Control and Prevention (CDC) [10]. This rate of prevalence is a historical record low for influenza in the United States and can be largely attributed to the general increase in personal health safety due to the COVID-19 pandemic.

3.1.2 Treatments over Time

Since the isolation of the influenza virus, scientists have been diligently working to uncover various methods of treatment. Historically, herbal medications have recurrently been used to treat infections and flu-like ailments [11]. Maoto, an herbal mixture synthesized from four different plants is used in Japan as a regular method of treatment for standard influenza infection [12]. Alike Maoto, numerous herbal remedies exist for the treatment of influenza; however, one of the more common modern treatments can be found in the form of the neuraminidase inhibitor oseltamivir. Since it received Food and Drug Administration approval in 1999, oseltamivir has been surrounded in controversy with numerous studies pointing out that the risks weigh heavier than the paltry benefits [13]. Even still, oseltamivir was largely at the forefront of influenza treatment between 1999 and the early twenty-first century. Before oseltamivir, the first neuraminidase inhibitor that was approved for use in the treatment of influenza had been zanamivir [14]. Numerous studies were conducted of the efficacy of zanamivir. An overarching review directed by Professor Carl James Heneghan in 2014 covered a total of 54 clinical trials and concluded that zanamivir is an effective treatment method for adults. However, the results of the review displayed insignificant results in the treatment of influenza for children. The herbal treatment using Maoto was compared to the effectiveness of oseltamivir or zanamivir through random trials in 2012 conducted by Shigeki Nabeshima [12]. These trials resulted in a surprising conclusion showing nearly equal results between both herbal and clinical methods.

3.1.3 Vaccine Landscape

Not only did 2021 record a low in infection rates but also a record high in immunizations. In 2021 alone, the CDC approximately distributed 198 million doses of the influenza vaccine, which was a record high, and as of February 25, 2022, another total of 174 million doses of the flu vaccine have been distributed by the CDC [10]. Currently, the World Health Organization recommends the use of quadrivalent vaccines. The recommendation includes egg-based or cell culture/recombinant-based vaccinations containing H1N1-like viruses, H3N2-like viruses, a Victoria lineage-type virus, and a Yamagata lineage-type virus [15].

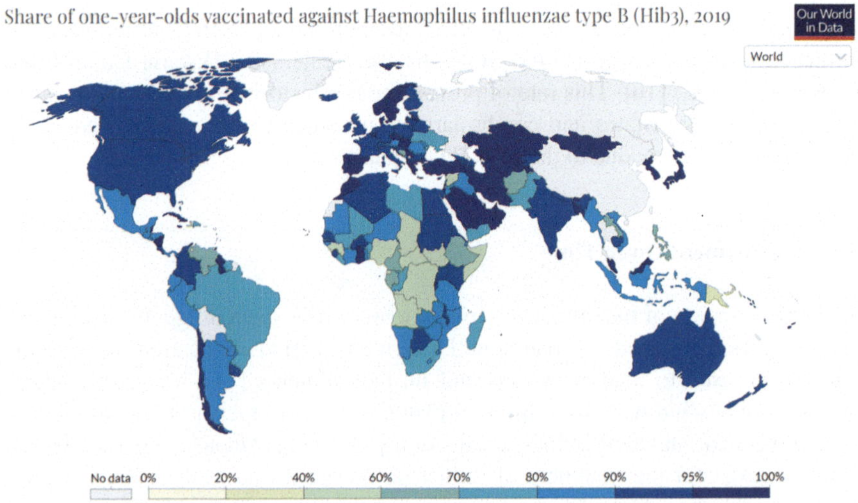

Source: World Health Organization (WHO); UNICEF. OurWorldInData.org/vaccination • CC BY. Note: Haemophilus influenza type B is a bacteria responsible for severe pneumonia, meningitis and other invasive diseases almost exclusively in children younger than 5 years

3.2 *HIV/AIDS*

HIV is a species of immune-attacking retrovirus characterized by long incubation periods in humans and mammals [16]. The virus primarily occurs as two types—HIV-1 and HIV-2—which differ in rates of progression and degrees of transmissibility [16]. Left untreated, however, both subtypes can cause AIDS, a condition marked specifically by a CD4 T lymphocyte count of less than 200 cells/mm^3 [17]. Because CD4 cells are the primary mechanisms behind cell-mediated immunity, a patient whose HIV infection has progressed to the AIDS stage becomes dangerously susceptible to fatal opportunistic infections, certain cancers, and even common diseases that an otherwise healthy individual can fight [18]. Left untreated, a person with HIV usually progresses through three stages: stage 1, acute HIV infection (mild influenza-like symptoms and high rates of transmission); stage 2, chronic HIV infection (asymptomatic for at least a decade until viral load increases and CD4 count decreases beyond repair); and stage 3, AIDS diagnosis [19]. It is critical to diagnose HIV in the early stages, before the viral load further impedes the immune system's ability to produce CD4 cells.

Early diagnosis and thereby efficient treatment of HIV is initially done with the availability of both at-home self-tests and rapid tests that produce same-day results in the lab. Following these tests, confirmatory testing at a healthcare center is necessary. These confirmatory tests measure levels of HIV antibodies in a patient's blood, which typically develop "within 28 days of infection". Due to the severity of this viral disease and its available treatments, the World Health Organization recommends people be retested prior to enrolling in treatment [18].

HIV is transmitted through bodily fluids such as blood, semen, and rectal or vaginal fluids. As such, it is most often contracted through unprotected anal or vaginal intercourse and sharing drug injection equipment such as syringes and needles [20]. The virus can also be passed perinatally from mother to baby during birth or breastfeeding [20]. Rarer forms of transmission include biting, oral sex, and transmission through pre-chewed food; usually, these modes involve a cut or abrasion in the mouth through which the virus can enter the bloodstream. Risk factors that increase an individual's likelihood of contracting HIV include having unprotected sex, receiving or performing unsafe injections, sharing contaminated needles, and accidental stick injuries in the healthcare setting [21].

3.2.1 Prevalence Worldwide/Rate of Infection

HIV is a major global health issue, having claimed over 36 million lives since the identification of the first case in 1981. At the end of 2020, there were "an estimated 37.7 million people" living with HIV [18]. Of these, 36 million were adults, and 53% were women and girls [22]; 25.4 million lived in the WHO African Region [18]. Since 2010, incident HIV rates have declined by 31%, thanks to increased access to treatment and educational resources worldwide [22]. In 2020, 1.5 million people became infected globally, and 680,000 died from AIDS-related complications [23]. This is a marked (52%) decrease in incidence since the peak in 1997, when three million people were diagnosed in that year alone [23]. As the figures below indicate, sub-Saharan Africa is home to just 10% of the world's population but comprises nearly two-thirds of the world's HIV prevalence. In this region, HIV has become a generalized epidemic, meaning that it spreads through the population at large rather than just through at-risk groups such as sex workers and users of injectable drugs [24, 25].

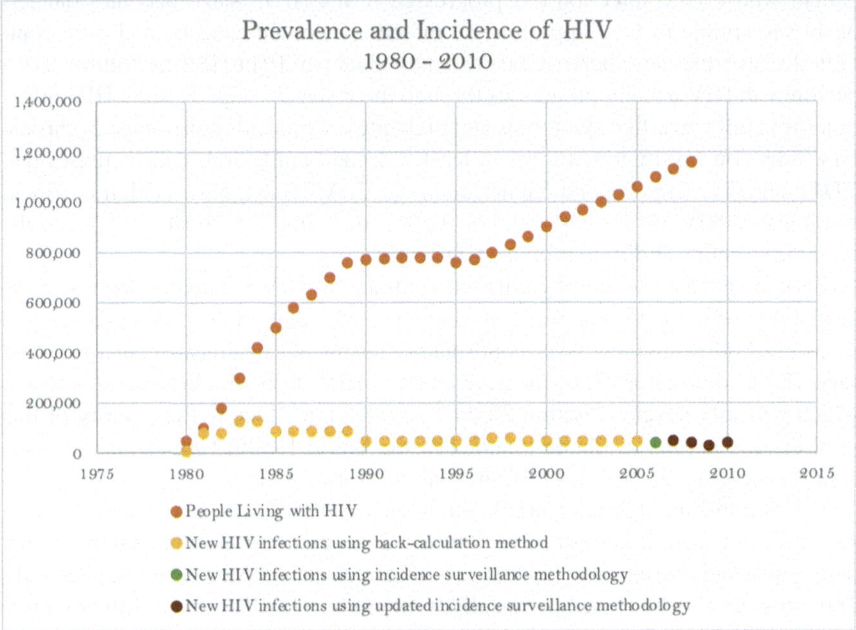

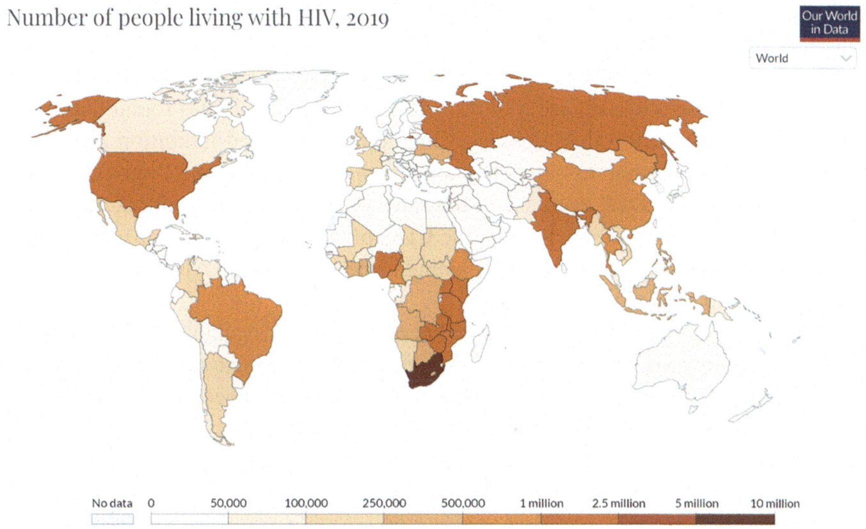

Source: Institute for Health Metrics and Evaluation, Global Burden of Disease (**2019**). OurWorldInData.org/hiv-aids • CC BY

In terms of AIDS, mortality rates have declined by 53% among women and girls [26] and by 41% among men and boys since 2010 [23]. Since the AIDS peak in 2004, mortality has reduced by 64% [23]. Still, the epidemic is not over, and its effects continue to be compounded by the prevalence of other viruses—most recently, COVID-19. Studies from several countries (South Africa, the United Kingdom) have found that the risk of mortality from a COVID-19 infection was doubled in those with HIV/AIDS compared to the general population [23]. Despite this statistic, the regions with highest rates of infection—regions in sub-Saharan Africa, namely, Mozambique and Zimbabwe—have had least access to COVID-19 vaccines [27].

3.2.2 Treatments over Time

Although there is currently no cure for HIV/AIDS, there are effective treatment options that enable patients to live healthy long lives with the virus as a manageable chronic health condition rather than a life-defining fatality. The current treatment regimen for HIV includes "three HIV medications from a minimum of two drug classes" [28]. Different drugs from different classes must be used because the life cycle of the virus is complex and multistage; targeting the virus at each stage of its life cycle thus decreases the likelihood of viral replication throughout the host [29]. This multistage treatment regimen is known as antiretroviral therapy (ART) and must be followed daily for life to be fully effective [30]. Common FDA-approved drug combinations for HIV include two nucleoside reverse transcriptase inhibitors and one non-nucleoside reverse transcriptase inhibitor, a protease inhibitor (PI), or integrase inhibitor (II) [28].

The first antiretroviral drug to reach FDA approval, in 1987, was zidovudine, a nucleoside reverse transcriptase inhibitor; in 1996, clinical studies found that combining zidovudine with other classes of medicines increased inhibition efficiency [28]. Since then, further research has shown that three-drug ART has led to an approximate 60% to 80% decrease in the rates of AIDs, hospitalization, and death [28]. In the United States specifically, the CDC plans to implement a 90-90-90 plan by 2030, wherein 90% of cases are HIV-diagnosed, 90% are on ART, and 90% have nearly fully suppressed the virus [28].

HIV prevention options have also become widely available. PrEP, or pre-exposure prophylaxis, is a medicine that at-risk individuals can take to prevent getting HIV through unprotected sex or use of contaminated needles [31]. There are currently two approved PrEP medications, Truvada and Descovy, the latter of which is not approved for those assigned female at birth [31]. Daily PrEP use can reduce the risk of sexually contracting HIV by 90% and the risk of contracting HIV through needles by 70% [32].

3.2.3 Vaccine Landscape

Research for an effective HIV vaccine is currently underway and has been for several decades. The difficulty scientists have faced in creating an effective HIV vaccine is the virus' rapid mutation rate and ability to evade the immune response. Yet as of March 2022, the National Institute of Allergy and Infectious Diseases launched a phase I clinical trial examining a potential mRNA vaccine for HIV, citing the effectiveness of the mRNA COVID-19 vaccine as support for the research [33].

3.3 HPV

Human papillomavirus, commonly known as HPV, is a common sexually transmitted disease that is associated with numerous cancers—cervical cancer, head and neck squamous cell carcinoma, and anal cancer [34]. There are over 200 types of the virus that exist. HPV infections can be categorized as low-risk infections and high-risk infections. Low-risk infections consist of causing genital warts, whereas high-risk infections may cause cervical and other types of cancers [35].

The biological structure of papillomaviruses can be described as "small, non-enveloped, epitheliotropic, double-stranded DNA viruses" [36]. Human papillomaviruses result in mucosal and cutaneous epithelial lesions and cancers [37]. Risk factors for HPV include having "multiple sex partners, sex at an early age, a history of sexually transmitted infections, and smoking" [35].

3.3.1 Worldwide Prevalence/Rate of Infection

Worldwide, HPV is found in 11–12% of women without cervical abnormalities [38]. Sub-Saharan Africa had the highest HPV prevalence of 24% [39] in cytologically healthy women. However, the highest prevalence of HPV infection in all women was in Asia; specifically, a large number of Eastern and Central and Southern Asian women were carriers [39]. Additionally, there appears to be a trend of higher HPV infection rates in developing regions rather than in developed regions [39]. In the United States, HPV is considered the "most commonly sexually transmitted infection" and has been associated with increased risk of cervical cancer and genital warts [40]. Adolescent girls and women under 25 are the most infected with HPV [39]. There is a peak prevalence of HPV in women around their early 20s [35]. Up to 79% of sexually active women are likely to acquire an HPV infection at some point during their lifetime [35]. On the other hand, genital warts is only prevalent in about 1% of sexually active Americans [37].

With more than 100 types of HPV strains, the HPV infection is categorized based as non-genital, mucosal or anogenital, and epidermodysplasia verruciformis [37]. Of these strains, HPV 6 and 11 are responsible for "approximately 90% of" genital warts, and HPV 16 and 18 are associated with "approximately 70% of cervical cancers" [40]. HPV 16 and 18 are considered the most prevalent strains of HPV [38]. HPV is transmitted when in contact with infected genital skin or mucosa [35]. Other possible routes of infection are orally or perinatally. While oral infection of HPV is possible, there is a low risk of transmission [35]. Similarly, perinatal transmission of HPV is rare [35].

3.3.2 Treatments over Time

Preventative care for HPV infection consists of two FDA-approved HPV vaccines. Gardasil is the quadrivalent recombinant HPV vaccine that was approved in 2006 and protects girls and women from ages 9 to 26 from HPV types 6, 11, 16, and 18 [35]. The second vaccine, Cervarix, was approved for girls and women from ages 10 to 25 and protects against HPV types 16 and 18 [35].

3.4 MMR

MMR, also known as measles, mumps, and rubella, are a group of single-stranded RNA viruses. Measles and mumps are negative sense paramyxoviruses, and rubella is a positive sense togavirus. This group of viruses has high transmissibility, with measles having a R0 of 12–18. This R0, also known as basic reproductive number, means that during the course of an infection, a person is likely to spread measles to 12–18 others. These specific viruses have had a spotlight in the public eye as of late, due to a now-defunct study linking the vaccine to autism. This negative press has resulted in many choosing to not vaccinate their children, thus leading to outbreaks [41].

Measles virus, also known as *morbillivirus*, replicates in the cytoplasm of host cells and is spread primarily through respiratory droplets. It is known for its viral prodrome of cough, coryza, conjunctivitis, and Koplik spots, as well as its characteristic cephalocaudal spreading maculopapular rash. Virulence factors include hemagglutinin and fusion protein, which aid in infectivity. Finally, measles has late-stage complications such as pneumonia and subacute sclerosing panencephalitis [42]. Mumps virus also replicates in the cytoplasm of host cells. It has the virulence factors of hemagglutinin, fusion protein, and neuraminidase. Mumps infection can be hinted at on physical exam because of its characteristic reproduction within salivary glands. It is known to also cause orchitis in males, which can lead to infertility. An additional known complication of mumps is meningitis [43].

Rubella, also known as German measles, can sometimes be mistaken for measles because of its maculopapular rash's similar cephalocaudal spread [44]. It is generally a disease of childhood and has different symptoms depending on the type of infection, congenital or acquired. Congenital rubella crosses the placenta and causes a myriad of symptoms. The most common presentation for congenital rubella is the triad of congenital cataracts, sensorineural deafness, and patent ductus arteriosus. Additional signs of congenital rubella are the "blueberry muffin rash," jaundice, microcephaly, and pulmonic stenosis [45]. Acquired childhood rubella signs include the cephalocaudal rash and posterior auricular and occipital lymphadenopathy. Rubella has been linked with arthritis. Rubella is also spread primarily through respiratory droplets.

3.4.1 Worldwide Prevalence/Rate of Infection

Measles is an extremely contagious disease and thus can have high rates of spread. It is estimated that there is a 90% chance of infection, if exposed to the virus [46]. Despite the high levels of contagion, the prevalence remains low due to the existence of MMR vaccine. The highest rates for measles can be seen in Africa and India; however, there are also large rates in Brazil. In 2019, there were an estimated 11,371 confirmed measles cases in Brazil. The primary age group that was affected by measles was from ages 15 to 29, which comprised 45% of the cases in Brazil. Somali was reported as having the greatest number of cases, nearing around 6000 cases as of 2022.

Mumps also has a high rate of transfer and is generally caused by the paramyxovirus. Since the introduction of the mumps vaccine, the rates dropped; however, recently, there has been a surge in the number of outbreaks and cases. The country most afflicted is China. The number of cases in China has reached around 130,000 by 2020 and has been increasingly seen in vaccinated patients. The number of cases of mumps in the United States has dropped greatly from 150,000 in 1967 to 2251 cases in 2018 (Lau, Roger, and Michael Turner) [47].

Rubella is an acute contagious disease that previously had a high prevalence but decreased by more than 95% from the vaccine. India has only recently seen a decline in the prevalence due to only launching the vaccine in 2017 [48]. Despite the introduction of the vaccine, there are still thousands of cases in India. Globally, however, the rates are extremely low, and it is only found in around 14 countries.

3.4.2 Treatments over Time

Measles virus, as it is known today, was first documented in the ninth century by a Persian doctor named Rhazes. Measles treatment has had some changes over the course of time, but much of its treatment remains the same. Historically, measles was treated supportively with antipyretics and hydration [49]. Today, treatment for measles is still largely supportive. Aimed at reducing complications, measles treatment largely consists of supportive fluids and analgesia. However, studies have demonstrated efficacy in vitamin A treatment. Vitamin A reduces mortality in measles patients less than 2 years old. This has now become a standard of care in this group. Vitamin A is efficacious due to its function as an immunomodulator, which increases antibody response against the virus. Post-exposure prophylaxis includes one dose of the MMR vaccine and immunoglobulin if the exposure was within 72 hours [50].

Mumps and rubella currently do not have any specific medications or drugs that can be taken as treatment; however, there are practices that can reduce the symptoms. Most symptoms should go away after 1–2 weeks, especially through the addition of hydration of fluids. There are some drugs, such as acetaminophen or ibuprofen, that can be taken to alleviate fever symptoms. Decreasing inflammation of the glands can be alleviated by utilizing heating or ice packs on the areas of inflammation [51].

3.4.3 Vaccine Landscape

Measles, mumps, and rubella vaccine is given in two doses. The first is given to children at 12–15 months of age, and the second is given from 4 to 6 years of age. The vaccine itself contains strains from each of the three viruses and has an extremely high efficacy rate, above 95%. The vaccine is slightly less effective for mumps but still results in an efficacy rate above 85%. Measles is currently labeled as eliminated in the United States due to the vaccine in 2016. However, low-income countries still have these diseases due to the lack of distribution of the vaccine. There have been speculations about autism being linked to the MMR vaccine; however, there is no evidence that can support this (Table 1).

Table 1 Viral diseases with their year of emergence and whether they have a treatment or vaccine available for use

Viral disease	Year emerged	Treatment available	Vaccine available
Chickenpox [52]	1691	Yes	Yes
Flu (influenza) [53]	1918	Yes	Yes
Herpes [54]	1.6 million years ago	Yes	No
HIV/AIDS [55]	1981	Yes	No
HPV [56]	500,000 years ago	No[a]	Yes
Infectious mononucleosis [57]	1880s	No	No
MMR [58]	Ninth century	No	Yes
Shingles [59]	1888[b]	Yes	Yes
Viral gastroenteritis (stomach flu) [60]	1972	Yes	Yes
Viral hepatitis [61]	100,000 years ago[c]	Yes	Yes
Viral meningitis [62]	1805	Yes	No
Viral pneumonia [63, 64]	1938[d]	No[e]	Yes

[a]Associated genital warts can be treated
[b]Although not distinguishable from smallpox, the emergence dates back to the fifteenth century
[c]HBV specifically has been estimated to originate between 3000 and 100,000 years ago
[d]Viral pneumonia has been common throughout human history along with most of the diseases in the table
[e]Most viruses responsible for pneumonia are not treatable (they resolve on their own), but there are some exceptions

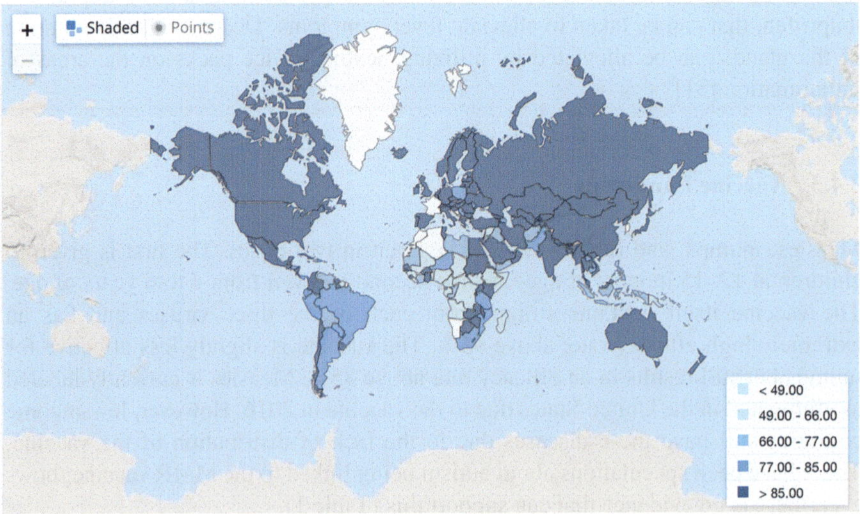

"Immunization, Measles (% of Children Ages 12–23 Months)." *The World Bank*, https://data.worldbank.org/indicator/SH.IMM.MEAS?end=2020&start=1980&view=chart

4 Conclusion

Global initiatives taken by organizations like the UNICEF and the World Health Organization have not been effectively implemented in several countries due to the dearth of the country's ownership and limited will, which is seen by their insufficient resources [65]. For the effect expansion of MMR treatment, there needs to be greater political support as well as a dependency on supplementary immunization activities (SIAs). A measles eradication plan should include the assurance of two measles-containing vaccine doses along with the SIAs. Using the MMR vaccine as a school entry requirement could also contribute to increasing the immunity withing target populations. Accurate reporting of cases on at least a weekly basis in affected countries could indicate areas that are spiking and need immediate containment or intervention. The strengthening of global immunization programs is crucial to achieving the eradication goals for measles, mumps, and rubella.

For all the aforementioned viral diseases including HPV and HIV, there must be adequate human, financial, and technical resources to investigate outbreaks and their causes. Since HPV presents major concerns for cervical cancer, routine screening for women with HPV must also be taken into consideration in target countries along with preventative transmission efforts (HPV vaccine) [41]. For both HPV and HIV, a national push toward breaking the stigma on STIs as well as providing educational programs in schools would prove beneficial as more people can recognize the symptoms and speak on the subject. Affordable STI screenings for those that are sexually active should also be highlighted as a global effort to further bring light to the diseases (and those affected) and help eliminate its transmission. There are affordable ways to prevent these viral diseases; however, they must be implemented and initiated at a global, national, and regional level in order to decrease or eliminate transmission.

References

1. Geldborn HR. Structure and classification of viruses. In: Medical microbiology. 4th ed. The University of Texas Medical Branch at Galveston; 1996. Chapter 41.
2. Crespo-Barreda A, Encabo-Berzosa MM, González-Pastor R, Ortíz-Teba P, Iglesias M, Serrano JL, Martín-Duque P. Viral and nonviral vectors for in vivo and ex vivo gene therapies. In: Laurence J, Baptista P, Atala A, Beusekom MV, editors. Translating regenerative medicine to the clinic. Zaragoza: Elsevier; 2016. p. 155–77.
3. Kobayashi K, Hisamatsu K, Suzui N, Hara A, Tomita H, Miyazaki T. A review of HPV-related head and neck cancer. J Clin Med. 2018;7(9):241. https://doi.org/10.3390/jcm7090241.
4. Bankamp B, Hickman C, Icenogle JP, Rota PA. Successes and challenges for preventing measles, mumps and rubella by vaccination. In: Current opinion in virology, vol. 34. Atlanta: Elsevier; 2019. p. 110–6.
5. Su S-B, Chang H-L, Chen K-T. Current status of mumps virus infection: epidemiology, pathogenesis, and vaccine. Int J Environ Res Public Health. 2020;17(5):1686. https://doi.org/10.3390/ijerph17051686.

6. Shimizu K. History of influenza epidemics and discovery of influenza virus. Nihon Rinsho. 1997;55(10):2505–11.
7. Van Epps HL. Influenza: exposing the true killer. J Exp Med. 2006;203(4):803. https://doi.org/10.1084/jem.2034fta.
8. Abrahão JS, de Arruda LB. Special issue "Emerging viruses: surveillance, prevention, evolution, and control". Viruses. 2020;12(3):306. https://doi.org/10.3390/v12030306. Published 11 Mar 2020.
9. Taubenberger JK, Morens DM. The pathology of influenza virus infections. Annu Rev Pathol. 2008;3:499–522. https://doi.org/10.1146/annurev.pathmechdis.3.121806.154316.
10. 2020–2021 flu season summary. Centers for Disease Control and Prevention. https://www.cdc.gov/flu/season/faq-flu-season-2020-2021.htm. Published 25 Oct 2021.
11. Mousa HA. Prevention and treatment of influenza, influenza-like illness, and common cold by herbal, complementary, and natural therapies. J Evid Based Complementary Altern Med. 2017;22(1):166–74. https://doi.org/10.1177/2156587216641831.
12. Nabeshima S, Kashiwagi K, Ajisaka K, et al. A randomized, controlled trial comparing traditional herbal medicine and neuraminidase inhibitors in the treatment of seasonal influenza. J Infect Chemother. 2012;18(4):534–43. https://doi.org/10.1007/s10156-012-0378-7.
13. Gupta YK, Meenu M, Mohan P. The Tamiflu fiasco and lessons learnt. Indian J Pharmacol. 2015;47(1):11–6. https://doi.org/10.4103/0253-7613.150308.
14. Heneghan CJ, Onakpoya I, Thompson M, Spencer EA, Jones M, Jefferson T. Zanamivir for influenza in adults and children: systematic review of clinical study reports and summary of regulatory comments. BMJ. 2014;348:g2547. https://doi.org/10.1136/bmj.g2547. Published 9 Apr 2014.
15. Recommended composition of influenza virus vaccines for use in the 2022–2023 Northern Hemisphere influenza season. World Health Organization. https://www.who.int/publications/m/item/recommended-composition-of-influenza-virus-vaccines-for-use-in-the-2022-2023-northern-hemisphere-influenza-season. Published 25 Feb 2022.
16. Acute HIV infection: NIH. HIV.gov. https://clinicalinfo.hiv.gov/en/glossary/acute-hiv-infection.
17. Acquired immunodeficiency syndrome (AIDS): NIH. HIV.gov. https://clinicalinfo.hiv.gov/en/glossary/acquired-immunodeficiency-syndrome-aids.
18. HIV/AIDS. World Health Organization. https://www.who.int/news-room/fact-sheets/detail/hiv-aids. Published 30 Nov 2021.
19. About HIV/AIDS. Centers for Disease Control and Prevention. https://www.cdc.gov/hiv/basics/whatishiv.html. Published 1 June 2021.
20. HIV.gov. How do you get or transmit HIV? n.d. Retrieved on 17 May 2021, from https://www.hiv.gov/hiv-basics/overview/about-hiv-and-aids/how-is-hiv-transmitted.
21. Centers for Disease Control and Prevention. HIV and youth. 2013. Retrieved 17 May 2021, from http://www.cdc.gov/hiv/risk/age/youth/index.html.
22. The global HIV/AIDS epidemic. HIV.gov. https://www.hiv.gov/hiv-basics/overview/data-and-trends/global-statistics. Published 30 Nov 2021.
23. Global HIV & AIDS statistics – fact sheet. UNAIDS. https://www.unaids.org/en/resources/fact-sheet.
24. Tanser F, de Oliveira T, Maheu-Giroux M, Bärnighausen T. Concentrated HIV subepidemics in generalized epidemic settings. Curr Opin HIV AIDS. 2014;9(2):115–25. https://doi.org/10.1097/COH.0000000000000034.
25. HIV prevalence and incidence, 1980–2010. Centers for Disease Control and Prevention. https://www.cdc.gov/nchhstp/newsroom/2012/hivincidencegraphics.html.
26. Global Burden of Disease Collaborative Network. Global burden of disease study 2019 (GBD 2019) results. Seattle: Institute for Health Metrics and Evaluation (IHME); 2021.
27. Covid-19 vaccines. World Health Organization. https://www.afro.who.int/health-topics/coronavirus-covid-19/vaccines.

28. Kemnic TR, Gulick PG. HIV antiretroviral therapy. [Updated 2022 Apr 30]. In: StatPearls [Internet]. Treasure Island (FL): StatPearls Publishing; 2022.
29. The HIV life cycle. National Institutes of Health. https://hivinfo.nih.gov/understanding-hiv/fact-sheets/hiv-life-cycle. Published 4 Aug 2021.
30. Antiretroviral Therapy (ART): NIH. HIV.gov. https://clinicalinfo.hiv.gov/en/glossary/antiretroviral-therapy-art.
31. About PrEP. Centers for Disease Control and Prevention. https://www.cdc.gov/hiv/basics/prep/about-prep.html. Published 20 Apr 2022.
32. Mermin J. Prep: a powerful prevention tool. HIV.gov. https://www.hiv.gov/blog/prep-a-powerful-prevention-tool. Published 8 Apr 2021.
33. NIH launches clinical trial of three mRNA HIV vaccines. National Institutes of Health. https://www.nih.gov/news-events/news-releases/nih-launches-clinical-trial-three-mrna-hiv-vaccines. Published 14 Mar 2022.
34. Cheng L, Wang Y, Du J. Human papillomavirus vaccines: an updated review. Vaccines (Basel). 2020;8(3):391. https://doi.org/10.3390/vaccines8030391. Published 16 July 2020.
35. Juckett G, Hartman-Adams H. Human papillomavirus: clinical manifestations and prevention. Am Fam Physician. 2010;82(10):1209–13.
36. IARC Working Group on the evaluation of carcinogenic risks to humans. Human papillomaviruses. Lyon (FR): International Agency for Research on Cancer; 2007. (IARC Monographs on the Evaluation of Carcinogenic Risks to Humans, No. 90.) 1, Human Papillomavirus (HPV) Infection.
37. Luria L, Cardoza-Favarato G. Human papillomavirus. [Updated 2022 Jan 24]. In: StatPearls [Internet]. Treasure Island (FL): StatPearls Publishing; 2022.
38. Forman D, de Martel C, Lacey CJ, et al. Global burden of human papillomavirus and related diseases. Vaccine. 2012;30(Suppl 5):F12–23. https://doi.org/10.1016/j.vaccine.2012.07.055.
39. Kombe Kombe AJ, Li B, Zahid A, et al. Epidemiology and burden of human papillomavirus and related diseases, molecular pathogenesis, and vaccine evaluation. Front Public Health. 2021;8:552028. https://doi.org/10.3389/fpubh.2020.552028. Published 20 Jan 2021.
40. Braaten KP, Laufer MR. Human papillomavirus (HPV), HPV-related disease, and the HPV vaccine. Rev Obstet Gynecol. 2008;1(1):2–10.
41. Guerra FM, et al. The basic reproduction number (R0) of measles: a systematic review. Lancet Infect Dis. 2017;17(12):e420–8.
42. Bellini WJ, Rota JS, Rota PA. Virology of measles virus. J Infect Dis. 1994;170(Suppl. 1):S15–23.
43. Rubin S, et al. Molecular biology, pathogenesis and pathology of mumps virus. J Pathol. 2015;235(2):242–52.
44. Frey TK. Molecular biology of rubella virus. Adv Virus Res. 1994;44:69–160.
45. Webster WS. Teratogen update: congenital rubella. Teratology. 1998;58(1):13–23.
46. Estofolete C, et al. Prevalence of measles antibodies in São José Do Rio Preto, São Paulo, Brazil: a serological survey model. Nat News, Nature Publishing Group, 20 Mar 2020, https://www.nature.com/articles/s41598-020-62151-3.
47. Viral mumps: increasing occurrences in the vaccinated population. Oral Surg Oral Med Oral Pathol Oral Radiol, Mosby, 27 June 2019, https://www.sciencedirect.com/science/article/pii/S2212440319305619?casa_token=aHAU_UCRHxQAAAAA%3AcmJ97ppdDYLcH7IHf9kAzY-tISGHZOaeIaoYBLHcxVc6ldddKx2K4rTxf96mOSn3KUylcBE31-g#fig0001.
48. Muliyil D, et al. Sero-prevalence of rubella among pregnant women in India, 2017. Vaccine, Elsevier, 15 Nov 2018, https://www.sciencedirect.com/science/article/pii/S0264410X18315263?casa_token=PB2bfHv7lHYAAAAA%3Af136rbYyUh1rbwXhr7h1JAd1d-iKTktD4hxBgKrTGhdHpaT3kJluzDOwS9W4txXDAt7glI6OsKQ.
49. D'Souza RM, D'Souza R. Vitamin A for the treatment of children with measles—a systematic review. J Trop Pediatr. 2002;48(6):323–7.
50. Griffin DE, Oldstone MBA, editors. Measles: history and basic biology, vol. 329. Berlin: Springer Science & Business Media; 2008.

51. Nofal A, et al. Treatment of recalcitrant warts with intralesional measles, mumps, and rubella vaccine: a promising approach. Int J Dermatol, U.S. National Library of Medicine. 2014. https://pubmed.ncbi.nlm.nih.gov/25070525/.
52. About chickenpox. In: Centers for Disease Control and Prevention. Centers for Disease Control and Prevention, 28 Apr 2021, https://www.cdc.gov/chickenpox/about/index.html.
53. Knobler SL, et al. The threat of pandemic influenza: are we ready? Workshop summary. Institute of Medicine (US) Forum on Microbial Threats, National Library of Medicine; 2005. https://www.ncbi.nlm.nih.gov/books/NBK22148/.
54. LaFee S, Romero J. Herpes infected humans before they were human. UC Health – UC San Diego, 10 June 2014, https://health.ucsd.edu/news/releases/pages/2014-06-10-herpes-origins-in-chimpanzees.aspx#:~:text=This%20level%20of%20divergence%20indicated,humans%20roughly%20200%2C000%20years%20ago.
55. A timeline of HIV and AIDS. HIV.gov, MHAF, 29 Apr 2022, https://www.hiv.gov/hiv-basics/overview/history/hiv-and-aids-timeline#:~:text=The%20HIV.gov%20Timeline%20reflects,risk%20for%2C%20HIV%20and%20AIDS.
56. Chen Z, et al. Niche adaptation and viral transmission of human papillomaviruses from archaic hominins to modern humans. PLoS Pathog. 2018;14(11). https://doi.org/10.1371/journal.ppat.1007352.
57. Evans AS. The history of infectious mononucleosis. Am J Med Sci. 1974;267(3):189–95. https://doi.org/10.1097/00000441-197403000-00006.
58. History of measles. Centers for Disease Control and Prevention, CDC, 5 Nov 2020, https://www.cdc.gov/measles/about/history.html.
59. Galetta KM, Gilden D. Zeroing in on zoster: a tale of many disorders produced by one virus. J Neurol Sci. 2015;358(1–2):38–45. https://doi.org/10.1016/j.jns.2015.10.004.
60. Oude Munnink BB, van der Hoek L. Viruses causing gastroenteritis: the known, the new and those beyond. Viruses. 2016;8(2):42. https://doi.org/10.3390/v8020042.
61. Littlejohn M, et al. Origins and evolution of hepatitis B virus and hepatitis D virus. Cold Spring Harb Perspect Med. 2016;6(1). https://doi.org/10.1101/cshperspect.a021360.
62. Edwards H. The history of meningitis. In: Meningitis research foundation. Meningitis Research Foundation; 2020. https://www.meningitis.org/blogs/the-history-of-meningitis.
63. Reimann HA. An acute infection of the respiratory tract with atypical pneumonia. J Am Med Assoc. 1938;111(26):2377. https://doi.org/10.1001/jama.1938.02790520033007.
64. Pneumonia. In: Johns Hopkins Medicine. Johns Hopkins Medicine; 2021. https://www.hopkinsmedicine.org/health/conditions-and-diseases/pneumonia.
65. Orenstein WA, Hinman A, Nkowane B, Olive JM, Reingold A. Measles and rubella global strategic plan 2012–2020 Midterm Review. Vaccine. Published 4 Jan 2018.

Drug Resistance in Antiviral Therapy

Seth Kwabena Amponsah and Benjamin Tagoe

Abstract Drug resistance in antiviral therapy is a major public health challenge. This resistance can occur among immunocompromised patients where persistent viral replication and protracted drug exposure result in selection of resistant strains. Consequences of drug resistance range from toxicity inherent in the use of second-line antiviral agents, severe disease, and even death (from progressive viral infection when no effective alternative treatment is available). Although scientific technology has identified a number of these proximal mechanisms of antiviral resistance, bigger evolutionary trends of the viruses remain obscure. New metrics for evaluating mutations, recombination rates, demographic histories of transmission, and selective forces during viral adaptation to antiviral drug treatment have been developed. Understanding levels of resistance and cross-resistance conferred by diverse mutations is required for accurate interpretation of genotypic assays. Identification of viral resistance to drug therapy is possible by linking distinctive viral alterations (phenotypic resistance) to a number of antiviral drugs. To reduce antiviral resistance, there is the need to optimize drug administration, select alternate therapy based on knowledge of resistance mechanisms, and develop novel antivirals. Experimental drugs with various viral targets are being investigated, and these agents may offer better treatment options. In this chapter, we delve into the mechanisms of viral resistance, manifestation of antiviral resistance, detection of antiviral resistance, and clinical implications of drug resistance in viruses.

Keywords DNA polymerase · Famciclovir · Hepatitis B virus · Lamivudine · Resistance · Varicella-zoster virus

S. K. Amponsah (✉)
Department of Medical Pharmacology, University of Ghana Medical School, Accra, Ghana

B. Tagoe
Fulfilment Operations and Academy, Zipline Ghana, Accra, Ghana

R. Shegokar, Y. Pathak (eds.), *Viral Drug Delivery Systems*,
https://doi.org/10.1007/978-3-031-20537-8_2

1 Resistance as an Evolutionary Process

Just like most cell-based organisms, viruses undergo evolution and natural selection. Currently, monitoring of viral evolutionary processes has become more streamlined due to availability of molecular data on strength of selection and nucleotide diversity [1, 2].

Over the last few decades, drugs that target mechanisms of viral replication have been widely used to treat viral infections. Generally, viral genomes are not likely to replicate if treatment is robust [3]. A less effective therapeutic regimen may result in successful replication of some viral genomes, which may ultimately lead to rapid adaptation towards resistance. Resistance to antiviral drugs is further worsened by high rates of mutations within viruses. Pharmaceutical companies and research laboratories have been forced to remain innovative in the process of drug discovery and development.

2 Clinical Definition of Antiviral Drug Resistance

Antiviral drug resistance occurs when viruses become less susceptibility to antiviral agents. This is usually confirmed by in vitro testing, with confirmation via genetic and biochemical analysis. In chronic hepatitis B, for example, definitions for first-line (main) and secondary antiviral drug failure exist [4]. First-line antiviral failure (sometimes referred to as non-responsiveness) occurs when there is the inability of an antiviral drug to decrease hepatitis B virus (HBV) deoxyribonucleic acid (DNA) viral load beyond 1.0 $\log_{10}$ IU/ml within the first 3 months of treatment. This is frequently caused by problems with efficacy of pharmacological or antiviral agents [5]. Second-line antiviral treatment failure occurs when there is an increase in viral load from nadir of at least 1.0 $\log_{10}$ IU/ml following two blood samples that are taken within 1 month. There is the possibility that patients who show efficacy to antiviral therapy may experience this second-line antiviral treatment failure sometime later. It is, thus, recommended that viral load is estimated at the beginning of therapy. In order to identify viral rebounds that may occur with antiviral resistance, an assay that is specific and sensitive (1000 IU/ml) to HBV DNA viral load is recommended. A recent international summit recommended that HBV DNA viral load testing be undertaken every 2–3 months, if health resources permit [4]. Patients on antiviral drugs should be monitored on a regular basis. This is especially important in individuals with advanced disease, since it is necessary to maintain viral suppression for an extended period.

This book chapter seeks to discuss variations in antiviral resistance mechanisms, resistance to common antiviral drugs, and identification of resistance. Additionally, information on infections caused by hepatitis B and C viruses, herpesvirus, human cytomegalovirus, varicella-zoster virus, and human immunodeficiency virus is reviewed.

3 Antiviral Drug Resistance

3.1 Variation in Viral Resistance Mechanisms

Due to differences in replication among viruses, development of antiviral drugs in modern times has primarily focused on targeting different stages of viral life cycle. This approach is relevant in reducing evolution of viruses toward resistance. Understanding the different mechanisms of viral resistance is key in antiviral drug development.

The mutation rate of hepatitis C virus (HCV) is relatively high, and this is further aided by repeated replication and poor censoring of encoded ribonucleic acid (RNA) polymerase [6]. Replication of HCV is close to the allowed maximum error rate before the loss of genomic integrity. Antiviral drugs that inhibit activity of protease or polymerase are usually used to treat HCV infections. A few mutations in HCV genome could result in resistance toward protease or polymerase inhibitors. It is likely resistant strains of HCV already exist in the population, considering its high mutation rate. Regardless, the HCV appears to be susceptible to new antiviral drug combinations such as ledipasvir and sofosbuvir. Research has shown that there is little to no cross-resistance between the two drugs; hence, these two agents could have promise in the future [7].

Herpesvirus (HSV) is known to have low diversity, partly due to its low rate of recombination [8]. HSV may occasionally have latent periods and also viral shedding while preserving its transmissibility even in asymptomatic individuals. Nucleoside inhibitors such as acyclovir are commonly used to manage HSV infections in immunocompromised patients. Therapy with acyclovir for systemic viral infections either target DNA polymerase or a thymidine kinase required for prodrug activation [9]. Infections caused by human cytomegalovirus (HCMV), which are HSV, could start in saliva (occupying a one compartment) before moving to other compartments. HCMV infections are generally asymptomatic; however, there is cause for concern among congenitally infected infants and immunocompromised hosts. Differentiation persists after compartmentalization to the point where viruses found in different compartments within a host may be very diverse [10].

The HCMV has polymorphism levels comparable to RNA viruses, despite the fact that usually DNA polymerases have high fidelity compared to RNA polymerases [11, 12]. Positive selection signatures can be found in a low percent (5%) of open reading frames of HCMV. Due to the fact that envelopes of HCMV assist to escape host immune defense, loci that are associated with envelope proteins appear to have elevated ratio of non-synonymous to synonymous substitutions (dN/dS). However, loci associated with replicative proteins are usually conserved. During treatment of HCMV infections, antivirals that act as nucleoside analogs, such as ganciclovir and cidofovir, are usually used. Resistance can occur in the viral kinase that is needed for the phosphorylation of prodrugs [13].

Neuraminidase (NA) and hemagglutinin (HA) are the virion-surface proteins found in *Influenza A virus* (IAV). These virion-surface proteins are encoded in the short genome of IAV. In IAV, HA and NA are used to bind and detach from the cell membrane of a host. The dN/dS ratio of HA and NA are substantially larger than those of the other viral proteins because they are surface antigens under diversifying selection [14, 15]. NA inhibitors, which have the tendency to disrupt viral envelope detachment from host cell membrane, are the most utilized antivirals in the treatment of IAV. IAV that have H274Y NA enzymes are known to be resistant to oseltamivir [16, 17]. It is possible that the difficulty of administering zanamivir, another NA inhibitor, has led to relatively few resistant viral strains [18]. As a recently synthesized antiviral agent, favipiravir is expected to induce mutagenesis in the IAV genome. So far, there has been little to no reports of viral resistance to favipiravir [19, 20].

Human immunodeficiency virus (HIV) is a retrovirus with an RNA genome. Once HIV infects a host, there is duplication of genome by reverse transcription that leads to the formation of a double-stranded DNA. This DNA is inserted into genome of the host and then reverted to RNA. During drug treatment, the process of reverse transcription is highly error-prone, and this can lead to a number of mutations and increased population diversity [21]. In fact, it is expected that at the start of pharmacotherapy, there exist mutations in the genome of HIV that are present in at least one infected cell [22, 23]. Multi-drug regimen is the most appropriate approach in HIV treatment today, where antivirals from varied classes, and without known cross-resistance mutations, are used [24]. A regimen like this decreases the likelihood for resistance to occur. Combinations with the various antiretroviral agents (integrase strand transfer inhibitors, non-nucleoside reverse transcriptase inhibitors, protease inhibitors, etc.) are recommended.

HBV is a DNA virus and replicates (by transcription) into intermediates of RNA. HBV then undergoes reverse transcription back to DNA. HBV is assumed to live as a quasi-species despite its DNA genome. In HBV, there is high level of population variation, which occurs due to lack of proofreading during the process of reverse transcription. Hence, there is often HBV with mutations that can lead to resistance to antiviral therapy [25, 26]. Lamivudine, a reverse transcriptase inhibitor, is typically used to treat HBV infections. However, a number of lamivudine-resistant HBV have been identified, often with low genetic barriers. As a result of the aforementioned, an additional reverse transcriptase inhibitor is frequently added [26]. Care must be taken when drug switching is done, especially if it is within same drug class because there is the likelihood of HBV-resistant mutations occurring. Although a few clinical trials have combination therapy in the treatment of HBV infections, this is often not the standard treatment [4, 26, 27].

3.2 Resistance to Common Antiviral Agents

Most drugs currently indicated for HSV infection treatment inhibit DNA polymerase in the virus. Nucleoside analogs such as ganciclovir and acyclovir are usually used to treat infections caused by HSV. Despite possessing weak activity against HCMV, famciclovir, valacyclovir, and acyclovir are used to manage other viral

Table 1 Mechanism of viral resistance to widely used drugs

Antiviral agent	Resistance mechanism	References
Cidofovir	Altered DNA polymerase	[33]
Famciclovir	Altered DNA polymerase or viral thymidine kinase	[34]
Acyclovir	Altered DNA polymerase or deficient viral thymidine kinase	[34]
Valacyclovir	Altered DNA polymerase or deficient viral thymidine kinase	[34]
Ganciclovir	Diminished drug phosphorylation or altered DNA polymerase	[33]
Vidarabine	Altered DNA polymerase	[35]
Foscarnet	Altered DNA polymerase	[33]
Zidovudine	Altered viral reverse transcriptase	[34]
Lamivudine	Altered viral reverse transcriptase	[36]
Didanosine	Altered viral reverse transcriptase	[36]
Indinavir	Altered viral protease	[36]
Nevirapine	Altered viral reverse transcriptase	[34]
Ritonavir	Altered viral protease	[36]
Saquinavir	Altered viral protease	[36]

infections such as varicella-zoster virus (VZV) infections. Other drugs like valganciclovir and ganciclovir are approved for HCMV infection. These drugs also possess in vitro activity against VZV and HSV [28–30]. The metabolism of acyclovir involves phosphorylation by viral thymidine kinase and subsequent conversion by host kinases to acyclovir triphosphate, which is the active form of the drug. Acyclovir triphosphate then inhibits viral (HSV and VZV) replication by chain termination. Ganciclovir is phosphorylated only once by viral kinases, and this makes the drug active against viruses [13, 28, 31].

Forscanet, a pyrophosphate analog, selectively binds to DNA polymerase of viruses. Eventually, DNA chain elongation is inhibited. Cidofovir, a nucleotide analog, requires phosphorylation by cellular enzymes. The activated form of cidofovir inhibits DNA polymerase of viruses. Second- and third-line HSV antiviral drugs, foscarnet and cidofovir, can be used when there is dose-limiting toxicities or suspected resistance to first-line drugs [32]. A summary of resistance mechanisms to the aforementioned antiviral drugs is outlined in Table 1.

4 Identification of Antiviral Drug Resistance

In order to discover drug resistance in viruses, mutations in their genome must be detected. These mutations should be validated via in vitro phenotypic assays, as being specifically related to drug resistance. A number of assays available in identifying resistance mutations include real-time polymerase chain reaction (PCR), allele-specific PCR, hybridization methods, and matrix-assisted laser desorption/ionization time-of-flight mass spectrometry (MALDI-TOF-MS) [27]. These highly sensitive tests can detect virus encoding the resistance mutations, which are present in only a small proportion of the overall viral quasi-species. A MALDI-TOF-MS-based technique may detect mutations that account for less than 1% of the viral quasi-species. On the contrary, direct PCR sequencing allows for the identification of all mutations, including any extra potential compensatory mutations, as well as new mutations that may have occurred [34, 35].

5 Factors Associated with Antiviral Drug Resistance

Depending on the antiviral drug target and/or class, the number and kind of substitutions required to confer resistance in viruses vary considerably; and this is known as the genetic barrier to resistance [21]. Usually, old antiviral classes have low genetic barriers, with viruses typically requiring few changes to achieve resistance [24]. The rate at which resistance mutations appear is influenced by genetic barriers. If resistance takes only one substitution while another requires many, the former is more likely to lead to resistance.

Data suggests that mutations that occur via transition are more prevalent than those requiring transversion; for example, genetic barriers should include not only the quantity of substitutions but also the kind of substitution [36]. In HCV, for example, mutations can occur that may lead to resistance; some require only a single nucleotide change [21]. With Fisher's geometric model, these mutations can be thought of as an adaptive mechanism [37, 38]. The distance to the ideal phenotype rises as the environment changes. In instances of new mutations, the viruses may have phenotypes that are closer to the optimal mutation and also keeping a distance between the phenotype and the ideal constant [39, 40].

Attempts to prevent drug resistance, however worthwhile, are not always successful. A different strategy is to target mutation rather than replicative processes, more precisely when population mutation rates increase and if most new mutations are detrimental to the virus. In non-recombining viruses, Muller's Ratchet is predicted to operate, lowering fitness over time [41]. Hill-Robertson interference (the process through which beneficial mutations are maintained and detrimental mutations often lost) will become stronger. The aforementioned can lead to population extinction [42].

Numerous RNA viruses replicate at or near the error threshold. New drugs exploit this by increasing the rate of base mutations in populations to the point where they exceed this threshold [43]. It is noteworthy, that, the exact mechanism of action of favipiravir is unknown. However, it is believed to work on RNA polymerase to reduce nucleoside incorporation fidelity. This would result in an increased production rate of mutant virion. Although resistance to other mutagenic drugs occurs as a result of mutations to restore polymerase fidelity, this has been observed less frequently with favipiravir [44]. Therefore, resistance to mutagenic therapy is not ruled out in the future.

6 Clinical Implications of Resistance in Viruses

6.1 Human Cytomegalovirus (HCMV)

As is the case with untreated HCMV infection, the clinical manifestations of drug-resistant HCMV infection can vary from asymptomatic to severe. There have been reports where there is asymptomatic infection with drug-resistant HCMV [45, 46]. Additionally, severe or fatal disease has been reported with drug-resistant HCMV. Data shows that host factors that predispose an individual to serious HCMV are almost same factors that aid in antiviral resistance. In suspected cases of antiviral drug resistance, laboratory-based tests should be conducted and immunosuppressive drugs avoided as much as possible. In the absence of acute, life-threatening, or blinding HCMV infection, antiviral drug selection should be guided by genotypic characterization. The extent of phenotypic resistance known to occur with certain gene mutations has major consequences for treatment selection. In non-life-threatening or vision-threatening illness, low-grade ganciclovir resistance may be addressed with higher intravenous doses [47, 48].

Furthermore, foscarnet is the best treatment option for ganciclovir-resistant viruses. Nephrotoxicity frequently complicates foscarnet use, and its long-term use can cause adverse effects. Cidofovir is an option for ganciclovir-resistant HCMV infections. Based on few in vitro data and case series, a combination of foscarnet and ganciclovir has been recommended for drug-resistant HCMV infections [32].

6.2 Varicella-Zoster Virus (VZV)

Drug-resistant VZV also has clinical consequences: direct impact of viral replication and adverse effects associated with alternative antiviral drugs. In immunocompromised hosts, a persistent verrucous variant of VZV can lead to drug resistance. Certain VZV DNA polymerase mutants selected in cell culture with foscarnet exhibited a sluggish growth phenotype, possibly indicating reduced virulence, although this has not been clinically verified [49].

Foscarnet is commonly used to treat suspected or confirmed acyclovir-resistant VZV infections. Resistance to foscarnet was identified in a small number of patients receiving treatment for acyclovir-resistant VZV infections and was related to mutation in viral DNA polymerase. Although there is paucity of data on cidofovir therapy for drug-resistant VZV, it is believed that cidofovir will preserve effectiveness against acyclovir-resistant VZV [49].

6.3 *Hepatitis B Virus (HBV)*

Drug-resistant HBV infection could lead to deteriorating liver histology, serum transaminase flares, and hepatic decompensation, among others. In drug resistance, there is a virologic breakthrough, where there is an increase in HBV DNA in serum, with a minimum of 1.0 $\log_{10}$ (tenfold) above nadir. There could also be a situation where there may be HBV reinfection with previously undetectable HBV DNA on at least two occasions (at least 1 month apart during treatment). Finally, a biochemical breakthrough occurs, as characterized by an increase in hepatic transaminase levels [4].

When there is evidence of antiviral drug-resistant mutation(s), the most appropriate course of action is to switch to another treatment or combination therapy. Tenofovir, a highly active antiviral agent with high efficacy against lamivudine-resistant viruses, seems to be a better option than adefovir monotherapy for HBV infections. Given the reported emergence of resistance to lamivudine, entecavir may not be the best alternative for lamivudine-resistant HBV [5].

The management of adefovir-resistant HBV is determined by antiviral treatment history of patient and the type of HBV mutation(s). Lamivudine is effective in adefovir-resistant HBV infections. Additionally, in vitro findings indicate that telbivudine might also be useful.

7 Conclusion

From a public health perspective, the unpredictability of viral evolution and drug resistance is a major concern to drug treatment in viral infections. Drugs that target mechanisms of viral (HBV, HCV, VZV, HCMV, HSV, and HIV) replication should be used to treat viral infections. Also, combinations with the various antiviral agents are recommended to reduce drug resistance. Where applicable, immune boosting agents should be used among patients with some form of drug-resistant viral infection. Furthermore, patients on antiviral drugs should be monitored (viral load testing) on a regular basis. This is especially important in individuals with advanced disease, since it is necessary to maintain viral suppression for an extended period.

References

1. Chevillotte M, von Einem J, Meier BM, Lin FM, Kestler HA, Mertens T. A new tool linking human cytomegalovirus drug resistance mutations to resistance phenotypes. Antiviral Res. 2010;85:318–27.
2. Irwin KK, Renzette N, Kowalik TF, Jensen JD. Antiviral drug resistance as an adaptive process. Virus Evol. 2016;2(1):vew014.
3. Alexander ME, Bowman CS, Feng Z, Gardam M, Moghadas SM, Röst G, et al. Emergence of drug resistance: implications for antiviral control of pandemic influenza. Proc R Soc B Biol Sci. 2007;274(1619):1675–84.

4. Reusser P, Cordonnier C, Einsele H, Engelhard D, Link H, Locasciulli A, et al. European survey of herpesvirus resistance to antiviral drugs in bone marrow transplant recipients. Bone Marrow Transplant. 1996;17(5):813–7.
5. Shaw T, Locarnini SA. Preclinical aspects of lamivudine and famciclovir against hepatitis B virus. J Viral Hepat. 1999;6(2):89–106.
6. Simmonds P. Genetic diversity and evolution of hepatitis C virus – 15 years on. J Gen Virol. 2004;85(Pt 11):3173–88.
7. Diana G, Gregory H. Ledipasvir/sofosbuvir (Harvoni): improving options for hepatitis C virus infection. P T. 2015;40(4):256–76.
8. Andrei G, Georgala A, Topalis D, Fiten P, Aoun M, Opdenakker G, et al. Heterogeneity and evolution of thymidine kinase and DNA polymerase mutants of herpes simplex virus type 1: implications for antiviral therapy. J Infect Dis. 2013;207(8):1295–305.
9. Griffiths A. Slipping and sliding: frameshift mutations in herpes simplex virus thymidine kinase and drug-resistance. Drug Resist Updat. 2011;14(6):251–9.
10. Renzette N, Gibson L, Bhattacharjee B, Fisher D, Schleiss MR, Jensen JD, et al. Rapid intra-host evolution of human cytomegalovirus is shaped by demography and positive selection. PLoS Genet. 2013;9(9):e1003735.
11. Lurain NS, Chou S. Antiviral drug resistance of human cytomegalovirus. Clin Microbiol Rev. 2010;23(4):689–712.
12. Renzette N, Bhattacharjee B, Jensen JD, Gibson L, Kowalik TF. Extensive genome-wide variability of human cytomegalovirus in congenitally infected infants. PLoS Pathog. 2011;7(5):e1001344.
13. Gilbert C, Boivin G. Human cytomegalovirus resistance to antiviral drugs. Antimicrob Agents Chemother. 2005;49(3):873–83.
14. Chen R, Holmes EC. Hitchhiking and the population genetic structure of avian influenza virus. J Mol Evol. 2010;70(1):98–105.
15. Bhatt S, Holmes EC, Pybus OG. The genomic rate of molecular adaptation of the human influenza A virus. Mol Biol Evol. 2011;28(9):2443–51.
16. Moscona A. Oseltamivir resistance—disabling our influenza defenses. N Engl J Med. 2005;353(25):2633–6.
17. Bloom JD, Gong LI, Baltimore D. Permissive secondary mutations enable the evolution of influenza oseltamivir resistance. Science (80-). 2010;328(5983):1272–5.
18. Thorlund K, Awad T, Boivin G, Thabane L. Systematic review of influenza resistance to the neuraminidase inhibitors. BMC Infect Dis. 2011;11:134.
19. Baranovich T, Wong S-S, Armstrong J, Marjuki H, Webby RJ, Webster RG, et al. T-705 (Favipiravir) induces lethal mutagenesis in influenza A H1N1 viruses in vitro. J Virol. 2013;87(7):3741–51.
20. Furuta Y, Gowen BB, Takahashi K, Shiraki K, Smee DF, Barnard DL. Favipiravir (T-705), a novel viral RNA polymerase inhibitor. Antiviral Res. 2013;100(2):446–54.
21. Götte M. The distinct contributions of fitness and genetic barrier to the development of antiviral drug resistance. Curr Opin Virol. 2012;2(5):644–50.
22. Goldberg DE, Siliciano RF, Jacobs WR. Outwitting evolution: fighting drug-resistant TB, Malaria, and HIV. Cell. 2012;148(6):1271–83.
23. Pennings PS, Kryazhimskiy S, Wakeley J. Loss and recovery of genetic diversity in adapting populations of HIV. PLoS Genet. 2014;10(1):e1004000.
24. Menéndez-Arias L. Molecular basis of human immunodeficiency virus type 1 drug resistance: overview and recent developments. Antiviral Res. 2013;98(1):93–120.
25. Khudyakov Y. Coevolution and HBV drug resistance. Antivir Ther. 2010;15(3 Pt B):505–15.
26. Bang KB, Kim HJ. Management of antiviral drug resistance in chronic hepatitis B. World J Gastroenterol. 2014;20(33):11641–9.
27. Shaw T, Bartholomeusz A, Locarnini S. HBV drug resistance: mechanisms, detection and interpretation. J Hepatol. 2006;44(3):593–606.
28. Elion GB, Furman PA, Fyfe JA, de Miranda P, Beauchamp L, Schaeffer HJ. Selectivity of action of an antiherpetic agent, 9-(2-hydroxyethoxymethyl) guanine. Proc Natl Acad Sci U S A. 1977;74(12):5716–20.

29. Wang D, Hicks CB, Goswami ND, Tafoya E, Ribeiro RM, Cai F, et al. Evolution of drug-resistant viral populations during interruption of antiretroviral therapy. J Virol. 2011;85(13):6403–15.
30. Strasfeld L, Chou S. Antiviral drug resistance: mechanisms and clinical implications. Infect Dis Clin North Am. 2010;24(2):413–37.
31. Gilbert C, Bestman-Smith J, Boivin G. Resistance of herpesviruses to antiviral drugs: clinical impacts and molecular mechanisms. Drug Resist Updat. 2002;5(2):88–114.
32. Deray G, Martinez F, Katlama C, Levaltier B, Beaufils H, Danis M, et al. Foscarnet nephrotoxicity: mechanism, incidence and prevention. Am J Nephrol. 1989;9(4):316–21.
33. Hasan MR, et al. Unusual accumulation of a wide array of antimicrobial resistance mechanisms in a patient with cytomegalovirus-associated hemophagocytic lymphohistiocytosis: a case report. BMC Infect Dis. 2020;20(1):237.
34. Kausar S, Said Khan F, Ishaq Mujeeb Ur Rehman M, Akram M, Riaz M, Rasool G, et al. A review: mechanism of action of antiviral drugs. Int J Immunopathol Pharmacol. 2021;35:20587384211002621.
35. Hong SP, Kim NK, Hwang SG, Chung HJ, Kim S, Han JH, et al. Detection of hepatitis B virus YMDD variants using mass spectrometric analysis of oligonucleotide fragments. J Hepatol. 2004;40(5):837–44.
36. Perno CF, Moyle G, Tsoukas C, Ratanasuwan W, Gatell J, Schechter M. Overcoming resistance to existing therapies in HIV-infected patients: the role of new antiretroviral drugs. J Med Virol. 2008;80(4):565–76.
37. Powdrill MH, Tchesnokov EP, Kozak RA, Russell RS, Martin R, Svarovskaia ES, et al. Contribution of a mutational bias in hepatitis C virus replication to the genetic barrier in the development of drug resistance. Proc Natl Acad Sci U S A. 2011;108(51):20509–13.
38. Foll M, Poh YP, Renzette N, Ferrer-Admetlla A, Bank C, Shim H, et al. Influenza virus drug resistance: a time-sampled population genetics perspective. PLoS Genet. 2014;10(2):e1004185.
39. Tenaillon O. The utility of Fisher's geometric model in evolutionary genetics. Annu Rev Ecol Evol Syst. 2014;45:179.
40. Hietpas RT, Bank C, Jensen JD, Bolon DNA. Shifting fitness landscapes in response to altered environments. Evolution (N Y). 2013;67(12):3512–22.
41. Lourenço JM, Glémin S, Galtier N. The rate of molecular adaptation in a changing environment. Mol Biol Evol. 2013;30(6):1292–301.
42. Muller HJ. The relation of recombination to mutational advance. Mutat Res – Fundam Mol Mech Mutagen. 1964;106:2–9.
43. Hill WG, Robertson A. The effect of linkage on limits to artificial selection. Genet Res (Camb). 2008;89(5–6).
44. Mullins JI, Heath L, Hughes JP, Kicha J, Styrchak S, Wong KG, et al. Mutation of HIV-1 genomes in a clinical population treated with the mutagenic nucleoside KP1461. PLoS One. 2011;6(1):e15135.
45. Pfeiffer JK, Kirkegaard K. A single mutation in poliovirus RNA-dependent RNA polymerase confers resistance to mutagenic nucleotide analogs via increased fidelity. Proc Natl Acad Sci U S A. 2003;100(12):7289–94.
46. Eid AJ, Arthurs SK, Deziel PJ, Wilhelm MP, Razonable RR. Emergence of drug-resistant cytomegalovirus in the era of valganciclovir prophylaxis: therapeutic implications and outcomes. Clin Transplant. 2008;22(2):162–70.
47. West P, Schmiedeskamp M, Neeley H, Oberholzer J, Benedetti E, Kaplan B. Use of high-dose ganciclovir for a resistant cytomegalovirus infection due to UL97 mutation. Transpl Infect Dis. 2008;10(2):129–32.
48. Rodriguez J, Casper K, Smallwood G, Stieber A, Fasola C, Lehneman J, et al. Resistance to combined ganciclovir and foscarnet therapy in a liver transplant recipient with possible dual-strain cytomegalovirus coinfection. Liver Transpl. 2007;13(10):1396–400.
49. Crassard N, Souillet AL, Morfin F, Thouvenot D, Claudy A, Bertrand Y. Acyclovir-resistant varicella infection with atypical lesions in a non-HIV leukemic infant. Acta Paediatr Int J Paediatr. 2000;89(12):1497–9.

Viral Diseases: Cellular Understanding of Diseases

Adithya Kaushal, Divya Kaushal, Ranjita Shegokar, and Yashwant Pathak

Abstract Viral diseases occur when an organism is invaded by pathogenic viruses or infectious virions, infectious virus particles containing an outer capsid and inner nucleic acid of a ribonucleic acid (RNA) or a deoxyribonucleic acid (DNA) that attach and enter susceptible host cells. Viral diseases share similar features, genomic type, and virion shape, yet their pathogenesis is different from one another. Viral diseases are prevalent, especially in the height of the SARS-CoV-2 pandemic. This pandemic has highlighted and continues to highlight viral diseases that are identical or similar to SARS-CoV-2. These diseases can be divided into three categories: DNA viruses, RNA viruses, and retroviruses. While many different viruses are associated with these three categories, three viral diseases were evaluated. The three prominently occurring viral diseases that are an ongoing public health threat worldwide are herpes simplex keratitis (HSK), Middle East respiratory syndrome coronavirus (MERS-CoV), and human immunodeficiency virus-1 (HIV-1). HSK is a double-stranded DNA virus (dsDNA) caused by the herpes simplex virus (HSV) infection of the cornea. HSK is one of the leading causes of blindness in the world. MERS is a single-stranded positive-sense RNA virus belonging to the *Coronaviridae* family, the most prominent family of RNA viruses. MERS-CoV can lead to symptoms ranging from upper respiratory infection, acute-to-severe lung infection, and multiorgan failure/death. HIV-1 is a retrovirus containing two single-stranded RNA molecules belonging to a family of heterogeneous lipid-enveloped RNA viruses. HIV-1 is the most widespread virus that severely damages the immune system, resulting in acquired immune deficiency syndrome (AIDS). The collective understanding of each pathogenesis and cellular comprehension is crucial to help identify the challenges and the opportunities to improve the current treatment for these viruses.

A. Kaushal · D. Kaushal · Y. Pathak (✉)
College of Pharmacy, University of South Florida Health, Tampa, FL, USA

R. Shegokar (✉)
CapnoPharm GmbH, Tübingen, Germany

R. Shegokar, Y. Pathak (eds.), *Viral Drug Delivery Systems*,
https://doi.org/10.1007/978-3-031-20537-8_3

Keywords DNA viruses · RNA viruses · Retroviruses · Herpes simplex keratitis (HSK) · Middle East respiratory syndrome coronavirus (MERS-CoV) · Human immunodeficiency virus-1 (HIV-1)

1 Introduction

Communicable diseases, specifically infectious diseases, have been present since the beginning of life on Earth. Moreover, infectious diseases have been extraordinarily prominent since the advent of human civilization. Historiographical methods of tracking and confirming the presence of infectious diseases have allowed for further clarification of the specificity of a particular type of infectious disease. Identification of types of infectious diseases is important for society writ large, but for scientific elucidation.

Infectious diseases can be novel in nature but can also be the result of previously dormant viruses that have been reintroduced to society through a plethora of factors [1]: environmental changes, lack of public health measures, and lack of public health education. Therefore, scientific elucidation of the said infectious diseases is essential because of the implications they have on society and also for further and continual scientific scholarship on understanding their circulation within populations.

Infectious diseases can be an entirely nebulous and all-encompassing classification of diseases that are predominant throughout the globe. There are an enumerable number of infectious diseases and various classifications of said diseases that could be described and expounded upon; however, the importance of this chapter is to uniquely highlight the prevalence of viral diseases and cellular understanding of the said viruses. Viral diseases significantly affect humans writ large and are singularly responsible for morbidity and mortality worldwide [2]. Therefore, comprehending viruses is crucial for intrinsically understanding the specific virus. However, for the hosts, particularly humans, for this chapter's discernment, they inhabit, affect, and ultimately impact disease manifestation [3].

The category of infectious diseases is wide ranging; to that effect, even the grouping of viral diseases falls into this group. While a variety of viruses could be discussed, the emphasis of this chapter will be to focus on DNA, RNA, and retroviruses uniquely. These three types of viruses were chosen not only because of their different viral biology and structure but also because of the preeminence of viral diseases in human populations that stem from these particular types of viruses.

This chapter examines three prevalent viral diseases to better understand pathogenesis, structure, biology, and relevant case studies. As mentioned, there are a variety of viruses that are based on DNA, RNA, and retroviruses. The three viral diseases that are further examined are herpes simplex keratitis (HSK), Middle East respiratory syndrome coronavirus (MERS-CoV), and human immunodeficiency virus-1 (HIV-1). HSK is a double-stranded DNA virus (dsDNA) caused by the herpes simplex virus (HSV), resulting in corneal infection. HSK is one of the world's most prominent causes of blindness. MERS is a single-stranded positive-sense RNA virus from the *Coronaviridae* family, an RNA virus family. MERS-CoV infection can cause various symptoms, such as upper respiratory infection, acute-to-severe lung infection, and multiorgan failure/death. Finally, HIV-1 is a retrovirus with two single-stranded RNA molecules that pertain to the lipid-enveloped RNA virus family. HIV-1 is the most prevalent virus that infects severe immune system damage, resulting in acquired immune deficiency syndrome (AIDS).

The relevance and importance of discussing viral diseases are to understand each disease cellularly better. Understanding the pathogenesis and cellular comprehension will help identify significant challenges and opportunities to improve therapeutics for these viruses. Through examining three distinct forms of viral diseases (DNA, RNA, and retrovirus), similarities and differences can be drawn between. Additionally, cellular understanding of viral diseases is constantly changing with time, and new findings are coming forth. Scientists are starting to understand disease processes that can help better understand and treat these diseases at a cellular level.

2 Background

Viruses are plentiful on Earth; it is estimated that there are 10^{31} unique viruses that are present [4]. Humans are primarily resistant to most of the present viruses; however, breakthroughs occur, and virus pathogenesis within humans can occur, as we have seen with COVID-19 [4]. While viruses are seemingly ubiquitous, one thing that should be kept in mind is that they are self-sustaining. While there is plenty of philosophical debate on whether viruses are living organisms, it remains scientifically steadfast, at the moment, that viruses are not living organisms [5]. In their essence, viruses are pathogens consisting of DNA or RNA as genetic material enclosed in a protein casing [6]. They are unique, infectious actors in terms of their lack of cellular and metabolic structures and their exclusivity in either having DNA or RNA as their primary genetic matter [6].

Although structurally and cellularly viruses are simplistic, the simplicity in their design is not a hindrance in their ability to be the ideal parasitic agent. Viruses, more specifically an individual pathogenic unit called a virion, begin their infection process through a fusion process, which results in viral fusion proteins binding to molecules on a cell membrane of the host organism [7]; for this chapter's purposes, assume a human host. Once encased in the cell membrane, viruses commandeer the cellular mechanisms of the host cell, allowing for the replication of viral genetic material [7]. Viruses repeat the process almost infinitely, producing many copies of their genetic material through its host's own cellular material.

The definition of a virus has changed over the SARS-CoV-2 pandemic. Many scientists and researchers are reexamining the definition to better understand viruses. Viruses are difficult to define due to the nature of their reproductive cycle, the intercellular stage of producing the viral particles (virions) and the extracellular stage of the virions escaping infected cells and reproducing in extracellular environments [8]. Viruses are the drivers of evolution and continuously adapt and grow to adjust to their environment. Viruses can be divided into three main categories, deoxyribonucleic acid (DNA), ribonucleic acid (RNA), and retroviruses (a combination of DNA and RNA). DNA, RNA, and retroviruses are similar yet have distinct cell cycles, replication, and expression strategies.

While this is a generalized and condensed account of viral functionality, viral machinations can differ depending on what type of virus is infecting a host cell. As mentioned, this chapter will examine the specificities in the viral design of DNA, RNA, and retroviruses (Table 1).

This table shows the similarities and differences of DNA, RNA, and retrovirus. Each was broken down to better understand the strand, size, sugar, replication, location, polymer chain length, bases associated, the types of each, and the common

Table 1 DNA, RNA and retrovirus background

Name	Deoxyribonucleic acid (DNA)	Ribonucleic acid (RNA)	Retrovirus
Scientific name	Deoxyribonucleic acid	Ribonucleic acid	Retroviridae
Strand	Double-stranded DNA (dsDNA)	Single-stranded RNA (ssRNA)	Two single-stranded RNA (ssRNA) enclosed in a lipid envelope
Size	0.6 nm	3000–7000 nt; capsid diameter 26–28 nm	7–12 kb
Sugar	Deoxyribose	Ribose	Ribose
Replication	Self-replicate	Cannot self-replicate; synthesized from DNA when needed	Reverse transcription – conversion of ssRNA genome copy to dsDNA
Location	Nucleus; small amount in mitochondria	Cytoplasm, nucleolus; sometimes a nucleus	Cytoplasm
Polymer chain	Long	Short	Short
Bases	Adenine, guanine, cytosine, thymine	Adenine, guanine, cytosine, uracil	Adenine, guanine, cytosine, thymine, uracil
Types	Double-stranded DNA viruses, single-stranded DNA viruses, and pararetroviruses	3 types: messenger RNA (mRNA), ribosomal RNA (rRNA), transfer RNA (tRNA)	Oncoviruses, lentiviruses, and spumaviruses
Common viruses	Herpes simplex keratitis (HSK), hepatitis B virus, human papillomavirus, adenovirus, Epstein–Barr virus (HHV4)	Middle Eastern Respiratory Syndrome Coronavirus (MERS-CoV), hepatitis C virus, Severe Acute Respiratory Syndrome (SARS), West Nile fever	Human immunodeficiency virus 1 (HIV-1), human immunodeficiency virus 2 (HIV-2), human T-lymphotropic virus-I (HTLV-I), HTLV-II, HTLV-V

viruses that are associated with each. This allows for a better understanding and breakdown of the DNA, RNA, and retrovirus.

2.1 *Deoxyribonucleic Acid (DNA)*

DNA is a molecule that contains the genetic blueprint for an organism's development, functioning, and reproduction [9]. Nuclear DNA is found in the nucleus of a cell. A complete nuclear DNA set is referred to as a genome [10]. DNA is composed of nucleotides, which are organic molecules that contain a nucleoside, a nitrogen-containing base, and phosphate and form a long chain of nucleotides and polymers. During reproduction, DNA is passed down from generation to generation. Furthermore, some viruses have a DNA core that allows them to survive in the nucleus of the cell they are infecting by replicating their own DNA using the host's biochemical makeup [11]. The virus DNA can be integrated with the DNA of the host cell. DNA viruses can be categorized into three categories of double-stranded, single-stranded, and pararetroviruses.

2.1.1 DNA Biology and Structure

As previously stated, DNA is a polymer composed of two distinct polynucleotide chains coiled around one another to form a double helix [12]. This polymer is made up of nucleotide repeats. The DNA structure is dynamic and can coil into tight loops like the helix or other shapes [13]. The two strands are coiled on the same axis and have the same pitch of 34 ångströms (Å) or 3.4 nm. Each of the pair chains contains a radius of 10 Å or 1.0 nm. An average DNA chain is measured to be 22–26 Å or 2.2–2.6 nm wide, whereas one nucleotide unit is measured at 3.3 Å or 0.33 nm.

DNA contains the genetic information required for organisms and viruses to develop, function, and reproduce. The DNA strands, or polynucleotides, are made up of monomer units called nucleotides. These nucleotides are organic molecules with nucleoside and phosphate on one end. DNA is made up of four nitrogen-containing nucleobases: adenine (A), thymine (T), guanine (G), and cytosine (C). These bases are classified into two types: pyrimidines and purines. Pyrimidines are aromatic heterocyclic organic compounds with a one-carbon nitrogen ring. Pyrimidines are composed of a single hydrogen-carbon ring and two nitrogen atoms. Thymine and cytosine are the pyrimidines for DNA. Purines are heterocyclic aromatic organic compounds composed of two-carbon nitrogen rings linked by an imidazole ring. Purines have two hydrogen-carbon rings and four nitrogen atoms. Adenine and guanine are the purines for DNA. Purines and pyrimidines perform the same functions and are DNA building blocks. Each pyrimidine and purine have a complementary nucleotide base held together by hydrogen bonds, allowing the base pairings to be separated for replication and translation. Adenine pairs with thymine via two hydrogen bonds, and guanine pairs with cytosine via three hydrogen bonds in DNA. To ensure complementary pairing between the bases, one strand of DNA will always be complementary to the second strand of DNA, as shown in Fig. 1.

In the process of compacting DNA molecules that are essential for gene regulation in living systems, DNA is tightly and orderly packed. Eukaryotic DNA is found in the cell nucleus and, to a lesser extent, in the mitochondria and chloroplasts. Prokaryotic DNA is found in nucleoids along with the cytoplasm. The genetic information in a genome is stored in the genes, and a complete set of genetic information within an organism is known as a genotype, which is sometimes referred to as alleles or carriers of a specific gene or genetic location [14]. This gene is a hereditary unit that influences specific characteristics of an organism. These genes can be transcribed and used as promoters or enhancers in regulatory sequences. In many species, only a small portion of the total genome sequence encodes proteins. Protein-coding axons make up about 1.5% of the human genome. Non-coding repetitive sequences make up more than half of the DNA in humans. Non-coding DNA (ncDNA) refers to DNA components that do not participate in protein sequences. As previously stated, the gene sequence of DNA contains genetic information that can influence an organism's phenotype. The messenger RNA sequence is defined by the bases in the DNA strand. Translation, the process by which ribosomes or the endoplasmic reticulum synthesize proteins in the nucleus of the cell, determines the relationship between the gene sequences and the amino acid sequences of the proteins. This is referred to as the genetic code or gene expression. This genetic code is made up of three-letter codons formed by three nucleotide sequences.

Additionally, transcription can occur when the codons of a gene are copied by RNA polymerase, an enzyme that synthesizes RNA using a DNA template into messenger RNA (mRNA) [12]. As aforementioned, the gene sequence of DNA contains genetic information that can influence an organism's phenotype. The mRNA sequence is defined by the bases in the DNA strand. Translation, the process by which ribosomes or the endoplasmic reticulum synthesize proteins in the nucleus of

Fig. 1 (**a**) DNA helix—Structure of the DNA helix containing two polynucleotide chains coiled around each other to form a helix shape. (**b**) DNA ladder—base pairs connected within the DNA helix. The DNA helix contains four nitrogenous bases: adenine, thymine, guanine, and cytosine. These bases are crucial and only pair up with the counter base. Adenine will pair with thymine, and guanine will pair with cytosine

the cell, determines the relationship between the gene sequences and the amino acid sequences of the proteins. This is referred to as the genetic code or gene expression. This genetic code is made up of three-letter codons formed by three nucleotide sequences. The specific enzyme creates a complementary strand of DNA through complementary base pairing from the original strand. The DNA polymerase extends DNA strands in a 5′ to 3′ direction, and the RNA polymerase will synthesize an RNA transcript complementary to move the template strand into a 3′ to 5′ direction. Using the DNA ligase, the 3′ of the DNA fragment will bind to the 5′ of the previous DNA.

2.2 *Ribonucleic Acid (RNA) Viruses*

RNA is a macromolecule found in all living cells that plays numerous biological roles [15]. In terms of structure, RNA is very similar to DNA, but the main difference is that RNA is single-stranded and has a backbone of alternating phosphate groups and a sugar ribose. RNA is responsible for converting genetic material from DNA into proteins. RNA is used in cellular protein synthesis via translation and carries genetic information that is translated into proteins by ribosomes. mRNA, ribosomal RNA (rRNA), and transfer RNA are three types of protein synthesis (tRNA). RNA viruses replicate their genomes, similarly to DNA, using RNA-dependent RNA polymerase (RdRp).

2.2.1 RNA Biology and Structure

RNA is a single-stranded polymeric molecule (ssRNA) that is required for gene coding, regulation, expression, and decoding [16]. RNA, like DNA, is a nucleic acid biopolymer that is required for life. Unlike DNA, which contains deoxyribose, RNA only contains one ribose sugar. The unique ribose backbone of RNA allows it to change at lower activation energy of hydrolysis than DNA. Each nucleotide in the RNA structure will have a ribose sugar with carbons labeled at 1′ to 5′ and a phosphate group attached at the 3′ position of the ribose and the 5′ position of the following. Because each phosphate group has a negative charge, RNA is a polyanion. Unlike DNA, RNA contains a hydroxyl group at the 2′ position of the ribose.

Adenine (A), uracil (U), guanine (G), and cytosine (C) are the four nitrogen-containing nucleobases that make up RNA [17]. There are two kinds of bases: pyrimidines and purines. The pyrimidines for DNA are uracil and cytosine. The purines for DNA are adenine and guanine. Except for thymine and uracil, three of the four bases are identical to the DNA bases. Both are pyrimidines, but in RNA, thymine is replaced by uracil (an unmethylated form of thymine). Purines and pyrimidines are RNA building blocks that perform the same functions. Each pyrimidine and purine has a complementary nucleotide base, which is held together by hydrogen bonds and allows the base pairings to be separated for replication and translation. Adenine is paired with uracil, while guanine is paired with cytosine. To ensure complementary pairing between the bases, one strand of RNA will always be complementary to the second strand of RNA, as shown in Fig. 2.

The RNA polymerase enzyme is catalyzed to synthesize RNA by using a DNA template through transcription. The process begins with the enzymes binding the promotor sequence in DNA. The DNA helix unwinds through helicase activity from the enzyme along the template strand in the 3′ to 5′ direction. During this process, the synthesis of a complementary RNA molecule in the 5′ to 3′ direction occurs [18]. After this process, the primary transcript of RNA is produced, yielding mature RNAs such as mRNA, tRNA, and rRNA. The specific primary transcript is used in mRNAs, which are modified for translation. mRNA is used to carry crucial information of protein sequences to the ribosomes [19]. This information is coded in codons that correspond to specific amino acids. The mRNA then moves from the nucleus to the cytoplasm where it is bound to ribosomes and translated with proteins into tRNA. The tRNA is a small chain of 80 nucleotides that transfer specific amino acids to a polypeptide chain when it is undergoing protein synthesis during translation. There the amino acid will attach, and the anticodon for the codon will bind through a specific sequence through hydrogen bonding. rRNA is the catalytic unit of the ribosome that hosts translation. The tRNA molecules are synthesized in the nucleolus. When out in the cytoplasm, the rRNA will combine with a protein to form a ribosome in which that ribosome may bind with an mRNA. In addition to the RNA polymerase enzyme, there are various RNA-dependent RNA polymerases (RdRp) that use the RNA as a template for the synthesis of a new RNA strand [20]. This enzyme uses an RNA template to catalyze a replication of a complementary RNA strand of the template. RdRP is a fundamental protein that is encoded in most RNA-containing genomes of viruses.

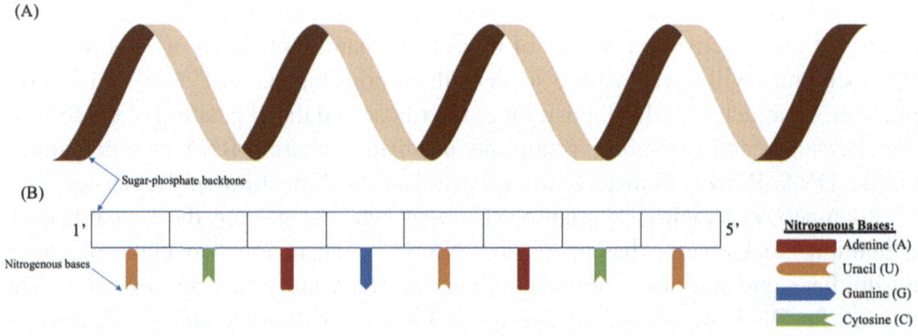

Fig. 2 (**a**) RNA helix—structure of the RNA that contains a single-stranded nucleotide. (**b**) RNA ladder—the four different nitrogenous bases of adenine, uracil, guanine, and cytosine. Similar to DNA, RNA has three of the same bases, adenine, guanine, and cytosine. The only difference is that the thymine from DNA is uracil in RNA. These bases are crucial and only pair up with the counter base. Adenine will pair with uracil, and guanine will pair with cytosine

2.3 *Retroviruses*

Retroviruses, which are members of the enveloped RNA virus family (*Retroviridae*), are common infectious agents found in a wide range of eukaryotes [21]. A retrovirus is a type of virus that uses and inserts a copy of the RNA genome into the DNA host cell that it infects, thereby altering the cell's genome. The retrovirus can infect either somatic cells or germline cells. When endogenous retroviruses infect the germlines, they become inherited (ERVs). Retroviruses are classified into three families: oncoretroviruses, lentiviruses, and spumaviruses. Each of these three groups is responsible for viruses that infect and kill humans, mammals, and birds.

2.3.1 Retrovirus Biology and Structure

Retroviruses, as previously stated, insert a copy of the RNA genome into the DNA host cell and alter the overall genome of the cell that they infiltrate. Once in the cell's cytoplasm, the retrovirus will use the reverse transcriptase (RT) enzyme to generate complementary DNA (cDNA) from an RNA template. The new DNA is integrated into the host cell genome via the integrase enzyme (IN), which forms covalent bonds between the host cell and the genetic information [22]. This occurs following the integration of double-stranded linear viral DNA from the RNA or DNA-dependent DNA polymerase reverse transcriptase. The IN's primary function is to insert viral DNA into the host chromosome, which is required for virus replication. During the integration process, the cell becomes a permanent carrier of the viral genome, and virus gene expression begins. The host cell's genome now contains the viral DNA and treats it as its own. The viral DNA will begin transcription and translation alongside the cell's genes. This process will begin to generate assembly proteins for virus copies (Fig. 3).

The average diameter of retrovirus virions is 80–100 nm. They have a lipid envelope surrounding an internal protein core that contains the viral genomic RNA. The core structure is made up of structural proteins such as matrix (MA), capsid (CA), and nucleocapsid (NC), as well as RT, integrase (IN), and protease (PR) enzymes. The viral envelope is formed when the virus particle buds from the host cell plasma membrane, into which viral glycoproteins are inserted. The glycoproteins are made up of disulfide-bonded transmembrane (TM) and surface (SU) subunits.

The retroviral genome contains dimers of single-stranded, positive-sense, linear RNA molecules that are packaged as viral particles. The structure of the retrovirus is a virus that contains genetic material that contains two concentric outer circles in which the envelope protein complex is embedded with capsid proteins represented as a hexagon. The layout of retroviruses in the RNA genome is 5′–gag–pro–pol–env–3′ [24]. Virions, independent particles of the retroviruses, are enveloped particles, ~100 nm in diameter. Virions contain two identical RNA molecules, ~7–10 kilobases in length [25]. These RNA molecules are dimers, formed by base pairings and complementary sequences. The envelope, RNA, and proteins are the three main

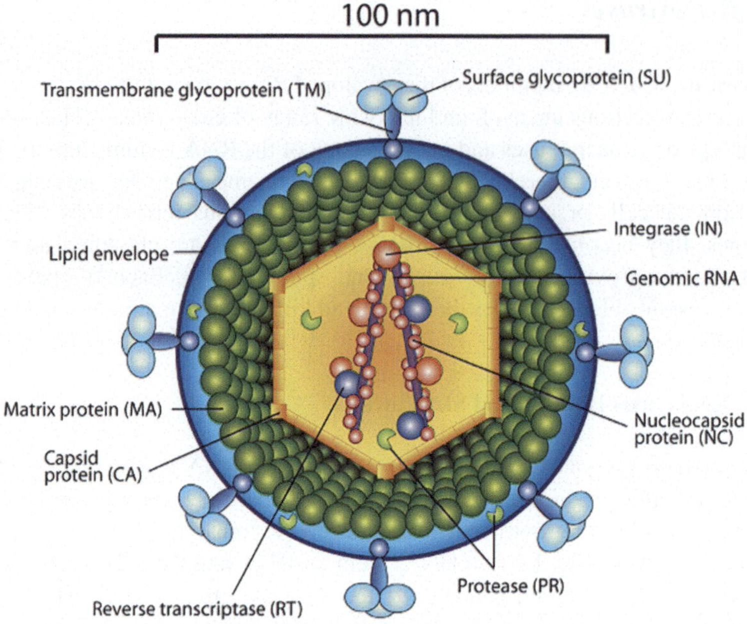

Fig. 3 Retrovirus Virion Structure [23]

components of all virions. The envelope is made up of lipids and glycoproteins that are encoded by the env spike protein, which forms the envelope and allows the retrovirus to target and attach to cells as well as infiltrate the cell membrane. The env spike protein is produced by the *env* gene, which also produces the surface protein (SU) and the transmembrane protein (TM). Both of these play an important role in the retrovirus's ability to bind to its target host via specific cell-surface receptors. The RNA in the virion is a dimer RNA with a cap at the 5′ end and a polymer at the 3′ end poly(A) tail. It contains noncoding terminal regions that are essential for replication and gene expression. The virion can contain five different proteins: group-specific antigen (gag), protease (pro), Pol proteins, Env proteins, and other RNA-associated proteins. Gag proteins are viral capsid components with two nucleic acid domains, the matrix (MA) and the nucleocapsid (NC) [26]. One of the most important functions of the Gag protein is to recognize, bind, and package retroviral genomic RNA into virions. In all retroviruses, the Gag protein serves as the precursor to the internal structural protein. Pro proteases are related to single domain proteases. Proteolytic cleavages produce mature gag and pol proteins during virion maturation. To produce mature proteins, the *pro* protein is cleaved by enzymes. Pol proteins are in charge of both the synthesis of viral DNA and the integration of the viral DNA into the host DNA cell after infection [27]. As previously stated, the env protein is critical in the entry and association of virions within the host cell. Retroviruses can become infectious, thanks to this specific protein. Other

Table 2 Organs affected by each virus

Virus	Primary organs affected
HSK	Eyes
MERS-CoV	Lungs and kidneys
HIV-1	Almost all bodily organs

retrovirus-dependent proteins can be found in virions. The NC protein, for example, coats the RNA (Table 2).

This table demonstrates the various viruses and the primary organs that are affected when infection takes place.

3 DNA Viruses: Herpes Simplex Keratitis (HSK)

HSK is an infection of the herpes simplex virus type 1 (HSV-1) or human herpesvirus type 1 (HHV-1) of the sensory ganglia, most commonly the trigeminal ganglion, and/or cornea – the transparent, protective outer layer of the eye covering the anterior chamber, iris, pupil, and lens [28]. HSK is caused by a reactivation of dormant HSV-1 within the body that may have been previously infected most likely through skin contact. It is estimated that approximately 90% of the human population is exposed to HSV-1 in childhood [29]. HSK has been classified as the most explicit cause of corneal blindness in developed countries across the globe. Global estimates have quantified that there are 1.5 million new cases of HSK each year, with 40,000 cases constituting extremely detrimental vision loss and/or blindness [30].

3.1 Virus Pathogenesis

As mentioned previously, HSV-1 typically occurs through close contact between persons. More specifically, HSV-1 penetrates the mucous membrane of the host, allowing for infection and replication to begin [31]. HSV-1 resulting in HSK can manifest itself through a variety of factors that are based on disease determinants of the particular host. Specifically, HSV-1 resulting in HSK can present itself in a host through re-activation of dormant HSV-1 that was previously active on a different bodily site on a host or simply through primary infection of HSK for the host that had no previous infection of HSV-1 [28]. For hosts that have been previously infected with HSV-1, the dormant virus has been shown through previous scientific studies and literature to lie in the mouth [28]. Moreover, the mouth has been identified as the primary site on hosts for the accumulation and spreading of HSV-1 [28]. For hosts where HSV-1 infection is novel and results in HSK, HSV-1 infection has been linked to droplet spread ocular entry via the tear film [28]. Through ocular entry, the conjunctiva of the host has been demonstrated as the primary site of infection and replication of HSV-1 [28]. HSV-1 entry either through the mouth or ocularly originates from the trigeminal ganglion, cornea, and brainstem. HSV-1, through neuronal axons, is carried via axonal retrograde transport and latent within nerve cell bodies at the sensory ganglia [28].

Once replication begins in nerve cell bodies, nascent parts of HSV-1 virions are transported anterogradely from the neuronal cell body to the periphery of the neuron, accompanied by further anterograde axonal spread to the cornea [28]. This movement of HSV-1 virions has been speculated to be a result of microtubule motors [28]. However, this movement is entirely contingent upon viral-DNA replication of HSV-1, which then results in anterograde transport of HSV-1 virions to the axonal chamber [32]. If the virus is unable to replicate, there is no transport and delivery to the axonal chamber and subsequently to the cornea. As a result, substantial infection of HSV-1 within the ganglion is vital for viral takeover.

In the cornea, HSV-1 primarily replicates within the epithelial sheaths [31]. While HSV-1 is anterogradely transported to and from axonal bodies resulting in HSK presenting itself within the cornea, HSV-1 can itself be present in the cornea via viral latency or novel infection. Thus, viral activation of HSV-1 within the cornea can serve as an additional genesis of HSK antigens that result in infection. While transport from the neuronal cell body to the periphery of the neuron can result in HSV-1 virions being transported to sensory ganglia resulting in HSK infection has been established, it is more likely and commonly believed that viral latency of HSV-1 within the cornea is responsible for HSK infection [31].

As with any type of viral infection, there is a substantial immune response that is present. In the case of particular diseases of the cornea, such as HSK, a substantial amount of native immune cells such as polymorphonuclear leukocytes (PMN), macrophages, and natural killer (NK) cells are introduced into the stroma [28]. In the leadup to corneal lesions presenting itself as a part of the latter stages of HSK infection, an influx of PMN and viral peaks after 4–5 days of initial infection [33]. A second immune inrush is noticed 7–8 days after initial infection and reaches an immune response apogee at around 15–21 days post-initial infection [33]. This second immune response can be uniquely contributed to the CD4+ T cells, particularly of the Th1 phenotype [33]. This secondary large immune response as a result of CD4+ T cells leads to a large inflammatory response, resulting in corneal lesions as a result of HSK manifestation [28].

3.2 Virus Biology and Structure

HSV-1 is a double-stranded DNA virus of significant size. HSV-1 has a genome of roughly 152 kb [34]. More specifically, HSV-1 is an alphaherpesvirus that is the first member of the human herpesvirus subset (HHV-1). An HSV-1 virion is approximately 120–300 nm and is constituted of a cloudy electron core that contains its viral genome [34]. The viral genome is encased in a 100 nm diameter capsid, tegument, and viral envelope, in the case of HSV-1 acquired from the host cell, encasing the entirety of viral genomic material. A distinct feature of herpesvirus is its unique capsid structure. In particular, it is composed of 162 capsomers, which are arranged to form an icosahedral [35]. The tegument is another critical part of the virion as this shapeless apparatus borders the capsid. It contains vital mRNAs and proteins, in particular the virion host shut-off (VHS) protein that necessitates viral genomic transfer from virion to host cell [35]. The envelope that encases the capsid is a lipid bilayer that is composed of glycoproteins. Glycoproteins, specifically gB, gC, gD, gH, and gL, are important for viral infection as they work in tandem with host cell receptors to initiate the interactivity of virion and host cell, resulting in the piercing of the viral capsid through the plasma membrane of the host cell [35]. Interaction of glycoproteins gB and/or gC glycoproteins with heparan sulfate precipitates the attachment of the virus to host cells [35]. Subsequently, gD binding to 3-OS heparan sulfate, herpes viral entry mediator (HVEM), or nectin-1 results in the viral capsid being transported to the cytoplasm of the host cell [35]. Moreover, it has been demonstrated that HSV-1 enters human host cells both through virion fusion with host cell surfaces and endocytosis [36]. Once the viral capsid is brought to the cytoplasm of the host cell, it is delivered to the host cell nucleus for viral replication via host cell DNA polymerase [35]. As viral replication takes place, new virions are released from that infected host cell, leading to further viral infection and subsequent replication [35]. In particular, the releasing of new virions is triggered by the removal of heparan sulfate moieties, through heparanase of the host enzyme, on the infected host cell [35]. As mentioned previously, HSV-1 can continue viral replication resulting in HSK manifestation, or it can become latent/dormant in the trigeminal ganglion via corneal nerves. Due to frequent oscillation from viral latency to viral activation, HSV-1 can be present within an infected host indefinitely.

3.3 *Clinical Cases*

HSV-1 infections, resulting in HSK, are the primary causes of corneal blindness in developed nations. For example, recent estimates have shown that 500,000 people within the United States have been infected with HSV-1, which subsequently results in HSK. Moreover, the prevalence of ocular HSV-1 infections is reflected in the costs spent on treatment. Estimates have also shown that the total costs of treatments, in the United States, for these specific types of infections are tens of millions of dollars annually [37]. However, it must be noted this may be a function of the healthcare system in the United States, as treatment decisions based on costs of said treatments are adjudicated on a case-by-case basis [38].

HSV-1 infections of the eye are typically clinically presented as unilateral infections, meaning either the right or left eye is infected [37]. While rare and typically presented in young children, approximately 1.3–12% of people who are infected can have a bilateral infection, meaning both eyes are infected [37]. The rarity of bilateral ocular HSV-1 infections is typically accompanied by the fact that such infections take place almost exclusively within patients who are immunocompromised [37]. Regardless of unilateral or bilateral infection, HSV-1 within the eye typically infects the corneal epithelium; however, it is possible that infections can take place in the anterior and posterior segments of the eye. The corneal epithelium is composed of five to seven layers of cells, approximately 50 μ in thickness, that serve as a protective barrier to the cornea [39]. As a barrier to the cornea, it prevents a variety of fluids, possibly resulting from tearing, and bacteria from penetrating the epithelium and corneal stroma. Ocular HSV-1 infections, known as HSK, which stands for herpes simplex keratitis or herpes stromal keratitis, typically present clinically in three types: epithelial, stromal, and endothelial [37]. Clinically, it has been demonstrated that epithelial keratitis is the most common type of HSK. Table 3 virus and related organ and tissue affected sites e.g. HIV- spleed or lympnodes, etc

Ophthalmologists through patient history and symptoms are able to initially presume, even before a slit-lamp examination, that an infection of HSK has taken place. HSK patients who are symptomatic typically report ocular irritation, redness, discharge, excessive tearing, light sensitivity, and/or pain. The diagnosis of HSK is then confirmed and completed by an ophthalmologist during their slit-lamp examination in order to rule out other ocular irritations such as allergic conjunctivitis or dry eye syndrome. The most prevalent form of HSK, epithelial keratitis, presents on the eye as rough, granulated spots that coalesce to form branching ocular lesions, classified as dendritic lesions [37]. The dendritic lesion may be initially difficult to detect via slit-lamp examination; therefore, an ophthalmologist will use fluorescein staining of the cornea to confirm the presence of dendritic lesions [37]. Stromal keratitis on the slit-lamp examination will show a cloudy, almost opaque, appearance of the corneal stroma, the middle layer of the cornea [37]. Stromal keratitis may result in corneal edema and angiogenesis and like epithelial keratitis will present itself through corneal lesions. Stromal keratitis has been associated with an

Table 3 Drug treatment for viruses

Virus	Preferred drug treatment
HSK	Acyclovir, valacyclovir, and trifluridine
MERS-CoV	No targeted drug treatments available Acetaminophen or nonsteroidal anti-inflammatory drug (NSAID) used to treat fevers and body fatigue
HIV-1	Antiretroviral therapy (ART) including at least two of the following drug classes: Nucleoside reverse transcriptase inhibitors Non-nucleoside reverse transcriptase inhibitors Protease inhibitors Fusion inhibitors CCR5 antagonists Integrase strand transfer inhibitor Attachment inhibitors Post-attachment inhibitors Pharmacokinetic enhancers

immune response to HSV-1 infection and can result in subsets of this particular type of keratitis manifesting itself: disciform, necrotizing, and immune stroma [37]. These three subsets of stromal keratitis are a result of the various immune structures that are triggered by HSV-1 infection. Disciform keratitis will present itself on slit-lamp examination as a disciform lesion that is disk-shaped in the presence of corneal stromal edema. There are typically accompanying Descemet membrane folds and a general decrease in corneal responsiveness as well. Additionally, while disciform keratitis is responsible for roughly 2% of HSK manifestations, there is significant recurrent infectivity, 20–48% [37]. Necrotizing keratitis is confirmed via slit-lamp examination due to an opaque corneal stroma and the presence of necrosis and ulcer formation. Immune stromal keratitis is similar to stromal keratitis in its clinical presentation. Lastly, endothelial keratitis upon slit-lamp examination is discerned by the presence of fine keratic precipitates, inflammatory cellular deposits typically of a circular shape, stromal edema, and acute iritis.

While HSK infections do not overtly compromise ocular function significantly, its latency and propensity for reactivation are why appropriate medications are of paramount concern to those who are infected. As a result, it is necessary that physicians caring for patients that are infected with HSK aggressively treat this viral infection in its infancy to reduce viral proliferation, increase virus dormancy, and most importantly prevent consequential vision loss. There are currently a variety of treatments that ophthalmologists have at their disposal. Before the prevalence of antiviral therapies to treat HSK infection, an eclectic range of treatments from chemical cautery, antiseptics, radiotherapy, and even snake venom were used to treat viral infection. However, as medicine evolved and continues to do so, more widely proven agents are used to treat HSK. Given the nature of the three types of HSK infection, treatments, although mostly similar, are slightly modified given the particular type of HSK a person may be infected with.

Epithelial keratitis, being the most common type of HSK, has a variety of treatments available. Typically, topical antiviral agents are the treatment of choice for these types of infections. More specifically, topical antiviral agents that are nucleoside analogues are chosen because of their ability to intersperse the production of viral DNA via the irrevocable binding of viral DNA polymerase within the host cell that has been infected [40]. Existing therapies included drugs whose main components are composed of idoxuridine, iododeoxyuridine, vidarabine, and trifluridine [34, 40]. While clinical efficacy was shown, it was noted that these drugs had a propensity for deactivating DNA synthesis of both normal and viral cells and also exhibited low body solubility, which required patients to apply nine droplets of solution into the affected eye or eyes, every 2 hours [34]. This resulted in ocular complications such as epithelial keratitis, ulcer formation, and prolonged ocular lesions [34]. As a result, most of these therapies were discontinued. However, trifluridine as a 1% solution remains a prevalent treatment, especially in the United States, for HSK despite its plentiful negative side effects [34, 40]. Current treatment of epithelial keratitis infections primarily surrounds the usage of acyclovir, ganciclovir, and valacyclovir [34, 40]. All three, derivations of one another, are also nucleoside analogs that are uniquely phosphorylated by the viral thymidine kinase, resulting in the drug being used as a substrate for viral DNA; these drugs essentially inhibit viral DNA production while simultaneously not interacting with normal cells of the host [40]. While new viral DNA is not able to be produced, these drugs are not a panacea for already infected cells. Acyclovir and ganciclovir are formulated as a 3% ointment or 400 mg tablets and 0.15% gel, respectively [34, 40]. However, their body solubilities within people are significantly low, acyclovir more than ganciclovir, resulting in higher dosages for those who are infected and increased side effects such as nausea, vomiting, diarrhea, other gastrointestinal disturbances, blurred vision, and general eye irritation [40]. Conversely, valacyclovir is taken orally and has shown an increased bio-solubility as compared to acyclovir and ganciclovir [40]. However, like the two previously mentioned drugs, it possesses side effects that are similar, including possible renal disease. In general, these drugs in combination with trifluridine 1% solution constitute the current treatments for epithelial keratitis. However, it must be noted the extended treatment for those who are infected, especially those who are immunocompromised, has not prevented newly formed viral cells from being eliminated.

Stromal keratitis is pathogenetically derived from an immune response to HSV-1 infection. As a result of a large inflammatory response, the three drugs, acyclovir, ganciclovir, and valacyclovir, are used, not as a direct counter to the inflammation but to target and reduce viral load to decrease the size of inflammation that may be present [40]. To treat the inflammatory response that is present due to stromal keratitis, ophthalmologists prescribe topical corticosteroids, primarily prednisolone acetate 1% [41], in conjunction with antiviral drugs, to dramatically reduce the proliferation of an inflammatory response seen in the corneal stroma. If stromal keratitis is present within someone who is infected but does not present any corneal ulcers, the typical treatment that is prescribed is a prophylactic dose of whichever antiviral drug coupled with a topical corticosteroid, the latter being tapered over 10 weeks or longer [40]. However, if corneal ulcers are present, a full treatment dose of an antiviral drug is given in conjunction with a topical corticosteroid, the latter still being tapered over 10 weeks or longer [40].

While a clinically present form of HSK, endothelial keratitis is not as commonly presented as compared to epithelial or stromal keratitis. However, clinical studies have demonstrated similar effectiveness in treatment style as was demonstrated with stromal keratitis that presents with corneal ulcers; a full treatment dose of an antiviral drug is given in conjunction with a topical corticosteroid, the latter being tapered over 10 weeks or longer [40]. Specifically, when an oral antiviral agent was given in conjunction with a topical corticosteroid, it resulted in a much quicker resolution of symptoms pertaining to infection that results in endothelial keratitis. Due to endothelial keratitis mainly having viral proliferation concentrated in the cornea posteriorly, therapeutics with the ability to permeate the anterior changer, such as the oral antiviral acyclovir, are able to also permeate the aqueous humor at full treatment levels. In a set of guidelines produced by The American Academy of Ophthalmology, prophylactic and full/therapeutic treatments for the various forms of HSK infection are explained in detail [42].

4 RNA Viruses: Middle East Respiratory Syndrome Coronavirus (MERS-CoV)

Middle East Respiratory Syndrome (MERS) is a viral respiratory infection caused by Middle East Respiratory Syndrome coronavirus (MERS-CoV), a lethal zoonotic pathogen first identified in Jeddah, Saudi Arabia, in June 2012 [26]. MERS-CoV, similar to the severe acute respiratory syndrome coronavirus (SARS-CoV) outbreak that emerged in 2002 in China and spread globally through human-to-human transmission and ended in 2004, was seen to induce a fatal severe lower respiratory infection. Despite the SARS-CoV outbreak between 2002 and 2004 (8906 cases, 774 deaths; reported by the World Health Organization (WHO)), the MERS outbreak of 2012 occurred a decade later [43]. The WHO confirmed 837 confirmed cases and 291 fatalities [43]. Currently, there is no vaccine or specific treatments for MERS-CoV other than human polyclonal immunoglobulin G (IgG) antibody (SAB-301) as a therapeutic agent [44].

4.1 Virus Pathogenesis

As stated previously, MERS-CoV was identified as a zoonotic virus, through camel to human transmission, in June 2012. MERS-CoV is a highly pathogenic virus leading to high rates of mortality [45]. This virus belongs to the same family of severe acute respiratory syndrome coronavirus (SARS-CoV). MERS-CoV emergence was of interest after the SARS-CoV 2002 outbreak. Similar to other SARS-CoV like SARS-CoV-2, patients and healthcare personnel had contracted a respiratory infection in April 2012, but it was not until June 2012 when the patients came in with severe pneumonia-like symptoms and kidney failure that a form of coronavirus was detached from his sputum [46]. MERS-CoV was determined after a patient in Saudi Arabia came to a hospital that had respiratory and pneumonia-like symptoms and later died. Another patient in the Middle East was detected to have similar symptoms of a severe respiratory infection and was found to be infected with the same virus as the patient that had died. Upon analysis, the coronavirus was referred to as human coronavirus Erasmus Medical Center (hCoV-EMC), until it was later classified as MERS-CoV [47]. From the detection of the virus, the spread of the virus has occurred through the Middle East and Europe. Between 2012 and 2018, the WHO received a total of 2279 confirmed cases of MERS-CoV and 806 associated deaths, resulting in a fatality rate of approximately 35.36% [48]. MERS-CoV is considered a major challenge to global health due to its high mortality rate (36%) and pathogenicity, as well as the lack of a vaccine or any other definite treatment, and presents a pressing need for research and development of definite therapeutic options and adequate management to prevent infection [49].

Virologists used a real-time reverse-transcription polymerase chain reaction (RT-qPCR) method to test the features of certain coronaviruses as well as screen for

RNA-dependent RNA polymerase (RdRp), a coronavirus gene that infects humans; MERS contains a positive RdRp [50]. Similarly to other coronaviruses, there was an inefficient spread between infected patients, suggesting a zoonotic transmission (transmitted between species from animals to humans or humans to animals). The origin is believed to come from Tylonycteris bat CoV HKU4 (Ty-BatCoV HKU4) and Pipistrellus bat CoV HKU5 (Pi-BatCoV HKU5), yet humans were infected by dromedary camels through direct or indirect contact [51, 52]. Transmission from humans to humans required close contact with the infected person.

MERS-CoV infection has an average incubation time of 5 days (2–14 days) [53]. During this time, the host exhibits no signs of infection. The disease's clinical symptoms range from mild upper respiratory infection symptoms like cough, fever, and myalgia to severe forms like pneumonitis and respiratory failure [54]. In addition, patients may experience abdominal pain, loss of appetite, nausea, diarrhea, vomiting, and gastrointestinal symptoms. Hemoptysis and diarrhea without any fever are less common symptoms [54].

Positive-stranded RNA viruses called coronaviruses mostly infect animals, especially bats, but a small subset can also infect humans and cause diseases [55]. Human coronaviruses are broadly classified into two types: α- and β-coronavirus [56]. MERS-CoV belongs to the coronavirus family. It has four major surface proteins that help the virus enter cells, envelope protein (E), spike protein (S), nucleocapsid protein (N), and membrane protein (M), as discussed in Sect. 4.2. The spike (S) protein is a transmembrane glycoprotein composed of two subunits.

SARS-CoV was deemed the most pathogenic coronavirus before MERS-CoV [57]. The higher pathogenicity of MERS-CoV was demonstrated by the greater number of deaths caused by this virus. MERS-CoV, like the SARS virus, infects and replicates in human airway epithelial cells and suppresses interferon production [58]. Unlike SARS-CoV, the MERS virus has a broad tissue tropism [59]. Compared to SARS-CoV, MERS-CoV can induce pro-inflammatory cytokines but not innate antiviral ones. MERS-CoV causes a delayed pro-inflammatory response and suppresses innate immunity, implying that it is more lethal than SARS-CoV [60]. After entering the respiratory tract, the MERS virus primarily interacts with the host DPP4 receptor via its spike (S) protein.

SDPP4 receptors can be found on the epithelial surface of many human organs, including the lungs, kidneys, liver, bone marrow, thymus, and intestines [61]. MERS-CoV infects macrophages, which are specialized cells that detect, phagocytose, and destroy bacteria and other harmful organisms. After infection, macrophages release pro-inflammatory chemokines and cytokines such as IL-1β, IL-6, and IL-8 [62]. In MERS, these virus-infected macrophages play a significant role in developing disease symptoms [63]. Infection of human epithelial cells with MERS-CoV causes the release of pro-inflammatory chemokines and cytokines from monocyte-derived macrophages. These chemokines/cytokines are thought to cause inflammatory changes and tissue injury by infiltrating immune cells in the lower respiratory tract [63]. MERS virus infection causes slow but significant IFN type I and II responses in epithelial cells [64].

4.2 Virus Biology and Structure

MERS-CoV is a single-stranded positive-sense RNA virus belonging to the *Coronaviridae* family. The *Coronaviridae* family members contain large, enveloped, single-stranded RNA viruses with genomes ranging 25–35 kb and virions of 118–140 nm in diameter [65]. MERS is an enveloped nidovirus, a large group of envelope-positive-stranded viruses. This nidovirus is decorated with homotrimers of the S glycoprotein that allows and mediates entry within the host cell. The MERS-CoV contains numerous proteins such as Spike (S), Envelope (E), Membrane (M), and Nucleocapsid (N) and in addition contain various virions, accessory, and non-structural proteins to allow for viral replication. As stated, MERS-CoV contains a Spike (S) protein, a type I transmembrane glycoprotein that is a crucial antigen located on the surface of the virus that can target and neutralize antibodies during the infection. The S protein comprises 1353 amino acids that are glycosylated and contain a large extracellular domain and short cytoplasmic terminal. The S protein virus binding, fusion, and entry can be divided into S1 and S2. These proteins are arranged using open reading frames (ORFs), encoded on the virion-sense strand (V) or on the complementary-sense strand (C) to determine the structural protein arrangement. The arrangement for MERS and other CoVs is 5′-ORF1a/b-S-E-M-N--poly(A)-3′ [44]. Like many other CoVs, MERS-CoV carries genes encoding accessory proteins that contribute to virulence and pathogenesis and antagonize host antiviral responses, such as a type I IFN response.

The S1 subunit has 240 residues, an N-terminal domain (NTD), and a receptor-binding domain (RBD). The internal subdomains of MERS-RBD are conserved across all CoVs, whereas the external subdomain structure varies considerably [66]. MERS-RBD external subdomains primarily participate in receptor binding, resulting in CoVs using different receptors. The RBD is in charge of dipeptidyl peptidase IV binding (DPP4, also known as human CD26). MERS-RBD has been shown to evoke strong antibody production in vivo, indicating that the S protein is immunogenic for the induction of neutralizing antibodies (nAbs) [67]. The S2 subunit participates in viral fusion by assembling heptapeptide repeats 1 and 2 (HR1 and HR2) into a typical six-helix bundle, fusion core. Three HR1 spirals are produced by three HR2 chains in the HR1 side groove core, resulting in an essential central coiled spiral core for membrane fusion [51, 68]. When S1 and DPP4 bind, HR1 binds HR2, forming a temporary intermediate structure. As a result, the viral and cell membranes move closer to fuse the membranes. The S protein plays a crucial role in the virus entry using proteases, enzymes that break proteins into single amino acids or smaller polypeptides resulting in new proteins.

MERS-CoV E protein is an inner membrane protein and a small structural protein. The length is 82 amino acids, including at least 1 transmembrane helix [69]. This protein is vital for intracellular trafficking, host recognition, viral assembly, and virus budding. By interacting with the N protein, the M protein plays a vital role in viral assembly, viral envelope formation, and viral core formation [70]. The N protein contains 413 amino acids and is classified as a phosphorylated essential protein. The N protein aids in the formation of a nucleocapsid by binding to the RNA genome, which is required for virus replication and assembly [71].

A comparison of MERS-CoV and other CoVs reveals that the N protein has low homology across the entire amino acid sequence. Overexpression of the SARS-CoV N protein can increase replication [72]. The nucleocapsid structure requires recognition of the viral RNA's characteristic sequences and binding to other structural viral proteins. The RNA is protected from nucleases in the host cell after the N protein forms a complex with the viral genomic RNA [73]. SSARS-CoV and MERS-CoV N proteins are ADP-ribosylated, a common post-translational modification.

CoV RNA synthesis is linked to replication organelles (ROs) made of a modified endoplasmic reticulum (ER) membrane. These are converted to double-membrane vesicles (DMVs) containing viral dsRNA and other membranes. MERS-CoV helicase, nsp13, is a critical viral replication enzyme that influences tropism and virulence [74].

4.3 *Clinical Cases*

Since 2012, MERS-CoV has represented a virulent pathogenic virus. Interestingly, MERS-CoV has been largely geographically contained to the Middle East. In addition, studies have shown that MERS-CoV pervasiveness may be contingent upon seasonal discrepancies, with summer months correlating to increased infectivity of MERS-CoV [75]. Eighty-eight percent of MERS-CoV infections have taken place in the Middle East, 11% have taken place in Asia, and 0.1% have taken place in the United States [75]. Moreover, many cases of MERS-CoV have been associated with travel to the Middle East [76]. While MERS-CoV is a zoonotic virus, human-to-human transmission of MERS-CoV has been demonstrated, especially between individuals who are in extremely close contact with one another [77]. However, larger-scale human-to-human transmission has been noted when individuals are in a hospital setting [78]. For example, in the Republic of Korea, from June 1 to July 31, 2015, at the Samsung Medical Center, 186 individuals, with 181 of the infections stemming from nosocomial transmission, and 25 healthcare employees were infected with MERS-CoV [76]. This significant transmission in a hospital setting was a result of a Korean national returning with a respiratory illness after traveling to various Middle Eastern states [76].

MERS-CoV is typically diagnosed via the use of real-time reverse transcriptase polymerase chain reaction (rRT-PCR) [79]. However, while rRT-PCR has been effective at detecting the presence of MERS-CoV, the time-till detection and determinations of viral load (VL) are constant sources of concern as it pertains to a timely and accurate rRT-PCR to determine the presence of MERS-CoV [79]. Cycle threshold values (Ct) represent the number of cycles where it takes to determine detection of, in this case, viral material [80]. Ct values are inversely related to VL; therefore, a lower Ct value is correlated to higher VL and vice versa [80]. However, in the case of MERS-CoV, scientific studies are in their infancy as it relates to Ct values and their connection to acute illness [81].

MERS-CoV can clinically manifest itself in a variety of ways. While the average incubation period for MERS-CoV is 5.2 days, it has been noted that it can be anywhere from 1.9 to 14.7 days [82]. Clinical manifestations of MERS-CoV infection can be on a spectrum ranging from asymptomatic to severe disease. Patients with mild infections have been documented as having low-temperature fevers, headaches, sore throat, and difficulty breathing. Patients with severe infections can have acute respiratory distress syndrome (ARDS), myocarditis, renal failure, and additional bacterial infection. Nassar et al. have documented a "classic" packaging of symptoms that are emblematic of MERS-CoV infection, which typically include but are not limited to fever with chills, shortness of breath, cough, and presentation of pneumonia [75].

Treatment for MERS-CoV is centered around the prevention of additional infections and respiratory and renal failure that may arise from milder to more severe cases [53]. Treatment is currently prescribed as such as there are no current vaccines or targeted therapeutics that have been created for MERS-CoV [53]. While a variety of targeted and combined drug therapies, such as antivirals, resveratrol, interferon, and monoclonal antibodies [82], are in their infancy in scientific and clinical studies, it is too early to determine the efficaciousness of the said care. At this point in time, MERS-CoV treatment is exclusively supportive care, where the body is provided with ancillary defenses, such as pain medication, in order to fight infection [83].

5 Retroviruses: Human Immunodeficiency Virus-1 (HIV-1)

Human immunodeficiency virus-1, more commonly known as HIV-1, is a retrovirus containing two single-stranded RNA molecules in a lipid-enveloped RNA virus. HIV-1 is part of a genus of retroviruses, *Lentivirus*, which have long incubation periods and long-duration illnesses. *Lentivirus* uses viral complementary DNA in the host cell DNA to infect the nondividing cells [84]. There are two subtypes of HIV, HIV-1 (more common) and HIV-2. HIV-1 is a more common pathogenic strain of the virus, which can be divided into a major group, Group M, and minor groups, Group N, O, and P [85]. The two subtypes of HIV originated from West-Central African non-human primates through zoonosis, an infectious pathogen transmitted from animal to humans. HIV-1 originated in Southern Cameroon through simian immunodeficiency virus (SIV) from chimpanzees. HIV can cause a variety of symptoms from asymptomatic to influenza-like symptoms. HIV-1 is the most widespread virus that severely damages the immune system resulting in acquired immune deficiency syndrome (AIDS), allowing pathogens to take advantage of a weakened immune symptom and thrive [86].

5.1 Virus Pathogenesis

As mentioned earlier, HIV-1 is a retrovirus with two ssRNA surrounded by a lipid envelope. The transmission of the virus starts as a single virus particle that can establish a new infection. HIV-1 is a slow degrading chronic infection that can lead to AIDS [87]. On June 5, 1981, the Centers for Disease Control and Prevention (CDC) first reported an article in its Morbidity and Mortality Weekly Report (MMWR) describing a rare lung infection, *Pneumocystis carinii* pneumonia (PCP), in five young, healthy gay men in Los Angeles, California. Immunologists and their colleagues reported that all men had an unusual infection in which the immune system was compromised [88]. By the time the article was published, two patients had died, and the others were dying. On the same day, in New York, a dermatologist reported a cluster of cases of a rare, aggressive cancer, Kaposi's sarcoma (KS), among gay men. KS is a rare type of cancer that can be classified through abnormal growth of lymph and blood cells, causing red or purple lesions – later discovered to be an AIDS-defining condition. A few days later, on June 16, a young white gay man presented and was admitted to the Clinical Center at the National Institutes of Health (NIH), exhibiting severe symptoms of immunodeficiency, the first patient with AIDS. The patient died a few months later. By August 28, there were over 100 cases, 94% of the cases resulting in patients whose orientation was gay/bisexual, and 40% of the cases resulted in death. By the end of the year, there were over 300 reported cases of individuals with severe immune deficiency, the majority of the cases being adults and a few children under the age of 13. By 1984, scientists had classified AIDS to be a retrovirus, HTLV-III. There has been substantial research to understand HIV-1 and AIDS better. As of 2020, over 37.7 million people have HIV – 36 million are adults, and 1.7 million are children between the ages of 0 and 14 [89]. With no current cure for HIV/AIDS, there is a treatment that can be taken, the antiretroviral therapy (ART), involving a combination of medicines for all patients living with HIV to live healthier and longer [90].

HIV has been classified as a sexually transmitted infection (STI). The transmission of HIV can come from five mucosal sources: blood, semen/pre-seminal fluid, rectal fluids, vaginal fluids, and breastmilk [91]. The epidemiological factor is the direct contact through human-to-human transmission (HHT), from one infected person to a susceptible person. The most common way that HIV is spread is through vaginal or anal sex with an infected person or through sharing of drug equipment such as needles. AIDS is a chronic and potentially life-threatening condition that damages the immune system and interferes with the ability of the body to fight infection. Patients who are infected with HIV develop flu-like illnesses. It has a 2–4-week incubation period. The symptoms that an infected person may go through include fever, headache, muscle aches/joint pain, rash, sore throat, mouth sores, swollen lymph glands, diarrhea, weight loss, cough, and night sweats [92]. HIV can lead to the progression of AIDS, at the point in which the immune system has been severely compromised, and opportunistic infections.

In most cases, HIV-1 infection begins with a single virion infecting a single target cell at the point of entry. Blood levels of the CD4+ T-cell target and antiviral antibodies; and viremia (virus in the blood), as measured by infectivity, immunoassay for viral proteins, and, most accurately, PCR for viral RNA [93]. The infection period is divided into two phases: eclipse (1–2 weeks) and acute/primary infection (2–4 weeks). The clinical latency period will follow these two infection periods (1–20 years). Virus replication occurs in the mucosa, submucosa, and draining lymphatics and possibly to a lesser extent in gut-associated lymphoid tissue (GALT) and systemic lymphatic tissues during the eclipse phase clinically silent [94]. Once detected in blood plasma, the virus multiplies exponentially due to explosive replication in GALT and peripheral lymphoid tissue compartments [95, 96]. The virus freely replicates and spreads from the site of infection to the numerous tissues and organs that provide replication sites. The acute infection phase is identified by relatively high levels of viremia and high proportions of infected CD4+ T cells in the blood and lymph nodes [97]. This stage is frequently, but not constantly, accompanied by "flu-like" symptoms such as fever and enlarged lymph nodes. The immune response begins to appear around peak viremia, both in the form of antibodies against all viral proteins and a CD8+ T-cell response against HIV-1 antigens expressed on infected cells. The high levels of viremia are most likely due to the absence of the early immune response and the generation of a large number of activated CD4+ cells as part of the host response, providing targets for viral replication [98]. This stage is characterized by a transitory decrease in the number of CD4+ T cells in the blood. The clinical latency phase (1–20 years) is differentiated by a constant or gradually increasing level of viremia, 1–100,000 copies/mL, and gradually declining levels of CD4+ T cells [98]. Individuals in this stage are asymptomatic and unaware that they have been infected. Numerous CD4+ T cells contract the virus and perish daily, indicating that it is far from latent.

HIV-1 has been isolated from bodily secretions, and the virus is expected to replicate in activated CD4+ T cells virtually anywhere in the body [99]. Natural isolates of HIV-1 require CD4 to infect cells. As a result, robust infection of cells is restricted to those expressing CD4. The normal function of CD4 is to act as a coreceptor in conjunction with the T-cell receptor binding to Class II MHC, found on antigen-presenting cells responsible for presenting heterologous peptides to CD4+ T helper cells [100]. Other types of cells, such as monocytes and macrophages, can express lower levels of CD4, and it has been reported that CD4 serves an additional function as an IL-16 receptor [101].

Despite no cure for HIV/AIDS, treatment can be taken to elongate the duration and quality of life. Therefore, it is crucial that proper testing be done in order to determine the diagnosis of HIV. As published by *The Lancet*, the findings provide comparable evidence on viral suppression and HIV transmission risk for gay men similar to heterosexual couples, implying that the risk of HIV transmission in gay couples through condomless sex is zero when the HIV viral load is suppressed. Their findings support the message of the U=U (undetectable equals untransmittable) complemented with early HIV testing and treatment [102].

5.2 Virus Biology and Structure

HIV-1 is a retrovirus containing two noncovalently linked single-stranded, positive-sense RNA molecules enclosed by a circular capsid lipid envelope composed of viral particle capsid (p24). HIV-1 is divided into four phylogenetic groups, M, N, O, and P, each of which represents a distinct introduction of simian immunodeficiency viruses (SIVs) from naturally infected great apes into humans. The M group caused the pandemic to spread [85]. The HIV virion is 100 nm in diameter and contains a single-stranded RNA (ssRNA), envelope glycoproteins, the capsid, and enzymes (reverse transcriptase, integrase, and proteases) [103]. The two copies of the unspliced RNA genome, 5′-capped and 3′-polyadenylated RNA, contribute to the HIV-1 recombination during reverse transcription of viral replication. Despite the presence of two copies of ssRNA genome within a virion, it produces a pseudodiploid, an essential component to viral reproduction giving rise to one DNA copy within the infected cells [104], otherwise known as a diploid with chromosomal translocations.

All retroviruses have the same viral structural proteins of *gag* (group-specific antigen), *pol* (gene or protein in retrovirus that codes for enzymes), and *env* (envelope) – all highlighted in Sect. 2.3.1. In addition, two main regulatory elements play a crucial role in HIV-1, *tat* (HIV trans-activator transcription) protein and *rev* (expression regulator of virion proteins) protein [105]. *Tat* is a protein containing 86–101 amino acids. It is vital in regulating viral genome RNA reverse transcription, viral mRNA synthesis, and virus release from infected cells. *Tat* binds to cellular proteins that facilitate their phosphorylation, which increases HIV gene transcription, resulting in a positive feedback cycle [106]. *Rev* is a transactivating protein, binding to the viral genome via an arginine-rich RNA-binding motif for regulating HIV-1 protein expression [107]. Using a nuclear localization signal in the *rev* gene, the *rev* protein is localized near the nucleus to ensure the export of unspliced and incomplete spliced mRNAs. This crucial synthesis of viral protein is crucial for viral replication. In addition, HIV-1 contains five accessory regulatory proteins: viral protein R (*vpr)*, Viral infectivity factor (*vif*), negative factor (*nef-*), Virus protein U (*vpu*), and *tev*. *Vpr* is a 96 amino acid 14-kDa protein that is required for virus replication in nondividing cells such as macrophages and plays a significant role in regulating HIV-1 pre-integration complex (PIC), nucleoprotein complex of viral genetic material, and viral and host proteins that insert viral genome to the host genome [108]. *Vif*'s function targets the human enzyme apolipoprotein B mRNA editing enzyme catalytic polypeptide (APOBEC) for ubiquitination and cellular degradation, thereby disrupting its antiviral activity [109]. This cytidine deaminase enzyme allows viral nucleic acids to be mutated. *Nef-* is a membrane-associated phosphoprotein with a myristoylated N-terminus. It is associated with functions during the virus's replication cycle. It is important in cell apoptosis and virus infectivity [110]. The *vpu* gene encodes a protein called *vpu*. *Vpu* protein is involved in the degradation of CD4 in the endoplasmic reticulum and the enhancement of virion release from infected cells' plasma membranes [111]. *Vpu* causes the degradation of the CD4 viral receptor, which contributes to the downregulation of CD4 expression during HIV infection. *Tev* is linked to a small number of HIV-1 isolates. It combines the *tat*, *env*, and *rev* genes and protein codes.

The HIV-1 envelope spikes, composed of trimers of noncovalently linked heterodimers of the surface gp120 and transmembrane gp41 glycoproteins, initiate a cascade of conformational changes that results in fusion of the viral and host cell membranes and release of the viral core into the cytoplasm [112]. HIV-1 infects mainly CD4-positive T lymphocytes and macrophages [113]. When gp120 interacts with the surface receptor CD4, a bridging sheet forms between the inner and outer domains of the gp120 monomer, exposing the binding site for a second cell surface molecule, typically the chemokine receptor CCR5 [114, 115]. Co-receptor engagement causes the fusion peptide at the N-terminus of gp41 to be inserted into the cell membrane. This causes significant rearrangements between the trimerized N- and C-terminal heptad repeat sequences within gp41, the formation of a six helical hairpin structure – additionally, the apposition and fusion of the viral and host cell membranes. At a depression formed between the inner and outer domains, CD4 binds gp120. These small molecules bind to and extend further into antiviral activity, increasing gp120 binding affinity and developing clinically valuable inhibitors. HIV-1 reverse transcriptase (HIV-1 RT) is a heterodimer composed entirely of the p66 and p51 subunits [116]. The p66 subunit contains two functional active sites, an N-terminal RNA- and DNA-dependent DNA polymerase and a C-terminal RNase H that digests the RNA component of RNA/DNA hybrids [117, 118]. During DNA polymerization, the subdomain catalytic residues Asp110, Asp185, and Asp186 activate the 3′ hydroxyls of the DNA primer and stabilize the hypothetical pentavalent-phosphorous intermediate state within the substrate dNTP, incorporating the nucleotide into the growing DNA chain and liberating free pyrophosphate [119]. Using DNA polymerization, catalytic residues of Asp110, Asp185, and Asp186 activate the 3′ hydroxyls of the DNA primer and stabilize the phosphorous intermediate state within the substrate dNTP integrating the nucleotide into the growing DNA chain and liberating unrestricted pyrophosphate.

The HIV core, which contains the replication enzymes RT and IN integrase (IN) and the viral genomic RNA, is surrounded by a cone-shaped shell made up of viral capsid (CA) protein [120]. Reverse transcription necessitates partial dissolution of the CA shell. CA, which is made up of independently folded N-terminal and C-terminal domains, is linked by a linker assembled into rings containing five or six protomers [121, 122]. The rings then join together to form a closed-shell molecule-like cone composed of hexamers and pentamers. This process allows for the integration of the virus into the host cell.

5.3 Clinical Cases

The AIDS epidemic is still prevalent as over 70 million people worldwide have acquired the infection and nearly 35 million people have died due to HIV-1 [123]. As of 2020, over 37.7 million people are living with HIV-1/AIDS worldwide, with 680,000 people who have died from HIV-related illnesses [124]. HIV infection is one of the leading causes of death among young adults between the ages of 25 and 44 in the United States [125]. Despite not being a cure for HIV-1, the virus is prevalent, which can be seen through funds allocated to antiretroviral therapy (ART) to allow those living with the virus to have a longer and healthier life. Various national and international groups such as the Centers for Disease Control and Prevention (CDC), NIH, and the WHO have created crucial information about the basics of HIV to prevent the disease. The essential prevention that is currently being used to prevent the HHT is abstinence, disposal of all drug equipment after singular use, and taking preexposure prophylaxis (PrEP) and post-exposure prophylaxis (PEP) [126]. PrEP is a prevention method used by HIV-negative individuals at high risk of being exposed to HIV. PEP is an ART drug prescribed to an HIV-negative individual who has encountered a single occurrence of high-risk exposure to HIV [127].

As discussed, there are two main types of human immunodeficiency virus, HIV-1 and HIV-2, with HIV-1 being discovered first and more prevalent worldwide. HIV-1 cases are clinically diagnosed using a series of diagnostic tests. HIV infection is detected by detecting HIV-specific antibodies in serum or plasma or the virus's presence through nucleic acid detection using polymerase chain reaction (PCR), p24 antigen testing, and virus growth in cell culture [128]. Antibody testing is the most frequently used method for determining HIV infection. Seroconversion can be detected in most cases within 2–3 weeks of infection using the highly sensitive HIV-1/HIV-2 enzyme immunoassay (EIA) tests currently on the market [129]. However, in a small proportion of early seroconverters still in the early stages of infection, the p24 antigen may become positive before detecting the antibody. As a result, for the laboratory to select appropriate testing, a clinical history that includes any recent high-risk behavior or symptoms consistent with seroconversion illness is required [130]. HIV diagnostic laboratories use another assay to confirm consistently positive EIA screen tests [129]. HIV diagnostic laboratories should repeatedly confirm positive EIA screen tests with another assay. The most commonly used confirmatory test, the Western blot, is a highly specific immunoblot that allows for the visualization of antibodies and the structural polypeptides of HIV. In cases where antibody testing is insufficient to determine an infected individual, DNA PCR, a nucleic acid amplification method that detects viral DNA integrated into the host cell's genomic DNA, is required [131]. The specificity of HIV-1 DNA PCR was 100% for all age groups, whereas the specificities of qualitative and quantitative RNA PCR assays were 96.1% and 95.5%, respectively [131].

In addition to diagnostic HIV testing, quantitative PCR (RNA) testing (viral load) may be performed to help determine the start of drug therapy and monitor its effectiveness [132]. HIV genotyping is a relatively new addition to patient management that is used to aid in the tracking of drug resistance and to guide the modification of antiretroviral drug selection [133]. Aside from diagnostic HIV testing, laboratories may also offer quantitative PCR (RNA) testing (viral load), which is used to help determine the initiation of drug therapy and monitor its effectiveness. HIV genotyping is a newer adjunct to patient management and is used to assist in tracking the development of drug resistance and guide the modification of antiretroviral drug selection [134].

A study was conducted in 2011 to examine the antiretroviral therapies to reduce the viral replication of HIV-1 [135]. The study was overseen by the National Institute of Allergy and Infectious Diseases (NIAID) through the NIH. Patients with HIV-1 and CD4 between 350 and 550 cells per cubic feet without prior use of antiretroviral therapies were eligible for this study. The drugs included in this study were the following: a combination of lamivudine and zidovudine (Combivir), efavirenz, atazanavir, nevirapine, tenofovir, lamivudine, zidovudine, didanosine, stavudine, a combination of lopinavir and ritonavir (Kaletra and Aluvia), ritonavir, and a combination of emtricitabine and tenofovir (Truvada). After enrollment, study participants were asked to attend three monthly visits, followed by quarterly visits, unless they became ill or needed additional antiretroviral medications. HIV-1-infected participants receiving antiretroviral therapy had one additional visit 2 weeks after starting therapy. HIV-1-uninfected partners were encouraged to return for all visits together for counseling on risk reduction and using condoms, treatment of sexually transmitted infections, and management of other medical conditions. HIV-1 serodiscordant couples were randomly assigned to either an early or delayed strategy for receiving antiretroviral therapy in a 1:1 ratio. Antiretroviral therapy was started in the partner with HIV-1 infection in the early-therapy group at enrollment. Therapy was started in the delayed-therapy group after two consecutive measurements with a CD4 count of 250 cells per cubic millimeter or less or after developing an AIDS-related illness. On a quarterly basis, HIV-1-uninfected partners were tested for HIV-1 seroconversion. A central laboratory evaluated samples from all seroconversion events, and an independent HIV endpoint committee reviewed the results. After beginning antiretroviral therapy, *virologic failure* was defined as two consecutive plasma HIV-1 RNA measurements of more than 1000 copies per milliliter at 16 weeks or later. The study's goal was to review and track the antiretroviral therapy recommendations.

This table shows the various drug treatments that are available to treat each respective virus. While all drug treatments have not been listed, preferred treatments have been noted.

6 Conclusion and Future Trends

Viral diseases constitute an almost indeterminable number of infectious diseases that are responsible for various medical ailments within a variety of living beings, especially humans. While these pathogenic viruses present an immense challenge to human civilization writ large [136], a further comprehension of pathogenic viruses has and continues to allow for the general edification of viral behavior. Moreover, this general edification corresponds with a scientific and clinical understanding of treating viral infection. This chapter has highlighted three prominent pathogenic viruses: HSK, MERS, and HIV-1. These three viruses were chosen because of their societal prevalence, distinct viral structures, and pathogenesis. The purpose was to highlight how the cellular machinations of each particular virus allow for viral replication.

Furthermore, it was showcased that because of the differing cellular compositions of each virus, viral infection took place in disparate fashions resulting in unique clinical presentation. Ultimately, treatment of each viral infection was shown to not simply target superficial clinical symptoms but target the root cellular machinery that powers each particular viral infection. Through these descriptions of viral comportment, this chapter strives to impress not only the necessity of understanding the cellular behavior of three frequently occurring viruses but stresses the importance of cellular apparatus in terms of using targeted and efficacious therapeutics to manage, reverse, and/or end viral infection.

The seemingly ubiquitous nature of viruses and subsequently viral diseases is often for the masses a matter of concern, but never pressing to the point of severe distress. Unless it is a vulnerable population (i.e., immunocompromised, elderly) and/or geographically prescient (i.e., Zika virus in South Florida, Ebola in Africa), there is a level of soft ignorance that covers society. However, due to the SARS-CoV-2 pandemic, viral awareness is quite possibly at an all-time high.

Given the availability and access to scientific and medical information, news, and personnel, society has been granted both a macro- and microlevel understanding of everything surrounding SARS-CoV-2. In particular, the viral structure, pathogenesis, and clinical manifestation of SARS-CoV-2 have been and continued to be expounded upon. Moreover, as the SARS-CoV 2 pandemic continues, novel scientific and clinical research has and continues to be done showing the wide-ranging impact SARS-CoV-2 has, especially on and with other viruses. In relation to this chapter's focus, recent scholarship has and continues to provide an insight of viral interplay between SARS-CoV-2 and HSK, MERS, and HIV-1. Majtanova et al. have shown that SARS-CoV-2 has been linked to development of HSK and/or serve as the genesis for the initiation of HSK [137]. As SARS-CoV-2 infection compromises the immune system, Majatanova et al. hypothesize that this results in a triggering of latent HSV-1 within the body, resulting in an ocular presentation, HSK [137]. Sajini et al. have shown that SARS-CoV-2 and MERS have the potential to form a recombinant virus [138]. During the transcription of genes, SARS-CoV-2 and MERS employ transcription regulatory sequences (TRSs), specific points during replication that facilitate recombination [138]. As a result, Sajini et al. hypothesize that due to the similar use of TRSs and a high frequency of recombination in SARS-CoV-2, there is a possibility of a recombinant SARS-CoV-2-MERS-CoV [138]. Ssentongo et al. suggest that individuals who are HIV-1 positive are at a higher risk of contracting SARS-CoV-2 compared to non-HIV-1-positive individuals [139]. It was noted that HIV-1-positive persons with SARS-CoV-2 experienced a variety of adverse effects such as increased severity of COVID-19 infection, decrease in T-cell production, and a generalized decrease in immune system responsiveness leading to tissue degradation [139]. Ssentongo et al. suggest that since the immune system is compromised within HIV-1-positive individuals, there is not a swift and effective immune response to a significant inflammatory response SARS-CoV-2 initiates [139]. Furthermore, the lack of a robust immune response is highlighted by the swift deterioration of CD4+ and CD8+ T cells, which are pivotal in combating infections [140].

As the SARS-CoV-2 pandemic rages on, examining the cellular background of a viral disease is crucial. Identifying cellular mechanics and current treatments while simultaneously creating opportunities for improving therapies remains a constant point of emphasis as it pertains to viral diseases. In conjunction with the SARS-CoV-2 pandemic, HSK, MERS, and HIV-1 are pressing public health concerns. Proper treatment and precautions must be taken to decrease infection rates and reduce the possibility of deleterious effects that SARS-CoV-2 may have on these already contagious viruses.

References

1. Institute of Medicine, Board on Global Health, Threats, F. O. M., Morse, S., Ph. D., E., M., D. Microbial evolution and co-adaptation: a tribute to the life and scientific legacies of Joshua Lederberg: workshop summary [E-book]. In: Emerging infections: condemned to repeat? 1st ed. National Academies Press; 2009. p. 195–6.
2. Chappell JD, Dermody TS. Biology of viruses and viral diseases. In: Mandell, Douglas, and Bennett's principles and practice of infectious diseases; 2015. p. 1681–1693.e4. https://www.ncbi.nlm.nih.gov/pmc/articles/PMC7152303/.
3. Sandaa RA, Bratbak G. Is the virus important? And some other questions. Viruses. 2018;10(8):442. https://www.ncbi.nlm.nih.gov/pmc/articles/PMC6116253/#:~:text=The%20importance%20of%20a%20virus,humans%2C%20animals%2C%20or%20crops.
4. Wu KJ. There are more viruses than stars in the universe. Why do only some infect us? National Geographic. 3 May 2021. Retrieved 19 May 2022, from https://www.nationalgeographic.com/science/article/factors-allow-viruses-infect-humans-coronavirus.
5. Are viruses alive? Microbiology Society. 10 May 2016. Retrieved 17 May 2022, from https://microbiologysociety.org/publication/past-issues/what-is-life/article/are-viruses-alive-what-is-life.html.
6. Taylor MW. What is a virus? In: Viruses and man: a history of interactions; 2014. p. 23–40. https://www.ncbi.nlm.nih.gov/pmc/articles/PMC7122971/.
7. Cohen FS. How viruses invade cells. Biophys J. 2016;110(5):1028–32. https://www.ncbi.nlm.nih.gov/pmc/articles/PMC4788752/#:~:text=Viruses%20initially%20stick%20to%20cell,two%20membranes%20would%20remain%20distinct.
8. Nasir A, Romero-Severson E, Claverie JM. Investigating the concept and origin of viruses. Trends Microbiol. 2020;28(12):959–67. https://doi.org/10.1016/j.tim.2020.08.003.
9. NHGRI. Deoxyribonucleic Acid (DNA) Fact Sheet. Genome.Gov. 9 Mar 2019. https://www.genome.gov/about-genomics/fact-sheets/Deoxyribonucleic-Acid-Fact-Sheet.
10. Nucleotide. Genome.Gov. 13 June 2022. https://www.genome.gov/genetics-glossary/Nucleotide.
11. van Etten JL, Lane LC, Dunigan DD. DNA viruses: the really big ones (giruses). Annu Rev Microbiol. 2010;64(1):83–99. https://doi.org/10.1146/annurev.micro.112408.134338.

12. Alberts B, Johnson A, Lewis J, et al. Molecular biology of the cell. 4th ed. New York: Garland Science; 2002. The structure and function of DNA. Available from: https://www.ncbi.nlm.nih.gov/books/NBK26821/.
13. Watson JD, Crick FHC. Molecular structure of nucleic acids: a structure for deoxyribose nucleic acid. Nature. 1953;171(4356):737–8. https://doi.org/10.1038/171737a0.
14. Wolfsberg TG, McEntyre J, Schuler GD. Guide to the draft human genome. Nature. 2001;409(6822):824–6. https://doi.org/10.1038/35057000.
15. Brosius J, Raabe CA. What is an RNA? A top layer for RNA classification. RNA Biol. 2016;13(2):140–4. https://doi.org/10.1080/15476286.2015.1128064.
16. Draper DE. A guide to ions and RNA structure. RNA. 2004;10(3):335–43. https://doi.org/10.1261/rna.5205404.
17. Leontis NB, Westhof E. Geometric nomenclature and classification of RNA base pairs. RNA. 2001;7(4):499–512. https://doi.org/10.1017/s1355838201002515.
18. Nudler E, Gottesman ME. Transcription termination and anti-termination in *E. coli*. Genes Cells. 2002;7(8):755–68. https://doi.org/10.1046/j.1365-2443.2002.00563.x.
19. Alberts B, Bray D, Lewis J, et al. Molecular biology of the cell. 3rd ed. New York: Garland Science; 1994. RNA synthesis and RNA processing. Available from: https://www.ncbi.nlm.nih.gov/books/NBK28319/.
20. Koonin E, Gorbalenya A, Chumakov K. Tentative identification of RNA-dependent RNA polymerases of dsRNA viruses and their relationship to positive strand RNA viral polymerases. FEBS Lett. 1989;252(1–2):42–6. https://doi.org/10.1016/0014-5793(89)80886-5.
21. Zheng J, Wei Y, Han GZ. The diversity and evolution of retroviruses: perspectives from viral "fossils". Virol Sin. 2022;37(1):11–8. https://doi.org/10.1016/j.virs.2022.01.019.
22. Beck B, Freudenreich O, Worth JL. Patients with human immunodeficiency virus infection and acquired immunodeficiency syndrome. In: Massachusetts general hospital handbook of general hospital psychiatry; 2010. p. 353–70. https://doi.org/10.1016/b978-1-4377-1927-7.00026-1.
23. Gifford R. Retroviral virion structure. 2013. https://doi.org/10.6084/m9.figshare.807677.v1.
24. Painter MM, Collins KL. HIV and retroviruses. In: Reference module in biomedical sciences; 2019. https://doi.org/10.1016/b978-0-12-801238-3.66202-5.
25. Coffin JM, Hughes SH, Varmus HE, editors. Retroviruses. Cold Spring Harbor: Cold Spring Harbor Laboratory Press; 1997. Genetic Organization. Available from: https://www.ncbi.nlm.nih.gov/books/NBK19370/.
26. Ball NJ, Nicastro G, Dutta M, Pollard DJ, Goldstone DC, Sanz-Ramos M, Ramos A, Müllers E, Stirnnagel K, Stanke N, Lindemann D, Stoye JP, Taylor WR, Rosenthal PB, Taylor IA. Structure of a spumaretrovirus gag central domain reveals an ancient retroviral capsid. PLoS Pathog. 2016;12(11):e1005981. https://doi.org/10.1371/journal.ppat.1005981.
27. Kim FJ, Battini JL, Manel N, Sitbon M. Emergence of vertebrate retroviruses and envelope capture. Virology. 2004;318(1):183–91. https://doi.org/10.1016/j.virol.2003.09.026.
28. Kaye S, Choudhary A. Herpes simplex keratitis. Prog Retin Eye Res. 2006;25(4):355–80. https://reader.elsevier.com/reader/sd/pii/S135094620600005X?token=4AE8ED6B9DF508DCD573765026A9FCEB32A882E59B6E84A5B211DC265146811ECC7303254576FF528506FAC502437CF6&originRegion=us-east-1&originCreation=20220612153745.
29. Al-Dujaili LJ, Clerkin PP, Clement C, McFerrin HE, Bhattacharjee PS, Varnell ED, Kaufman HE, Hill JM. Ocular herpes simplex virus: how are latency, reactivation, recurrent disease and therapy interrelated? Future Microbiol. 2011;6(8):877–907. https://www.ncbi.nlm.nih.gov/pmc/articles/PMC3403814/.
30. Nijm L. Herpes simplex epithelial keratitis – EyeWiki. American Academy of Ophthalmology EyeWiki. 19 Jan 2022. Retrieved 15 May 2022, from https://eyewiki.aao.org/Herpes_Simplex_Epithelial_Keratitis.
31. Farooq AV, Shukla D. Herpes simplex epithelial and stromal keratitis: an epidemiologic update. Surv Ophthalmol. 2012;57(5):448–62. https://www.ncbi.nlm.nih.gov/pmc/articles/PMC3652623/.

32. Dai X, Zhou ZH. Structure of the herpes simplex virus 1 capsid with associated tegument protein complexes. Science. 2018;360(6384). https://www.science.org/doi/10.1126/science.aao7298.
33. Deshpande SP, Zheng M, Lee S, Rouse BT. Mechanisms of pathogenesis in herpetic immunoinflammatory ocular lesions. Vet Microbiol. 2002;86(1–2):17–26. https://reader.elsevier.com/reader/sd/pii/S0378113501004874?token=1143E6938548CA10544B51EEF771A5C36381DDF6A96F8D2F2E08E17CC0C881E3BC231283C7D0E0321685B4634C5C7132&originRegion=us-east-1&originCreation=20220612165725.
34. Tsatsos M, MacGregor C, Athanasiadis I, Moschos MM, Hossain P, Anderson D. Herpes simplex virus keratitis: an update of the pathogenesis and current treatment with oral and topical antiviral agents. Clin Exp Ophthalmol. 2016;44(9):824–37. https://onlinelibrary.wiley.com/doi/pdf/10.1111/ceo.12785?saml_referrer.
35. Lobo AM, Agelidis AM, Shukla D. Pathogenesis of herpes simplex keratitis: the host cell response and ocular surface sequelae to infection and inflammation. Ocul Surf. 2019;17(1):40–9. https://www.ncbi.nlm.nih.gov/pmc/articles/PMC6340725/.
36. Rahn E, Petermann P, Hsu MJ, Rixon FJ, Knebel-Mörsdorf D. Entry pathways of herpes simplex virus type 1 into human keratinocytes are dynamin- and cholesterol-dependent. PLoS One. 2011;6(10):e25464. https://journals.plos.org/plosone/article?id=10.1371/journal.pone.0025464#:~:text=Our%20results%20suggest%20that%20HSV,impact%20during%20entry%20into%20keratinocytes.
37. Azher T, Yin XT, Tajfirouz D, Huang A, Stuart P. Herpes simplex keratitis: challenges in diagnosis and clinical management. Clin Ophthalmol. 2017;11:185–91. https://www.ncbi.nlm.nih.gov/pmc/articles/PMC5261835/#b7-opth-11-185.
38. Lairson DR. Prevention of herpes simplex virus eye disease. Arch Ophthalmol. 2003;121(1):108. https://jamanetwork.com/journals/jamaophthalmology/fullarticle/415043.
39. Sridhar M. Anatomy of cornea and ocular surface. Indian J Ophthalmol. 2018;66(2):190. https://www.ncbi.nlm.nih.gov/pmc/articles/PMC5819093/.
40. Chodosh J, Ung L. Adoption of innovation in herpes simplex virus keratitis. Cornea. 2020;39(1):S7–S18. https://www.ncbi.nlm.nih.gov/pmc/articles/PMC7664964/.
41. Das S, Roy A, Fernandes M. How much clinical practice is aligned with the Herpetic Eye Disease Study! Indian J Ophthalmol. 2021;69(5):1339. https://www.ncbi.nlm.nih.gov/pmc/articles/PMC8186589/.
42. Herpes simplex virus keratitis: a treatment guideline – 2014. American Academy of Ophthalmology EyeWiki. 1 June 2014. Retrieved 15 May 2022, from https://www.aao.org/clinical-statement/herpes-simplex-virus-keratitis-treatment-guideline#REFERENCES.
43. Summary of probable SARS cases with onset of illness from 1 November 2002 to 31 July 2003. World Health Organization. 24 July 2015. Retrieved 23 May 2022, from https://www.who.int/publications/m/item/summary-of-probable-sars-cases-with-onset-of-illness-from-1-november-2002-to-31-july-2003.
44. Li YH, Hu CY, Wu NP, Yao HP, Li LJ. Molecular characteristics, functions, and related pathogenicity of MERS-CoV proteins. Engineering. 2019;5(5):940–7. https://www.sciencedirect.com/science/article/pii/S2095809918307598?via%3Dihub.
45. van den Brand JM, Smits SL, Haagmans BL. Pathogenesis of Middle East respiratory syndrome coronavirus. J Pathol. 2014;235(2):175–84. https://pubmed.ncbi.nlm.nih.gov/25294366/.
46. Al-Abdallat MM, Payne DC, Alqasrawi S, Rha B, Tohme RA, Abedi GR, al Nsour M, Iblan I, Jarour N, Farag NH, Haddadin A, Al-Sanouri T, Tamin A, Harcourt JL, Kuhar DT, Swerdlow DL, Erdman DD, Pallansch MA, Haynes LM, Gerber R, Jordan MERS-CoV Investigation Team. Hospital-associated outbreak of Middle East respiratory syndrome coronavirus: a serologic, epidemiologic, and clinical description. Clin Infect Dis. 2014;59(9):1225–33. https://academic.oup.com/cid/article/59/9/1225/419021.

47. Shapiro M, London B, Nigri D, Shoss A, Zilber E, Fogel I. Middle East respiratory syndrome coronavirus: review of the current situation in the world. Disaster Mil Med. 2016;2(1). https://www.ncbi.nlm.nih.gov/pmc/articles/PMC5329956/.
48. Middle East respiratory syndrome coronavirus (MERS-CoV) – Saudi Arabia 2018. World Health Organization. 28 Dec 2018. Retrieved 21 May 2022, from https://www.who.int/emergencies/disease-outbreak-news/item/28-december-2018-mers-saudi-arabia-en.
49. Fehr AR, Channappanavar R, Perlman S. Middle East respiratory syndrome: emergence of a pathogenic human coronavirus. Annu Rev Med. 2017;68(1):387–99. https://pubmed.ncbi.nlm.nih.gov/27576010/.
50. Lu G, Liu D. SARS-like virus in the Middle East: a truly bat-related coronavirus causing human diseases. Protein Cell. 2012;3(11):803–5. https://link.springer.com/article/10.1007/s13238-012-2811-1.
51. Zumla A, Hui DS, Perlman S. Middle East respiratory syndrome. Lancet. 2015;386(9997):995–1007. https://www.thelancet.com/journals/lancet/article/PIIS0140-6736(15)60454-8/fulltext.
52. Choudhry H, Bakhrebah MA, Abdulaal WH, Zamzami MA, Baothman OA, Hassan MA, Zeyadi M, Helmi N, Alzahrani F, Ali A, Zakaria MK, Kamal MA, Warsi MK, Ahmed F, Rasool M, Jamal MS. Middle East respiratory syndrome: pathogenesis and therapeutic developments. Futur Virol. 2019;14(4):237–46. https://www.futuremedicine.com/doi/10.2217/fvl-2018-0201.
53. Ramadan N, Shaib H. Middle East respiratory syndrome coronavirus (MERS-CoV): a review. Germs. 2019;9(1):35–42. https://pubmed.ncbi.nlm.nih.gov/31119115/.
54. MERS Symptoms & Complications. Centers for Disease Control and Prevention. 2 Aug 2019. Retrieved 25 May 2022, from https://www.cdc.gov/coronavirus/mers/about/symptoms.html.
55. Mohd HA, Al-Tawfiq JA, Memish ZA. Middle East Respiratory Syndrome Coronavirus (MERS-CoV) origin and animal reservoir. Virol J. 2016;13(1). https://pubmed.ncbi.nlm.nih.gov/27255185/.
56. Chan JFW, Lau SKP, To KKW, Cheng VCC, Woo PCY, Yuen KY. Middle East respiratory syndrome coronavirus: another zoonotic betacoronavirus causing SARS-like disease. Clin Microbiol Rev. 2015;28(2):465–522. https://pubmed.ncbi.nlm.nih.gov/25810418/.
57. Chan JFW, Chan KH, Choi GKY, To KKW, Tse H, Cai JP, Yeung ML, Cheng VCC, Chen H, Che XY, Lau SKP, Woo PCY, Yuen KY. Differential cell line susceptibility to the emerging novel human betacoronavirus 2c EMC/2012: implications for disease pathogenesis and clinical manifestation. J Infect Dis. 2013;207(11):1743–52. https://academic.oup.com/jid/article/207/11/1743/797450.
58. Kindler E, Jónsdóttir HR, Muth D, Hamming OJ, Hartmann R, Rodriguez R, Geffers R, Fouchier RAM, Drosten C, Müller MA, Dijkman R, Thiel V. Efficient replication of the novel human betacoronavirus EMC on primary human epithelium highlights its zoonotic potential. MBio. 2013;4(1). https://pubmed.ncbi.nlm.nih.gov/23422412/.
59. Zielecki F, Weber M, Eickmann M, Spiegelberg L, Zaki AM, Matrosovich M, Becker S, Weber F. Human cell tropism and innate immune system interactions of human respiratory coronavirus EMC compared to those of severe acute respiratory syndrome coronavirus. J Virol. 2013;87(9):5300–4. https://pubmed.ncbi.nlm.nih.gov/23449793/.
60. Menachery VD, Eisfeld AJ, Schäfer A, Josset L, Sims AC, Proll S, Fan S, Li C, Neumann G, Tilton SC, Chang J, Gralinski LE, Long C, Green R, Williams CM, Weiss J, Matzke MM, Webb-Robertson BJ, Schepmoes AA, Shukla AK, Metz TO, Smith RD, Waters KM, Katze MG, Kawaoka Y, Baric RS. Pathogenic influenza viruses and coronaviruses utilize similar and contrasting approaches to control interferon-stimulated gene responses. MBio. 2014;5(3). https://pubmed.ncbi.nlm.nih.gov/24846384/.
61. Wang N, Shi X, Jiang L, Zhang S, Wang D, Tong P, Guo D, Fu L, Cui Y, Liu X, Arledge KC, Chen YH, Zhang L, Wang X. Structure of MERS-CoV spike receptor-binding domain complexed with human receptor DPP4. Cell Res. 2013;23(8):986–93. https://www.nature.com/articles/cr201392.

62. Lau SKP, Lau CCY, Chan KH, Li CPY, Chen H, Jin DY, Chan JFW, Woo PCY, Yuen KY. Delayed induction of proinflammatory cytokines and suppression of innate antiviral response by the novel Middle East respiratory syndrome coronavirus: implications for pathogenesis and treatment. J Gen Virol. 2013;94(12):2679–90. https://pubmed.ncbi.nlm.nih.gov/24077366/.
63. Zhou J, Chu H, Li C, Wong BHY, Cheng ZS, Poon VKM, Sun T, Lau CCY, Wong KKY, Chan JYW, Chan JFW, To KKW, Chan KH, Zheng BJ, Yuen KY. Active replication of Middle East respiratory syndrome coronavirus and aberrant induction of inflammatory cytokines and chemokines in human macrophages: implications for pathogenesis. J Infect Dis. 2013;209(9):1331–42. https://pubmed.ncbi.nlm.nih.gov/24065148/.
64. Yin Y, Wunderink RG. MERS, SARS and other coronaviruses as causes of pneumonia. Respirology. 2017;23(2):130–7. https://pubmed.ncbi.nlm.nih.gov/29052924/.
65. Payne S. Family coronaviridae. Viruses. 2017:149–58. https://www.ncbi.nlm.nih.gov/pmc/articles/PMC7149805/.
66. Du L, Kou Z, Ma C, Tao X, Wang L, Zhao G, Chen Y, Yu F, Tseng CTK, Zhou Y, Jiang S. A truncated receptor-binding domain of MERS-CoV spike protein potently inhibits MERS-CoV infection and induces strong neutralizing antibody responses: implication for developing therapeutics and vaccines. PLoS One. 2013;8(12):e81587. https://journals.plos.org/plosone/article?id=10.1371/journal.pone.0081587.
67. Lu G, Hu Y, Wang Q, Qi J, Gao F, Li Y, Zhang Y, Zhang W, Yuan Y, Bao J, Zhang B, Shi Y, Yan J, Gao GF. Molecular basis of binding between novel human coronavirus MERS-CoV and its receptor CD26. Nature. 2013;500(7461):227–31. https://www.nature.com/articles/nature12328.
68. Lu L, Liu Q, Zhu Y, Chan KH, Qin L, Li Y, Wang Q, Chan JFW, Du L, Yu F, Ma C, Ye S, Yuen KY, Zhang R, Jiang S. Structure-based discovery of Middle East respiratory syndrome coronavirus fusion inhibitor. Nat Commun. 2014;5(1). https://www.nature.com/articles/ncomms4067.
69. Surya W, Li Y, Verdià-Bàguena C, Aguilella VM, Torres J. MERS coronavirus envelope protein has a single transmembrane domain that forms pentameric ion channels. Virus Res. 2015;201:61–6. https://www.sciencedirect.com/science/article/pii/S0168170215001136?via%3Dihub.
70. Liu J, Sun Y, Qi J, Chu F, Wu H, Gao F, Li T, Yan J, Gao G. The membrane protein of severe acute respiratory syndrome coronavirus acts as a dominant immunogen revealed by a clustering region of novel functionally and structurally defined cytotoxic T-lymphocyte epitopes. J Infect Dis. 2010;202(8):1171–80. https://academic.oup.com/jid/article/202/8/1171/926918.
71. Lin SC, Ho CT, Chuo WH, Li S, Wang TT, Lin CC. Effective inhibition of MERS-CoV infection by resveratrol. BMC Infect Dis. 2017;17(1). https://bmcinfectdis.biomedcentral.com/articles/10.1186/s12879-017-2253-8.
72. Hu Y, Li W, Gao T, Cui Y, Jin Y, Li P, Ma Q, Liu X, Cao C. The severe acute respiratory syndrome coronavirus nucleocapsid inhibits type I interferon production by interfering with TRIM25-mediated RIG-I ubiquitination. J Virol. 2017;91(8). https://journals.asm.org/doi/10.1128/JVI.02143-16.
73. Grunewald ME, Fehr AR, Athmer J, Perlman S. The coronavirus nucleocapsid protein is ADP-ribosylated. Virology. 2018;517:62–8. https://www.sciencedirect.com/science/article/pii/S0042682217303999?via%3Dihub.
74. Zhang R, Li Y, Cowley TJ, Steinbrenner AD, Phillips JM, Yount BL, Baric RS, Weiss SR. The nsp1, nsp13, and M proteins contribute to the hepatotropism of murine coronavirus JHM. WU. J Virol. 2015;89(7):3598–609. https://journals.asm.org/doi/10.1128/JVI.03535-14.
75. Nassar MS, Bakhrebah MA, Meo SA, Alsuabeyl MS, Zaher WA. Middle East Respiratory Syndrome Coronavirus (MERS-CoV) infection: epidemiology, pathogenesis and clinical characteristics. Eur Rev Med Pharmacol Sci. 2018;22(15):4956–61. https://www.european-review.org/article/15635.

76. Azhar EI, Hui DS, Memish ZA, Drosten C, Zumla A. The Middle East Respiratory Syndrome (MERS). Infect Dis Clin N Am. 2019;33(4):891–905. https://www.ncbi.nlm.nih.gov/pmc/articles/PMC7127753/.
77. Middle East respiratory syndrome coronavirus (MERS-CoV). World Health Organization. 8 Nov 2019. Retrieved 27 May 2022, from https://www.who.int/health-topics/middle-east-respiratory-syndrome-coronavirus-mers#tab=tab_1.
78. Omrani AS, Memish ZA. Therapeutic options for middle east respiratory syndrome coronavirus (MERS-CoV) infection: how close are we? Curr Treat Options Infect Dis. 2015;7(3):202–16. https://www.ncbi.nlm.nih.gov/pmc/articles/PMC7100761/.
79. al Johani S, Hajeer AH. MERS-CoV diagnosis: an update. J Infect Public Health. 2016;9(3):216–9. https://www.ncbi.nlm.nih.gov/pmc/articles/PMC7102781/.
80. Waudby-West R, Parcell BJ, Palmer CN, Bell S, Chalmers JD, Siddiqui MK. The association between SARS-CoV-2 RT-PCR cycle threshold and mortality in a community cohort. Eur Respir J. 2021;58(1):2100360. https://erj.ersjournals.com/content/58/1/2100360.
81. Feikin DR, Alraddadi B, Qutub M, Shabouni O, Curns A, Oboho IK, Tomczyk SM, Wolff B, Watson JT, Madani TA. Association of higher MERS-CoV virus load with severe disease and death, Saudi Arabia, 2014. Emerg Infect Dis. 2015;21(11). https://www.ncbi.nlm.nih.gov/pmc/articles/PMC4622256/.
82. Mann R, Perisetti A, Gajendran M, Gandhi Z, Umapathy C, Goyal H. Clinical characteristics, diagnosis, and treatment of major coronavirus outbreaks. Front Med. 2020;7. https://www.ncbi.nlm.nih.gov/pmc/articles/PMC7691433/.
83. Middle East Respiratory Syndrome (MERS). Johns Hopkins Medicine. 19 Nov 2019. Retrieved 29 May 2022, from https://www.hopkinsmedicine.org/health/conditions-and-diseases/middle-east-respiratory-syndrome-mers.
84. Cockrell AS, Kafri T. Gene delivery by lentivirus vectors. Mol Biotechnol. 2007;36(3):184–204. https://link.springer.com/article/10.1007/s12033-007-0010-8.
85. Sharp PM, Hahn BH. Origins of HIV and the AIDS pandemic. Cold Spring Harb Perspect Med. 2011;1(1):a006841. https://www.ncbi.nlm.nih.gov/pmc/articles/PMC3234451/.
86. Powell MK, Benková K, Selinger P, Dogoši M, Kinkorová Luňáčková I, Koutníková H, Laštíková J, Roubíčková A, ŠPůrková Z, Laclová L, Eis V, ŠAch J, Heneberg P. Opportunistic infections in HIV-infected patients differ strongly in frequencies and spectra between patients with low CD4+ cell counts examined postmortem and compensated patients examined antemortem irrespective of the HAART era. PLoS One. 2016;11(9):e0162704. https://journals.plos.org/plosone/article?id=10.1371/journal.pone.0162704.
87. Simonetti F. Diagnosis of human immunodeficiency virus infection. In: Dewar R, Maldarelli F, editors. Diagnosis of human immunodeficiency virus infection. 9th ed. Elsevier; 2020. p. 1619–41.
88. A timeline of HIV and AIDS. HIV.Gov. 29 Apr 2022. Retrieved 23 May 2022, from https://www.hiv.gov/hiv-basics/overview/history/hiv-and-aids-timeline.
89. Global Statistics. HIV.Gov. 30 Nov 2021. Retrieved 23 May 2022, from https://www.hiv.gov/hiv-basics/overview/data-and-trends/global-statistics#:%7E:text=According%20to%20UNAIDS%20%3A,children%20aged%200%2D14%20years.
90. HIV treatment: the basics | NIH. HIVinfo.NIH.Gov. 16 Aug 2021. Retrieved 25 May 2022, from https://hivinfo.nih.gov/understanding-hiv/fact-sheets/hiv-treatment-basics#:%7E:text=treatment%20for%20HIV%3F-,The%20treatment%20for%20HIV%20is%20called%20antiretroviral%20therapy%20(ART).,HIV%20live%20longer%2C%20healthier%20lives.
91. How is HIV transmitted? HIV.Gov. 26 Jan 2021. Retrieved 21 May 2022, from https://www.hiv.gov/hiv-basics/overview/about-hiv-and-aids/how-is-hiv-transmitted.
92. HIV/AIDS – symptoms and causes. Mayo Clinic. 26 Mar 2022. Retrieved 19 May 2022, from https://www.mayoclinic.org/diseases-conditions/hiv-aids/symptoms-causes/syc-20373524.
93. Swanstrom R, Coffin J. HIV-1 pathogenesis: the virus. Cold Spring Harb Perspect Med. 2012;2(12):a007443. https://www.ncbi.nlm.nih.gov/pmc/articles/PMC3543077/.

94. Haase AT. Targeting early infection to prevent HIV-1 mucosal transmission. Nature. 2010;464(7286):217–23. https://www.nature.com/articles/nature08757.
95. Ribeiro RM, Qin L, Chavez LL, Li D, Self SG, Perelson AS. Estimation of the initial viral growth rate and basic reproductive number during acute HIV-1 infection. J Virol. 2010;84(12):6096–102. https://pubmed.ncbi.nlm.nih.gov/20357090/.
96. Guadalupe M, Reay E, Sankaran S, Prindiville T, Flamm J, McNeil A, Dandekar S. Severe CD4$^+$ T-cell depletion in gut lymphoid tissue during primary human immunodeficiency virus type 1 infection and substantial delay in restoration following highly active antiretroviral therapy. J Virol. 2003;77(21):11708–17. https://pubmed.ncbi.nlm.nih.gov/14557656/.
97. The stages of HIV infection. HIV.Info.NIH.Gov. 20 Aug 2021. Retrieved 24 May 2022, from https://hivinfo.nih.gov/understanding-hiv/fact-sheets/stages-hiv-infection#:~:text=Acute%20HIV%20infection%20is%20the,and%20spreads%20throughout%20the%20body.
98. Miller WC, Rosenberg NE, Rutstein SE, Powers KA. Role of acute and early HIV infection in the sexual transmission of HIV. Curr Opin HIV AIDS. 2010;5(4):277–82. https://pubmed.ncbi.nlm.nih.gov/20543601/.
99. Anderson JA, Ping LH, Dibben O, Jabara CB, Arney L, Kincer L, Tang Y, Hobbs M, Hoffman I, Kazembe P, Jones CD, Borrow P, Fiscus S, Cohen MS, Swanstrom R. HIV-1 populations in semen arise through multiple mechanisms. PLoS Pathog. 2010;6(8):e1001053. https://pubmed.ncbi.nlm.nih.gov/20808902/.
100. Cheng-Mayer C, Tasca S, Ho SH. Coreceptor switch in infection of nonhuman primates. Curr HIV Res. 2009;7(1):30–8. https://pubmed.ncbi.nlm.nih.gov/19149552/.
101. Liu Y, Cruikshank WW, O'Loughlin T, O'Reilly P, Center DM, Kornfeld H. Identification of a CD4 domain required for Interleukin-16 binding and lymphocyte activation*. J Biol Chem. 1999;274(33):23387–95. https://pubmed.ncbi.nlm.nih.gov/10438516/.
102. Rodger AJ, Cambiano V, Bruun T, Vernazza P, Collins S, Degen O, Corbelli GM, Estrada V, Geretti AM, Beloukas A, Raben D, Coll P, Antinori A, Nwokolo N, Rieger A, Prins JM, Blaxhult A, Weber R, van Eeden A, Brockmeyer NH, Clarke A, Guerrero JDR, Raffi F, Bogner JR, Wandeler G, Gerstoft J, Gutiérrez F, Brinkman K, Kitchen M, Ostergaard L, Leon A, Ristola M, Jessen H, Stellbrink H-J, Phillips AN, Lundgren J, PARTNER Study Group, Janeiro N. Risk of HIV transmission through condomless sex in serodifferent gay couples with the HIV-positive partner taking suppressive antiretroviral therapy (PARTNER): final results of a multicentre, prospective, observational study. Lancet. 2019;393(10189):2428–38. https://www.thelancet.com/journals/lancet/article/PIIS0140-6736(19)30418-0/fulltext.
103. Lu K, Heng X, Summers MF. Structural determinants and mechanism of HIV-1 genome packaging. J Mol Biol. 2011;410(4):609–33. https://pubmed.ncbi.nlm.nih.gov/21762803/.
104. King SR, Duggal NK, Ndongmo CB, Pacut C, Telesnitsky A. Pseudodiploid genome organization aids full-length human immunodeficiency virus type 1 DNA synthesis. J Virol. 2008;82(5):2376–84. https://pubmed.ncbi.nlm.nih.gov/18094172/.
105. Debaisieux S, Rayne F, Yezid H, Beaumelle B. The ins and outs of HIV-1 tat. Traffic. 2011;13(3):355–63. https://onlinelibrary.wiley.com/doi/10.1111/j.1600-0854.2011.01286.x.
106. Haseltine WA. Molecular biology of the human immunodeficiency virus type 1. FASEB J. 1991;5(10):2349–60. https://faseb.onlinelibrary.wiley.com/doi/10.1096/fasebj.5.10.1829694.
107. Bukrinsky M, Adzhubei A. Viral protein R of HIV-1. Med Virol. 1999;9(1). https://onlinelibrary.wiley.com/doi/10.1002/(SICI)1099-1654(199901/03)9:1%3C39::AID-RMV235%3E3.0.CO;2-3.
108. da Costa KS, Leal E, dos Santos AM, Lima e Lima AH, Alves CN, Lameira J. Structural analysis of viral infectivity factor of HIV type 1 and its interaction with A3G, EloC and EloB. PLoS One. 2014;9(2):e89116. https://journals.plos.org/plosone/article?id=10.1371/journal.pone.0089116.

109. Feinberg MB, Greene WC. Molecular insights into human immunodeficiency virus type 1 pathogenesis. Curr Opin Immunol. 1992;4(4):466–74. https://doi.org/10.1016/s0952-7915(06)80041-5.
110. Bour S, Schubert U, Strebel K. The human immunodeficiency virus type 1 Vpu protein specifically binds to the cytoplasmic domain of CD4: implications for the mechanism of degradation. J Virol. 1995;69(3):1510–20. https://doi.org/10.1128/jvi.69.3.1510-1520.1995.
111. Benko DM, Schwartz S, Pavlakis GN, Felber BK. A novel human immunodeficiency virus type 1 protein, tev, shares sequences with tat, env, and rev proteins. J Virol. 1990;64(6):2505–18. https://doi.org/10.1128/jvi.64.6.2505-2518.1990.
112. Zhu P, Liu J, Bess J, Chertova E, Lifson JD, Grisé H, Ofek GA, Taylor KA, Roux KH. Distribution and three-dimensional structure of AIDS virus envelope spikes. Nature. 2006;441(7095):847–52. https://doi.org/10.1038/nature04817.
113. Liu J, Bartesaghi A, Borgnia MJ, Sapiro G, Subramaniam S. Molecular architecture of native HIV-1 gp120 trimers. Nature. 2008;455(7209):109–13. https://doi.org/10.1038/nature07159.
114. Kwong PD, Wyatt R, Robinson J, Sweet RW, Sodroski J, Hendrickson WA. Structure of an HIV gp120 envelope glycoprotein in complex with the CD4 receptor and a neutralizing human antibody. Nature. 1998;393(6686):648–59. https://doi.org/10.1038/31405.
115. Rizzuto CD, Wyatt R, Hernández-Ramos N, Sun Y, Kwong PD, Hendrickson WA, Sodroski J. A conserved HIV gp120 glycoprotein structure involved in chemokine receptor binding. Science. 1998;280(5371):1949–53. https://doi.org/10.1126/science.280.5371.1949.
116. Hu WS, Hughes SH. HIV-1 reverse transcription. Cold Spring Harb Perspect Med. 2012;2(10):a006882. https://doi.org/10.1101/cshperspect.a006882.
117. Kohlstaedt LA, Wang J, Friedman JM, Rice PA, Steitz TA. Crystal structure at 3.5 Å resolution of HIV-1 reverse transcriptase complexed with an inhibitor. Science. 1992;256(5065):1783–90. https://doi.org/10.1126/science.1377403.
118. Jacobo-Molina A, Ding J, Nanni RG, Clark AD, Lu X, Tantillo C, Williams RL, Kamer G, Ferris AL, Clark P. Crystal structure of human immunodeficiency virus type 1 reverse transcriptase complexed with double-stranded DNA at 3.0 A resolution shows bent DNA. Proc Natl Acad Sci. 1993;90(13):6320–4. https://doi.org/10.1073/pnas.90.13.6320.
119. Huang H, Chopra R, Verdine GL, Harrison SC. Structure of a covalently trapped catalytic complex of HIV-1 reverse transcriptase: implications for drug resistance. Science. 1998;282(5394):1669–75. https://doi.org/10.1126/science.282.5394.1669.
120. Gamble TR, Vajdos FF, Yoo S, Worthylake DK, Houseweart M, Sundquist WI, Hill CP. Crystal structure of human cyclophilin a bound to the amino-terminal domain of HIV-1 capsid. Cell. 1996;87(7):1285–94. https://doi.org/10.1016/s0092-8674(00)81823-1.
121. Pornillos O, Ganser-Pornillos BK, Kelly BN, Hua Y, Whitby FG, Stout CD, Sundquist WI, Hill CP, Yeager M. X-ray structures of the hexameric building block of the HIV capsid. Cell. 2009;137(7):1282–92. https://doi.org/10.1016/j.cell.2009.04.063.
122. Blair WS, Pickford C, Irving SL, Brown DG, Anderson M, Bazin R, Cao J, Ciaramella G, Isaacson J, Jackson L, Hunt R, Kjerrstrom A, Nieman JA, Patick AK, Perros M, Scott AD, Whitby K, Wu H, Butler SL. HIV capsid is a tractable target for small molecule therapeutic intervention. PLoS Pathog. 2010;6(12):e1001220. https://doi.org/10.1371/journal.ppat.1001220.
123. Why the HIV epidemic is not over. World Health Organization. 2022. https://www.who.int/news-room/spotlight/why-the-hiv-epidemic-is-not-over.
124. HIV/AIDS. World Health Organization. 30 Nov 2021. https://www.who.int/data/gho/data/themes/hiv-aids.
125. Selik RM. HIV infection as leading cause of death among young adults in US cities and states. JAMA. 1993;269(23):2991. https://doi.org/10.1001/jama.1993.03500230073032.
126. Choi KH, Coates TJ. Prevention of HIV infection. AIDS. 1994;8(10):1371–90. https://doi.org/10.1097/00002030-199410000-00003.

127. PrEP and PEP | Preventing new HIV infections | Clinicians | HIV | CDC. Centers for Disease Control and Prevention. 10 Sept 2020. https://www.cdc.gov/hiv/clinicians/prevention/prep-and-pep.html.
128. Bystryak S, Acharya C. Detection of HIV-1 p24 antigen in patients with varying degrees of viremia using an ELISA with a photochemical signal amplification system. Clin Chim Acta. 2016;456:128–36. https://doi.org/10.1016/j.cca.2016.02.022.
129. Sahni A, Nagendra A, Roy P, Patrikar S. Usefulness of enzyme immunoassay (EIA) for screening of anti HIV antibodies in urinary specimens: a comparative analysis. Med J Armed Forces India. 2014;70(3):211–4. https://doi.org/10.1016/j.mjafi.2013.10.011.
130. Fearon M. The laboratory diagnosis of HIV infections. Can J Infect Dis Med Microbiol. 2005;16(1):26–30. https://doi.org/10.1155/2005/515063.
131. Cunningham CK, Charbonneau TT, Song K, Patterson D, Sullivan T, Cummins T, Poiesz B. Comparison of human immunodeficiency virus 1 DNA polymerase chain reaction and qualitative and quantitative RNA polymerase chain reaction in human immunodeficiency virus 1-exposed infants. Pediatr Infect Dis J. 1999;18(1):30–5. https://doi.org/10.1097/00006454-199901000-00009.
132. Zhang M, Versalovic J. HIV Update. Pathol Patterns Rev. 2002;118(Suppl_1):S26–32. https://doi.org/10.1309/lr1b-mhm7-gdwg-t1eu.
133. Oelemann WMR, Lowndes CM, VeríSsimo Da Costa GC, Morgado MG, Castello-Branco LRR, Grinsztejn B, Alary M, Bastos FI. Diagnostic detection of human immunodeficiency virus type 1 antibodies in urine: a Brazilian Study. J Clin Microbiol. 2002;40(3):881–5. https://doi.org/10.1128/jcm.40.3.881-885.2002.
134. Hirsch MS, Brun-Vezinet F, Clotet B, Conway B, Kuritzkes DR, D'Aquila RT, Demeter LM, Hammer SM, Johnson VA, Loveday C, Mellors JW, Jacobsen DM, Richman DD. Antiretroviral drug resistance testing in adults infected with human immunodeficiency virus type 1: 2003 recommendations of an International AIDS Society–USA Panel. Clin Infect Dis. 2003;37(1):113–28. https://doi.org/10.1086/375597.
135. Cohen MS, Chen YQ, McCauley M, Gamble T, Hosseinipour MC, Kumarasamy N, Hakim JG, Kumwenda J, Grinsztejn B, Pilotto JH, Godbole SV, Mehendale S, Chariyalertsak S, Santos BR, Mayer KH, Hoffman IF, Eshleman SH, Piwowar-Manning E, Wang L, Makhema J, Mills LA, de Bruyn G, Sanne I, Eron J, Gallant J, Havlir D, Swindells S, Ribaudo H, Elharrar V, Burns D, Taha TE, Nielsen-Saines K, Celentano D, Essex M, Fleming TR, HPTN 052 Study Team. Prevention of HIV-1 infection with early antiretroviral therapy. N Engl J Med. 2011;365(6):493–505. https://doi.org/10.1056/nejmoa1105243.
136. Artika IM, Wiyatno A, Ma'roef, C. N. Pathogenic viruses: molecular detection and characterization. Infect Genet Evol. 2020;81:104215. https://www.sciencedirect.com/science/article/pii/S1567134820300472?via%3Dihub.
137. Majtanova N, Kriskova P, Keri P, Fellner Z, Majtan J, Kolar P. Herpes simplex keratitis in patients with SARS-CoV-2 infection: a series of five cases. Medicina. 2021;57(5):412. https://www.ncbi.nlm.nih.gov/pmc/articles/PMC8146587/.
138. Sajini AA, Alkayyal AA, Mubaraki FA. The recombination potential between SARS-CoV-2 and MERS-CoV from cross-species spill-over infections. J Epidemiol Glob Health. 2020;11(2):155. https://www.ncbi.nlm.nih.gov/pmc/articles/PMC8242116/.
139. Ssentongo P, Heilbrunn ES, Ssentongo AE, Advani S, Chinchilli VM, Nunez JJ, Du P. Epidemiology and outcomes of COVID-19 in HIV-infected individuals: a systematic review and meta-analysis. Sci Rep. 2021;11(1). https://www.ncbi.nlm.nih.gov/pmc/articles/PMC7973415/#CR40.
140. CD4-CD8 Ratio – Health Encyclopedia – University of Rochester Medical Center. University of Rochester Medical Center. 2022. Retrieved 29 May 2022, from https://www.urmc.rochester.edu/encyclopedia/content.aspx?contenttypeid=167&contentid=cd4_cd8_ratio.

Viral Infections: Current Treatment Options

Sagar Salave, Dhwani Rana, Arti Bodar, Dignesh Khunt, Bhupendra Prajapati, and Jayvadan Patel

Abstract Viruses are small tiny droplets of microbes and mostly cause and also create a pandemic situation in the world such as COVID-19. Numerous viruses are reported up to date, and still new ones are coming, which leads to continuous research on finding new diagnostic tests with a novel treatment option. Numerous bioinformatics sites continue updating virus genomes with their cell member structure protein to assist the drug developer in finding the safest treatment with good efficacy. The current chapter is focused on the comprehensive review of different types of viruses with an available treatment option.

Keywords Virus · Viral treatment · Zika virus · Coronavirus · Herbal remedies · CNS virus

1 Introduction

Viruses are considered to be plentiful biological species of all known ecosystems on our planet; an estimated 1×10^{31} viruses exist on earth [1, 2]. The diverse habitat of viruses expands in an extensive range of surroundings from deep sea, marine

S. Salave · D. Rana
Department of Pharmaceutics, NIPER-Ahmedabad, Gandhinagar, Gujarat, India

A. Bodar · D. Khunt
Graduate School of Pharmacy, Gujarat Technological University, Gandhinagar, Gujarat, India

B. Prajapati (✉)
Shree S K Patel College of Pharmaceutical Education and Research, Ganpat University, Mahesana, Gujarat, India

J. Patel
Nootan Pharmacy College, Sankalchand Patel University, Visnagar, Gujarat, India

R. Shegokar, Y. Pathak (eds.), *Viral Drug Delivery Systems*,
https://doi.org/10.1007/978-3-031-20537-8_4

sediments, acidic (pH 3) hot springs (>80 °C), hypersaline lakes, solar salterns, deep ice (>30 m) in polar lakes, alkaline (pH 8) lakes, terrestrial subsurface, to freshwater [3, 4]. These vast biogeographical distributions cause their ecological dynamics. Viruses are obligate parasites, having the ability to replicate only inside the living cells due to not being absent of encoding all the proteins required for the replication in their genomes [1, 5]. These sub-microscopic, intracellular, mobile genetic elements are incompetent of carrying out their life-sustaining functions outside the host cell and hence exist as potentially active but inert entities outside of cells [6, 7]. The viral genome consists of nucleic acid, single- or double-stranded RNA, or DNA, inside the protein coat called a capsid. Capsid embedded with nucleic acid is known as nucleocapsid [6]. Certain viruses possess an additional membrane covering the symmetrically arranged nucleocapsid called an envelope, typically made up of phospholipid bilayer and viral glycoproteins [8, 9]. The proteins entrenched in the outer membrane coat function as cellular receptors or ligands, which are required for cellular binding and access for the virus into the host cell [10].

In its infective form, the fully assembled virus is termed as virion, which functions to deliver its genetic material (DNA or RNA) into the host cell for its expression (transcription and translation) [9, 11]. As viruses are devoid of ribosomes, they cannot synthesize their proteins and hence depend on the host cell's machinery for translating their mRNA into proteins. By parasitizing the host cells, they carry out metabolic functions as they cannot generate ATP on their own [6]. When infected, the host cell generally dies as the virus forces and refrains the host cell to carry out its normal functions. Upon death, the host cell releases mature virions, which communicate to other cells. Occasionally, certain viruses modify the cell's functions, while many times, the sick cell loses control of cell division, which leads to cancer cells. Viruses typically infect a particular type of cell or a particular species of plants or animals. Certain viruses cause infection in a specific age group, while some viruses lead to chronic infections [12]. With the purpose of evading the host's defenses to facilitate the process of replication and propagation, viruses have evolved several unique sophisticated methods [10]. Even the closely related virus strains having small genetic differences vary greatly in pathogenicity, virulence, and transmissibility [13]. This chapter aims at highlighting diverse kinds of viral infection and the available therapeutic interventions along with the emerging therapies for the same.

2 Virus Classification

The International Committee on Taxonomy of Viruses (ICTV) takes the charge of taxonomic assignment and classification of viruses. They develop, refine, and maintain the universal viral taxonomy; the five defined hierarchical ranks are the order, family, subfamily, genus, and species. There are more than 9000 virus

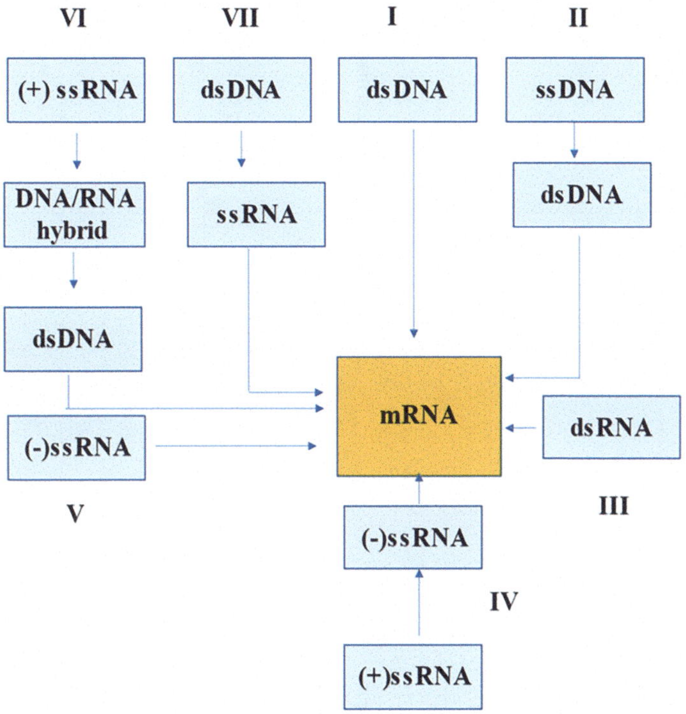

Fig. 1 Classification of viruses based on the mRNA by Baltimore classification

species [14, 15]. However, viruses are broadly categorized into three classes based on their genome, i.e., virus containing DNA as their genome, virus containing RNA as their genome, and lastly virus containing reverse transcriptase (RT) enzyme [16]. The Baltimore classification scheme proposed by David Baltimore in 1971 is based on mRNA synthesis (Fig. 1) [17].

2.1 DNA Viruses

This class includes poxviruses, herpesviruses, poxviruses, papillomaviruses, and adenoviruses among many others. The genetic material, i.e., DNA, can be classified as single stranded (ss) and double stranded (ds). Conditional to the virus, during infection, the viral DNA is transcribed by cellular and/or viral RNA polymerases to produce mRNAs, which get translated into viral proteins. The DNA polymerase replicating DNA can be of viral or cellular origin [16]. Table 1 summarizes the classification of viruses.

Table 1 Classification of viruses

Family	Subfamily	Genus	Genetic material	Place of replication of the genomes and assembly of progeny viruses (site of capsid assembly)	Natural hosts
DNA viruses [9, 16, 18]					
Herpesviridae	*Alphaherpesvirinae*	*Antivirus, Mardivirus* *Scutavirus, Simplexvirus* *Varicellovirus*	dsDNA, linear	Nucleus	Vertebrates
	Betaherpesvirinae	*Cytomegalovirus* *Muromegalovirus* *Roseolovirus* *Proboscivirus*			
	Gammaherpesvirinae	*Lymphocryptovirus* *Rhadinovirus* *Macavirus* *Percavirus*			
Parvoviridae	*Parvovirinae*	*Amdoparvovirus* *Aveparvovirus* *Bocaparvovirus* *Copiparvovirus* *Erythroparvovirus* *Dependoparvovirus* *Protoparvovirus* *Tetraparvovirus*	ssDNA, linear	Nucleus	Vertebrates, insects
	Densovirinae	*Ambidensovirus* *Iteravirus* *Brevidensovirus* *Hepandensovirus* *Penstyldensovirus*			

Poxviridae	*Chordopoxvirinae*	*Orthopoxvirus* *Parapoxvirus* *Avipoxvirus* *Capripoxvirus* *Leporipoxvirus* *Suipoxvirus* *Molluscipoxvirus* *Yatapoxvirus*	dsDNA, linear	Cytoplasm	Humans, vertebrates, arthropods
	Entomopoxvirinae	*Alphaentomopoxvirus* *Betaentomopoxvirus* *Gammaentopoxvirus*			
Iridoviridae	*Alphairidovirinae*	*Lymphocystivirus* *Megalocytivirus* *Ranavirus*	dsDNA, linear	Cytoplasm	Amphibia Fish Invertebrates
	Betairidovirinae	*Iridovirus* *Chloriridovirus*			
Adenoviridae		*Atadenovirus* *Aviadenovirus* *Ichtadenovirus* *Mastadenovirus* *Siadenovirus*	ssDNA, linear	Nucleus	Vertebrates
Hepadnaviridae		*Orthohepadnavirus* *Avihepadnavirus*	dsDNA, circular	Cytoplasm/nucleus	Humans, apes, birds
Polyomaviridae		*Alphapolyomavirus* *Betapolyomavirus* *Deltapolyomavirus* *Gammapolyomavirus*	dsDNA, circular	Nucleus	Mammals and birds

(continued)

Table 1 (continued)

Family	Subfamily	Genus	Genetic material	Place of replication of the genomes and assembly of progeny viruses (site of capsid assembly)	Natural hosts
Papillomaviridae		*Alphapapillomavirus* *Betapapillomavirus* *Chipapillomavirus* *Deltapapillomavirus* *Dyodeltapapillomavirus* *Dyoepsilonpapillomavirus* *Dyoetapapillomavirus* *Dyoiotapapillomavirus* *Dyokappapillomavirus* *Dyolambdapapillomavirus* *Dyomupapillomavirus* *Dyonupapillomavirus* *Dyoomikronpapillomavirus* *Dyopipapillomavirus* *Dyorhopapillomavirus* *Dyosigmapapillomavirus* *Dyothetapapillomavirus* *Dyoxipapapillomavirus* *Dyozetapapillomavirus* *Epsilonpapillomavirus* *Etapapillomavirus* *Gammapapillomavirus* *Iotapapillomavirus* *Kappapapillomavirus* *Lambdapapillomavirus* *Mupapillomavirus* *Nupapillomavirus* *Omegapapillomavirus* *Omikronpapillomavirus* *Phipapillomavirus* *Pipapillomavirus* *Psipapillomavirus* *Rhopapillomavirus* *Sigmapapillomavirus* *Taupapillomavirus* *Thetapapillomavirus* *Upsilonpapillomavirus* *Xipapillomavirus* *Zetapapillomavirus*	dsDNA, circular	Nucleus	Vertebrate

Circoviridae		*Circovirus* *Cyclovirus*	ssDNA, circular	Nucleus	Birds and mammals
Asfarviridae		*Asfivirus*	dsDNA, linear	Cytoplasm	Pigs, warthogs, bushpigs Vector: Argasid ticks
Inoviridae		*Fibrovirus* *Habenivirus* *Inovirus* *Lineavirus* *Saetivirus*	ssDNA	Cytoplasm	Bacteria
Microviridae	*Bullavirinae*	*Alpha3microvirus* *G4microvirus* *Phix174microvirus*	ssDNA, circular	Cytoplasm	Microvirus: enterobacteria
	Gokushovirinae	*Chlamydiamicrovirus* *Bdellomicrovirus* *Spiromicrovirus*			*Clamydiamicrovirus* and *Bdellomicrovirus*: intracellular parasitic bacteria *Spiromicrovirus*: *Spiroplasma*
Geminiviridae		*Mastrevirus* *Capulavirus* *Curtovirus* *Begomovirus* *Grablovirus* *Topocuvirus* *Eragrovirus* *Turncurtovirus* *Becurtovirus*	ssDNA, circular	Nucleus	Plants

(continued)

Table 1 (continued)

Family	Subfamily	Genus	Genetic material	Place of replication of the genomes and assembly of progeny viruses (site of capsid assembly)	Natural hosts
Baculoviridae		*Alphabaculovirus* *Betabaculovirus* *Deltabaculovirus* *Gammabaculovirus*	dsDNA, circular	Nucleus	Arthropods: Lepidoptora Hymenoptera Diptera Crustacean: decapoda (shrimps)
RNA viruses [9, 16, 19–22]					
Coronaviridae	*Coronavirinae*	*Alphacoronavirus* *Betacoronavirus* *Gammacoronavirus* *Deltacoronavirus*	ssRNA(+), linear	Cytoplasm	Vertebrates
	Torovirinae	*Bafinivirus* *Torovirus*			
Arteriviridae	*Crocarterivirinae*	*Muarterivirus*	ssRNA(+), linear	Cytoplasm	Vertebrates
	Equarterivirinae	*Alphaarterivirus*			
	Heroarterivirinae	*Lambdaarterivirus*			
	Simarterivirinae	*Deltaarterivirus* *Etaarterivirus* *Epsilonarterivirus* *Iotaarterivirus* *Thetaarterivirus* *Zetaarterivirus*			
	Variarterivirinae	*Gammaartevirus* *Betaarterivirus*			
	Zealarterivirinae	*Kappaarterivirus*			
Togaviridae		*Alphavirus* *Rubivirus*	ssRNA(+), linear	Cytoplasm	Humans, mammals, marsupials, birds, mosquitoes

Flaviviridae		*Flavivirus* *Hepacivirus* *Pegivirus* *Pestivirus*	ssRNA(+), linear	Cytoplasm	Humans, mammals. Vector: ticks or mosquitoes
Picornaviridae		*Aphthovirus, Aquamavirus, Avihepatovirus, Ampivirus, Avisivirus, Cardiovirus, Dicipivirus, Cosavirus Enterovirus (includes rhinoviruses), Erbovirus Gallivirus, Harkavirus Hepatovirus, Hunnivirus Kobuvirus, Kunsagivirus Limnipivirus, Megrivirus Mischivirus, Mosavirus Oscivirus, Parechovirus Pasivirus, Passerivirus Potamipivirus, Rabovirus Rosavirus, Sakobuvirus Salivirus, Sapelovirus Senecavirus, Sicinivirus Teschovirus, Torchivirus, Tremovirus*	ssRNA(+), linear	Cytoplasm	Vertebrates
Astroviridae		*Avastrovirus* *Mamastrovirus*	ssRNA(+), linear	Cytoplasm	Vertebrates
Caliciviridae		*Lagovirus* *Nebovirus* *Norovirus* *Sapovirus* *Vesivirus*	ssRNA(+), linear	Cytoplasm	Vertebrates
Hepeviridae		*Orthohepevirus* *Piscihepevirus*	ssRNA(+), linear	Cytoplasm	Humans, pig, wild boar, monkey, some rodents, chicken

(continued)

Table 1 (continued)

Family	Subfamily	Genus	Genetic material	Place of replication of the genomes and assembly of progeny viruses (site of capsid assembly)	Natural hosts
Filoviridae		*Ebolavirus* *Marburgvirus* *Cuevavirus*	ssRNA(−), linear	Cytoplasm	Reservoir: Bats Occasional: humans and primates
Rhabdoviridae		*Almendravirus, Cytorhabdovirus, Curiovirus, Dichorhavirus, Ephemerovirus* *Hapavirus, Ledantevirus* *Lyssavirus, Novirhabdovirus* *Nucleorhabdovirus, Perhabdovirus, Sigmavirus* *Sprivivirus, Sripuvirus, Tibrovirus, Tupavirus, Varicosavirus, Vesiculovirus*	ssRNA(−), linear	Cytoplasm	Vertebrates Invertebrates Plants
Bornaviridae		*Orthobornavirus* *Carbovirus*	ssRNA(−), linear	Nucleus	Natural: horses, sheep, cattle, rodents, birds. Occasional: humans
Orthomyxoviridae		*Alphainfluenzavirus* *Betainfluenzavirus* *Gammainfluenzavirus* *Deltainfluenzavirus* *Isavirus* *Quaranjavirus* *Thogotovirus*	ssRNA(−), linear	Nucleus	Aquatic birds, humans, pig, horse, seals

Arenaviridae		*Antennavirus* *Hartmanivirus* *Mammarenavirus* *Reptarenavirus*	ssRNA(−), linear	Cytoplasm	Reservoir: rodents Occasional: humans
Reoviridae	*Spinareovirinae*	*Aquareovirus* *Coltivirus* *Cypovirus* *Dinovernavirus* *Fijivirus* *Idnoreovirus* *Mycoreovirus* *Orthoreovirus* *Oryzavirus*	dsRNA	Cytoplasm	Vertebrate Invertebrate Plant Fungi
	Sedoreovirinae	*Cardoreovirus* *Mimoreovirus* *Orbivirus* *Phytoreovirus* *Rotavirus* *Seadornavirus*			
Birnaviridae		*Aquabirnavirus* *Avibirnavirus* *Blosnavirus* *Entomobirnavirus*	dsRNA	Cytoplasm	Salmonid fish, young sexually immature chickens, insects

(continued)

Table 1 (continued)

Family	Subfamily	Genus	Genetic material	Place of replication of the genomes and assembly of progeny viruses (site of capsid assembly)	Natural hosts
RT viruses [9, 19]					
Caulimoviridae		*Caulimovirus* *Badnavirus* *Cavemovirus* *Petuvirus* *Rosadnavirus* *Solendovirus* *Soymovirus* *Tungrovirus*	dsDNA	Nucleus/cytoplasm	Plants, insects
Hepadnaviridae		*Orthohepadnavirus* *Avihepadnavirus*	dsDNA, circular		Human, apes, birds
Retroviridae	*Orthoretrovirinae*	*Alpharetrovirus* *Betaretrovirus* *Gammaretrovirus* *Deltaretrovirus* *Epsilonretrovirus* *Lentivirus*	ssRNA(+), linear	Nucleus	Arthropods: Lepidoptora Hymenoptera Diptera Crustacean: decapoda (shrimps)
	Spumaretrovirinae	*Simiispumavirus* *Bovispumavirus* *Equispumavirus* *Felispumavirus* *Prosimiispumavirus*			

2.2 *RNA Viruses*

The RNA genome is directly translated from RNA to RNA. They are divided into plus-strand RNA, having genome that is a messenger RNA, and minus strand, having genome that is an anti-messenger sense. Certain viruses have double-strand RNA as their genome. Protein loaded in the virus is needed for the replication of RNA viral due to the cells not containing RNA to DNA copying enzymes. The minus-strand and double-strand RNA viruses are composed of enzymes (RNA synthesizing) that produce mRNA. RNA intermediates matching to the genome aids in replication, and the process is similar to that of DNA replication [16]. Table 1 represents viruses belonging to this class.

2.3 *Reverse Transcriptase (RT) Viruses*

RT viruses such as retroviruses and SARS-CoV-2 RNA are embedded with the RT enzyme, which produces DNA copy from the template of RNA [16]. The examples of viruses belonging to this class are mentioned in Table 1.

3 Different Types of Viral Infection

Viral infection results when the virus invades the host cell, emptying its genetic material, and forces the host cell to replicate it. This process releases new viruses, which further infect new cells [16]. Different types of viral infections have been depicted in Fig. 2. The following section describes the various types of viral infections in detail.

3.1 *Respiratory Tract Infection*

This type of infection normally occurs in the upper respiratory tract, i.e., the upper airways, nose, lungs, and throat. Upper respiratory infections may include sinusitis, sore throat, and the common cold. Another respiratory viral infection also comprises pneumonia, influenza, and coronaviruses. In pediatric patients, viruses can cause inflammation of the upper and lower airways called laryngotracheobronchitis [16]. However, lower airway inflammation is called bronchiolitis. Commonly, more severe respiratory symptoms are observed in elderly patients, infants, and people with lung disorders.

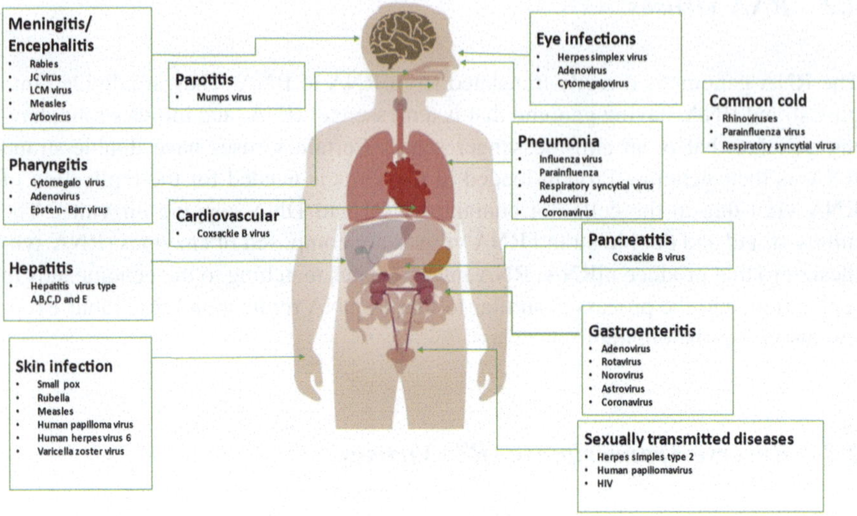

Fig. 2 Overview of various types of viral infections in humans

3.2 *Gastrointestinal Tract Infection*

Viral infection of the gastrointestinal tract produces a variety of clinical symptoms and eventual disease progression. Table 2 summarizes viral infection of the gut with causative agents.

Gastroenteritis is a condition where the gastrointestinal tract gets infected and results in the inflammation of its inside lining. Causative agents for this type of clinical condition are rotaviruses, noroviruses, cytomegalovirus, herpes simplex virus, adenovirus, Epstein-Barr virus, and human papillomavirus (HPV). Notable symptoms of viral gastroenteritis are nausea, watery diarrhea, and vomiting. Dehydration is the most solemn complication of this illness [23].

3.3 *Hepatic Infection*

Hepatotropic viruses replicate in the liver and can cause infection. These mainly comprise hepatitis A, hepatitis B, hepatitis C, and hepatitis E viruses. Hepatitis and viral damage arise due to the immune response into the liver [24]. In some pieces of evidence, it has been found that the liver can be affected as a part of generalized host infection with viruses that mainly target the other tissues, particularly the upper

Table 2 Types of viral infection and causative agents

Causative virus	Infection
Viral skin infection [26, 27]	
Chickenpox in children	Varicella-zoster virus (VZV)
Molluscum contagiosum	Poxvirus
Herpes zoster or shingles	Varicella-zoster virus (VZV)
HSV infection	Herpes simplex virus
CMV infections	Cytomegalovirus
Warts	Human papillomavirus (HPV)
Measles	Paramyxovirus
Lower respiratory tract infection [28–30]	
Bronchiolitis	Adenovirus, coronaviruses, influenza viruses, rhinovirus, bocavirus
Exacerbations of wheezing/asthma	Rhinovirus, adenovirus, parainfluenza virus (PIV), coronaviruses, influenza viruses, bocavirus
Croup	Influenza, adenovirus
Pneumonia	Influenza, PIV, adenovirus, RSV, hMPV, human respiratory syncytial virus (HRSV)
Viral infection of the gastrointestinal tract [23]	
Acute gastroenteritis	Caliciviruses, rotaviruses, astroviruses, adenoviruses, toroviruses
Ulcerative mucosal disease, mainly in immunocompromised hosts	Cytomegalovirus, herpes simplex viruses
Mass lesions and malignancies that are worse in immunocompromised hosts	Epstein-Barr virus, human papillomaviruses, human herpesvirus-8
Motility disorders	Rotavirus, caliciviruses, varicella-zoster virus, Epstein-Barr virus, cytomegalovirus
Viruses causing acute CNS disease [31]	
Meningoencephalitis	Herpes simplex viruses-1
Encephalomyelitis	Rabies virus
Meningoencephalomyelitis	West Nile virus
Encephalitis	Nipah virus
Meningitis, encephalitis	Equine encephalitis viruses
Meningitis, encephalitis, myelitis	Mumps virus
Meningitis, encephalitis	Rubella virus
Meningitis, meningoencephalitis, myelitis	Coxsackievirus, echovirus
Meningitis, myelitis	Poliovirus
Meningitis, encephalitis	California encephalitis virus
Viruses causing chronic CNS disease [31]	
Encephalitis, meningitis, myelitis	Human immunodeficiency virus (HIV)
Myelitis	Human T-cell leukemia virus-1 and virus-2
Progressive multifocal, leukoencephalopathy	John Cunningham virus (JC virus)
Leukoencephalitis, cerebellitis, meningitis, myelitis	Varicella-zoster virus
Encephalitis	Measles virus
Encephalitis	Cytomegalovirus
Encephalitis, meningitis, myelitis	Epstein-Barr virus (EBV)

respiratory tract. Causative agents for this clinical condition are herpes viruses, cytomegalovirus, herpes simplex virus, adenovirus, parvovirus, and SARS-coronaviruses [25].

3.4 Skin Infection

Clinical conditions like warts and other blemishes are a result of some viral infection. Several other viruses also affect the other parts of the body and produce skin rashes including chickenpox viral infection [26]. Table 2 describes the commonly found skin infections and their causative virus.

3.5 Placental and Fetal Infection

Cytomegalovirus, rubella virus, and Zika virus are currently known for infection of the placenta and fetus in pregnant women [32].

3.6 Viral Infection in the Nervous System

More than 100 different viruses are recognized which infect the brain and spinal cord. Viral causes of central nervous system disease have been linked to a variety of pathways, including localised infection, replication (as in the case of encephalitis) and systemic infection (as in the case of acute disseminated encephalomyelitis), as well as immune-related processes (as in the case of encephalomyelitis) [33]. Viruses such as the West Nile virus and rabies virus infect the brain and cause encephalitis. Some viruses cause infection on layers of tissues that cover the brain and spinal cord, which cause meningitis.

4 Diagnosis of Viral Infection

Rapid diagnosis aids in disease management by allowing timely therapy and preventing complications, and precise viral diagnosis helps indecisive forecast and sometimes allows discontinuation of antibiotic therapy [22, 23]. Some of the well-known diagnostic techniques are complement fixation test; hemagglutination inhibition test; next-generation sequencing; radioimmunoassay; chemiluminescent immunoassay; enzyme immunoassays such as fluorescence polarization immunoassay, chemiluminescent immunoassay, and microparticle enzyme immunoassay; nucleic acid-based amplification test including real-time polymerase chain reaction; transcription-mediated amplification; quantitative polymerase chain reaction; and nucleic acid sequence-based amplification [23]. Mass spectrometric-based methods are also explored in clinical virology for diagnosis, structural investigation of biomolecules, and genotype detection. These methods are widely used, but specific and accurate diagnosis remains a challenge. Microfluidic technology using lab-on-a-chip (LOC) devices allows point-of-care diagnostic by carrying out immediate reactions within the chip or in a portable device [23, 24]. These LOC-based sophisticated sample-to-result devices illustrate robustness, economic operational costs, and opportunities for diagnostic innovations [25].

5 Antiviral Drug Development Strategies

There has been considerable development in the methods of drug discovery over the past few decades. Antiviral drugs are generally directed against specific viral enzymes and inhibit the phases of viral entry or release. However, information regarding the structure and function of viral proteins serves to be beneficial for rational antiviral drug designing. Molecular mechanisms underlying virus and host interactions are also widely studied upon, and the resultant understanding of virus life cycles at the molecular level has proposed several targets for therapeutic intervention [26, 55]. Newer advances in this regard are based on targeting host factors important for virus replication, gene silencing strategies intended at interfering with viral gene expression, and exploitation of the innate immune response system. Virostatics aiming to interfere with viral replication steps are leveraged for treating certain viral infections [26].

The antiviral drug discovery paradigm can be apportioned into target identification, lead generation, lead optimization, and leadership development using high-throughput screenings and yielding clinical candidates. Lead generation involves certain approaches like substrate-based approaches, screening approaches, and biostructural approaches to arrive at a workable and novel lead. Several antiviral agents such as HIV proteinase inhibitors and influenza virus neuraminidase inhibitors are designed using the substrate-based approach, whereas HIV RT inhibitors and HIV proteinase inhibitors and influenza virus polymerase inhibitors are screened from the libraries of compounds. High-resolution structural data for the generation of leads- structural approach has yielded few HIV proteinase inhibitors [55]. Alternative approaches to antiviral therapies have also been developed such as immunomodulators, antisense oligonucleotides, ribozymes, and aptamer. Figure 3a, b depict the strategies for the development of an antiviral drug.

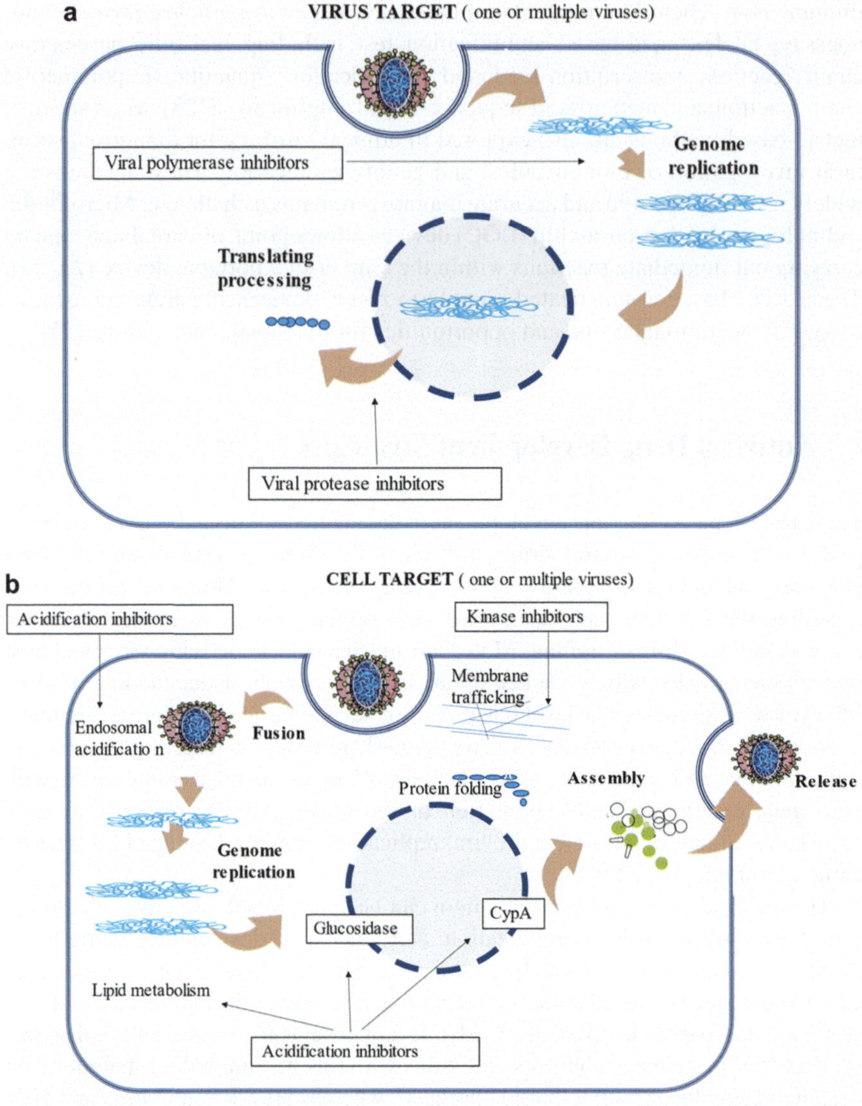

Fig. 3 (**a**) Strategies for development of antiviral drugs (virus target). (**b**) Strategies for development of antiviral drugs (organ target)

6 Current Treatment of Viral Infection

6.1 Method to Treat Viral Infection

Treatment of viral infection is based on two approaches.

6.1.1 Treatment of Symptoms

Symptoms associated with viral infection can be rectified by common medication. These medications can be general therapy or over-the-counter drugs. These symptoms include fever and aches, nausea, vomiting, runny nose, sore throat, diarrhea, and dehydration [8].

- *Anti-hypercytokinemia as Treatment Modality*

The high fatality rate of virus infection is called hypercytokinemia, which is also referred to as a cytokine storm in immunocompromised individuals. This is a condition characterized by swiftly proliferating and highly activated cells, macrophages, and NK cells and associated with overproduction and release of more than 150 inflammatory cytokines and chemical mediators by the immune system [56]. If this condition is left untreated, it may cause severe pathological complications, even death of individuals. Cytokine storm is closely linked with several viruses including dengue fever (DF) viruses, Ebola, Marburg, Lassa, and Crimean-Congo hemorrhagic fever [57]. Several strategies have been adopted to treat this medical condition, which are enlisted in Table 3.

6.1.2 Antiviral Therapy

Several antiviral drugs have been well established due to the successful employment of basic science to generate an effective chemical moiety for severe viral infection [42]. Antiviral therapy is one of the most exhilarating aspects of viral science. The development of novel antiviral therapy is very much a work in progress including the drug discovery program for different types of viruses such as coronaviruses, dengue filoviruses, and other viruses [58]. In Table 3, various well-established antiviral drugs have been enlisted including their mechanism of action.

6.2 Herbal Treatment as an Antiviral Therapy

Medicinal plants have played a crucial role in the treatment of various infectious and non-infectious ailments. A variety of herbal medicines possess broad-spectrum antiviral infection. Very early, Boots drug company (Nottingham, England) has screened around 288 plants for anti-influenza activity [59]. Later studies have stated the inhibitory effects of medicinal herbal plants on the replication of a variety of viruses.

Research itself evidenced that viral infections such as severe acute respiratory syndrome (SARS) virus and poxvirus [60], herpes simplex virus type 2 (HSV-2) [61], hepatitis B virus [62], and HIV [63] were inhibited by several medicinal plants. Table 3 enlist reported herbal plants for antiviral therapy.

Table 3 Different treatment options for viral diseases

Summary and selected examples of therapeutic options for cytokine storms induced by emerging viruses				
Treatment strategy	**Mode of action**	**Specific drug/ product**	**Virus**	**Reference**
Anti-inflammatory drugs	Direct anti-inflammatory effects resulting in, among others, reduction of inflammatory cytokines	NSAIDs	Pandemic influenza Ebola	[34, 35]
Biological or chemical modifiers	Suppression levels of inflammatory cytokines by affecting various signaling pathways	SIP1R agonists SERP-1 p38 and MAPK inhibitors	Human pandemic influenza A Ebola	[36, 37]
Downregulation of inflammatory genes	Silencing of inflammatory master genes	DNAzyme DZ 13 for silencing of the *c-Jun* gene	Avian influenza A/H5N1 virus	
Heat shock treatment	Induction of HSPs that play a role in regulating innate and adaptive immunity	Short-term heat shock (39 °C for 4 h) for induction of HSP70	Avian influenza A/H5N1 virus	[38]
Combination therapy	Antiviral plus anti-inflammatory drugs	Zanamivir plus COX-2 inhibitors Oseltamivir plus S1P1R agonists	Avian and pandemic influenza	[39, 40]
Supportive therapy	Fluid, oxygen, and electrolytes	IV fluid resuscitation, crystalloids, and oxygen therapy	Ebola	[41]
Antiviral therapy [42]				
Virus family	**Specific virus (disease)**	**Example of drug**	**Mechanism of action**	
Retrovirus	HIV (AIDS)	Zidovudine (AZT)	Reverse transcriptase inhibitor; nucleoside analog; prevents the synthesis of DNA transcripts	
		Nevirapine	Reverse transcriptase inhibitor; nonnucleoside analog; prevents the synthesis of DNA transcripts	
		Atazanavir	Protease inhibitor; blocks processing of viral proteins	
		Maraviroc	Entry inhibitor; binds host cell CCR5 to inhibit binding of R5-tropic HIV to this coreceptor	
		Raltegravir	Integrase strand transfer inhibitor; blocks integration of linear dsDNA reverse transcript	

(continued)

Table 3 (continued)

Orthomyxovirus	Influenza virus (influenza)	Amantadine	Binds and blocks the H+ ion channel formed by the viral M2 proteins, prevents RNA uncoating; type A viruses only
		Oseltamivir	Binds the enzymatic site on the viral neuraminidase, prevents cleavage of terminal sialic acid residues, and release of virions from infected cells; all influenza type A and B viruses
Poxvirus	Variola (smallpox)	Brincidofovir	Viral polymerase inhibitor; cytosine derivative; prevents the synthesis of DNA transcripts
Adenovirus	Adenovirus viremia		
Polyomavirus	BK virus in renal transplant patients		
Herpesvirus	Herpes simplex (encephalitis)	Acyclovir	Viral DNA polymerase inhibitor; guanine derivative; prevents the synthesis of DNA transcripts
	Cytomegalovirus (retinitis)	Ganciclovir, valganciclovir	
Hepadnavirus	Hepatitis B virus (chronic hepatitis)	Tenofovir, emtricitabine	HBV DNA polymerase inhibitor as well as HIV reverse transcriptase inhibitor; nucleotide analog; prevents the synthesis of viral DNA
Hepacivirus	Hepatitis C virus (chronic hepatitis)	Sofosbuvir	Nucleoside analog inhibitor of viral RNA polymerase (NS5)
		Simeprevir	Protease NS3 inhibitor—blocks processing of viral polypeptide
		Ledipasvir	Viral NS5A inhibitor—targets viral protein essential for replication but whose function is incompletely characterized
Herbs as antiviral therapy for viral infection			
Virus	**Medicinal plant used**	**Antiviral effect**	**Reference**
Dengue virus type-2 (DEN-2)	*Azadirachta indica* Juss. (neem)	Aqueous extract inhibited DEN-2 both in vitro and in vivo	[43]
Human adenovirus type 1	Black soybean extract	Inhibition of human adenovirus type 1 and coxsackievirus B1 in a dose-dependent manner	[44]

(continued)

Table 3 (continued)

Vesicular stomatitis virus (VSV)	*Trichilia glabra* L.	Leaves extract inhibits VSV	[45]
Human immunodeficiency virus	*Phyllanthus amarus* Schum. & Thonn.	Inhibits HIV replication both in vitro and in vivo	[46]
	Olive leaf extract (OLE)	Inhibits acute infection and cell-to-cell transmission of HIV-1	[47]
Severe acute respiratory syndrome-associated coronavirus (SARS-CoV)	*Lycoris radiata*	Lycorine, isolated from *Lycoris radiate*, possesses anti-SARS-CoV	[48]
Viral hemorrhagic septicemia virus (VHSV)	*Olea europaea* L.	Leaf extract inhibited viral replication	[49]
Hepatitis C virus (HCV)	*Saxifraga melanocentra* Engl. & Irmsch.	A compound, namely, 1,2,3,4,6-penta-*O*-galloyl-beta-d-glucoside, isolated from *Saxifraga melanocentra*	[50]
Poliovirus	*Guazuma ulmifolia* Lam.	Both plants extract inhibited poliovirus replication, as well as blocking the synthesis of viral antigens in infected cell cultures	[51]
Hepatitis B virus	*Polygonum cuspidatum* Sieb. & Zucc	Inhibits hepatitis B virus in a stable HBV-producing cell line	[52]
Influenza virus	*Geranium sanguineum* L.	A medicinal plant reducing the infectivity of various influenza virus strains in vitro and in vivo	[53]
Herpes simplex virus (HSV)	*Phyllanthus urinaria* L.	1346TOGDG and geraniin inhibited HSV-1 and HSV-2, respectively	[54]

7 Conclusion

Viruses are considered to be plentiful biological species among all known ecosystems on our planet; an estimated 1×10^{31} viruses exist on earth. The diverse habitat of viruses expands in an extensive range of surroundings from deep sea to acidic (pH 3) hot springs. Antiviral drugs are generally directed against specific viral enzymes and inhibit the phases of viral entry or release. Herbal medicine possesses broad-spectrum antiviral infection. Studies have shown inhibitory effects of medicinal herbal plants on the replication of a variety of viruses. Virus infections such as SARS, HSSV-2, and HIV were inhibited by several medicinal plants. The current chapter highlighted diverse kinds of viral infections and their available therapeutic interventions along with emerging therapies for the same.

References

1. Wu KJ. There are more viruses than stars in the universe. Why do only some infect us? Science. 2020;15.
2. Microbiology by numbers. Nat Rev Microbiol. 2011;9(9):628.
3. Breitbart M, Rohwer F. Here a virus, there a virus, everywhere the same virus? Trends Microbiol. 2005;13(6):278–84.
4. Sano E, Carlson S, Wegley L, et al. Movement of viruses between biomes. Appl Environ Microbiol. 2004;70(10):5842–6.
5. Greber UF, Way M. A superhighway to virus infection. Cell. 2006;124(4):741–54.
6. Chaitanya KV. Structure and organization of virus genomes. In: Genome and genomics. Nature Publishing Group; 2019. p. 1–30.
7. Summers WC. Virus infection. In: Encyclopedia of microbiology. Elsevier Inc.; 2009. p. 546–52.
8. Lodish H, Berk A, Zipursky SL, Matsudaira P, Baltimore D, Darnell J. Viruses: structure, function, and uses. Molecular cell biology. 4th ed. WH Freeman; 2000.
9. Gelderblom HR. Structure and classification of viruses. 4th ed. University of Texas Medical Branch at Galveston; 1996.
10. Herman P, Pauwels K. Biosafety recommendations on the handling of animal cell cultures. In: Al-Rubeai M, editor. Animal cell culture. Cell engineering, vol. 9; 2015. p. 689–716.
11. Molecular Expressions Cell Biology: Virus Structure. https://micro.magnet.fsu.edu/cells/virus.html. Date Accessed 2021-12-07.
12. Werth BJ. Overview of antibiotics – infections – MSD manual consumer version. University of Washington School of Pharmacy; 2018.
13. Payne S. Virus transmission and epidemiology. In: Viruses. Academic Press; 2017. p. 53–60.
14. Lefkowitz EJ. Taxonomy and classification of viruses. In: Manual of clinical microbiology. Wiley; 2015. p. 1390–404.
15. Walker PJ, Siddell SG, Lefkowitz EJ, et al. Changes to virus taxonomy and to the International Code of Virus Classification and Nomenclature ratified by the International Committee on Taxonomy of Viruses (2021). Arch Virol. 2021;166(9):2633–48.
16. Strauss JH, Strauss EG. Overview of viruses and virus infection. Viruses Hum Dis. 2008:1–33.
17. Baltimore D. Expression of animal virus genomes. Bacteriol Rev. 1971;35(3):235–41.
18. Strauss JH, Strauss EG. DNA-containing viruses. In: Viruses and human disease. Elsevier; 2008. p. 261–323.
19. SIB Swiss Institute of Bioinformatics. ViralZone. https://viralzone.expasy.org/. Date Accessed 2021-12-07.
20. Strauss JH, Strauss EG. Plus-strand RNA viruses. In: Viruses and human disease. Elsevier; 2008. p. 63–136.

21. Strauss JH, Strauss EG. Minus-strand RNA viruses. In: Viruses and human disease. Elsevier; 2008. p. 137–91.
22. Ryu WS. Other negative-strand RNA viruses. In: Molecular virology of human pathogenic viruses. Elsevier; 2017. p. 213–24.
23. Goodgame RW. Viral infections of the gastrointestinal tract. Curr Gastroenterol Rep. 1999;1(4):292–300.
24. Rehermann B, Nascimbeni M. Immunology of hepatitis B virus and hepatitis C virus infection. Nat Rev Immunol. 2005;5(3):215–29.
25. Adams DH, Hubscher SG. Systemic viral infections and collateral damage in the liver. Am J Pathol. 2006;168(4):1057–9.
26. Thandi CS, Whittam L. Diagnosis and management of common viral skin infections. Prescriber. 2021;32(4):10–4.
27. Bansal R, Tutrone WD, Weinberg JM. Viral skin infections in the elderly: diagnosis and management. Drugs Aging. 2002;19(7):503–14.
28. Pavia AT. Viral infections of the lower respiratory tract: old viruses, new viruses, and the role of diagnosis. Clin Infect Dis. 2011;52(SUPPL. 4):S284.
29. Skappak C, Ilarraza R, Wu YQ, et al. Virus-induced asthma attack: the importance of allergic inflammation in response to viral antigen in an animal model of asthma. PLoS One. 2017;12(7):e0181425.
30. Boncristiani HF, Criado MF, Arruda E. Respiratory viruses. Encycl Microbiol. 2009:500–18.
31. Power C, Noorbakhsh F. Central nervous system viral infections: clinical aspects and pathogenic mechanisms. Neurobiol Dis. 2007:485–99.
32. Arora N, Sadovsky Y, Dermody TS, et al. Microbial vertical transmission during human pregnancy. Cell Host Microbe. 2017;21(5):561–7.
33. Bookstaver PB, Mohorn PL, Shah A, et al. Management of viral central nervous system infections: a primer for clinicians. J Cent Nerv Syst Dis. 2017;9:117957351770334.
34. Zhao Y, Ren J, Harlos K, et al. Toremifene interacts with and destabilizes the Ebola virus glycoprotein. Nature. 2016;535(7610):169–72.
35. Carter MJ. A rationale for using steroids in the treatment of severe cases of H5N1 avian influenza. J Med Microbiol. 2007;56(7):875–83.
36. Johnson JC, Martinez O, Honko AN, et al. Pyridinyl imidazole inhibitors of p38 MAP kinase impair viral entry and reduce cytokine induction by Zaire ebolavirus in human dendritic cells. Antivir Res. 2014;107(1):102–9.
37. Chen H, Zheng D, Abbott J, et al. Myxomavirus-derived serpin prolongs survival and reduces inflammation and hemorrhage in an unrelated lethal mouse viral infection. Antimicrob Agents Chemother. 2013;57(9):4114–27.
38. Oldstone MBA, Rosen H. Cytokine storm plays a direct role in the morbidity and mortality from influenza virus infection and is chemically treatable with a single sphingosine-1-phosphate agonist molecule. Curr Top Microbiol Immunol. 2014;378:129–47.
39. Walsh KB, Teijaro JR, Wilker PR, et al. Suppression of cytokine storm with a sphingosine analog provides protection against pathogenic influenza virus. Proc Natl Acad Sci U S A. 2011;108(29):12018–23.
40. Xue J, Fan X, Yu J, et al. Short-term heat shock affects host-virus interaction in mice infected with highly pathogenic avian influenza virus H5N1. Front Microbiol. 2016;7:924.
41. Zheng BJ, Chan KW, Lin YP, et al. Delayed antiviral plus immunomodulator treatment still reduces mortality in mice infected by high inoculum of influenza A/H5N1 virus. Proc Natl Acad Sci U S A. 2008;105(23):8091–6.
42. Bah EI, Lamah M-C, Fletcher T, et al. Clinical presentation of patients with Ebola virus disease in Conakry, Guinea. N Engl J Med. 2015;372(1):40–7.
43. Richman DD, Nathanson N. Antiviral therapy. In: Viral pathogenesis. Academic Press; 2016. p. 271–87.
44. Gelderblom HR. Structure and classification of viruses. In: Baron S, editor. Medical microbiology. 4th ed. Galveston: University of Texas Medical Branch at Galveston; 1996.

45. Yamai M, Tsumura K, Kimura M, et al. Antiviral activity of a hot water extract of black soybean against a human respiratory illness virus. Biosci Biotechnol Biochem. 2003;67(5):1071–9.
46. Cella M, Riva DA, Coulombié FC, et al. Virucidal activity presence in Trichilia glabra leaves. Rev Argent Microbiol. 2004;36(3):136–8.
47. Notka F, Meier G, Wagner R. Concerted inhibitory activities of Phyllanthus amarus on HIV replication in vitro and ex vivo. Antivir Res. 2004;64(2):93–102.
48. Lee-Huang S, Zhang L, Huang PL, et al. Anti-HIV activity of olive leaf extract (OLE) and modulation of host cell gene expression by HIV-1 infection and OLE treatment. Biochem Biophys Res Commun. 2003;307(4):1029–37.
49. Li SY, Chen C, Zhang HQ, et al. Identification of natural compounds with antiviral activities against SARS-associated coronavirus. Antivir Res. 2005;67(1):18–23.
50. Micol V, Caturla N, Pérez-Fons L, et al. The olive leaf extract exhibits antiviral activity against viral haemorrhagic septicaemia rhabdovirus (VHSV). Antivir Res. 2005;66(2–3):129–36.
51. Zuo GY, Li ZQ, Chen LR, et al. In vitro anti-HCV activities of Saxifraga melanocentra and its related polyphenolic compounds. Antivir Chem Chemother. 2005;16(6):393–8.
52. Felipe AMM, Rincão VP, Benati FJ, et al. Antiviral effect of Guazuma ulmifolia and Stryphnodendron adstringens on poliovirus and bovine herpesvirus. Biol Pharm Bull. 2006;29(6):1092–5.
53. Chang JS, Liu HW, Wang KC, et al. Ethanol extract of Polygonum cuspidatum inhibits hepatitis B virus in a stable HBV-producing cell line. Antivir Res. 2005;66(1):29–34.
54. Pantev A, Ivancheva S, Staneva L, et al. Biologically active constituents of a polyphenol extract from Geranium sanguineum L. with anti-influenza activity. Zeitschrift fur Naturforsch – Sect C J Biosci. 2006;61(7–8):508–16.
55. Yang CM, Cheng HY, Lin TC, et al. The in vitro activity of geraniin and 1,3,4,6-tetra-O-galloyl-β-d-glucose isolated from Phyllanthus urinaria against herpes simplex virus type 1 and type 2 infection. J Ethnopharmacol. 2007;110(3):555–8.
56. Jones PS. Strategies for antiviral drug discovery. Antivir Chem Chemother. 1998;9(4):283–302.
57. Ragab D, Salah Eldin H, Taeimah M, et al. The COVID-19 cytokine storm; what we know so far. Front Immunol. 2020;11:1446.
58. Turner MD, Nedjai B, Hurst T, et al. Cytokines and chemokines: at the crossroads of cell signalling and inflammatory disease. Biochim Biophys Acta, Mol Cell Res. 2014;1843(11):2563–82.
59. Meganck RM, Baric RS. Developing therapeutic approaches for twenty-first-century emerging infectious viral diseases. Nat Med. 2021;27(3):401–10.
60. Mukhtar M, Arshad M, Ahmad M, et al. Antiviral potentials of medicinal plants. Virus Res. 2008;131(2):111–20.
61. Kotwal GJ, Kaczmarek JN, Leivers S, et al. Anti-HIV, anti-poxvirus, and anti-SARS activity of a nontoxic, acidic plant extract from the Trifollium species Secomet-V/anti-Vac suggests that it contains a novel broad-spectrum antiviral. Ann N Y Acad Sci. 2005;1056:293–302.
62. Schnitzler P, Schneider S, Stintzing FC, et al. Efficacy of an aqueous Pelargonium sidoides extract against herpesvirus. Phytomedicine. 2008;15(12):1108–16.
63. Kwon DH, Kwon HY, Kim HJ, et al. Inhibition of hepatitis B virus by an aqueous extract of Agrimonia eupatoria L. Phyther Res. 2005;19(4):355–8.

Mucosal Targeting Strategies for Antiviral Drug Delivery

Tayo Alex Adekiya, Mumuni Sumaila, Raphael Taiwo Aruleba, and Yahya E. Choonara

Abstract Antiviral drug efficacy and bioavailability are critical factors in the treatment of viral infections. The existence of new viral infections globally, as well as the presence of multidrug-resistant viruses and their transmission, has created significant challenges to the use of antiviral drugs. Numerous studies have explored the limitations that include poor bioavailability due to low drug permeability and/or solubility, as well as short half-life and mucosal shielding that necessitates the administration of larger doses that may result in side effects. This chapter therefore provides a concise incursion into the various mucosal delivery systems for antiviral drugs and their mode of actions using a nano-enabled mucosal targeting approach. Mucoadhesive polysaccharide-based nanoparticles as a potential antiviral drug delivery strategy is also discussed, including the use of lipoidal systems for site-specific mucosal antiviral drug delivery. Furthermore, mucosal delivery of antiviral drugs via the oral, nasal, and inhalation routes are assimilated and discussed in terms of current and future trends in this field.

Keyword Antiviral · Viral infections · Mucosal · Mucoadhesive · Drug delivery systems · Nanoparticles

T. A. Adekiya
Department of Pharmaceutical Sciences, Howard University, Washington, DC, USA

M. Sumaila
Pharmacy Department, Faculty of Health Science, Southern Africa Nazarene University, Manzini, Eswatini

R. T. Aruleba
Department of Molecular and Cell Biology, Faculty of Science, University of Cape Town, Cape Town, South Africa

Y. E. Choonara (✉)
Wits Advanced Drug Delivery Platform Research Unit, Department of Pharmacy and Pharmacology, School of Therapeutic Science, Faculty of Health Sciences, University of the Witwatersrand, Johannesburg, South Africa

R. Shegokar, Y. Pathak (eds.), *Viral Drug Delivery Systems*,
https://doi.org/10.1007/978-3-031-20537-8_5

1 Introduction

The COVID-19 pandemic has highlighted the great impact viral infections have on public health and the global economy. Viruses can replicate rapidly in host cells and have the ability to attack any component of the host cell [1]. As a result, the clinical effectiveness and bioavailability of antiviral drugs are highly essential factors for the treatment of viral infections. The emergence of new infections across the world, as well as the existence of multidrug-resistant forms of viruses and their transmission, has posed serious limitations in the clinical value of antiviral drugs [2], for instance, low antiviral drug bioavailability due to poor permeability or solubility and/or short half-life that necessitate the administration of larger doses, which may result in side effects. Certain antiviral drugs have also been shown to interact with conventional drugs, resulting in harmful drug-drug interactions [3]. Furthermore, side effects are frequent, resulting from chronic therapy, which may further impede the patient's ability to adhere to the drug regimen. Some viruses, such as Ebola, Zika, and human immunodecifiency virus (HIV), may spread into difficult-to-reach anatomical areas such as lymphatic system, the central nervous system (CNS), and synovial fluid, this making it difficult to optimally treat such infections [3]. Unfortunately, the oral or parenteral routes of antiviral drug administration have significant limitations that emphasize the need for superior drug delivery systems [1]. Various drug delivery systems have been proven to optimize the efficacy and patient compliance and reduce the side effects of antiviral drugs. Drug delivery systems with improved dosing frequency and treatment durations thereby make antiviral therapy more cost-effective.

The use of nanoparticles to optimize the delivery of antiviral drugs has been the subject of many experimental studies over the last few decades [4]. As nanocarriers for antiviral drugs, nanoparticles are capable of decreasing mucociliary clearance and evading macrophage phagocytosis, hence increasing drug absorption [4]. As such, nanocarriers have the capacity to pass the mucosal barrier and the ability to improve drug bioavailability through particle absorption mechanisms [5]. Mucosal adhesives, also known as mucoadhesives, were first used in the early 1980s to regulate the delivery of drugs [6]. Mucoadhesive drug delivery systems (MDDSs) use the bioadhesivity of particular polymers that become adhesive during hydration and could therefore be utilized to deliver a drug to a particular mucosal site in the body over prolonged periods of time [7]. Owing to its large surface area as well as strong blood flow, MDDSs provide rapid absorption and excellent bioavailability. MDDSs bind to the layer of mucus that covers the epithelial surface of the mucosa as well as mucin molecules, where it increases the residence time of the dose concentration at the absorption site [7]. For example, polyethylene glycol (PEG) has been reported to improve nanoparticle mucoadhesion by interpenetration and entanglement with mucin fibers [8]. In addition, mucolytic agents such as N-acetyl-l-cysteine disrupt the mesh-like architecture of mucus by hydrolyzing the disulfide bonds to increase mucosal drug delivery [9].

Conventional MDDS, such as via oral and nasal/pulmonary routes, have received the most attention, although rectal, sublingual, vaginal, ocular, and transcutaneous routes have also been investigated [7]. The delivery of drugs via the oral mucosa offers numerous advantages when compared to other drug delivery systems, such as excellent accessibility, bypassing hepatic first-pass metabolism, influx of drug in one direction, as well as enhanced barrier permeability when compared to intact skin [7].

A large number of viruses, bacteria, protozoa, yeast, and multicellular parasites, which cause human disease invade or enter through mucosal tissues. Mucin glycoproteins (in mucus), which are released in large quantity by mucosal epithelia, can function as a physical barrier, a viscous material, or lubricant based on their structure, thereby protecting tissue surfaces against infection and injury [10]. Using a mucoadhesive drug delivery system for antiviral drug intervention can localize the drug in a specific site of the body with enhanced contact time and bioavailability . This chapter therefore provides a concise incursion into the delivery of antiviral drugs and their mode of actions using a nano-enabled mucosal targeting drug delivery approach. Mucoadhesive polysaccharide-based nanoparticles as a potential antiviral drug delivery strategy are discussed, including the use of lipoidal systems for site-specific mucosal antiviral drug delivery. Furthermore, mucosal delivery of antiviral drugs via the oral and nasal routes is assimilated and discussed in terms of current and future trends in this field.

2 Drug Treatment of Viral Infections

Various drugs have been developed to inhibit the growth of viruses in humans, and many have proven effective [11]. Nevertheless, with a multitudinous virus population infecting humans, developing novel, efficient antiviral drugs is still needed. According to Tompa et al. [11], an efficient antiviral drug should (a) evade drug resistance even after long-term clinical use, (b) fight off integrated viral DNA in the human host genome, (c) remedy co-infections by different viruses, (d) avoid drug interactions when co-administered with another drug, and (e) be cheap and low toxic in patients. A proposed strategy for rational drug identification, design, and administration of viral infections is summarized in Fig. 1.

Three strategies can be explored in the quest for a new drug regimen tackling various viral infections. The combination of available drugs will delay drug resistance development while achieving an additive/synergistic curative effect. For example, nanotechnology is an attractive option due to the dynamic physicochemical properties of nanomaterials that can even aid host-directed treatment (HDT). Noteworthy, the synergistic effect of nanomaterials can evade side effects and antiviral resistance and potentially reduce the therapeutic dose. Augmenting the host cellular and immune response via HDT via the use of cytokines, antibodies, small molecules, synthetic nucleic acids and repositioned drugs is also a key area for antiviral treatment intervention.

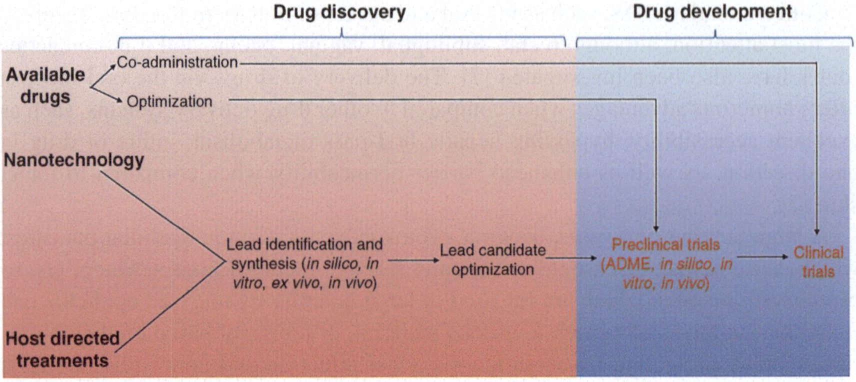

Fig. 1 Potential pipeline for identifying and developing a new antiviral drug

2.1 Hepatitis B Virus Infection

Hepatitis B virus (HBV) is a member of the *Hepadnaviridae* family and is transmitted by contact with infected body fluids and blood, which in turn triggers cirrhosis and hepatocellular cancer [12]. In 2015, WHO's global estimated cases of chronic HBV showed that 257 million persons live with chronic HBV [13]. Despite this, existing drugs cannot clear HBV infection and viral rebounds when the regimen is stopped because covalently closed circular DNA remains stable in the nucleus of hepatocytes infected by the virus even after treatment [14].

For the treatment of chronic HBV, antiviral drugs such as entecavir, adefovir dipivoxil, lamivudine, tenofovir alafenamide, tenofovir disoproxil fumarate, and telbivudine have been used. Lamivudine was the first drug approved as an oral agent against HBV, but the emergence of drug resistance has limited the use of this drug for long-term therapy [15]. Hence, it is no longer recommended as a first-line treatment option for chronic HBV. Noteworthy, lamivudine is also anti-HIV-1, and it has been suggested to have activity against SARS-CoV-2 based on its binding affinity to the virus main protease [16, 17]. Among the aforementioned anti-HBV agents, the nucleoside analogues entecavir and the two prodrugs of tenofovir (tenofovir alafenamide and tenofovir disoproxil fumarate) are the preferred first-line choices. In chronological order, these three drugs showed a high suppression of the virus in over 95% of the patients, with high profile safety and effectiveness up to 5 years of treatment [18, 19]. These drugs are associated with a high barrier to resistance [20]. Hence, they offer better therapy for successful long-term treatment. Precisely, patients (HBeAg-positive) who received entecavir (ETV)-022 at a dose of 0.5 mg for 48 weeks showed superior virologic suppression and histological improvement over patients (HBeAg-positive) who received lamivudine [19]. Tenofovir is another drug that remains active against lamivudine-resistant HBV [21]. This drug has been shown to be potent in treating both HBeAg-negative and HBeAg-positive HBV infections with no safety problems [22]. This drug is also used to treat patients co-infected with HIV-1 and HBV [23].

Adefovir dipivoxil is another well-tolerated drug and produced long-term virological, serological, and histological improvement in HBeAg-positive patients [24]. However, it is not recommended as the first-line choice because entecavir and tenofovir are more potent options. Drug delivery systems can provide a better avenue to increase the efficacies of these drugs and improve patient compliance. For example, studies have shown that a mucoadhesive carrier can extend the residence time of antigen in the nasal cavity and triggers a strong immune adjuvant capability [25, 26]. This was supported by a more recent study that highlighted that mucoadhesive glycol chitosan NPs induced strong immunity (cell-mediated and humoral) after nasal immunization of BALB/c mice against HBV antigen [27]. Indeed, both the humoral and cell-mediated responses to HBV antigens are essential for viral clearance by the host. Hence, this suggests that mucosal administration of drugs or vaccines will trigger clearance of viral infection.

2.2 *Influenza Virus Infection*

Influenza virus is a century-old virus, a member of the *Orthomyxoviridae* family, with about 250,000–500,000 annual death reported from its seasonal and pandemic flu. Influenza is a highly contagious disease that can resolve within 3–7 days. It can result in secondary infections or develop into a severe disease such as pneumonia or acute respiratory syndrome [28]. The latter can be fatal, especially in elderly patients. Indeed several influenza pandemics have hit the world such as the "Spanish flu, the calamitous influenza pandemic caused by H1N1 in 1918," "Asian influenza pandemic caused by the H2N2 virus in 1957," "Hong Kong pandemic caused by the H3N2 virus in 1968," and "the 2009 swine flu caused by H1N1 virus" [29, 30]. Against influenza, clinicians have depended on more conventional therapy techniques, and the drugs used are key for prophylaxis or early containment of this infectious disease.

Anti-influenza drugs such as neuraminidase inhibitors (zanamivir and oseltamivir) and M2 ion channel inhibitors (amantadine and rimantadine), among others, are antiviral drugs approved for influenza, but their efficacy is now threatened by the growing resistance caused by drug-induced selective pressure [31]. Notably, the M2 ion channel inhibitors are not recommended for influenza B and C viruses due to the lack of M2 protein [28]. Hence, therapy with M2 ion channel inhibitors is limited to influenza A but still faced with a high rate of resistance in this virus strain. For example, over 90% global resistance rate was reported in 2005–2006, nearly 50% in Europe, and over 80% in Asia and the USA (2006–2007) [32]. Adding to the drawbacks, amantadine is associated with the CNS side effects. Indeed, various new amantadine and rimantadine derivatives have been shown to have significant activity against influenza type A. These derivatives require more study to ascertain if they offer superior potency, selectivity, and resistance profile over the parent compounds.

However, neuraminidase (NA) inhibitors are the most widely used drug treatment for controlling influenza because they block the activity NA. Zanamivir is well tolerated with no adverse events in patients. However, its absolute oral bioavailability is poor (averaging 2%) [33]. Hence, this calls for the development of alternate delivery routes for zanamivir. A dose of 10 mg/25 mg is recommended for an adult to inhale twice a day for influenza. This NA inhibitor is more effective than oseltamivir. Oseltamivir is administered orally for 5 days at a daily dose of 75 mg (twice) in adults, but in severe cases of influenza, duration of treatment can be extended. This drug presents more negative effects than zanamivir and is capable of inducing resistant viral strains [29]. Indeed, observational studies have suggested that oseltamivir is associated with improved clinical outcomes and reduced mortality in severe influenza [34, 35]. Resistance to this drug has been reported in children and adults. Further, the active metabolite of oseltamivir is significantly less active against the NA in influenza B than in influenza A.

Other NA inhibitors are laninamivir, a drug powder administered at a single inhalation dose of 40 mg. Peramivir is another NA inhibitor but administered via drip infusion at a dose of 300 mg for 15mins. It shows high activity against both influenza A and B viruses, respectively [29]. Several studies have used combination therapy approaches to remedy the various influenza virus types. Indeed, combination therapy warrants more study to ensure dose specificity, efficacy, drug-drug interaction, and adverse events. Overall, in the fight against influenza, drug-resistant virus variants pose a major threat, and it is high time that new mucosal drug delivery approaches are investigated. Moreover, mucoadhesive carriers hold great potential for drug delivery and vaccines due to extended residence time of carrier, penetration, and mucoadhesion. Chitosan is an example of delivery vehicle with a mucoadhesive property that can overcome the mucociliary clearance. Sawaengsak and co-workers [36] showed that a chitosan encapsulated hemagglutinin (HA) split influenza virus vaccine induced higher systemic and mucosal antibody titers than the HA-split influenza virus alone in mice. More so, the vaccine-induced cell-mediated immunity as mice spleen secreted high numbers of IFN-y.

2.3 Herpes Simplex Virus Infection

The herpes simplex virus (HSV) is divided into two types, HSV-1 and HSV-2. They are a member of the *Simplexvirus* family, their infection has been with mankind since time immemorial, and they remain one of the most common infectious agents affecting humans [37]. HSV-1 is more prevalent than HSV-2; however, symptoms associated with both are skin lesions, whitlow, keratitis, and genital sores. Often, the disease caused by these viruses is subclinical, and infection can be mild to severe, particularly in immunosuppressed patients. In light of these, effective drugs are needed to provide a better quality of life for patients, particularly those who are severely affected.

Three classes of drugs targeting the viral DNA replication are licensed for use against HSV, which are guanosine analogues (acyclovir and ganciclovir), pyrophosphate analogue foscarnet, and acyclic nucleotide analogues (cidofovir) [38]. Acyclovir remains the gold standard used for the treatment of HSV. Acyclovir has been phosphorylated thrice and acts by terminating viral DNA chain and competitive inhibition of the DNA polymerase [39]. The drug is very potent and safe and has high bioavailability, particularly the esterified analogue. However, the widespread use of this drug has resulted in HSV-resistant strains. Additionally, it requires long-term treatment administration in immunosuppressed individuals, and this in turn exacerbates drug resistance [40]. Other drugs such as cidofovir and the pyrophosphate analogue forcarnet are nephrotoxic and have poor oral bioavailability [38]. Moreover, this highlights the need for a novel approach to the delivery of antiviral agents for HSV or a new drug that is effective against both the parent viruses and drug-resistant virus strains.

Overall, more research time and effort are needed in virology and pharmaceutical science niches for new molecular entities or drug delivery interventions against HSV. In order to aid in the local and systemic delivery of acyclovir, Chaudhary and Sherma [41] developed an oral mucosal drug carrier for HSV. They suggested that the high mucoadhesive property of the vehicle could result in extending the retention and enhance absorption across mucosal tissue.s

3 The Role of Mucosal Targeting in Antiviral Drug Delivery

As highlighted above, the existing drug treatment options available in the war against viruses is limited and hence causing a hurdle to their efficacy and resulting in drug resistance. The role of mucus (or mucosal linings of tissues) together with nanotechnology could provide a solution to address antiviral drug efficacy and resistance based upon the chemical, biological, and physical benefits of mucus targeting nanomaterials (Fig. 2). Mucosal drug delivery systems can achieve superior drug penetration and permeation through absorption membranes [42]. The action of the absorption membrane is dependent on the mucosal layer and the epithelial tissue. Various regions of the human body (e.g., eyes, mouth, gastrointestinal system, urogenital cavity) are protected by mucosal membranes, hence making mucosal targeted drug delivery a significant option [43]. Small drug molecules and viruses can penetrate through the mucosal membranes easily. Mucoadhesives are designed to allow longer administration of a drug by allowing a longer contact time between the delivery device and the mucosa. Several mucoadhesives are commercially available in the form of tablets and pastes [44]. The oral mucosal cavity easy accessibility offers a great route for local and systemic drug delivery. The oral mucosa has an excellent blood supply, ability for quick repair after injury, and relative

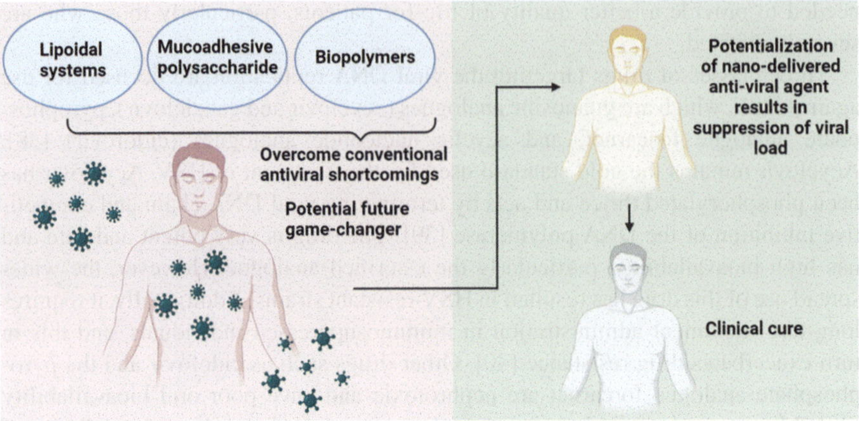

Fig. 2 Diagram depicts how drug delivery can alleviate and cure viral infection

permeability. Although some challenges need to be overcome, mucosal drug delivery is a potential game-changer technology for the delivery of antiviral drugs and HDT.

3.1 Advantages of Nano-enabled Mucosal Delivery Carriers

Mucosal nano-enabled antiviral drug delivery is a promising strategy attracting much attention due to the mucoadhesive and mucopenetrating properties of nanosized materials, particularly polysaccharide nanomaterials. Given the noninvasiveness, painlessness, and patient friendliness of the mucosal delivery route, mucosal drug delivery may offer a convenient antiviral delivery pathway with a corresponding improvement in patient adherence to antiviral therapy. Moreover, mucosal delivery, for example, buccal, ocular, nasal, and vaginal delivery, presents a variety of therapeutic benefits such as the ability to overcome hepatic first-pass metabolism, target precise tissue, and improve bioavailability. Despite the numerous appealing features of mucosal targeted drug delivery, the route offers a significant barrier to vectorization of conventional antiviral therapeutic. Firstly, the mucus gel layer on the mucosal epithelia, apically bound to the cell surface, plays a protective role to the cells underneath, preventing the permeation of compounds into the cells [42]. Secondly, mucins, component of mucus and high-molecular-weight glycoproteins, form an interpenetrating network restricting the free passage of components within and across the mucus [45]. Thirdly, the hydrophilicity and anionic properties of mucus create a steric hindrance to molecular diffusion through mucus layer [45]. Fourthly, the dynamic nature of mucus in regard to nonstop secretion and shedding from the mucosal surfaces poses additional challenge to mucosal drug delivery [45]. Hence, mucosally delivered antivirals must diffuse upstream to allow access to and crossing the epithelium. Finally, the epithelia layer below the mucus, with cells interconnected by tight junctions, offers a serious biological tissue-based barrier to antiviral delivery. Drug properties, such as hydrophilicity, lipophilicity, molecular size, and others, impact transepithelial movement of drugs intended for systemic action [46].

Water-soluble compounds (e.g., protein/peptide drugs) preferentially use the paracellular path as the main route of absorption. However, the presence of intercellular tight junctions in this route hinders drug transport. Transcellular route, on the other hand, favors lipid-soluble compounds because the latter possess cell-membrane partitioning capacity. Drug compounds with small molecular size passively diffuse through the paracellular pathways; however, largely sized compounds (such as proteins and peptides) are hindered, highlighting the sieving property of the intercellular tight junctions [46].

The application of nanotechnology to enhance mucosal delivery of antiviral drugs has provided numerous benefits with regard to solubility/permeability enhancement, modulation of drug biodistribution/disposition, bioactive protection against degradation in biological milieu, biomimetic properties, targeted delivery, and others [42, 47, 48]. Nanosizing imparts unique properties into materials, and this has been explored by researchers to improve the biopharmaceutical performance of various compounds with mucosal delivery challenge. The small particulate size (1–100 nm) allows for easy delivery of loaded drug compounds into anatomically privileged sites unreachable to conventional drugs. The large particulate surface area to volume ratios accommodate large bioactive payloads [47]. Furthermore, the possibility of drug entrapment within nanoparticle architecture and ease of structural modification (with polymers, such as poly(ethylene glycol)) [47] could give rise to dose optimization and improvement in drug delivery capacity. This can be possible due to nanocarrier-mediated increase in retention time and polymer-mediated stability enhancement of the delivery system [4]. Additionally, nanoparticles can be engineered with tunable surface chemistry and functionalized with targeting moieties to enable mucosal permeation/cellular entry and cell-type specific delivery, respectively [48–50].

Recently, Sanna and colleagues [48] designed targeted nanoparticles incorporating a mixture of the hydrophobic polymer, poly(epsilon-caprolactone), and the amphiphilic block copolymers, poly(D,L-lactic-co-glycolic acid)-block-poly(ethylene glycol), for site-specific delivery of the anti-SARS-CoV-2 drug, remdesivir (RDV) [48]. Using a Vero E6 cells infection model, authors demonstrated that angiotensin-converting enzyme-2 (ACE2) receptor ligand-decorated nanoparticles with encapsulated RDV show a significant improvement in antiviral efficacy (EC_{50}, 0.67 mM) compared to unformulated RDV (EC_{50}, 0.92 mM). Additionally, they reported a competitive particulate-ACE2 receptor interaction between the viral particles and the polymeric nanoparticles as the latter demonstrated a basal antiviral property when tested without the encapsulated RDV [48].

4 Mucoadhesive Polysaccharide-Based Nanoparticles

Mucoadhesion, the ability of nanomaterials to undergo interfacial attractive interaction with the mucus or mucosal membrane, permits prolonged nanocarrier residence time at the site of absorption following administration. In addition, bio-adhesive phenomenon can be used to extend drug release rate from mucoadhesive formulations, reduce frequency of dosing, enhance drug bioavailability, afford targeted delivery to body sites/tissues, and, overall, improve therapeutic outcomes [51]. Mucoadhesivity has been largely employed in the development of polymer-based dosage forms for vaginal, gastrointestinal, ocular, buccal, and nasal drug delivery [51]. Given the numerous benefits of mucoadhesiveness to drug delivery, a wide array of polysaccharides (naturally derived mucoadhesive polymers with desirable safety) [46] have been explored for the design of efficient nanocarriers for antiviral drug delivery.

For example, chitosan nanoparticles are polysaccharide-based nanocarriers that have shown wide benefit in antiviral drug delivery capacity stemming from their high drug entrapment, sustained-release capacity, and minimal cytotoxicity [52–54]. Szymańska and co-worker [55] fabricated zidovudine-encapsulated mucoadhesive chitosan glutamate particulates and demonstrated the vaginal delivery potential for the treatment of herpes (HSV-2) infection. These particulate carriers, with more than 80% average zidovudine entrapment, displayed a significant muco-retaining property on isolated human vaginal epithelium, hence, suggesting that the zidovudine-loaded mucoadhesive chitosan particulate can remain attached to the vaginal mucosa resisting the washing effect of vaginal fluid and allowing sufficient time for systemic absorption via the vaginal epithelium. However, permeation studies, as reported by authors, revealed that the glutamated chitosan-based carrier failed to facilitate drug absorption across the human tissue as there was a drop (~25 μg/cm^2) in the permeated amount of zidovudine over 24 h period compared to the unformulated drug dispersion [55]. Other reports have highlighted the permeability-enhancing properties of chitosan-based carriers, which were attributed to increase in epithelial tight junction opening and interaction with extracellular matrix components [56, 57]. This is contrary to the report from [55], hence, the need for further studies to establish the penetration-enhancing behavior of chitosan.

In another study, Ekama et al. [58] employed the ionic gelation technique in designing a sodium tripolyphosphate-crosslinked chitosan particulate carrier for intravaginal co-delivery of antiretroviral drugs, maraviroc and tenofovir. The polysaccharide-based carrier progressively reduced the degree of HIV-1BaL infectivity (up to 1.0 μg/mL particulate concentration) during in vitro assay using the TZM-bl indicator cell model. Further, authors reported a sustained-release kinetics for the encapsulated antiretroviral drugs; however, while the onset of tenofovir release was within 1.0 h into the release study, the release of maraviroc commenced 11 h later. The authors reported that hydrophilicity variation between both antiretrovirals might have influenced the difference in release behavior [58]. The hydration dynamics of polysaccharide carriers has been reported to also impact tenofovir release behavior from the biopolymer architecture [59].

Martín-Illana and associate [59] demonstrated the influence of rapid water permeation on tenofovir release characteristic from polysaccharide-based ethylcellulose/chitosan-containing trilayer vaginal films. Rapid film hydration favors polyelectrolyte complexes formation through ionic and/or hydrogen bonding, hence, favoring sustained release of tenofovir. Drug delivery kinetics can be modulated by incorporating inorganic drug release regulator, which can weaken the polysaccharide chains and consequently increase ionic mobility through the formation of low-molecular-weight polyelectrolyte complexes [59].

Intestinal absorption of orally administered antiviral drugs determines the bioavailability and system efficacy of the drug toward viral clearance. Antivirals with high water solubility/low permeability (e.g., zidovudine, zanamivir, and oseltamivir) face gastrointestinal absorption challenge. Various mucoadhesive polysaccharide nanoparticulate strategies have been used to conquer this drawback. Pedreiro et al. [60] synthesized mucoadhesive zidovudine-loaded sodium starch glycolate/hypromellose phthalate particulate carrier using co-precipitation/solvent evaporation method to increase gastrointestinal absorption and bioavailability of encapsulated drug. Using the everted gut sac model of permeability study, authors reported a twofold increase in zidovudine intestinal absorption from the synthesized polysaccharide particulate carrier in comparison with the free, unformulated drug. This result may be attributed to the increase in amorphous properties of the zidovudine in polysaccharide carrier, which altered the intestinal epithelia-carrier interaction as opposed to its free form, hence, the resultant improvement in permeability. Additionally, the increase in lipophilicity and gastrointestinal residence of the polysaccharide particulate carrier may have contributed to the observed increase in intestinal absorption [60]. Other studies that demonstrated the permeability-enhancing and bioavailability-improving properties of mucoadhesive polysaccharide-based carrier systems have been reported [52, 54].

Thiolation of chitosan, which is the conjugation of protein-derived thioether functionality to chitosan, is a strategy that has been employed to improve mucoadhesiveness, cohesiveness, permeation enhancement, and efflux-pump inhibition of chitosan-based nanocarriers [61]. This approach is increasingly gaining extended application in mucosal antiviral drug delivery. Rajawat et al. [62] covalently conjugated N-Acetyl cysteine to chitosan through a carbodiimide-mediated coupling reaction producing chitosan-N-acetyl cysteine particles for ocular delivery of acyclovir. Thioether-modified chitosan was reported to improve mucoadhesiveness by 1.45-fold compared to the unmodified chitosan. Dissolution study (in a simulated tear fluid) displayed early burst release followed by an extended release for 12 h. Overall, authors posited that the thiolated polysaccharide particulate may offer a promising strategy for efficient delivery of acyclovir for the treatment of ocular HSV-1 and HSV-2 infection.

In another study, Kapanigowda and colleagues [63] showed that ganciclovir-loaded chitosan-based mucoadhesive particulate significantly increases the intraocular bioavailability (AUC = ~5.0-fold, C_{max} = ~2.7-fold, in aqueous humor) of ganciclovir in comparison to the unformulated drug solution using a *Wistar* rat eyes model. The mechanism of bioavailability improvement could be attributed to carrier-mediated increase in corneal permeation, resistance to nasolacrimal drainage, and improvement in drug stability [63].

5 Lipoidal Systems for Mucosal Delivery

Among the nanocarriers that can be used in nanomedicines, lipid-based carriers (LBCs), which comprise liposomes, solid lipid nanoparticles (SLNs), nanostructured lipid capsules (NLCs) in particular, lipid micelles, lipopolymers, and lipoproteins have received increased attention in the industry due to a plethora of characteristics [64]. LBCs have high encapsulation efficiency for amphiphilic and hydrophobic bioactive molecules and the ability to entrap proteins and peptides, oligonucleotide (such as mRNA), and lipid-drug conjugates (Fig. 3). Others include controlled drug release through lipid composition, ability to tailor the properties to the specific application, easy surface modification, amphiphilic nature, lower plasma circulation, biodegradable and biocompatible, non-immunogenic, and less or no toxicity [64, 65]. Furthermore, it has the ability to overcome solubility and

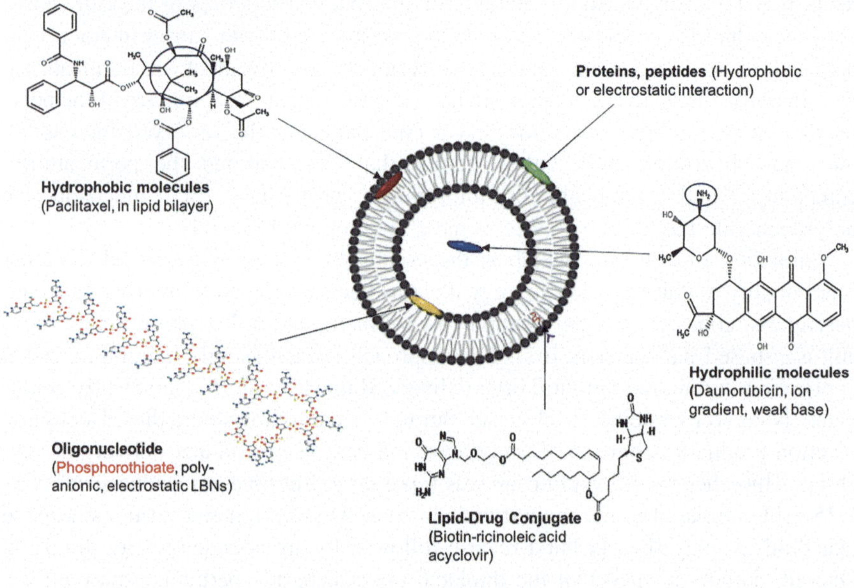

Fig. 3 Schematic representation of LBNs as a highly versatile carrier for the delivery of several therapeutic agents. The hydrophobic molecules are entrapped in the lipid bilayer; the hydrophilic molecules are embedded within the LBNs due to their ion gradient and weak base; oligonucleotide uses the polyanionic to form electrostatic interactions with the LBCs. There is hydrophobic or electrostatic interaction between proteins and peptides with LBNs; lipid-drug conjugate could be sandwich within the hydrophobic head and within the lipid bilayer

bioavailability problem faced with therapeutic molecules with poor permeability and/or solubility; LBCs could bypass multidrug resistance mechanisms [65]. LBCs can mimic the chylomicrons formation through which the entrapped therapeutic molecules can be effectively transported along with the carrier during the classical transcellular lipid absorption mechanisms [66]. The PEGylation of lipoidial nanodelivery systems could improve their mucoadhesive property and thereby help in the interpenetration and entanglement with mucin fibers [8]. More so, the dispersion of lipoidal nanosystems in mucoadhesive polymeric gel formulation possesses the additional importance of increasing the mucus penetration of the drug through oral administration as well as control and extend the release rate of drugs. According to Du and colleagues [67], the addition of oleic acid to liquid crystal systems increases the pH sensitivity of lipid drug nanocarrier internal nanostructure and triggers the generation of mucoadhesive hexosomes for better drug delivery to the buccal mucosa.

Currently, there are over 20 liposomal drugs (e.g., liposomal doxorubicin, liposomal amphotericin B, patisiran, etc.) in the markets. Recently, two lipid-based nanoparticles (LNPs) COVID-19 vaccines (e.g., COVID-19 mRNA-Lipid based nanoparticles vaccine) were authorized to be used by the Food and Drug Administration (FDA) as well as the European Medicines Agency (EMA) with over 50 additional ones in clinical trials.

5.1 The Use of Micro- and Nanoemulsions as Mucosal Targeted Delivery System

Microemulsions are thermodynamically stable oil (such as castor oil, soybean oil, and peanut oil)-and-water dispersions stabilized with a surfactant or a cosurfactant, whereas nanoemulsions are single-phase thermodynamically stable with particle size range of 20–500 nm [68]. For their preparation, many techniques such as high-pressure homogenization, phase inversion temperature technique, low-energy emulsification, ultrasonication, or microfluidization can be utilized. The advantages of nanoemulsion and microemulsion include greater water solubility, high loading capacity, extended time of residence in the GIT, improved bioavailability and absorption, as well as lymphatic absorption. Abdou et al. [69] demonstrated that the addition of 0.3% chitosan as a mucoadhesive agent to the nanoemulsion-loaded zolmitriptan increased its residence duration and zeta potential while having no impact on globule size. They also demonstrated that the designed mucoadhesive nanoemulsion had better permeability coefficients across the nasal mucosa than the free zolmitriptan solution. In vivo investigations revealed that the mucoadhesive nanoemulsion containing zolmitriptan formulation had a shorter T_{max} and a greater AUC_{0-8} in the brain as opposed to the bear nasal or intravenous solution [69]. Similarly, Kumar et al. developed a mucoadhesive nanoemulsion-based carrier system with the inclusion of PEG-400 as a mucoadhesive agent as an alternative approach for the transportation of olanzapine via a quick nose-to-brain route for improved distribution and transport into and within the brain [70].

Thus, the formulation of mucoadhesive microemulsion or nanoemulsion can be of help in mucosal delivery of antiviral drugs. In many studies, the bioavailability of insoluble antiviral drugs such as acyclovir and nevirapine has been enhanced by keeping them in a molecular dispersion in the GIT and so extending the absorption window present in the GI lumen [71, 72]. Mahboobian et al. observed that the gel-nanoemulsion loaded acyclovir generated utilizing the low-energy approach indicated that acyclovir penetration of the optimal gel emulsion was approximately 2.8-fold greater than the acyclovir uncoated [73]. This can be employed as an effective topically applied delivery strategy for viral ophthalmic disease treatment.

5.2 *Liposomes*

Liposomes are small spherical vesicles ranging from 15 to 1000 nm in size synthesized through the encloses of phospholipids in an aqueous core through which the drug is distributed. For site specificity, a number of targeted ligands can be conjugated to the liposome surface. Liposomes can entrap and be used for the delivery of both lipophilic and hydrophilic drugs. Liposome lipid layers possess the ability to prevent the encapsulated drug from GI degradation as well as aid in its prolonged release [74]. Liposomes are frequently prepared using techniques such as passive and active loading, solvent dispersion, detergent removal, and mechanical dispersion method through the use of supercritical fluid technology, sonication, membrane contactor technology, dual asymmetric centrifugation, freeze-drying, and cross-flow filtration technology. Polyethylene glycol (PEG)-modified liposomal systems have mucus-penetrating particles for systemic absorption of drug delivery via oral administration.

Mucoadhesive liposomes could enable long-term absorption in the GIT mucosal layer; a drug-loaded liposome may still be removed during the turnover of mucus first before liposomes infiltrate via the layer of mucosal to reach epithelial cells [75]. Intestinal mucus-penetrating characteristics of pluronic F127-coated liposomal system have been demonstrated to enhance oral absorption of lipophilic drugs [76]. Ramana et al. developed a liposomal delivery system from egg phospholipids loaded with nevirapine, which demonstrated maximal stability at physiological pH, prevented systemic toxic side effects, and increased availability at the targeted location [77]. Alsarra et al. created nanoliposome-loaded acyclovir using the lipid film hydration technique, which allowed the drug to stay in touch with the absorptive regions in the nasal cavity for longer periods of time and enhanced direct absorption via the nasal mucosa [78].

5.3 Solid Lipid Nanoparticles

Over the past decade, solid lipid nanoparticles (SLNs) have gained popularity as adaptable nanosized vehicles for therapeutic agents, and they have been extensively studied worldwide. SLNs are nanospheres or nanosized spherical lipid carriers, which are composed of solid lipid matrix such as glycerides, waxes, or fatty acids as lipid constituents and tween, polyxythylene ethers, bile salts, phospholipids, and/or polyvinyl alcohol as physiological-compatible emulsifiers for stabilization [79]. In water or aqueous surfactant solution, the spherical particle size of SLNs ranges between 50 and 1000 nm in diameter. Meanwhile, at room temperature, the solid lipids, which could be regarded of as perfect crystal lipid matrices, absorb pharmaceuticals or other bioactive molecules between the fatty acid chains of the SLNs solid hydrophobic core [79, 80]. Additionally, drug laden can be attached, particularly to the surface matrix of the carrier instead of being dispersed or embedded into the core of the solid matrix [79, 80].

SLNs possess several advantages over other nanocarriers, which include biocompatibility and biodegradability. SLNs can be produced without the utilization of organic solvents, and they are highly stable in physical conditions and highly reproducible and have safe cost [79]. Other advantages include the ability to encapsulate lipophilic and hydrophilic bioactive molecules with several physiological and pharmacological features, easy to produce in large scale, controlled drug release, can be used to improve the stability of the incorporated active substances, and can be sterilized [79, 80]. In addition, SLNs as a nanocarrier systems have been shown to be safe and effective delivery methods for therapeutic agents and capable of enhancing the pharmacokinetic profile and efficacy of the entrapped therapeutic molecules [81]. It has been proven to be a highly efficacious skin delivery carrier treatment [82, 83], which also has the potential of delivering gene [84] and other therapeutic agents for cancer and several other diseases. SLNs have been used as a targeted drug delivery through the blood-brain barrier (BBB) into the CNS for neurological diseases treatment and other psychological manifestations [85, 86].

PEGylated SLNs have gained intensive research for therapeutic agent delivery into the brain due to their lipidic nature, which makes them easily and readily taken up by the brain tissues [87], as well as improve the mucoadhesive property of SLNs. The surface of SLNs may be modified, and chitosan can be utilized to enhance mucoadhesion and nanoparticle transport to the pulmonary mucosa, as well as enhance drug delivery to alveolar macrophages [88]. Vieira et al. found that chitosan-coated SLNs laden with rifampicin, an anti-tuberculosis drug, have greater mucoadhesive characteristics as well as higher permeability in A549 alveolar epithelial cells compared to non-coated SLNs in an in vitro study. This suggested that the generated chitosan-SLNs could be employed as a potential carrier for safer and more effective anti-tuberculosis intervention [88]. It has been reported that SLN can improve the bioavailability and reduce the toxicity and sustain the release profile of antiviral agents. SLN-loaded antiviral drugs can actively sustain the inhibition of HIV virus generation; moreover, SLNs have been shown to enhance the essential oil accumulation in the skin, thereby enhancing antiherpetic efficacy. Thus, SLNs that cover and penetrate mucosal barriers quickly and evenly can be employed as an effective topically applied delivery strategy for viral disease treatment.

5.4 *Nanostructured Lipid Carriers*

In comparison to SLNs, nanostructured lipid carriers (NLCs) are second-generation SLNs, which employ liquid lipids to provide higher loading capacity, controlled release pattern, and stability. More so, they might also be surface functionalized to attain target specificity. It has been reported in an in vitro mucoadhesive investigation that PEG-NLCs as well as polyvinyl alcohol (PVA)-NLCs were more than twofold mucoadhesive to newly porcine intestinal mucosa than chitosan (CS)-NLCs and uncoated-NLCs [89]. NLC has been utilized to improve antifungal activity through the delivery of Miconazole to the oral mucosa [90]. When compared to plain drug, NLCs exhibited a ninefold enhancement in ease of penetration through the porcine nasal mucosa, and NLC had no adverse impact on the nasal mucosa, making it safe for intranasal delivery in mice [91]. Owing to these studies, and the attributes of NLCs, they can be employed as a mucosal drug delivery approach for antiviral intervention.

6 The Use of Mucoadhesive Oral Delivery Nanosystems

After antiviral drugs are ingested, they have to move through the gut until they reach the small intestine where absorption and transport to systemic circulation take place. Several factors influence the oral bioavailability of antiviral drugs. Examples include (i) bioaccessibility of the antiviral drug (percentage of absorbable drug present in the intestinal fluids), (ii) antiviral drug stability (percentage of drug that possess stability against enzymatic/chemical degradation), and (iii) antiviral drug absorption (percentage of drug that can permeate the intestinal mucosa/epithelium). Antiviral drugs with high lipophilic molecules and poor aqueous solubility, such as efavirenz, zidovudine, and dolutegravir, and antiviral essential oils and some phytochemicals encounter dissolution challenge in the gastrointestinal fluids, hence, having minimal bioaccessibility [92]. Mucoadhesive polysaccharide nanoparticles can be used to improve the solubility of lipophilic antivirals in the aqueous gastrointestinal fluid. Belgamwar and collaborators [93] improved the aqueous solubility of efavirenz (a BCS-II classified antiretroviral with poor oral bioavailability) using a novel chitosan-HPβCD copolymer-grafting strategy. The synthesized chitosan-HPβCD tremendously increased efavirenz solubilization by 380-fold compared to the free drug, demonstrating the bioavailability-enhancing potential of the polysaccharide hybrid system following oral administration. The solubility-improving effect of the hybrid polysaccharide nanoparticles could be attributed to an inclusion complex formation between the hydrophobic openings of chitosan-HPβCD copolymer and efavirenz [94]. Moreover, there could be hydrogen bonding occurring between the N-H and "O" atoms of the antiretroviral drug and the polysaccharides, permitting drug solubilization.

The gastric mucosa presents a major barrier to the oral delivery of antiviral drugs impacting the oral bioavailability of these drugs. Permeating the mucus layer and intestinal epithelium before gaining access to systemic circulation has not been an easy task for some antivirals, particularly hydrophilic and high-molecular-weight drugs. In addition, the presence of specialized cells of the intestinal epithelium, such as Paneth cells, enterocytes, goblet cells, M cells, and endocrine cells, can lessen the uptake of antiviral drugs [92].

Miller and co-workers employed ion-pairing strategy to improve the gastrointestinal absorption of zanamivir heptyl ester and guanidino oseltamivir, two highly polar antiviral drugs [53]. Using the counterion 1-hydroxy-2-naphthoic, authors produced antivirals with improved lipophilicity and permeability. Apparent permeability of both drug formulations through Caco-2 cell monolayers was considerably increased by 0.8–3.0 × 10^{-6} cm/s compared to formulations without the counterion. The effect of counterion inclusion on effective permeability (P_{eff}) of the antiviral drug formulations was studied using the rat jejunal perfusion assay model. While the counterion failed to increase the P_{eff} of guanidino oseltamivir, the P_{eff} of zanamivir heptyl ester was enhanced by 4.0 × 10^{-5} cm/s compared to the counterion-free drug formulations. Authors suggested that the observed variation in P_{eff} between the experimental antivirals may be due to dissociation and ion-exchange phenomena largely affecting the guanidino oseltamivir formulation compared to zanamivir heptyl ester formulation. Hence, the latter exhibited more endogenous stability during membrane permeation. Overall, the combined influence of mucoadhesiveness, muco-penetrativeness, and solubilization-enhancing ability of polysaccharide nanosystems could explain the oral bioavailability-improving property of the nanocarriers [93, 94].

7 The Use of Mucoadhesive Nasal Delivery Systems

The human nose has a unique structure coated with very vascularized mucosa for performing its fundamental function as air conditioner and filter protecting the subordinate airways. The functional and structural characteristics of the nasal mucosa are critical for elucidating the physiological nasal defense systems and laying the foundation for the biopharmaceutical idea of nasal drug delivery. Nasal drug administration has been employed as an alternate approach for the systemic distribution of drugs that were previously only available through intravenous delivery [95]. This is owing to the vast surface area, permeable endothelium membrane, lack of first-pass metabolism, high total blood flow, and ease of access. More so, the delivery of drugs that cannot be delivered via the oral route to the brain can be achieved through the nasal route, due to its direct and noninvasive method.

Recently, there has been a lot of attention on nasal delivery of drugs for systemic therapy, including a lot of chemical, peptide, and protein therapies. Following intranasal administration, drugs are discharged rapidly from the nasal cavity, which results in a quick systemic drug absorption [95, 96]. Meanwhile, numerous ways of improving the residence period of drug products in the nasal mucosa have been documented, which results in an increased nasal drug absorption. The intranasal drug delivery system provides a noninvasive alternative method for drug administration to obtain efficient and effective drug levels, via direct absorption into the bloodstream circulation across the nasal mucosal membranes [95, 96]. In the past, mucoadhesive systems for both peroral and oral delivery have been developed. The nasal mucosa offers an excellent location for the bioadhesive delivery systems for drug.

In order to address the challenges inherent with the use of antiviral agents for viral infection treatment, nasal administration of antiviral drugs through the use of innovative devices of mucoadhesive drug delivery systems through nasal\pulmonary is therefore proposed as an alternative route. It is widely understood that following nasal administration, drug-escaping mucociliary elimination as well as enzymatic degradation may not only enter into blood circulation, but it can also penetrate the cerebrospinal fluid (CSF) or brain tissue via the olfactory area and/or the nasal trigeminal route [97]. Thus, the nasal approach also appears to be a viable technique for obtaining antiviral agent absorption in the CNS, as it has the ability to transport drugs effectively into the CNS from the nasal cavity. To address CNS viral infections, antiviral drugs such as zidovudine have been administered intranasally using thermosensitive hydrogels [97]. Ribavirin is an antiviral drug that has the potential in treating viral infections in both animals and humans. In vivo nasal delivery of α-cyclodextrin-containing agglomerates to rats resulted in a complete accumulation of ribavirin in all the regions of the brain investigated compared to micronized ribavirin delivered without excipient microparticles [98]. Alsarra et al. demonstrated that an intranasal liposomal delivery method coupled with a mucoadhesive gel technology generates therapeutically relevant plasma concentration of low-molecular-mass hydrophilic drugs, acyclovir. In their study, it was discovered that the combination of liposomal systems, acyclovir, as well as mucoadhesive gel did not only enhance longer interaction between the absorptive regions of the nasal cavity and drug, but it also enabled direct uptake across the nasal mucosa [78].

8 Mucoadhesive Inhalation Delivery Systems

Inhalational delivery of therapeutic agents is a widely accepted and ideal route for treating many pulmonary diseases like asthma, chronic obstructive pulmonary disease, lung cancer, pulmonary infections, and others. Most inhalational formulations are designed to produce direct local effect on the lung tissue; hence, such delivery system encourages minimal systemic bioavailability and reduced adverse drug effect [99]. Given the large surface area of the human lungs alongside the highly vascularized and penetrable epithelia, drug given through this route can achieve rapid absorption and high pulmonary bioavailability [100]. Additionally, inhaled drugs can evade hepatic first-pass metabolism, produce rapid onset of action, allow for reduced therapeutic dose, and minimize systemic toxicity [100]. Despite these attractive benefits of inhalational route of drug delivery, inhalable drugs are faced with several drawbacks, which impact the pulmonary delivery of therapeutic agents.

The biological obstacles that exist in the pulmonary airways such as mucus, ciliated cells, and resident macrophages effectively hinder drug localization, permeation, and adsorption in the lung [101]. Drugs entering the lung airways are cleared by dual mechanisms based on the location of deposition. Ciliated cells of the pulmonary airways aid in the removal of drugs that enter the upper airways, while the resident alveolar macrophages engulf and digest particles localized in the lower airways [101]. In order to harness the benefits offered by inhalational drug delivery, several nano-based strategies have been tested for efficacy in overcoming the highlighted disadvantage of pulmonary route. Most of these methods were designed to improve on drug properties such as aqueous solubility, dissolution rate, and drug efflux/clearance. Nanocarriers with neutral surface charge can effectively permeate the airway mucosa; however, cationic carriers favorably adhere to the anionic mucus mesh in a typical inhalational delivery model [102]. Mucoadhesive carriers based on positively charged polysaccharides, like trimethyl chitosan, positively charged cellulose and positively charged starch, show an extended residence in the airway epithelium, thus encouraging optimum absorption of drug at diseased site.

Jamali et al. [103] designed a positively charged chitosan-based nanoparticles for intranasal delivery of small interfering RNA (siRNA). The authors investigated the inhibiting effect of chitosan/siRNA nanoparticle complex on influenza virus replication and nanocarrier-mediated prophylactic efficacy post-exposure to lethal influenza virus. In vitro study, using Vero cell infection model and hemagglutination assay, showed efficient gene transfection and chitosan/siRNA nanoparticles significantly inhibited influenza virus replication, producing about 17-fold decrease in viral titer at least 72 h in comparison with untreated samples. Further, authors assessed the in vivo performance of the polysaccharide-based gene delivery system in an influenza virus-infected BALB/c mice. Mice received intranasal dose of chitosan/siRNA and challenged, intranasally, with lethal PR-8 (H1N1) virus. While greater than half of mice treated with the gene-carrying polysaccharide nanoparticles were protected against PR8 virus challenge, all the mice in the control group were killed after receiving same dose of the influenza virus as the test group. Additionally, the chitosan-based gene carrier significantly reduced PR8-mediated morbidity as the observed weight reduction was smallest in the chitosan/siRNA-treated groups in comparison with the control groups. The authors suggested that polymer mucoadhesivity, influenced by the polysaccharide-mucus ionic interaction, contributed largely to the observed delivery efficacy of chitosan/siRNA nanoparticle complex. Also, given that the polysaccharide-based nanocarrier deposits the antiviral gene directly into the lungs (infection site), the nanosystem could reduce loss of siRNA into systemic circulation, hence, minimizing drug toxicity.

Very recently, the targeted delivery benefit offered by inhalational delivery of mucoadhesive nanocarriers was presented as a potentially effective therapeutic approach for COVID-19 infection. Hanafy and colleagues [104] reported the efficacy of chitosan-coated bovine serum albumin (CT-BSA) nanoparticles toward the intranasal co-delivery of silymarin/curcumin (SL/CUR) for COVID-19 intervention. In order to study the anti-inflammatory and anti-COVID-19 efficacy of the SL/CUR-encapsulated CT-BSA nanoparticles, authors employed the oleic acid mice model where they created acute respiratory distress syndrome (ARDS) in the experimental animal similar to the inflammatory process encountered in a COVID-19-infected patient [104]. The synthesized mucoadhesive, inhalable, CT-BSA-SL/CUR nanoparticles significantly reduced the expression of interleukin-6 and c-reactive protein by 2.3-fold and 1.7-fold, respectively, compared to the group treated with the free capsule. Additionally, plaque reduction assay, using Vero E6 cell (for SARS-CoV-2 virus) model, revealed that SL-loaded CT-BSA nanoparticles, at inhalational dose 25 μg/mL, exhibited anti-COVID-19 activity (Fig. 4); hence, authors posited that inhalational CT-BSA-SL/CUR nanocarrier could improve the histopathological effect of COVID-19 infection as well as inhibit viral growth.

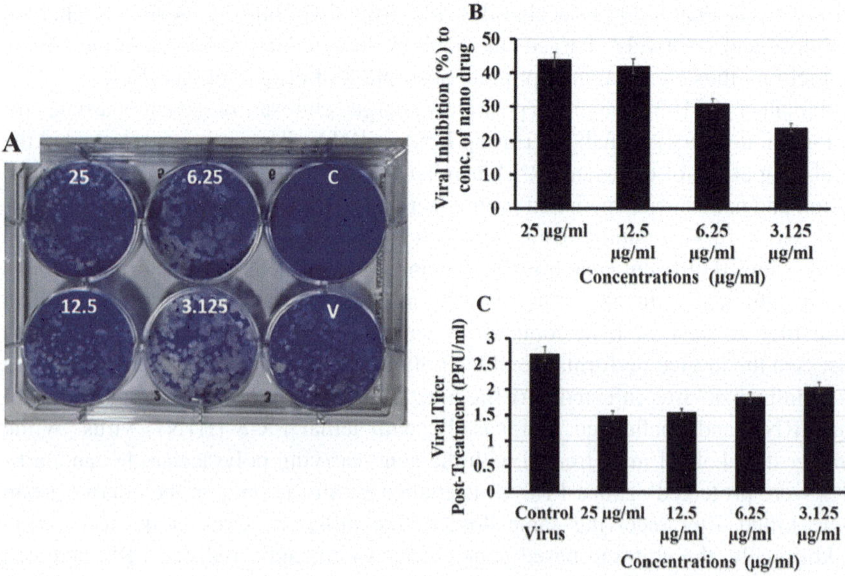

Fig. 4 Illustration of the anti-COVID-19 activity of SL-loaded CT-BSA nanoparticles showing (**a**) plates for plaque reduction using Vero E6 cell model. (**b**) Percentage inhibition versus sample concentration plot. (**c**) Viral titer versus sample concentration following treatment. Plates allowed to solidify before incubating at 37 °C until viral plaques is formed (3–4 days). Control sample = untreated virus + Vero E6 cells. (Reprinted with permission from Ref. [104]. Copyright (2021) Elsevier Science Publisher B. V)

The application of inhalable mucoadhesive nanocarriers for the delivery of antivirals into the central nervous system (CNS) is gaining more attention. Despite the presence of the blood-brain barrier (BBB), viral infection can still spread to the CNS and produce significant neurotropic effects. Canine distemper virus and measles virus can infect the CNS, causing demyelinating disease in dogs and humans, respectively. Eastern equine encephalitis virus is capable of inducing death or long-term and serious neurological sequel. HSV-1 has been reported to cause severe encephalitis [97]. HIV-1 can also cross the BBB into the CNS and, in more severe infection, can cause dementia [105, 106]. There are several potent antiviral agents facing serious challenges crossing the BBB into the CNS; thus, the therapeutic efficacy of these agents is hampered due to poor central bioavailability. Bioavailability-improving properties of mucoadhesive polysaccharide nanoparticles have been employed when the delivery of antiviral across the blood-brain barrier (BBB) is required for central activity [53, 107]. In this case, nanoformulation administered intranasally can deliver bioactivity to the brain via the olfactory or trigeminal route [107].

Hansen and co-workers [108] demonstrated that the cellulosic polysaccharide, cationic-hydroxyethylcellulose (cat-HEC), is viable for application in nasal drug delivery. They showed that cat-HEC significantly increased acyclovir penetration across porcine nasal mucosa. In a similar study [107], hydroxypropyl-β-cyclodextrin (HPβCD) nanoparticles, prepared by cross-linking diphenyl carbonate to cyclodextrin, was reported to improve the transnasal delivery of dolutegravir sodium (DGS) for the treatment of central retroviral infection. The nanosystem produced a 1.51-fold increase in porcine nasal mucosa permeability compared to the unformulated DGS solution. Furthermore, animal study using albino male Wistar rats, showed that intranasally administered nanoparticles significantly increased cerebrospinal fluid concentration of DGS by 4.13-fold and 2.86-fold compared to intravenously and intranasally administered DGS solutions, respectively [107]. The observed increase in permeability can be ascribed to the improvement in DGS solubility and nanonization imparted by the HPβCD nanoparticles. This consequently led to an increase in drug absorption into the CNS.

9 Mucoadhesive Nasal Sprays

Nasal sprays are widely used and usually preferred by both patients and physicians. Sprays that are liquid (solution and suspension) are more convenient than powder sprays since powder sprays can easily irritate the nasal mucosa [109]. The administration of the nanocarrier formulation through nasal spray can lead to the biodistribution of the formulation homogeneously across the nasal mucosa, which allows for higher nanoparticles containing drug to come into contact with the membrane of nasal mucosal for an extended duration of time. Saindane et al. [110] reported that this type of technology would not only extend the period of contact between nasal mucosa and drug but will also maintain the release of the drug over a longer period of time. Saindane et al. [110] created a nanosuspension-based carvedilol nasal gelling spray in situ using a formulation of in situ gelling nasal spray that included gellan gum as an ion-activated transporter. The created formulation was sprayed into the nasal cavity as a low-viscosity solution. When a mist of the solution comes into contact with nasal fluid, it transforms into a sprayed gel on the mucosa of the nose [110]. These in situ gelling intranasal spray formulations have several advantages, including the ability to combine specific dose concentration of the drug with wider distribution all through the nasal mucosa, resulting in increased bioavailability, as well as ease of use because drugs may be administered from any place at any time without skill.

Aref et al. [111] investigated the therapeutic effectiveness of the intranasal spray of ivermectin mucoadhesive nanosuspension in COVID-19 patients' treatment. The study's outcomes demonstrated that the local administration of nasal spray of mucoadhesive nanosuspension containing ivermectin is safe and efficient for the treatment of individuals with mild COVID-19 and results in quick viral clearance and a shorter period of anosmia [111]. Exploring this delivery approach could be a promising technology in the mucosal delivery of drugs in antiviral intervention of treating viral infections.

10 Conclusion

Several drugs have been designed to limit viral replication, and many have proven successful. Nonetheless, with a diverse virus population infecting humans, the development of novel, effective antiviral drugs is still required. However, the approach to antiviral drug delivery has proven to be a viable strategy to increase antiviral efficacy, improve patient compliance, and reduce the side effect profile of existing antiviral drugs in the absence of new drug molecules. In particular, mucosal targeted antiviral drug delivery systems can improve bioavailability, leading to a more cost-effective approach to antiviral treatment. Using mucoadhesive drug delivery systems can achieve site-specific drug delivery and increase contact time to establish a form of HDT linking to mucosal immunity.

References

1. Sharma P, Chawla A, Arora S, Pawar P. Novel drug delivery approaches on antiviral and antiretroviral agents. J Adv Pharm Technol Res. 2012;3(3):147.
2. Strasfeld L, Chou S. Antiviral drug resistance: mechanisms and clinical implications. Infect Dis Clin. 2010;24(3):809–33.
3. Chakravarty M, Vora A. Nanotechnology-based antiviral therapeutics. Drug Deliv Transl Res. 2021;11(3):748–87.
4. Suk JS, Xu Q, Kim N, Hanes J, Ensign LM. PEGylation as a strategy for improving nanoparticle-based drug and gene delivery. Adv Drug Deliv Rev. 2016;99:28–51.

5. Rizvi SA, Saleh AM. Applications of nanoparticle systems in drug delivery technology. Saudi Pharm J. 2018;26(1):64–70.
6. Yadav VK, Gupta AB, Kumar R, Yadav JS, Kumar B. Mucoadhesive polymers: means of improving the mucoadhesive properties of drug delivery system. J Chem Pharm Res. 2010;2(5):418–32.
7. Shaikh R, Singh TRR, Garland MJ, Woolfson AD, et al. Mucoadhesive drug delivery systems. J Pharm Bioallied Sci. 2011;3(1):89.
8. Maisel K, Reddy M, Xu Q, Chattopadhyay S, et al. Nanoparticles coated with high molecular weight PEG penetrate mucus and provide uniform vaginal and colorectal distribution in vivo. Nanomedicine. 2016;11(11):1337–43.
9. Lock JY, Carlson TL, Carrier RL. Mucus models to evaluate the diffusion of drugs and particles. Adv Drug Deliv Rev. 2018;124:34–49.
10. Linden SK, Sutton P, Karlsson NG, Korolik V, et al. Mucins in the mucosal barrier to infection. Mucosal Immunol. 2008;1(3):183–97.
11. Tompa DR, Immanuel A, Srikanth S, Kadhirvel S. Trends and strategies to combat viral infections: a review on FDA approved antiviral drugs. Int J Biol Macromol. 2021;172:524–41.
12. Iannacone M, Guidotti LG. Immunobiology and pathogenesis of hepatitis B virus infection. Nat Rev Immunol. 2021;2021:1–14.
13. World Health Organization. Global hepatitis report, 2017. Geneva: World Health Organization; 2017. http://apps.who.int/iris/bitstream/handle/10665/255016/9789241565455-eng.pdf?sequence=1.
14. Hu J, Protzer U, Siddiqui A. Revisiting hepatitis B virus: challenges of curative therapies. J Virol. 2019;93(20):e01032–19.
15. Chang TT, Lai CL, Chien RN, Guan R, et al. Four years of lamivudine treatment in Chinese patients with chronic hepatitis B. J Gastroenterol Hepatol. 2004;19:1276–82.
16. Quercia R, Perno CF, Koteff J, Moore K, et al. Twenty-five years of lamivudine: current and future use for the treatment of HIV-1 infection. J Acquir Immune Defic Syndr. 2018;78(2):125.
17. Fadaka AO, Aruleba RT, Sibuyi NRS, Klein A, et al. Inhibitory potential of repurposed drugs against the SARS-CoV-2 main protease: a computational-aided approach. J Biomol Struct Dyn. 2020;2020:1–13.
18. Marcellin P, Gane E, Buti M, Afdhal N, et al. Regression of cirrhosis during treatment with tenofovir disoproxil fumarate for chronic hepatitis B: a 5-year open-label follow-up study. Lancet. 2013;381(9865):468–75.
19. Chang TT, Lai CL, Kew Yoon S, Lee SS, et al. Entecavir treatment for up to 5 years in patients with hepatitis B e antigen–positive chronic hepatitis B. Hepatology. 2010;51(2):422–30.
20. Seto WK, Lo YR, Pawlotsky JM, Yuen MF. Chronic hepatitis B virus infection. Lancet. 2018;392(10161):2313–24.
21. Lada O, Benhamou Y, Cahour A, Katlama C, et al. In vitro susceptibility of lamivudine-resistant hepatitis B virus to adefovir and tenofovir. Antivir Ther. 2004;9(3):353–63.
22. Marcellin P, Heathcote EJ, Buti M, Gane E, et al. Tenofovir disoproxil fumarate versus adefovir dipivoxil for chronic hepatitis B. N Engl J Med. 2008;359(23):2442–55.
23. Peters MG, Andersen J, Lynch P, Liu T, et al. Randomized controlled study of tenofovir and adefovir in chronic hepatitis B virus and HIV infection: ACTG A5127. Hepatology. 2006;44(5):1110–6.
24. Marcellin P, Chang TT, Lim SGL, Sievert W, et al. Long-term efficacy and safety of adefovir dipivoxil for the treatment of hepatitis B e antigen–positive chronic hepatitis B. Hepatology. 2008;48(3):750–8.
25. Jaganathan KS, Singh P, Prabakaran D, et al. Development of a single-dose stabilized poly(d,l-lactic-co-glycolic acid) microspheres-based vaccine against hepatitis B. J Pharm Pharmacol. 2004;56:1243–50.
26. Felt O, Buri P, Gurny R. Chitosan: a unique polysaccharide for drug delivery. Drug Dev Ind Pharm. 1998;24:979–93.

27. Pawar D, Jaganathan KS. Mucoadhesive glycol chitosan nanoparticles for intranasal delivery of hepatitis B vaccine: enhancement of mucosal and systemic immune response. Drug Deliv. 2016;23(1):185–94.
28. Amarelle L, Lecuona E, Sznajder JI. Anti-influenza treatment: drugs currently used and under development. Arch Bronconeumol (English Edition). 2017;53(1):19–26.
29. Shie JJ, Fang JM. Development of effective anti-influenza drugs: congeners and conjugates–a review. J Biomed Sci. 2019;26(1):1–20.
30. Garten RJ, Davis CT, Russell CA, Shu B, et al. Antigenic and genetic characteristics of swine-origin 2009 A (H1N1) influenza viruses circulating in humans. Science. 2009;325(5937):197–201.
31. Sarker A, Gu Z, Mao L, Ge Y, et al. Influenza-existing drugs and treatment prospects. Eur J Med Chem. 2022;2022:114189.
32. Hayden F. Developing new antiviral agents for influenza treatment: what does the future hold? Clin Infect Dis. 2009;48:S3–S13.
33. Cass LM, Efthymiopoulos C, Bye A. Pharmacokinetics of zanamivir after intravenous, oral, inhaled or intranasal administration to healthy volunteers. Clin Pharmacokinetics. 1999;36(1):1–11.
34. Lee N, Choi KW, Chan PKS, Hui DSC, et al. Outcomes of adults hospitalised with severe influenza. Thorax. 2010;65(6):510–5.
35. McGeer A, Green KA, Plevneshi A, Shigayeva A, et al. Antiviral therapy and outcomes of influenza requiring hospitalization in Ontario, Canada. Clin Infect Dis. 2007;45(12):1568–75.
36. Sawaengsak C, Mori Y, Yamanishi K, Mitrevej A, et al. Chitosan Nanoparticle Encapsulated Hemagglutinin-Split Influenza Virus Mucosal Vaccine. AAPS PharmSciTech. 2014;15:317–25.
37. Greco A, Diaz JJ, Thouvenot D, Morfin F. Novel targets for the development of anti-herpes compounds. Infect Disord Drug Targets. 2007;7(1):11–8.
38. Zinser E, Krawczyk A, Mühl-Zürbes P, Aufderhorst U, et al. A new promising candidate to overcome drug resistant herpes simplex virus infections. Antivir Res. 2018;149:202–10.
39. Morfin F, Thouvenot D. Herpes simplex virus resistance to antiviral drugs. J Clin Virol. 2003;26(1):29–37.
40. Christophers J, Clayton J, Craske J, Ward R, et al. Survey of resistance of herpes simplex virus to acyclovir in Northwest England. Antimicrob Agents Chemother. 1998;42(4):868–72.
41. Chaudhary B, Verma S. Preparation and evaluation of novel in situ gels containing acyclovir for the treatment of oral herpes simplex virus infections. Sci World J. 2014;2014:1.
42. Laffleur F, Bernkop-Schnürch A. Strategies for improving mucosal drug delivery. Nanomedicine. 2013;8(12):2061–75.
43. Cojocaru FD, Botezat D, Gardikiotis I, Uritu CM, et al. Nanomaterials designed for antiviral drug delivery transport across biological barriers. Pharmaceutics. 2020;12(2):171.
44. Hearnden V, Sankar V, Hull K, Juras DV, et al. New developments and opportunities in oral mucosal drug delivery for local and systemic disease. Adv Drug Deliv Rev. 2012;64(1):16–28.
45. Boegh M, Hanne MN. Mucus as a barrier to drug delivery – understanding and mimicking the barrier properties. Basic Clin Pharmacol Toxicol. 2015;116(3):179–86.
46. Sumaila M, Thashree M, Kumar P, Choonara YE. Lipopolysaccharide nanosystems for the enhancement of oral bioavailability. AAPS PharmSciTech. 2021;22(7):1–25.
47. Singh L, Kruger HG, Maguire GE, Govender T, et al. The role of nanotechnology in the treatment of viral infections. Ther Adv Infect Dis. 2017;4(4):105–31.
48. Sanna V, Satta S, Hsiai T, Sechi M. Development of targeted nanoparticles loaded with antiviral drugs for SARS-CoV-2 inhibition. Eur J Med Chem. 2022;13:114121.
49. Petros RA, DeSimone JM. Strategies in the design of nanoparticles for therapeutic applications. Nat Rev Drug Discov. 2010;9(8):615–27.
50. Caron J, Reddy LH, Lepêtre-Mouelhi S, Wack S, et al. Squalenoyl nucleoside monophosphate nanoassemblies: new prodrug strategy for the delivery of nucleotide analogues. Bioorg Med Chem Lett. 2010;20(9):2761–4.

51. Khutoryanskiy VV. Advances in mucoadhesion and mucoadhesive polymers. Macromol Biosci. 2011;11(6):748–64.
52. Joshy KS, Susan MA, Snigdha S, Nandakumar K, et al. Encapsulation of zidovudine in PF-68 coated alginate conjugate nanoparticles for anti-HIV drug delivery. Int J Biol Macromol. 2018;107:929–37.
53. Miller JM, Dahan A, Gupta D, Varghese S, et al. Enabling the intestinal absorption of highly polar antiviral agents: ion-pair facilitated membrane permeation of zanamivir heptyl ester and guanidino oseltamivir. Mol Pharm. 2010;7(4):1223–34.
54. Belgamwar A, Khan S, Yeole P. Intranasal chitosan-g-HPβCD nanoparticles of efavirenz for the CNS targeting. Artif Cells Nanomed Biotechnol. 2018;46(2):374–86.
55. Szymańska E, Krzyżowska M, Cal K, Mikolaszek B, et al. Potential of mucoadhesive chitosan glutamate microparticles as microbicide carriers–antiherpes activity and penetration behavior across the human vaginal epithelium. Drug Deliv. 2021;28(1):2278–88.
56. Sonaje K, Chuang EY, Lin KJ, Yen TC, et al. Opening of epithelial tight junctions and enhancement of paracellular permeation by chitosan: microscopic, ultrastructural, and computed-tomographic observations. Mol Pharm. 2012;9(5):1271–9.
57. Frank LA, Chaves PS, D'Amore CM, Contri RV, et al. The use of chitosan as cationic coating or gel vehicle for polymeric nanocapsules: increasing penetration and adhesion of imiquimod in vaginal tissue. Eur J Pharm Biopharm. 2017;114:202–12.
58. Ekama SO, Ilomuanya MO, Azubuike CP, Bamidele TA, et al. Mucoadhesive microspheres of maraviroc and tenofovir designed for pre-exposure prophylaxis of HIV-1: an in vitro assessment of the effect on vaginal lactic acid bacteria microflora. HIV/AIDS (Auckland, NZ). 2021;13:399.
59. Martín-Illana A, Chinarro E, Cazorla-Luna R, Notario-Perez F, et al. Optimized hydration dynamics in mucoadhesive xanthan-based trilayer vaginal films for the controlled release of tenofovir. Carbohydr Polym. 2022;278:118958.
60. Pedreiro LN, Cury BS, Chaud MV, Gremião MP. A novel approach in mucoadhesive drug delivery system to improve zidovudine intestinal permeability. Braz J Pharm Sci. 2016;52:715–25.
61. Werle M, Bernkop-Schnürch A. Thiolated chitosans: useful excipients for oral drug delivery. J Pharm Pharmacol. 2008;60(3):273–81.
62. Rajawat GS, Shinde UA, Nair HA. Chitosan-N-acetyl cysteine microspheres for ocular delivery of acyclovir: synthesis and in vitro/in vivo evaluation. J Drug Deliv Sci Technol. 2016;35:333–42.
63. Kapanigowda UG, Nagaraja SH, Ramaiah B, Boggarapu PR. Improved intraocular bioavailability of ganciclovir by mucoadhesive polymer based ocular microspheres: development and simulation process in Wistar rats. DARU J Pharm Sci. 2015;23(1):1–1.
64. Soares S, Sousa J, Pais A, Vitorino C. Nanomedicine: principles, properties, and regulatory issues. Front Chem. 2018;6:360.
65. Adekiya TA, Kondiah PP, Choonara YE, Kumar P, et al. A review of nanotechnology for targeted anti-schistosomal therapy. Front Bioeng Biotechnol. 2020:32.
66. Adekiya TA, Kumar P, Kondiah PP, Pillay V, et al. Synthesis and therapeutic delivery approaches for praziquantel: a patent review (2010-present). Expert Opin Ther Pat. 2021;31(9):851–65.
67. Du JD, Liu Q, Salentinig S, Nguyen TH, et al. A novel approach to enhance the mucoadhesion of lipid drug nanocarriers for improved drug delivery to the buccal mucosa. Int J Pharm. 2014;471(1–2):358–65.
68. Gupta PK, Bhandari N, Shah H, Khanchandani V, et al. An update on nanoemulsions using nanosized liquid in liquid colloidal systems. In: Nanoemulsions-properties, fabrications and applications. Intechopen Publisher; 2019.
69. Abdou EM, Kandil SM, El Miniawy HM. Brain targeting efficiency of antimigrain drug loaded mucoadhesive intranasal nanoemulsion. Int J Pharm. 2017;529(1–2):667–77.

70. Kumar M, Misra A, Mishra AK, Mishra P, et al. Mucoadhesive nanoemulsion-based intranasal drug delivery system of olanzapine for brain targeting. J Drug Target. 2008;16(10):806–14.
71. Peira E, Chirio D, Carlotti ME, Spagnolo R, et al. Formulation studies of microemulsions for topical applications of acyclovir. J Drug Deliv Sci Technol. 2009;19(3):191–6.
72. Manyarara TE, Khoza S, Dube A, Maponga CC. Formulation and characterization of a paediatric nanoemulsion dosage form with modified oral drug delivery system for improved dissolution rate of nevirapine. MRS Adv. 2018;3(37):2203–19.
73. Mahboobian MM, Mohammadi M, Mansouri Z. Development of thermosensitive in situ gel nanoemulsions for ocular delivery of acyclovir. J Drug Deliv Sci Technol. 2020;55:101400.
74. Akbarzadeh A, Rezaei-Sadabady R, Davaran S, Joo SW, et al. Liposome: classification, preparation, and applications. Nanoscale Res Lett. 2013;8(1):1–9.
75. Yamazoe E, Fang JY, Tahara K. Oral mucus-penetrating PEGylated liposomes to improve drug absorption: differences in the interaction mechanisms of a mucoadhesive liposome. Int J Pharm. 2021;593:120148.
76. He H, Lu Y, Qi J, Zhu Q, et al. Adapting liposomes for oral drug delivery. Acta Pharm Sin B. 2019;9(1):36–48.
77. Ramana LN, Sethuraman S, Ranga U, Krishnan UM. Development of a liposomal nanodelivery system for nevirapine. J Biomed Sci. 2010;17(1):1–9.
78. Alsarra IA, Hamed AY, Alanazi FK. Acyclovir liposomes for intranasal systemic delivery: development and pharmacokinetics evaluation. Drug Deliv. 2008;15(5):313–21.
79. Mukherjee S, Ray S, Thakur RS. Solid lipid nanoparticles: a modern formulation approach in drug delivery system. Indian J Pharm Sci. 2009;71(4):349.
80. Naseri N, Valizadeh H, Zakeri-Milani P. Solid lipid nanoparticles and nanostructured lipid carriers: structure, preparation and application. Adv Pharm Bull. 2015;5(3):305.
81. ud Din F, Aman W, Ullah I, Qureshi OS, et al. Effective use of nanocarriers as drug delivery systems for the treatment of selected tumors. Int J Nanomed. 2017;12:7291.
82. Souto EB, Baldim I, Oliveira WP, Rao R, et al. SLN and NLC for topical, dermal, and transdermal drug delivery. Expert Opin Drug Deliv. 2020;17(3):357–77.
83. Liu M, Wen J, Sharma M. Solid lipid nanoparticles for topical drug delivery: mechanisms, dosage form perspectives, and translational status. Curr Pharm Des. 2020;26(27):3203–17.
84. Gómez-Aguado I, Rodríguez-Castejón J, Vicente-Pascual M, Rodríguez-Gascón A, et al. Nucleic acid delivery by solid lipid nanoparticles containing switchable lipids: plasmid DNA vs. Messenger RNA. Molecules. 2020;25(24):5995.
85. He H, Yao J, Zhang Y, Chen Y, et al. Solid lipid nanoparticles as a drug delivery system to across the blood-brain barrier. Biochem Biophys Res Commun. 2019;519(2):385–90.
86. Gastaldi L, Battaglia L, Peira E, Chirio D, et al. Solid lipid nanoparticles as vehicles of drugs to the brain: current state of the art. Eur J Pharm Biopharm. 2014;87(3):433–44.
87. Puri A, Loomis K, Smith B, Lee JH, et al. Lipid-based nanoparticles as pharmaceutical drug carriers: from concepts to clinic. Crit Rev Ther Drug Carrier Syst. 2009;26(6):523.
88. Vieira AC, Chaves LL, Pinheiro S, Pinto S, et al. Mucoadhesive chitosan-coated solid lipid nanoparticles for better management of tuberculosis. Int J Pharm. 2018;536(1):478–85.
89. Chanburee S, Tiyaboonchai W. Mucoadhesive nanostructured lipid carriers (NLCs) as potential carriers for improving oral delivery of curcumin. Drug Dev Ind Pharm. 2017;43(3):432–40.
90. Mendes AI, Silva AC, Catita JA, Cerqueira F, et al. Miconazole-loaded nanostructured lipid carriers (NLC) for local delivery to the oral mucosa: improving antifungal activity. Colloids Surf B Biointerfaces. 2013;111:755–63.
91. Agbo CP, Ugwuanyi TC, Ugwuoke WI, McConville C, et al. Intranasal artesunate-loaded nanostructured lipid carriers: a convenient alternative to parenteral formulations for the treatment of severe and cerebral malaria. J Control Release. 2021;334:224–36.
92. Delshadi R, Bahrami A, McClements DJ, Moore MD, et al. Development of nanoparticle-delivery systems for antiviral agents: a review. J Control Release. 2021;331:30–44.

93. Belgamwar A, Khan S, Rathi L. Synthesis of chitosan-graft-HPβCD copolymer by novel one pot technique and its application for solubility enhancement of Efavirenz. J Young Pharm. 2017;9(2):172.
94. Joshy KS, George A, Snigdha S, Joseph B, et al. Novel core-shell dextran hybrid nanosystem for anti-viral drug delivery. Mater Sci Eng. 2018;93:864–72.
95. Türker S, Onur E, Ózer Y. Nasal route and drug delivery systems. Pharm World Sci. 2004;26(3):137–42.
96. Keller LA, Merkel O, Popp A. Intranasal drug delivery: opportunities and toxicologic challenges during drug development. Drug Deliv Transl Res. 2021;2021:1–23.
97. Dalpiaz A, Pavan B. Nose-to-brain delivery of antiviral drugs: a way to overcome their active efflux? Pharmaceutics. 2018;10(2):39.
98. Giuliani A, Balducci AG, Zironi E, Colombo G, et al. In vivo nose-to-brain delivery of the hydrophilic antiviral ribavirin by microparticle agglomerates. Drug Deliv. 2018;25(1):376–87.
99. Abdellatif AA, Tawfeek HM, Abdelfattah A, Batiha GE, et al. Recent updates in COVID-19 with emphasis on inhalation therapeutics: nanostructured and targeting systems. J Drug Deliv Sci Technol. 2021;63:102435.
100. Eedara BB, Alabsi W, Encinas-Basurto D, Polt R, et al. Inhalation delivery for the treatment and prevention of COVID-19 infection. Pharmaceutics. 2021;13(7):1077.
101. Lee WH, Loo CY, Traini D, Young PM. Inhalation of nanoparticle-based drug for lung cancer treatment: advantages and challenges. Asian J Pharm Sci. 2015;10(6):481–9.
102. Prasher P, Sharma M. Mucoadhesive nanoformulations and their potential for combating COVID-19. Nanomedicine. 2021;16(28):2497–501.
103. Jamali A, Mottaghitalab F, Abdoli A, Dinarvand M, et al. Inhibiting influenza virus replication and inducing protection against lethal influenza virus challenge through chitosan nanoparticles loaded by siRNA. Drug Deliv Transl Res. 2018;8(1):12–20.
104. Hanafy NA, El-Kemary MA. Silymarin/curcumin loaded albumin nanoparticles coated by chitosan as muco-inhalable delivery system observing anti-inflammatory and anti COVID-19 characterizations in oleic acid triggered lung injury and in vitro COVID-19 experiment. Int J Biol Macromol. 2022;198:101–10.
105. Kaul M. HIV's double strike at the brain: neuronal toxicity and compromised neurogenesis. Front Biosci J Virtual Libr. 2008;13:2484.
106. Kolson DL, González-Scarano F. HIV and HIV dementia. J Clin Invest. 2000;106(1):11–3.
107. Belgamwar AV, Khan SA, Yeole PG. Intranasal dolutegravir sodium loaded nanoparticles of hydroxypropyl-beta-cyclodextrin for brain delivery in neuro-AIDS. J Drug Deliv Sci Technol. 2019;52:1008–20.
108. Hansen K, Kim G, Desai KG, Patel H, et al. Feasibility investigation of cellulose polymers for mucoadhesive nasal drug delivery applications. Mol Pharm. 2015;12(8):2732–41.
109. Kumar A, Pandey AN, Jain SK. Nasal-nanotechnology: revolution for efficient therapeutics delivery. Drug Deliv. 2016;23(3):671–83.
110. Saindane NS, Pagar KP, Vavia PR. Nanosuspension based in situ gelling nasal spray of carvedilol: development, in vitro and in vivo characterization. AAPS PharmSciTech. 2013;14(1):189–99.
111. Aref ZF, Bazeed SE, Hassan MH, Hassan AS, et al. Clinical, biochemical and molecular evaluations of ivermectin mucoadhesive nanosuspension nasal spray in reducing upper respiratory symptoms of mild COVID-19. Int J Nanomedicine. 2021;16:4063.

Micro- and Nanoemulsions in Antiviral Treatment

Nidhi Mishra, Neelu Singh, and Poonam Parashar

Abstract Most of the antiviral agents offer challenges of high toxicity, poor aqueous solubility, and compromised bioavailability, resulting in unsatisfactory clinical outcome. Microemulsions are soft nanocarriers bestowed with thermodynamic stability, easy formulation technique, high entrapment efficiency, and modifying release. Microemulsions have the ability to improve the solubility and stability, reducing dose-dependent toxicity of the drug and thus enhancing overall bioavailability of the drug. Further microemulsion can be surface engineered for delivering payload at the desired site. This chapter gives an insight of improved drug delivery and stability of antivirals through microemulsions, citing works of various researchers. The findings of experiments stating encasing of drug into soft nanocarriers give strong evidence of their drug delivery potential, hence a significant boost in therapeutic outcomes.

Keywords Antivirals · Drug delivery · Nanotechnology · Soft nanocarriers · Toxicity

1 Introduction

Even though the world has advanced in several domains, viral infections continue to thrive and add to human morbidity, along with its multiple social, economic, and cultural ramifications. Coronavirus, Ebola virus, Nipah virus, Zika virus, dengue

---

N. Mishra · N. Singh
Department of Pharmaceutical Sciences, Babasaheb Bhimrao Ambedkar University (A Central University), Lucknow, UP, India

P. Parashar (✉)
Amity Institute of Pharmacy, Amity University Uttar Pradesh Lucknow Campus, Lucknow, UP, India

R. Shegokar, Y. Pathak (eds.), *Viral Drug Delivery Systems*,
https://doi.org/10.1007/978-3-031-20537-8_6

virus, chikungunya virus, and various influenza virus strains—H5N1 (avian flu), H1N1 (swine flu), and H3N2—have all triggered crises in past as well as current years. Lately, a devastating pandemic spurred by the new coronavirus (nCoV) has took the lives of nearly 5.13 million people worldwide, with substantial socioeconomic repercussions [1, 2]. In West Africa, a major outbreak of Ebola virus infection killed 11,315 people out of 28,616 documented cases between 2014 and 2016. In the first quarter of 2019, Australia received 27,540 influenza cases. Even though influenza prevalence has plummeted around the globe, specific types of the influenza virus are being observed in different areas of the world, with the seasonal influenza A virus predominating. Over the last few years, Zika virus transmission has reached epidemic proportions in various parts of the world. Dengue fever is now afflicting South East Asia 17 times more than other viral diseases, worsening the epidemic [3–5]. As a logical consequence, the economic repercussions of viral maladies have been enormous.

Infectious diseases are believed to be responsible for around 15 million (>25%) of the 57 million annual fatalities globally; this estimate does not encompass the millions of deaths that occur as a result of previous infections (for instance, streptococcal rheumatic heart disease) or complications linked with chronic infections like liver failure and hepatocellular carcinoma in humans infected with hepatitis B or C viruses. People in impoverished countries, primarily infants and children, bear the brunt of the morbidity and mortality attributable to viral diseases (about three million children die each year from malaria and diarrheal diseases alone). In prosperous nations, indigenous and impoverished minorities are severely affected by viral diseases fatalities [6, 7].

Viruses are obligatory intracellular parasites that are made up of either double- or single-stranded DNA or RNA wrapped in a protein covering known as the capsid. There are two types of antiviral medications: nonretroviral and retroviral drugs. Infections with the non-human immunodeficiency virus (HIV) are treated using nonretroviral inhibitors. Antiretroviral drugs, on the other hand, reduce HIV replication, delay the onset of AIDS, and extend the patient's life. Viruses are categorized into families based on their genome type and replication technique, according to the Baltimore classification system [8–10]. On the basis of replication method, viruses are classified into seven classes (Baltimore system of classification) as shown in Fig. 1. Further, on the basis of envelop present/absent, viruses are classified into two classes:

I. Enveloped viruses; example: dengue viruses, hepatitis B and C viruses, human immunodeficiency virus, herpesviruses
II. Non-enveloped viruses; example: human papilloma viruses [12]

The preponderance of viral infections is subclinical, implying that the body's defensive mechanisms stop the infection from progressing before clinical symptoms appear. Infections like these are important epidemiologically because they allow the virus to spread through populations. The following stages of viral disease pathogenesis are involved:

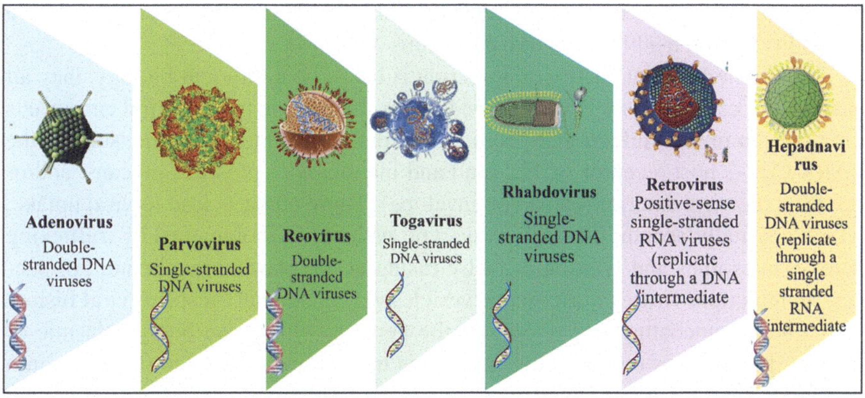

Fig. 1 Schematic representation of various kind of viruses [10, 11]. (Author's original)

(i) Virus attachment at the point of entry
(ii) Virus penetration into the host cell
(iii) Virus uncoating
(iv) Virus replication via transcription and translation leading to the synthesis of virus-specific proteins
(v) Nucleocapsid assembly
(vi) Virion release, resulting in further infection spread [13]

Accessibility of tissue to the pathogenic virus, cell vulnerability to viral reproduction, and virus resistance to host defensive mechanisms are all factors that affect pathogenic pathways [14]. Blocking cellular synthesis of biomolecules, which restricts cellular energy, is one of the virus's mechanisms for causing host cell demise. The indirect approach to cell destruction is the integration of the viral genome with the host genome, which causes mutations in the host genome. The infection pathway is investigated in terms of virulence, virus-dependent variables, virulence genes, inoculum amount, replication speed, and viral infection dissemination [15]. Viral DNA enters the nucleus of the host cell, where it is then converted to mRNA by the host cell RNA polymerase, followed by translation of the mRNA into virus-specific proteins. The proteins produced include enzymes that aid in the production of additional viral DNA, as well as coat and envelope proteins. Virions are released by budding or cell lysis following complete construction of coat proteins around viral DNA. The virion's mRNA is synthesized by enzymes present in the virion, or the viral RNA serves as its own mRNA and is translated into numerous enzymes, including RNA polymerase and structural proteins. Virions are released after they have been assembled [16].

The nucleus of the host cell has no role in virus propagation. Reverse transcriptase is found in the retrovirus virion. This reverse transcriptase copies the viral RNA to DNA. The DNA copy is subsequently incorporated into the host cell's genome, forming a provirus, which is transcribed into new genomic RNA and mRNA before being translated into viral proteins. Budding releases the viruses that have developed. HIV is a retrovirus that replicates in RNA. Some RNA retroviruses have the ability to turn healthy cells into cancerous cells [17].

Despite the fact that each viral infection is based on a different biology, they all share a few key processes that could be used as therapeutic targets. Viral entrance is the first step in viral infection. Specific receptors or attachment factors, such as the SARS-CoV-2 host receptor ACE2, bind and interact with proteins, glycans, and/or lipids on the viral exterior (capsid or envelope). These contacts lead to viral uptake, which is usually accomplished by endocytic pathways accompanied by trafficking across endosomes and lysosomes or by fusing at the plasma membrane. For successful infection, viruses must escape vesicles via endosomal escape or viral fusion, subsequently uncoating and release of the genome, which can be cytoplasmic or nuclear. If mRNA is required, the genome is transcribed, translated (typically into enormous complex polyproteins that are processed by viral or host proteases), and replicated [18–20]. Although many new RNA viruses encode their own polymerases, others rely on host polymerases. Viruses employ the host translation machinery in the same way that cells do, but viral translation is regulated differently. Following the creation of structural proteins and genomes, viral particles are assembled, followed by viral egress or cell lysis, and the cycle is restarted [9]. Therapeutics can be used to target each of these phases in the overall viral life cycle. Antiviral therapies are classified as either direct acting or host factor, depending on whether they target viral or cellular factors [21]. Antibodies that bind the receptor, endocytosis inhibitors, host protease inhibitors, lipidomic reprogramming medicines, kinase inhibitors (e.g., baricitinib), and other host-factor antivirals target parts of the host cell essential for successful viral infection or pathogenesis [17, 22, 23].

The antiviral targets conserved with genes/motifs within a virus family (such as the coronavirus RNA-dependent RNA polymerase or 3CL protease, which have greater than 60% and 40% sequence identity, respectively) or if diverse virus family members co-opt the same host pathways to promote virus replication and/or pathogenesis, both direct-acting and host-factor therapeutics can be broadly applicable (such as the furin protease). Clustered regularly interspaced short palindromic repeats (CRISPR) screens are effective techniques for identifying putative targets in necessary host components [7, 24–26]. Furthermore, disruption of biological pathways in both infected and normal cells may make host-factor interventions more hazardous. There is a definite disease progression within an organism as infection progresses, in addition to the viral life cycle. The viral phase of the disease is when symptoms begin, and incubation durations range from 1 to 14 days. The viral phase is replaced by an inflammatory phase as the disease progresses, in which the body's antiviral inflammatory responses begin to restrict viral replication, ultimately inflicting harm to the body [27–29]. Antiviral therapy intervention windows are usually only effective during the viral phase of infection, beyond which they are rendered ineffective. In outbreak scenarios, testing is critical because of the short treatment windows. In the inflammatory phase, anti-inflammatory and immunomodulatory medications, such as corticosteroids, are used to decrease inflammatory damage by dampening the host immune response. The corticosteroid dexamethasone was currently being used to treat COVID-19 patients who were hospitalized [30]. Baricitinib, a kinase inhibitor in the JAK/STAT pathway that has been approved as an emergency COVID-19 treatment, reduces cytokine release. Another option is to use cytokines like interferon, which have antiviral properties without causing tissue damage, earlier in the disease [31]. Therapeutics based on RNA come in a variety of forms. RNA-based therapies are a relatively new therapeutic class with a lot of potential. Multiple RNA-based treatments have been approved by the FDA in the previous 20 years, and many more are currently in various phases of clinical testing. Because they are more modular and easier to create than traditional medications, these therapies have a lot of promise. To have a therapeutic effect, RNA can function in a variety of ways. Antisense oligonucleotides (ASOs) bind to an mRNA and alter translation, splicing, or the availability of RNA-binding proteins by base pairing. Anti-microRNAs (anti-miRs) bind directly to microRNAs (miRNAs) and prevent them from functioning. To cause translational repression or mRNA degradation, miRNAs or small interfering RNAs (siRNAs) can be utilized as treatments. Small activating RNAs (saRNAs) are similar to microRNAs (miRNAs), except they are nuclear and activate transcription. mRNAs can be employed as a treatment or as vaccines to create protein products [32, 33].

In a similar way to classic small-molecule drugs, RNA aptamers can be selected to bind proteins or small molecules with high affinity, but with activity across a wider spectrum of substrates. Finally, CRISPR effectors and guide RNAs (gRNAs) can be utilized to target cellular RNAs for destruction or to modify genes using CRISPR effectors and guide RNAs (gRNAs). Overall, RNA-based products can be employed for a wide range of purposes, like gene expression regulation, splicing, and translation [34]. The greatest drawback with addressing viral diseases is that it's difficult to predict how the virus will interact with the host's defenses. It can either stop the virus from spreading or induce an immune response in the afflicted tissue [35]. The various approaches for inhibition of viral infection are listed in Fig. 2.

2 Challenges Encountered with Antiviral Drugs

The quality of life of individuals battling with viral infections has enhanced as a consequence of continued research into antiviral therapeutics. However, the emergence of novel viral infections around the world, as well as the advent of multidrug-resistant viruses and their transmission, has posed increased difficulties to antiviral therapy's clinical efficacy.

Some of the common problems associated with antiviral drugs are:

- Some antiviral treatments have been documented to cause negative drug-drug interactions when taken with other prescribed drugs. Furthermore, toxic side effects are a regular result of long-term therapy modules, which may make it even more difficult for patients to adhere to their entire treatment regimen [36].

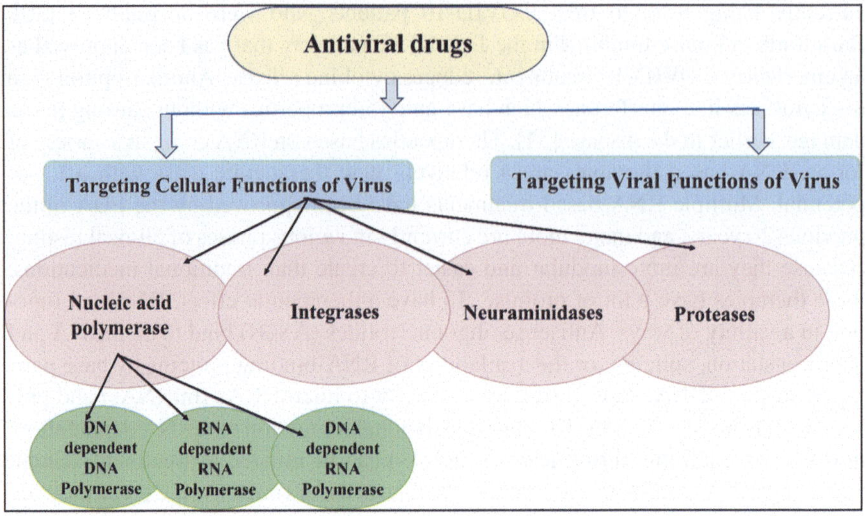

Fig. 2 Various approaches for combatting viral infections

- Several antiviral drugs have a short half-life, resulting in increased dosage frequency and suboptimal patient compliance [37].
- Drug resistance is expected to emerge as a consequence of chronic drug administration, especially in immunocompromised patients [38].
- Limited bioavailability due to inadequate solubility/permeability may necessitate the administration of a greater dose, resulting in toxicity [39].
- Viruses such as HIV, Zika, and Ebola disseminate to impenetrable anatomical sites such as the CNS, synovial fluid, and lymphatic system, rendering difficulties for drugs to reach, therefore reducing clinical efficiency [40].
- Numerous viral infections can persist in the latent state for a long time, complicating screening and treatment [41].
- Antiviral drugs' selectivity toward the virus over the host cell and the identification of the target that is exclusive to the virus life cycle are two additional obstacles in their design and development [42].
- The structure and function of each virus are distinct, rendering the discovery of a broad-spectrum antiviral drug challenging [43].

The challenges encountered can be defeated using nanocarrier approach as shown in Fig. 3. The various nanocarriers such as liposomes, niosomes, microemulsion, nanoparticles, nanostructured lipid carriers, and micelles have shown marked improvement in therapeutic outcome of antiviral drugs. The antiviral agents when loaded into nanocarriers exhibited improved therapeutic efficacy ascribing to improved solubility, enhanced bioavailability, enhanced permeation, cellular uptake, improved accumulation and retention, and targeting specificity.

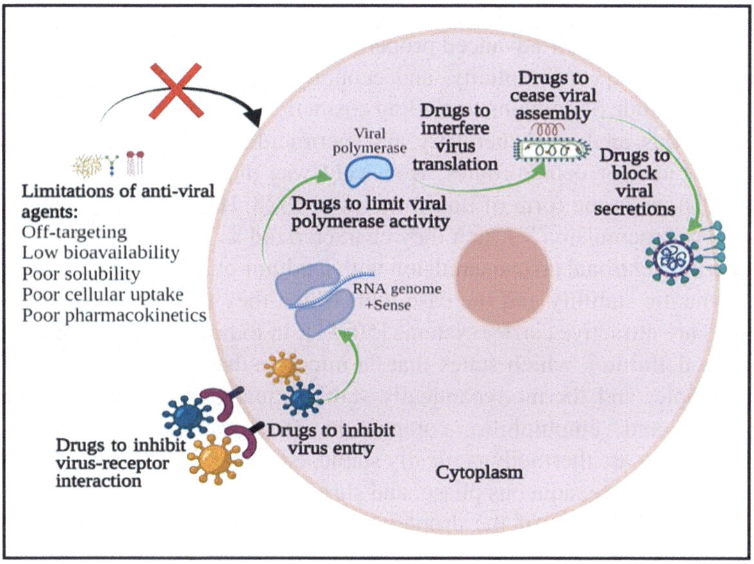

Fig. 3 An illustration of steps involved in virus replication and involvement of antiviral drug-loaded nanocarrier to cease distinctive stages of viral replication nanocarriers

3 Micro-/Nanoemulsions as Drug Carriers

The holistic picture for the establishment of novel drug delivery systems is improving by the day, resulting in the formation of various revolutionary antiviral delivery systems. Antiviral medication development can be hampered by several issues, including limited efficacy of antiviral agents, low solubility of the chemical, low bioavailability when delivered in standard dosage forms, short half-life of certain compounds, and systemic toxic side effects. By enhancing the design, formulation, and administration of antiviral medicines while taking into account the aforesaid parameters, innovative drug delivery techniques can be used to produce successful therapy [44, 45].

With the development of nanotechnology, it has been feasible to better understand the biological mechanics of live cells and establish technologies that aid in the early detection and management of several viral infections. Drug and gene delivery; the use of fluorescent biological markers; the recognition of proteins, pathogens, and malignancies; the separation and purification of biomolecules; tissue engineering; MRI contrast intensification; and pharmacokinetic investigations are a few of the applications to list out. With its potential to efficiently address with viral maladies and overcome the challenges met by conventional antivirals, it has opened up a huge area of research and application [46–49].

The scope of this chapter is limited to microemulsions and nanoemulsions; therefore, these two novel delivery systems will only be discussed in this chapter.

Microemulsions have attracted the interest of formulation scientists since their invention, owing to their advanced properties in terms of stability, solubility, facile formulation aspect simplicity, and economic viability. Microemulsions are used in a multitude of domains, including cosmetics, immunology, sensor devices, coating, textiles, analytical chemistry, and spermicide, in addition to drug delivery via oral, topical, or ocular routes. Rodewald was the first scientist to discover microemulsions in the form of liquid waxes in 1928. Hoar and Schulman coined the term "microemulsion," which they characterized as a clear solution made by titrating a conventional coarse emulsion with medium-chain alcohols. Due to their thermodynamic stability and the ease with which they may be prepared, microemulsions are attractive carrier systems [50, 51]. In today's reality, we can ponder Attwood's definition, which states that "a microemulsion is a transparent, optically isotropic, and thermodynamically stable liquid dosage form made up of water, oil, and amphiphilic compounds (surfactant and cosurfactant)." Microemulsions are thermodynamically stable, isotropic, and transparent systems made up of oil phase, aqueous phase, and surfactant, usually in combination with a cosurfactant. The size of the droplets might range from 10 to 200 nm. O/W microemulsions, W/O microemulsions, and a bicontinuous microemulsion with excellent solubilizing potential are the three types of the same. Microemulsions have a distinctive mode of action for skin penetration because they react with lipids on the skin, causing the intercellular space to shift and the drug to be delivered. The size of the dispersed phase droplets is the major difference between an emulsion and a microemulsion [52, 53].

Microemulsions are intriguing delivery approaches as they offer for both controlled and sustained drug release for a range of routes of administration. As a delivery mechanism, microemulsions have a few distinct characteristics, the most notable of which are that they are less toxic, enable greater drug absorption, and modulate drug release rates. Microemulsions are used in a wide range of drug targeting and controlled drug release applications. Owing to their propensity to solubilize lipophilic drugs, they offer distinguishable characteristics like increased bioavailability. Microemulsions can transport water-soluble drugs into the aqueous phase, demonstrating that they can transport both lipophilic and hydrophilic molecules. Microemulsions have a diverse spectrum of applications since they can be administered via all basic drug delivery routes. When compared to other biphasic dosage forms, microemulsions have a longer shelf life [54]. Microemulsions are developed with the intent of maximizing their peculiar qualities, such as minimizing toxic side effects and decreasing the amount of the carrier vehicle. Due to their ease of administration, they are significantly superior to conventional dosage forms. They offer resistance to hydrolysis and oxidation. They make it much easier for patients to comply [55].

Initially, nanoemulsion and microemulsion were assumed indistinguishable terminologies; however, with advancements in research, the controversy has been addressed, and the terminologies are now clearly distinct because microemulsions entail more surfactant, i.e., >20%, and are much more thermodynamically stable. Additionally, microemulsion seems to have a much smaller hydrodynamic diameter, yet microemulsion's ability to be deployed for internal purposes is limited due to its high surfactant content. Nanoemulsions with a low surfactant content, on the other hand, are better for any medication that is required to be administered internally [56].

Nanoemulsion is a lipid-based formulation that is commonly utilized for systemic delivery of drug to enhance bioavailability. In other words, nanoemulsions are lipid-based drug delivery systems that are thermokinetically stable and are entirely comprised of oil, surfactant, cosurfactant, and water with droplet sizes in the nanometer range [57]. Also, nanoemulsion can be defined as a two-phase system in which one phase is dispersed as a small nano-sized (10–200 nm) droplet in a continuous phase attributable to the surfactant (emulsifier) molecules' protective coating. For preparation of nanoemulsions, many techniques such as low-energy emulsification, phase inversion temperature method, ultrasonication, high-pressure homogenization, or microfluidization are used. As lipids, several oils such as soyabean oil, castor oil, and peanut oil have been employed. Enhanced water solubility, drug loading efficiency, GIT retention time, absorption and bioavailability, and lymphatic uptake are all advantages inherited by nanoemulsions [58–61]. Another class of emulsion includes multiple emulsion also known as double emulsion or emulsion of emulsion, which is a complex heterogeneous dispersion system. Multiple emulsions are basically of two types, water-oil-water (W/O/W) multiple emulsions and oil-water-oil (O/W/O) multiple emulsions, and have wide applications in drug delivery.

Pharmaceutical scientists primarily employ oil-in-water (O/W) nanoemulsions, in which oil is the dispersed phase and water is the continuous phase, for systemic delivery of hydrophobic drugs to elevate efficacy and bioavailability [62, 63]. Irrespective of their thermodynamic stability and propensity to form nanosized droplets, when used topically, nanoemulsions have a plethora of potential merits over unstable dispersions [64]. Nanoemulsions have exhibited potential over traditional topical formulations (e.g., gels and emulsions) due to superior percutaneous permeation. Nanoemulsion formation is a non-spontaneous ($G > 0$) process, and scientists propose an external energy source while preparing nanoemulsion as they are only stable kinetically [65]. Specific physicochemical characteristics of nanoemulsion systems may make them more appropriate for antiviral drug therapy with better cure rates. The drug molecule is loaded in nano-sized oil droplets that are dispersed in a continuous aqueous phase in an O/W nanoemulsion. Drug distribution over impermeable barriers is facilitated by these nanosized drug-loaded oil droplets, which increases bioavailability. The antiviral activity of loaded drugs can be improved by increasing the surface area-to-volume ratio of the system. Certain attributes of nanoemulsion systems make them ideal for drug delivery via mobilization too [49, 66]. The use of a nanoemulsion system to

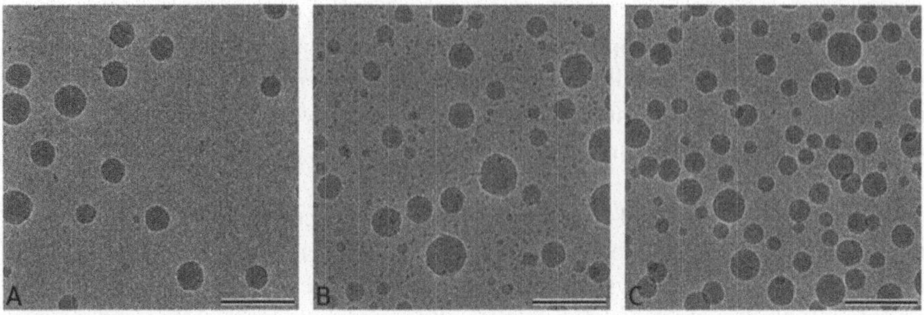

Fig. 4 Cryo-electron microscopic images of formulations made with (**a**) blank span, (**b**) 10% sucrose esterase monoester and (**c**) 75% sucrose esterase monoester. (Reproduced with permission)

deliver antiviral drugs via nebulization could augment antiviral drug efficacy even against COVID-19 [67]. The SEM images of acyclovir-loaded nanoemulsion are shown in Fig. 4.

Microemulsion-related research studies: Taking into consideration the numerous afore-mentioned advantages, Zhu et al. formulated microemulsion of penciclovir for dermal delivery and compared the permeation potential of the microemulsion with commercial cream (penciclovir content 1.0%, w/w). The result demonstrated 3.5 times enhanced permeation of formulation when compared with commercial cream [68]. Likewise, Alkhatib et al. prepared microemulsions using variable concentration of Span 20, Tween 80, and IPM (isopropyl myristate) and evaluated it against *Aspergillus niger*, *Aspergillus flavus*, *Bacillus cereus*, *Candida albicans*, *Candida glabrata*, herpes simplex virus type 2 (HSV-2), and the VERO cell line. The results showed significant antimicrobial activities against *A. niger* and herpes simplex virus type 2 (HSV-2) when exposed to microemulsion having composition 0.166, 0.242, and 0.506 of Span 20, Tween 80, and IPM respectively. The microemulsion was also capable of destroying HSV-2 virus at a 200-fold dilution in Dulbecco's modified Eagle medium [69]. Another study by Shishu et al. reported development of acyclovir-loaded microemulsion for treating cutaneous herpetic infections [70]. The prepared microemulsion demonstrated 1.7-fold enhanced permeation when compared with marketed formulation (Herpex, 5%®). Further in v vivo studies in herpes simplex virus-1 infection-induced female BALB/c mice showed depressed herpetic skin lesions in topical formulation administered topical formulation subsequent to post 24 hours. Similarly, Carvalho et al. attempted transdermal delivery of zidovudine through microemulsion and lamellar liquid crystalline systems [71]. The result of in vitro permeation studies over pig ear skin displayed significantly higher (twofold) permeation through microemulsion when compared with lamellar liquid crystalline systems and found to be non-irritant to the skin. On the contrary, lamellar liquid crystalline systems served as a drug reservoir, resulting in retarded partitioning of acyclovir and thus reduced permeation into the skin.

One more study performed by Sasivimolphan et al. described formulation and assessment of oxyresveratrol-loaded microemulsion for topical delivery for treatment of herpes simplex virus [72]. The prepared oxyresveratrol-loaded microemulsion showed substantially higher permeation potential up to 93.04 times through shed snake skin at the end of 6 hours when compared with Vaseline (20% w/w concentration). Further, 20%, 25%, and 30%w/w microemulsion was applied topically in HSV-1-infected mice (dose = 5 times/day for 7 days), subsequent to infection development. The results of the experiment revealed that the microemulsion (25% and 30% w/w) was significantly efficient in deferring skin lesion development and reducing mortality when compared with untreated control. Comparably, Lee et al. formulated itraconazole-loaded microemulsion for intranasal delivery for rhinovirus infection therapy [73]. The microemulsion displayed significantly higher drug release in comparison to drug suspension. Further subsequent to microemulsion administration, there was a marked reduction in inflammatory, namely, IL-1, IL-6, CXCL1, CCL2, and TNF-α, post 8-hour HRV1B infection when compared with itraconazole suspension administered groups.

Nanoemulsion-related research studies: Prabhakar et al. formulated indinavir-loaded nanoemulsion (w/o/w) for brain delivery against HIV infection [74]. The results of pharmacokinetic studies revealed 2.44-fold increase in brain-specific accumulation when assessed with naïve drug solution indicating higher brain uptake. A new study executed by Nabila et al. described formulation of curcumin-encased nanoemulsion and its assessment against four dengue virus serotypes (DENV-1, DENV-2, DENV-3, and DENV-4). The results demonstrated that curcumin-encased nanoemulsion exhibited reduced CC_{50} value of 52.97 μg/mL when compared with DMSO curcumin solution that holds a CC_{50} value of 61.51 μg/mL [75]. Similarly, curcumin-encased nanoemulsion presented reduced IC_{50} value of 0.96 μg/mL, 2.61 μg/mL, 22.62 μg/mL, and 15.13 μg/mL against DENV-1, DENV-2, DENV-3, and DENV-4 strains, respectively, which was significantly lower than that of DMSO curcumin solution (1.12 μg/mL, 4.03 μg/mL, 35.9 μg/mL, and 17.24 μg/mL). Another study executed by Jena et al. showed improved efficacy of lamivudine when formulated as multiple emulsion stabilized through biopolymer gum odina. Some more studies pertaining to development of microemulsion and nanoemulsion for delivery of antiviral agents are given in Table 1.

All the above studies give concrete evidence of enhanced therapeutic potential of drug(s) when formulated as nanoemulsion/microemulsion in terms of enhanced permeation, elevated cytotoxicity, improved uptake, and stability of drugs. However, still methods and optimization are required to develop effective formulation that can reach to clinical trial phases and may rise as successful products for clinical applications.

Table 1 Micro- and nanoemulsions for antiviral drug delivery

Formulation	Drug loaded	Main excipients	Activity	Route	Major in vitro/in vivo studies	Purpose of study	References
Microemulsion	Penciclovir	Oleic acid, Cremophor EL, ethanol, and water	Antiviral activity (herpes simplex virus, varicella-zoster virus, Epstein-Barr virus, hepatitis virus, and cytomegalovirus)	Topical	Permeability studies	To increase permeability of penciclovir	[68]
	–	Sorbitan monolaurate (Span 20), polyoxyethylene sorbitan monooleate (Tween 80), sodium pentane sulfonic acid, isopropyl myristate (IPM), and ethanol	Antifungal activity, Antiviral activity (herpes simplex virus type 2)	Topical	Cytotoxicity study, cellular respiration, cytopathic effect (CPE) inhibition assay	To evaluate the antifungal and antiviral activities of microemulsion formulations: o/w with hydrophilic nanodroplet, w/o with hydrophobic nanodroplet, and bicontinuous	[69]
	Acyclovir	Isopropyl myristate (IPM), Captex 355, Labrafac CC, Span 20, Tween 20, and water	Cutaneous HSV-I infection	Topical	In vitro skin permeation studies through mice skin, skin retention study, histopathological examination	To develop, evaluate, and optimize acyclovir-loaded microemulsion	[70]
	Acyclovir	Oleic acid, castor oil, Tween 80, and ethanol	–	Topical	Ex vivo skin permeation studies Pharmacokinetic studies of the plain drug and prepared microemulsion	To topically deliver acyclovir	[76]
	Nevirapine	Liquid paraffin, chitosan	–	Oral	Stability assessment via phase separation and creaming, globule size, and pH	To evaluate the stability of emulsion prepared after adding aqueous dispersion of liquid paraffin oil and chitosan without any surfactant	[77]

	Acyclovir	Labrasol, Plurol Oleique, and Labrafac	Antiviral	Oral	In vitro intestinal permeability studies In vivo absorption study Pharmacokinetic studies	To improve the absorption and the overall bioavailability of acyclovir	[78]
	Zidovudine	IPM, Labrasol and Plurol Oleique	Anti-retroviral	Transdermal	In vitro skin permeation and retention studies, dermal irritation test	To investigate, compare, and optimize zidovudine microemulsion for transdermal administration in antiretroviral therapy	[71]
	Tetrahydrocur-cumin (THC)	Glycerol monolaurate, Cremophor EL, Solutol HS 15, Capmul MCM, Capmul MCM C8, Capmul MCM C10, Carbopol, Labrafac PG, Labrafac WL 1349, Transcutol P, Peceol, Capryol 90, Lauroglycol 90, Labrafil M 2125 CS, Tween 20, Tween 80, Span 20, Span 80, PEG 400, and oleic acid	HIV/AIDS prophylaxis	Topical	In vitro permeability study, in vitro cell viability assay, in vitro anti-HIV activity using β-galactosidase assay and p24 antigen assay, cell uptake studies, and lactobacilli viability assay	To develop, characterize, and optimize THC-loaded microemulsion (gel based) to increase solubility, adhesivity, and retention time in the vaginal mucosa	[79]
	Oxyresveratrol	IPM, oleic acid, olive oil, Tween 80, Span 80, Cremophor RH 40, propylene glycol, butanol, and isopropanol	Cutaneous HSV-1 infection	Topical	In vitro skin permeation study, in vivo animal studies using BALB/c mice as animal model to monitor the development of skin lesions and mortality after application of microemulsion	To prepare oxyresveratrol-loaded microemulsion and evaluate its physicochemical properties and therapeutic efficacy	[72]

	Itraconazole (ITZ)	Benzyl alcohol, Cremophor EL, Solutol HS15, and water	Rhinovirus infection	Intranasal	In vivo animal studies using BALB/c mice as animal model; levels of cytokines and chemokines were measured using ELISA, RT-PCR	To develop and evaluate ITZ microemulsion for treating rhinovirus infection	[73]
	Acyclovir	IPM, Tween 20, Span 20, Water, DMSO, menthol, propylene glycol	Cutaneous HSV-I infection	Topical	Ex vivo skin permeation study Histopathological studies, in vivo antiviral studies using murine model of cutaneous HSV-I infection, cytotoxicity assay, viral plaque assay	To develop and evaluate microemulsion with enhanced penetration ability for better therapeutic efficacy	[80]
Nanoemulsion	Saquinavir mesylate	Capmul MCM, Tween 80, PEG 400, and water	Anti-HIV	Intranasal	Ex vivo permeation studies Nasal cilio toxicity, biodistribution studies, gamma scintigraphy	Intranasal administration of saquinavir for CNS targeting in neuro-AIDS, to reduce the viral load at anatomical reservoir sites	[81]
	Saquinavir	Lipoid®-80, deoxycholic acid, water	Anti-HIV	Oral and intravenous	Biodistribution study following oral and intravenous administration Analysis of absorption and disposition of saquinavir, non-compartmental pharmacokinetic analysis	To enhance oral bioavailability and brain deposition	[82]

	Nevirapine	Soyabean oil, ethyl oleate, and oleic acid	Anti-HIV	Oral	In vitro drug permeability studies	To reduce drug efflux and enhance release profile and permeability	[83]
	Indinavir	Lecithin, Tween 80	Anti-HIV	Intravenous	Brain uptake studies by fluorescence microscopy, pharmacokinetic and tissue distribution studies	To improve brain delivery and uptake via proposed mechanism (LDL-mediated endocytosis and Tween 80-mediated P-gp inhibition)	[84]
	Acyclovir	Cyanoacrylates, polyacrylic acid, sodium carboxymethyl cellulose, hyaluronic acid, hydroxypropyl cellulose (HPMC), polycarbophil, chitosan, and gellan	Chicken pox, herpes simplex virus (HSV)	Oral	Permeability studies	To reduce dose, minimize side effects, and improve oral bioavailability and blood circulation time	[85]
	Acyclovir	Solutol HS 15 Span 80	–	Topical	In vitro skin studies using Franz diffusion cell via HPLC	To optimize phase inversion method for developing W/O/W nanoemulsions for the dermal delivery of hydrophilic acyclovir	[74]
	Nelfinavir mesylate	Polyvinylpyrrolidone, HPMC, sodium carboxymethyl cellulose, and methyl cellulose	Anti-HIV	Oral	Ex vivo drug release, in vivo absorption study, pharmacokinetic study	To increase biodistribution and bioavailability, maximize therapeutic efficacy, enhance solubility and drug-loading	[86]
	Indinavir (lactoferrin modified)	Oleic acid, Span 80, polysorbate 80 (P80), glycerol	Anti-HIV	Intravenous	In vitro hemolysis study, neuropharmacokinetic studies	To enhance brain penetration and residence time of indinavir	[87]

	Coumestrol	IPM, oleylamine, Dioleyl phosphocholine, polysorbate 80 Distearoyl phosphocholine	Anti HSV-1 and HSV-2	Topical	Esophageal mucosa permeation/retention study, mucosal microscopic analysis (histopathology) Confocal fluorescence microscopy In vitro anti-herpes activity on Vero cell lines	To enhance the antiviral effects of coumestrol	[88]
	Curcumin	Hidden (patent protection)	Vulvar intraepithelial neoplasia (VIN) associated with human papilloma virus (HPV) infection	–	Cytotoxicity study (MTT assay), cellular uptake assay Phototoxicity assay, caspase-3/7 enzymatic activity measurement	To treat VIN by curcumin loaded in nanoemulsion or in combination with photodynamic therapy	[89]
	Curcumin	Castor oil, Cremophor RH 40, and PEG 400	Dengue virus	–	Cell viability assay (MTT) Antiviral activity value determination (plaque assay)	To evaluate the potential antiviral activity of curcumin nanoemulsion against four Indonesian-derived dengue serotypes	[75]
	Genistein	IPM, oleylamine, Dioleoyl phosphocholine, Polysorbate 80 Distearoyl phosphocholine	Anti-herpes	Topical	Esophageal mucosa permeation/retention study, histological and confocal fluorescence microscopy studies, in-vitro evaluation of anti-HSV-1 (29R strain) on Vero cell lines	To evaluate distribution of genistein on excised porcine esophageal mucosa and anti-HSV-1 activity against acyclovir-resistant viral strain	[90]

4 Conclusion

Nanotechnology has transformed the world by providing new answers to a variety of challenges that plague today's healthcare. Nanomedicines have a variety of advantages over traditional approaches for preventing, diagnosing, and treating viral infections, thanks to recent advancements in nanomedicine design and engineering. Nanomedicine techniques are advantageous because they have unique properties such as small particle size, a high area-to-volume ratio, the ability to modify the surface to achieve desired selectivity, and biocompatibility. Furthermore, these unique techniques have significant potential in antiviral therapies by assisting in the management of issues such as drug resistance, limited solubility and bioavailability, burst release, and short duration of action. Future research could focus on achieving "multi-functionalization" of nanomaterials to accomplish site-specific, concurrent drug delivery, as well as "multiplexing" to treat a wide range of diseases and associated symptoms in a diversified population. In the realm of viral infections, the advent of theranostics that can provide precise diagnosis, effective therapy, and real-time monitoring is becoming increasingly important. Various technologies, such as nanotraps, nanodiamonds, and nanofibers, are being used in persisting research against the influenza and HIV-1 viruses and can be applied to other viral illnesses. The use of nanomaterials as an adjuvant to antiviral vaccinations and studies to improve immune response has shown promise in the prevention and treatment of viral infections. The complexity involved in their fabrication and characterization, as well as their large-scale production, are among the few roadblocks in the development of these advanced kinds of nanosystems. However, with significant advances in the fields of polymer chemistry, biology, and nanotechnology, there is reason to believe that the rate at which new viral infections originate can be regulated and that global viral infection management can be addressed. In an essence, microemulsions and nanoemulsions developed to deliver antiviral drugs could be a powerful therapeutic agent for treating COVID-19 and other viral disorders, necessitating further research in these areas.

References

1. Chauhan D, Arya K, Saxena VL. NIPAH virus: a review article. 2018.
2. Epstein JH, Anthony SJ, Islam A, Kilpatrick AM, Khan SA, Balkey MD, et al. Nipah virus dynamics in bats and implications for spillover to humans. Proc Natl Acad Sci. 2020;117(46):29190–201.
3. Bloom DE, Cadarette D. Infectious disease threats in the twenty-first century: strengthening the global response. Front Immunol. 2019;10:549.
4. Sahanaa C, Mishra AK, Bazroy J. Trend of morbidity and mortality of dengue in Tamil Nadu and Puducherry, South India. Int J Community Med Public Health. 2018;5(1):322–5.
5. Chakravarty M, Vora A. Nanotechnology-based antiviral therapeutics. Drug Deliv Transl Res. 2021;11:748–87.

6. De Clercq E, Li G. Approved antiviral drugs over the past 50 years. Clin Microbiol Rev. 2016;29(3):695–747.
7. Yao R, Ianevski A, Kainov D. Safe-in-man broad spectrum antiviral agents. In: Antiviral drug discovery and development. Springer; 2021. p. 313–337.
8. Ita K. Transcutaneous permeation of antiviral agents. J Drug Deliv Sci Technol. 2017;41:293–302.
9. Delshadi R, Bahrami A, Mcclements DJ, Moore MD, Williams L. Development of nanoparticle-delivery systems for antiviral agents: a review. J Control Release. 2021;331:30.
10. Sharma P, Chawla A, Arora S, Pawar P. Novel drug delivery approaches on antiviral and anti-retroviral agents. J Adv Pharm Technol Res. 2012;3(3):147.
11. Lodish H, Berk A, Zipursky SL, Matsudaira P, Baltimore D, Darnell J. Viruses: structure, function, and uses. In: Molecular cell biology. 4th ed. WH Freeman; 2000.
12. Franklyne JS, Gopinath PM, Mukherjee A, Chandrasekaran N. Nanoemulsions: the rising star of antiviral therapeutics and nano-delivery system-current status and prospects. Curr Opin Colloid Interface Sci. 2021:101458.
13. Chinchar VG. Replication of viruses. Encyclopedia of virology, vol. 1471. Elsevier; 1999.
14. Baron S. Viral pathogenesis–medical microbiology. 1996.
15. Manjarrez-Zavala ME, Rosete-Olvera DP, Gutiérrez-González LH, Ocadiz-Delgado R, Cabello-Gutiérrez C. Pathogenesis of viral respiratory infection. Respir Dis Infect New Insight. 2013;1:3–32.
16. Rampersad S, Tennant P. Replication and expression strategies of viruses. Viruses. 2018:55–82.
17. Gentile G, Micozzi A. Speculations on the clinical significance of asymptomatic viral infections. Clin Microbiol Infect. 2016;22(7):585–8.
18. García-Cárceles J, Caballero E, Gil C, Martínez A. Kinase inhibitors as underexplored antiviral agents. J Med Chem. 2021;65:935.
19. Yang K-C, Lin J-C, Tsai H-H, Hsu C-Y, Shih V, Hu C-MJ. Nanotechnology advances in pathogen-and host-targeted antiviral delivery: multipronged therapeutic intervention for pandemic control. Drug Deliv Transl Res. 2021:1–18.
20. Yang G, Chen S, Zhang J. Bioinspired and biomimetic nanotherapies for the treatment of infectious diseases. Front Pharmacol. 2019;10:751.
21. Poduri R, Joshi G, Jagadeesh G. Drugs targeting various stages of the SARS-CoV-2 life cycle: exploring promising drugs for the treatment of COVID-19. Cell Signal. 2020;74:109721.
22. Maus A, Strait L, Zhu D. Nanoparticles as delivery vehicles for antiviral therapeutic drugs. Engin Regen. 2021;2:31–46.
23. Payne S. Chapter 17: Family coronaviridae. In: Viruses: from understanding to investigation. Elsevier Inc. All rights reserved. Academic Press; 2017.
24. Parvez MK, Parveen S. Evolution and emergence of pathogenic viruses: past, present, and future. Intervirology. 2017;60(1–2):1–7.
25. Andersen PI, Ianevski A, Lysvand H, Vitkauskiene A, Oksenych V, Bjørås M, et al. Discovery and development of safe-in-man broad-spectrum antiviral agents. Int J Infect Dis. 2020;93:268–76.
26. Mourya DT, Yadav PD, Ullas P, Bhardwaj SD, Sahay RR, Chadha MS, et al. Emerging/re-emerging viral diseases & new viruses on the Indian horizon. Indian J Med Res. 2019;149(4):447.
27. Adamson CS, Chibale K, Goss RJ, Jaspars M, Newman DJ, Dorrington RA. Antiviral drug discovery: preparing for the next pandemic. Chem Soc Rev. 2021;50:3647–55.
28. Kausar S, Said Khan F, Ishaq Mujeeb Ur Rehman M, Akram M, Riaz M, Rasool G, et al. A review: mechanism of action of antiviral drugs. Int J Immunopathol Pharmacol. 2021;35:20587384211002621.
29. Ryu W-S. Virus life cycle. In: Molecular virology of human pathogenic viruses. Academic Press; 2017. p. 31–45.
30. Ranjbar K, Moghadami M, Mirahmadizadeh A, Fallahi MJ, Khaloo V, Shahriarirad R, et al. Methylprednisolone or dexamethasone, which one is superior corticosteroid in the treatment

of hospitalized COVID-19 patients: a triple-blinded randomized controlled trial. BMC Infect Dis. 2021;21(1):1–8.
31. Zhang X, Zhang Y, Qiao W, Zhang J, Qi Z. Baricitinib, a drug with potential effect to prevent SARS-COV-2 from entering target cells and control cytokine storm induced by COVID-19. Int Immunopharmacol. 2020;86:106749.
32. Walker PJ, Siddell SG, Lefkowitz EJ, Mushegian AR, Dempsey DM, Dutilh BE, et al. Changes to virus taxonomy and the International Code of Virus Classification and Nomenclature ratified by the International Committee on Taxonomy of Viruses (2019). Arch Virol. 2019;164(9):2417–29.
33. Moelling K, Broecker F. Viruses and evolution–viruses first? A personal perspective. Front Microbiol. 2019;10:523.
34. Roos W, Ivanovska I, Evilevitch A, Wuite G. Viral capsids: mechanical characteristics, genome packaging and delivery mechanisms. Cell Mol Life Sci. 2007;64(12):1484–97.
35. Rouse BT, Sehrawat S. Immunity and immunopathology to viruses: what decides the outcome? Nat Rev Immunol. 2010;10(7):514–26.
36. Oshikoya KA, Oreagba IA, Ogunleye OO, Lawal S, Senbanjo IO. Clinically significant interactions between antiretroviral and co-prescribed drugs for HIV-infected children: profiling and comparison of two drug databases. Ther Clin Risk Manag. 2013;9:215.
37. Gerber JG. Using pharmacokinetics to optimize antiretroviral drug-drug interactions in the treatment of human immunodeficiency virus infection. Clin Infect Dis. 2000;30(Supplement_2):S123–9.
38. Strasfeld L, Chou S. Antiviral drug resistance: mechanisms and clinical implications. Infect Dis Clin. 2010;24(3):809–33.
39. Singh R, Lillard JW Jr. Nanoparticle-based targeted drug delivery. Exp Mol Pathol. 2009;86(3):215–23.
40. Vyas S, Subhedar R, Jain S. Development and characterization of emulsomes for sustained and targeted delivery of an antiviral agent to liver. J Pharm Pharmacol. 2006;58(3):321–6.
41. Chaudhuri A, Kennedy P. Diagnosis and treatment of viral encephalitis. Postgrad Med J. 2002;78(924):575–83.
42. Bule M, Khan F, Niaz K. Antivirals: past, present and future. In: Recent advances in animal virology. Springer; 2019. p. 425–446.
43. Adalja A, Inglesby T. Broad-spectrum antiviral agents: a crucial pandemic tool. Expert Rev Anti-Infect Ther. 2019;17(7):467–70.
44. Saini R, Saini S, Sharma S. Nanotechnology: the future medicine. J Cutan Aesthet Surg. 2010;3(1):32–3.
45. Joo K-I, Lei Y, Lee C-L, Lo J, Xie J, Hamm-Alvarez SF, et al. Site-specific labeling of enveloped viruses with quantum dots for single virus tracking. ACS Nano. 2008;2(8):1553–62.
46. Blecher K, Nasir A, Friedman A. The growing role of nanotechnology in combating infectious disease. Virulence. 2011;2(5):395–401.
47. Mendes PM. Cellular nanotechnology: making biological interfaces smarter. Chem Soc Rev. 2013;42(24):9207–18.
48. Villanueva-Flores F, Castro-Lugo A, Ramírez OT, Palomares LA. Understanding cellular interactions with nanomaterials: towards a rational design of medical nanodevices. Nanotechnology. 2020;31(13):132002.
49. Cojocaru F-D, Botezat D, Gardikiotis I, Uritu C-M, Dodi G, Trandafir L, et al. Nanomaterials designed for antiviral drug delivery transport across biological barriers. Pharmaceutics. 2020;12(2):171.
50. Sharma AK, Garg T, Goyal AK, Rath G. Role of microemulsions in advanced drug delivery. Artif Cells Nanomed Biotechnol. 2016;44(4):1177–85.
51. Gibaud S, Attivi D. Microemulsions for oral administration and their therapeutic applications. Expert Opin Drug Deliv. 2012;9(8):937–51.
52. Katiyar BS, Katiyar SS, Mishra PS, Sailaja DL. Microemulsions: a novel drug carrier system. Int J Pharm Sci Rev Res. 2013;20(2):138–48.

53. Kanwar R, Rathee J, Patil MT, Mehta SK. Microemulsions as nanotemplates: a soft and versatile approach. In: Microemulsion—chemical nanoreactor. InTech; 2019.
54. Peira E, Chirio D, Carlotti ME, Spagnolo R, Trotta M. Formulation studies of microemulsions for topical applications of acyclovir. J Drug Deliv Sci Technol. 2009;19(3):191–6.
55. Chudasama A, Patel V, Nivsarkar M, Vasu K, Shishoo C. Investigation of microemulsion system for transdermal delivery of itraconazole. J Adv Pharm Technol Res. 2011;2(1):30.
56. Ahmad N, Alam MA, Ahmad FJ, Sarafroz M, Ansari K, Sharma S, et al. Ultrasonication techniques used for the preparation of novel eugenol-Nanoemulsion in the treatment of wounds healings and anti-inflammatory. J Drug Deliv Sci Technol. 2018;46:461–73.
57. Sivakumar M, Tang SY, Tan KW. Cavitation technology–a greener processing technique for the generation of pharmaceutical nanoemulsions. Ultrason Sonochem. 2014;21(6):2069–83.
58. Gupta PK, Bhandari N, Shah H, Khanchandani V, Keerthana R, Nagarajan V, et al. An update on nanoemulsions using nanosized liquid in liquid colloidal systems. In: Nanoemulsions – properties, fabrications and applications. InTech; 2019.
59. Hobson JJ, Edwards S, Slater RA, Martin P, Owen A, Rannard SP. Branched copolymer-stabilised nanoemulsions as new candidate oral drug delivery systems. RSC Adv. 2018;8(23):12984–91.
60. Sutradhar KB, Amin ML. Nanoemulsions: increasing possibilities in drug delivery. Eur J Nanomed. 2013;5(2):97–110.
61. Rajpoot P, Pathak K, Bali V. Therapeutic applications of nanoemulsion based drug delivery systems: a review of patents in last two decades. Recent Pat Drug Deliv Formul. 2011;5(2):163–72.
62. Kumar M, Bishnoi RS, Shukla AK, Jain CP. Techniques for formulation of nanoemulsion drug delivery system: a review. Prev Nutr Food Sci. 2019;24(3):225.
63. Qadir A, Faiyazuddin M, Hussain MT, Alshammari TM, Shakeel F. Critical steps and energetics involved in a successful development of a stable nanoemulsion. J Mol Liq. 2016;214:7–18.
64. Tadros T, Izquierdo P, Esquena J, Solans C. Formation and stability of nano-emulsions. Adv Colloid Interf Sci. 2004;108:303–18.
65. McClements DJ. Nanoemulsions versus microemulsions: terminology, differences, and similarities. Soft Matter. 2012;8(6):1719–29.
66. Shah K, Chan LW, Wong TW. Critical physicochemical and biological attributes of nanoemulsions for pulmonary delivery of rifampicin by nebulization technique in tuberculosis treatment. Drug Deliv. 2017;24(1):1631–47.
67. Kumar M, Jain CP. Possible benefits of reformulating antiviral drugs with nanoemulsion system in the treatment of novel coronavirus infection. Curr Drug Ther. 2020;2:3.
68. Zhu W, Yu A, Wang W, Dong R, Wu J, Zhai G. Formulation design of microemulsion for dermal delivery of penciclovir. Int J Pharm. 2008;360(1–2):184–90.
69. Alkhatib MH, Aly MM, Rahbeni RA, Balamash KS. Antimicrobial activity of biocompatible microemulsions against Aspergillus niger and herpes simplex virus type 2. Jundishapur J Microbiol. 2016;9(9).
70. Shishu SR. Development of novel microemulsion-based topical formulations of acyclovir for the treatment of cutaneous herpetic infections. AAPS PharmSciTech. 2009;10(2):559.
71. Carvalho ALM, da Silva JA, Lira AAM, Conceição TMF, de Souza Nunes R, de Albuquerque Junior RLC, et al. Evaluation of microemulsion and lamellar liquid crystalline systems for transdermal zidovudine delivery. J Pharm Sci. 2016;105(7):2188–93.
72. Sasivimolphan P, Lipipun V, Ritthidej G, Chitphet K, Yoshida Y, Daikoku T, et al. Microemulsion-based oxyresveratrol for topical treatment of herpes simplex virus (HSV) infection: physicochemical properties and efficacy in cutaneous HSV-1 infection in mice. AAPS PharmSciTech. 2012;13(4):1266–75.
73. Lee J-J, Shim A, Jeong JY, Lee SY, Ko H-J, Cho H-J. Development of intranasal nanovehicles of itraconazole and their immunological activities for the therapy of rhinovirus infection. Colloids Surf B: Biointerfaces. 2016;143:336–41.
74. Schwarz JC, Klang V, Karall S, Mahrhauser D, Resch GP, Valenta C. Optimisation of multiple W/O/W nanoemulsions for dermal delivery of acyclovir. Int J Pharm. 2012;435(1):69–75.

75. Nabila N, Suada NK, Denis D, Yohan B, Adi AC, Veterini AS, et al. Antiviral action of curcumin encapsulated in nanoemulsion against four serotypes of dengue virus. Pharm Nanotechnol. 2020;8(1):54–62.
76. Kumar B, Jain SK, Prajapati SK, Mahor A, Kumar A. Development and characterization of transdermal microemulsion gel for an antiviral drug. Int J Pharm Sci Res. 2010;1(6):57–74.
77. Bajaj H, Bisht S, Yadav M, Singh V, Singh M. Design and development of nevirapine loaded surfactant free chitosan microemulsion. Acta Poloniae Pharm. 2011;68(6):981–8.
78. Ghosh PK, Majithiya RJ, Umrethia ML, Murthy RS. Design and development of microemulsion drug delivery system of acyclovir for improvement of oral bioavailability. AAPS PharmSciTech. 2006;7(3):E172–E7.
79. Mirani A, Kundaikar H, Velhal S, Patel V, Bandivdekar A, Degani M, et al. Tetrahydrocurcumin-loaded vaginal nanomicrobicide for prophylaxis of HIV/AIDS: in silico study, formulation development, and in vitro evaluation. Drug Deliv Transl Res. 2019;9(4):828–47.
80. Kaur A, Sharma G, Gupta V, Ratho RK, Shishu, Katare OP. Enhanced acyclovir delivery using w/o type microemulsion: preclinical assessment of antiviral activity using murine model of zosteriform cutaneous HSV-1 infection. Artif Cells Nanomed Biotechnol. 2018;46(2):346–54.
81. Mahajan HS, Mahajan MS, Nerkar PP, Agrawal A. Nanoemulsion-based intranasal drug delivery system of saquinavir mesylate for brain targeting. Drug Deliv. 2014;21(2):148–54.
82. Vyas TK, Shahiwala A, Amiji MM. Improved oral bioavailability and brain transport of Saquinavir upon administration in novel nanoemulsion formulations. Int J Pharm. 2008;347(1–2):93–101.
83. Manyarara TE, Khoza S, Dube A, Maponga CC. Formulation and characterization of a paediatric nanoemulsion dosage form with modified oral drug delivery system for improved dissolution rate of nevirapine. MRS Adv. 2018;3(37):2203–19.
84. Prabhakar K, Afzal SM, Surender G, Kishan V. Tween 80 containing lipid nanoemulsions for delivery of indinavir to brain. Acta Pharm Sin B. 2013;3(5):345–53.
85. Hassan H, Adam SK, Othman F, Shamsuddin AF, Basir R. Antiviral nanodelivery systems: current trends in acyclovir administration. J Nanomater. 2016;2016:1.
86. Patel A, Shelat P, Lalwani A. Development and optimization of solid self-nanoemulsifying drug delivery system (S-SNEDDS) using Scheffe's design for improvement of oral bioavailability of nelfinavir mesylate. Drug Deliv Transl Res. 2014;4(2):171–86.
87. Karami Z, Saghatchi Zanjani MR, Rezaee S, Rostamizadeh K, Hamidi M. Neuropharmacokinetic evaluation of lactoferrin-treated indinavir-loaded nanoemulsions: remarkable brain delivery enhancement. Drug Dev Ind Pharm. 2019;45(5):736–44.
88. Argenta DF, Bidone J, Koester LS, Bassani VL, Simões CMO, Teixeira HF. Topical delivery of coumestrol from lipid nanoemulsions thickened with hydroxyethyl cellulose for antiherpes treatment. AAPS PharmSciTech. 2018;19(1):192–200.
89. Bonfim CMD, Monteleoni LF, Calmon MDF, Cândido NM, Provazzi PJS, Lino VDS, et al. Antiviral activity of curcumin-nanoemulsion associated with photodynamic therapy in vulvar cell lines transducing different variants of HPV-16. Artif Cells Nanomed Biotechnol. 2020;48(1):515–24.
90. Argenta D, Bidone J, Misturini F, Koester L, Bassani V, Simoes C, et al. In vitro evaluation of mucosa permeation/retention and antiherpes activity of genistein from cationic nanoemulsions. J Nanosci Nanotechnol. 2016;16(2):1282–90.

Novel Formulation Approaches for Treatment of Ebola Virus

Sankha Bhattacharya, Shambhavi Singh, Sambuddha Chakraborty, Bhupendra G. Prajapati, Mahavir Chougule, and Jayvadan K. Patel

Abstract The segment of the United Nations that deals with the medical aspects of all the different parts of the world, the World Health Organization, on the 3rd of May 2021, professed the ongoing Ebola outbreak to be over. Ebola virus is acknowledged to be the causative agent of Ebola viral disease (EVD) or Ebola fever, which is of an hemorrhagic nature, a viral hemorrhagic disease found in humans in addition to some arboreal primates. The diagnosis is made using recognition of viral RNA through a reverse transcriptase-polymerase chain reaction in real time and even through fast diagnostic tests, the basis of which are detection of antigens using an enzyme-linked immunosorbent assay. Although vaccines are still important in reducing deaths due to EVD, other strategic approaches to working for its prevention and management are required. Several patented agents, biotherapeutics, etc., are required for prophylactic or therapeutic purposes. This article deals with different medicines and vaccines, etc., which are emerging or alternatives to the treatment of EVD. A shift in paradigm toward improvement in medical and clinical health structures had a quick impact on EVD containment while also laying the groundwork for introducing new medicines and therapies to the afflicted countries once the said medicines combined with the therapies became obtainable.

S. Bhattacharya · S. Singh
Department of Pharmaceutics, School of Pharmacy & Technology Management, SVKM'S NMIMS Deemed-to-be University, Shirpur, Maharashtra, India

S. Chakraborty
Department of Microbiology, Tripura University, Agartala, Tripura, India

B. G. Prajapati (✉)
Department of Pharmaceutics and Pharmaceutical Technology, Shree S K Patel College of Pharmaceutical Education and Research, Ganpat University, Mahesana, Gujarat, India

M. Chougule
Department of Pharmaceutical Sciences, Mercer University College of Pharmacy, Atlanta, GA, USA

J. K. Patel
Nootan Pharmacy College, Sankalchand Patel University, Mehsana, Gujarat, India

R. Shegokar, Y. Pathak (eds.), *Viral Drug Delivery Systems*,
https://doi.org/10.1007/978-3-031-20537-8_7

Keywords Ebola virus · Vaccine · Diagnostics · Theranostics · Viral genomics · Pathophysiology

Abbreviations

EBOV	Ebola virus
EVD	Ebola viral disease
GP	Glycoprotein
NP	Nucleoprotein

1 Introduction

Nature certainly has an enigmatic way of acting, as it has implanted a sense of survival in a virus particle as small as 100 nm that can shut down the entire human civilization! [1, 2]

Between July and September 1976, a hemorrhagic fever outbreak occurred in southern Sudan and 800 km away in Bumba, the Democratic Republic of Congo, which was earlier recognized as northern Zaire, with a very high fatality of around 30–80% in Sudan and 89% in Zaire [3]. Blood and tissue samples were taken from clinically ill patients and then sent to classified laboratories in the UK, Belgium, and the Centers for Disease Control and Prevention in Atlanta, USA, owing to the severity of the diseases [4]. The samples were further processed and inoculated in cultured Vero cells, with a distinct cytopathic effect observed on the third day after inoculation. The supernatants from the cell culture were then collected and analyzed using electron microscopy. The electron micrograph revealed some filamentous virus particles that morphologically resembled the Marburg virus, a hemorrhagic fever-causing virus from Marburg, Germany. Indirect immunofluorescence revealed a serological difference between the Marburg virus and African clinical isolates, indicating that African hemorrhagic fever is caused by a different virus [5]. On the river Ebola in Zaire, Prof. S. R. Pattyn of the Institute of Tropical Medicine in Antwerp, Belgium, and Mr. E. T. W. Bowen of the Microbiological Research Establishment in Porton Down, UK, named this viral agent Ebola virus. *Filoviridae* are filament-like, an incased virus with linear, negative-sense, non-segmented ribonucleic acid (RNA). The first member of the *Filoviridae* family was the Marburg virus, which was isolated from clinical samples of hemorrhagic fever outbreaks in the German town of Marburg in 1967. In 1976, Ebola was classified as the *Zaire ebolavirus*, but it was then renamed the Ebola virus (EBOV) in the 2010 by the International Committee on Virus Taxonomy [6]. There are in all six known *ebolavirus* species, namely: *Zaire ebolavirus*, *Sudan ebolavirus*, *Taï Forest ebolavirus*, *Bundibugyo ebolavirus*, *Reston ebolavirus*, and *Bombali ebolavirus* (Fig. 1a) [7]. All viruses can be the causative agent for viral hemorrhagic fever in humans. This virus causes a zoonotic disease, and each of its outbreaks is triggered by its

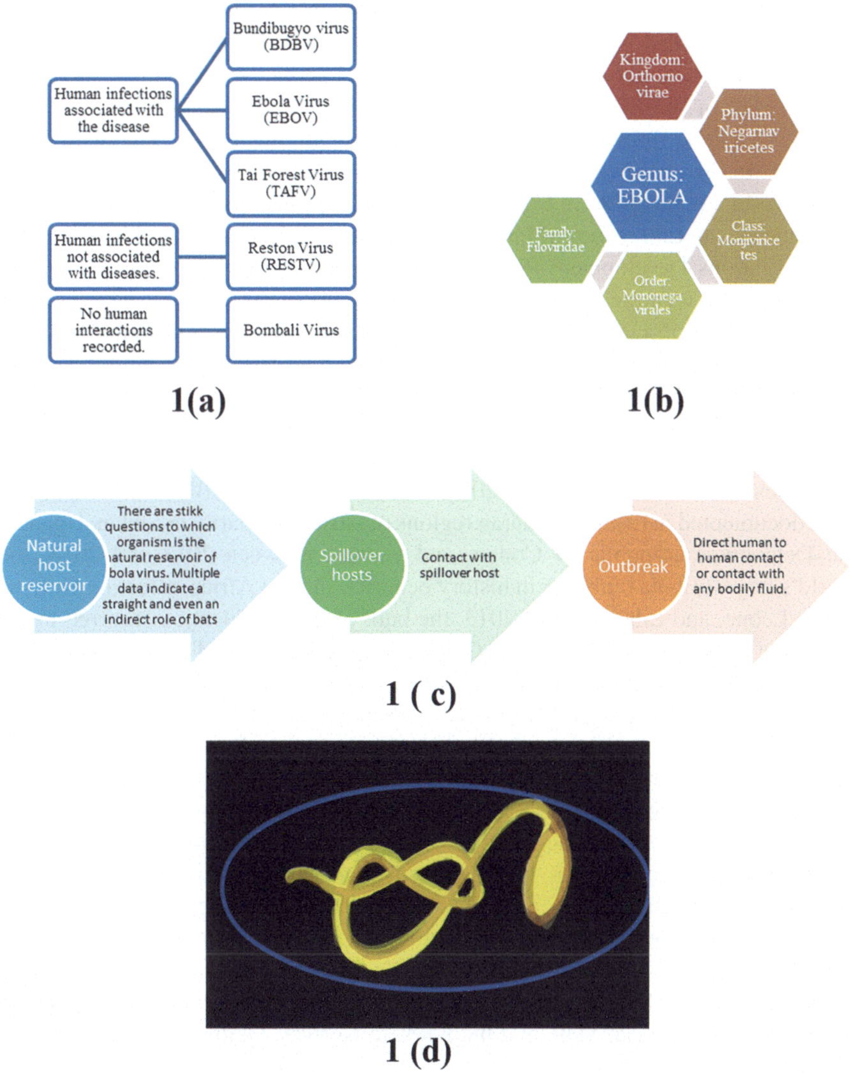

Fig. 1 (**a**) Different species within the genus *Ebola*, which are associated with infection. (**b**) Ebola virus disease configuration and genetics. (**c**) General steps that pave the way for an outbreak. (**d**) Microscopic view of Ebola virus as observed

introduction into an animal (Fig. 1b). It has now been found out that bats are most probably the natural reservoir as the EBOV is primarily an animal-borne disease [8].

The EBOV has infected humans because of spillover events. Spillover occurs when many new species meet the reservoir of a viral pathogen that belongs to a different species, such as humans and fruit bats. The disease can be passed from parent to child through uninterrupted interaction with purulent blood, body fluids, or the patient's skin. From the time the very first patient with the Ebola virus disease (EVD) was recorded, the World Health Organization (WHO), a specialized part of the United Nations that deals with public health at an international level, has registered more than 20 EBOV outbreaks. Few of them, such as those in Guinea, Liberia, and Sierra Leone from 2014 to 2016, had such high death rates. The 2014 EVD outbreak was the first-ever recorded large-scale introduction of the EBOV to the rest of the world's population; in May 2014, the WHO pronounced this epidemic to be a "Public Health Emergency of International Concern". In research published by the United Nations Development Program, many related variables of the 2014–2016 pandemic were identified, among which a few highlighted factors are denial or dishonesty; refusal of quarantine and dishonesty about infection contributed to about 74% of viral spread, traditional practices involved high-risk behavior during funerals, 60% were linked aside from that, and the epidemic was exacerbated by a lack of awareness, religious belief, stigmatization, poverty, poor diet, poor health care services, and mobilization [9]. After a drawn-out battle that began in 1976, the outbreak was declared over on 3 May 2021. Since then, over 20 EBOV outbreaks have been documented in the Sub-Saharan regions of Africa, primarily in Sudan, Uganda, the Democratic Republic of Congo, and Gabon. Between 2013 and 2016, the world's largest Ebola outbreak in history occurred in west Africa, affecting Guinea, Sierra Leone, and Liberia. Until 2013, the bulk of EVD epidemics occurred in the Democratic Republic of the Congo, Gabon, and the Republic of the Congo.

The outbreak that occurred in West Africa from 2013 through 2016 was thought to have started in December 2013 in the Guinean village of Meliandou. Liberia was the first country to declare the 2014–2016 Ebola outbreak over on 9 May 2015. Sierra Leone and Guinea also saw a decline in new infections over time [10]. The average fatality rate in an EVD case is close to 50%. Case mortality statistics in preceding epidemics have vacillated from 25% through 90%. Studying statistics, it has been determined that there has been medical advancement that has resulted in a decrease in the fatality rate over the years from 1976 to the present day. Through the years of progression, this illness even witnessed a 100% mortality rate, with 318 recorded cases, to 42%, with just 130 recorded cases. The majority of cases were seen in the region of the Democratic Republic of Congo, with some spillovers seen in the region of Liberia, Mali, and the USA, to Norway, Germany, France, and the Netherlands, etc.

The Ebola outburst in the western parts of Africa officially ended; however, this was only recorded in June 2016, when the organization proclaimed that Liberia and Guinea were Ebola free for the fourth and second times respectively. The disease was initially found in 1976 during two simultaneous epidemics; the first was in the South Sudanese town of Nzara, while the second was in the Democratic Republic of the Congo in Yambuku, a community along the river Ebola, the origin of its name [11].

The EBOV is a zoonotic illness, which means it may spread to people, apes, and other animals. Whether through a direct or indirect mode, the infection spread through interaction with bats or by handling the dead, or even by taking care of ill humans, resulting in those most susceptible becoming infected [12]. Even though the natural reservoir is yet to be asserted, bats have found their way to being the most plausible suspect (Fig. 1c). There are three different varieties of fruit bats *Epomops franqueti*, *Hypsignathus monstrosus*, and *Myonycteris torquata*, that have been discovered to be the most acceptable for carrying the virus without becoming ill. It is unknown whether other animals are implicated in the propagation of the disease as of 2013 [13]. There are a few speculations about plants, arthropods, rodents, and birds being potential virus reservoirs. Uninterrupted contact with purulent persons' blood, saliva, and other bodily fluids can cause secondary human-to-human transmission to occur because of this. The danger of nosocomial infections is particularly great when tending to the sick or even handling infected dead bodies, which is why they are so common in hospitals. EBOV belongs to the family *Filoviridae* of the order *Mononegavirales* (Fig. 1b). The virus particles belong to the genus, which has negative polarity to the single RNA strand and resembles long stretched filaments when seen under an electron microscope, with some particles curving into the shape of the number six (Fig. 1d).

With fever generally ranging over 38.3 °C (101 °F) alongside fatigue, weakness, loss of appetite, muscular pain, and sore throat, the symptoms follow the initial influenza-like stage. Severe dehydration is frequently caused by the combination of severe vomiting and diarrhea, which are common side effects. Shortness of breath along with chest pain, as well as swelling, headaches, and confusion, usually follows. Approximately 5–7 days from the onset of symptoms, the skin at times fosters a maculopapular rash, which is a highly inflamed patch together with many minute irregularities (Fig. 2a).

The genome of the EBOV is 19 kb long and contains seven open reading frames, encodes structural proteins such as the virion envelope glycoprotein (GP), nucleoprotein (NP), and matrix proteins VP24 and VP40 [14]. *Bundibugyo ebolavirus* (Bundibugyo virus), *Reston ebolavirus* (Reston virus), *Sudan ebolavirus* (Sudan virus), *Taï Forest ebolavirus* (Taï Forest virus), and *Zaire ebolavirus* are the five species of the genus *Ebolavirus*. EBOV genomes contain seven genes: 3′-UTR-NP-VP35-VP40-GP-VP30-VP24-L-5′-UTR, 3′-UTR-NP-VP35-VP40-GP-VP30-VP24-L-5′-UTR, 3′-UTR-NP-VP35-VP40-GP-VP sequencing of the pentameric ebolavirus genomes (Bundibugyo virus, EBOV, Reston virus, Sudan virus, and Taï Forest virus) is distinct, as are the number and location of gene overlays. Ebolavirus virions, like all filoviruses, are filament-like elements that can take the shape of a shepherd's crook, a "U," or a "6," and can be curved torus or branched [15].

Among the proteins encoded by the genome are NP [16], polymerase cofactor (VP35), matrix protein (VP40), GP, soluble GP (sGP), small soluble GP (ssGP), transcription activator (VP30), minor matrix protein (VP24), and RNA-dependent RNA polymerase (L) (Fig. 2b). GP allows the infection to enter monocytes and additionally macrophages, where cell injury or viral molecule openness can advance the arrival of connected cytokines, which leads to aggravation and fever, and also endothelial cells, which can cause vascular harm. Twelve different filoviruses have been found. The seven filoviruses discovered to cause infection in humans belong to the *Ebolavirus* genus (Bundibugyo virus, EBOV, Reston virus, Sudan virus, and Taï Forest virus) (Fig. 1) or the *Marburgvirus* genus (Marburg virus and Ravn virus) [17].

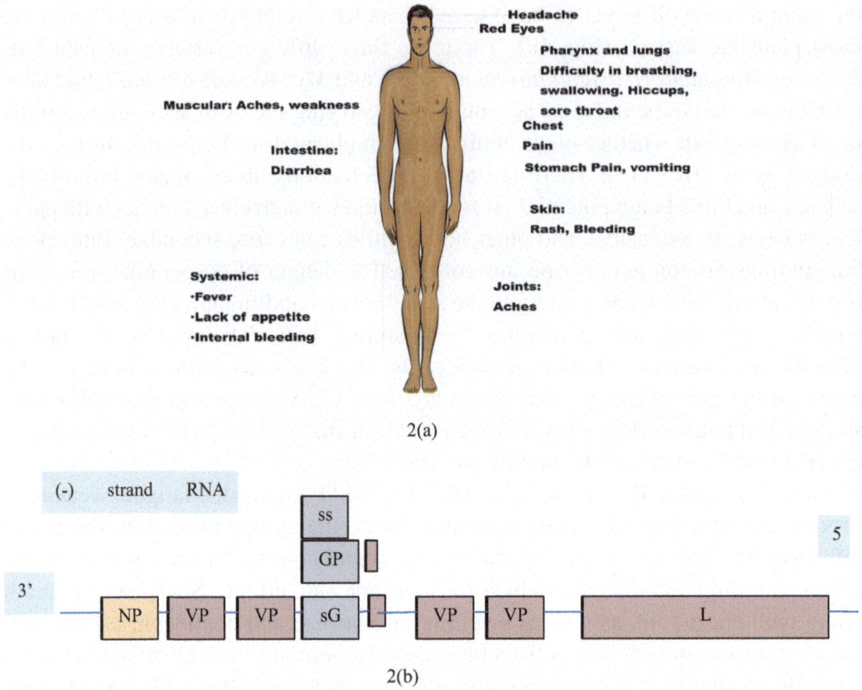

Fig. 2 (**a**) Signs and symptoms of Ebola virus (EBOV) presentation in a human. (**b**) The EBOV has a 19-kb linear negative-sense RNA genome with seven structural and nonstructural proteins produced by seven genes

Ebola disease, caused by Bundibugyo virus, EBOV, Sudan virus, or Taï Forest virus, and Marburg disease, caused by Marburg virus or Ravn virus, are the two principal subtypes of filovirus disease recognized by the WHO International Classification of Diseases Revision 11 of 2018. This filovirus disease subcategorization is mostly based on the growing body of evidence of molecular differences between the Ebola and Marburg viruses [18]. These are the differences that might influence inadvertent primate hosts: virus–host reservoir tropism, pathogenesis, and disease phenotype. Highly variable filoviruses, often with unidentified pathogenic potential, are found to be dispersed broadly through the African, Asian, and European continents in a wide range of host reservoirs, according to genomics.

A nanocarrier is a type of nanomaterial that has been used to convey another chemical, such as medicine. Micelles, polymers, carbon-based materials, liposomes, and other substances are often utilized as nanocarriers. Bioavailability and therapeutic efficacy are improved by nanocarriers, which provide preferential accumulation at the target site [19]. Although several nanocarriers have been created, only a few have received clinical approval. Colloidal drug carrier systems with particle sizes fewer than 500 nm are known as nanocarriers. Because of their potential in the realm of drug delivery, these have been investigated extensively in recent decades. Nanocarriers can change the basic characteristics and bioactivity of medications because of their extensive surface area-to-volume ratio. To name a few advantages, nanocarriers can improve pharmacokinetics, even biodistribution, as well as reducing toxicities, improving the ability of the active pharmaceutical ingredient to become more soluble and turn more stabilized to control drug release, and target-specified therapeutic medication. In addition to what has been stated above, a small change to the composition between organic, inorganic, or hybrid, and even to its size and shapes with the surface, which are the external properties (attachment of targeted moieties, surface charge, functional groups, PEGylation, or another coating) often results in changes to the physicochemical properties of nanocarriers [20]. The ultimate objective of utilizing nanocarriers to carry medicine is to cure a disease effectively while minimizing adverse effects.

Experimental postexposure treatments for EBOV focus on (1) preventing filovirus-associated coagulopathies (recombinant nematode anticoagulant protein and recombinant human activated protein C); (2) inhibiting viral replication or translation (nucleotide analogs [Toyama Chemical, Tokyo, Japan], BCX4430 [BioCryst Pharmaceuticals, Durham, NC, USA], brincidofovir [Chimerix, Durham, NC, USA], and brinci [mAb cocktails]). Clinical studies for some of these postexposure treatment possibilities are now underway.

Ebola virus multiplies rapidly in monocytes, macrophages, dendritic cells, and other cells such as liver cells, fibroblasts, and adrenal gland cells, resulting in enormous quantities of viruses. As a result of this viral replication, a surge in inflammatory chemical signals is seen, which often leads to a septic condition. There are usually three phases to an illness: a generic fever, headache, and myalgia lasting a few days, which is often followed by the gastric phase, which mostly includes diarrhea and vomiting, abdominal discomfort, and dehydration. Liver and kidney function deteriorates in the advanced and last stages of the disease, resulting in relentless metabolic deterioration, convulsions, shock, and mortality, which results in the occurrence of bleeding through the mucosa. Another set of characteristic symptoms seen is bloody diarrhea with multiple organ failure within the 16-day window from the onset of symptoms.

Ebola virus proteins interfere with the capacity of cells to produce and respond to interferon proteins, including interferon-α, interferon-β, and interferon-γ, which restrict the response of the human immune system to viral infections (Fig. 3) [21]. Ebola vaccines of many forms are being developed to combat EVD, including inactivated vaccines, viral vector vaccines, subunit vaccinations, and DNA vaccines. EBOV is now being treated with a licensed vaccine that has been named Ervebo, and Ebanga and Inmazeb are the two licensed and approved monoclonal antibody medicines; the latter is a cocktail, but the former is a single monoclonal antibody [22]. These unquestionably played a prime part in limiting the recent outbreaks. However, no vaccines or medicines have been authorized to prevent or even treat infections that have occurred in humans caused by other *Ebolavirus* genus members (e.g., Sudan virus; Bundibugyo virus, Taï Forest virus), *Marburgvirus* genus associates, or diseases caused by developing filoviruses.

A monoclonal antibody formulation Mab114, which was produced at the facility of Cook pharmacy, with reference to a study conducted at the National Institutes of

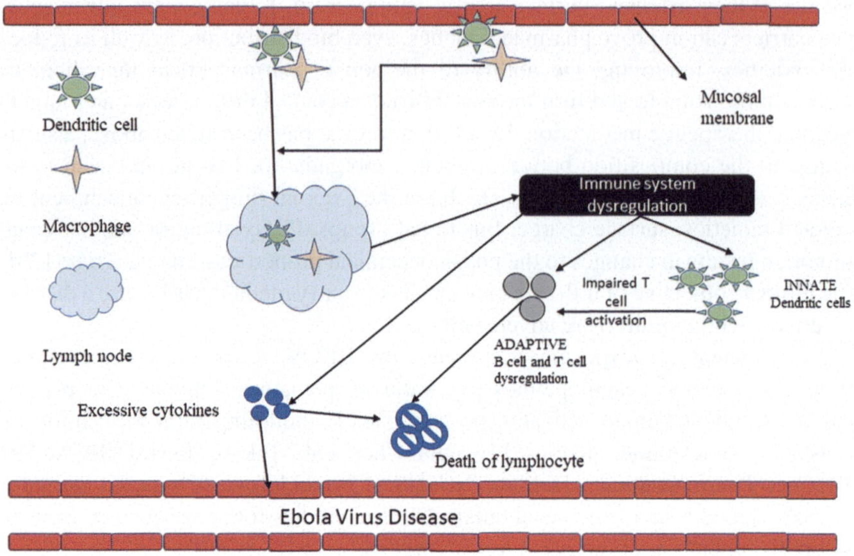

Fig. 3 Pathogenesis of Ebola virus disease

Health (NIH). The phase I study was the open-label dosage intensification trial with the primary target of making sure that the formulation was safe and tolerable. Here, the GP of different EBOV strains was targeted by the human IgG1 monoclonal antibody. The infusions of Mab114 were found to be safe as well as well tolerated. Some associated adverse effects, such as malaise, myalgia, joint pains, and nausea, have been observed.

A miracle drug called remdesivir by Gilead sciences was developed. *Zaire*, *Sudan*, and *Bundibugyo ebolavirus* strains were those targeted with the help of remdesivir. It is a small nucleotide molecule, which, in its chemical nature, is a prodrug with action ranging over a broad spectrum. The study was conducted in a placebo-controlled randomized trial to test for its efficacy. It was observed that there was an increase in levels of serum amylase, but no negative side effects were observed.

REGN-3470-3471-3479 was developed at the Illinois-based Regeneron Pharmaceuticals. The developed formulation manufactured was a cocktail of three different monoclonal antibodies that targeted the non-overlapping epitopes that are specific to the Zaire strain of the Ebola virus. The study was conducted in a placebo-controlled randomized phase, with the most prevalent side effect or symptom observed being a headache.

ZMapp was a cocktail of three different chimeric monoclonal antibodies that were specific to the epitopes on the GP of the Zaire strain of Ebola virus. This was developed in collaboration between MappBio, Leafbio, and Defyrus Inc. An open-label multicenter randomized safety and effectiveness trial with an adaptive design was conducted. The PREVAIL trial provided the most convincing evidence of the efficacy of an Ebola treatment. Even though the study's findings were found to be quite promising, the numbers did not provide sufficient data to be definitive about the efficacy of ZMapp.

Ervebo, a vaccine developed by Merck, was the foremost EBOV vaccine to be permitted for use in the USA and the European Union (EU). In 2019, the *Zaire ebolavirus* vaccine, which is live-attenuated and recombinant, was authorized for use in anyone aged 18 or over to prevent illness caused by the virus. Since the start of the Ebola outbreak, more than 80 vaccine candidates have entered the development stage. Moderna, GlaxoSmithKline (GSK), and Merck were among the developers. Most of the candidates, on the other hand, did not proceed to the clinical stage.

The evolution of science across time is discussed. We are retracing our steps from 2020 at snail's pace. With the FDA's approval of Ebanga, a human monoclonal antibody permitted for use in people of all ages in the case of infection, we were on the verge of a breakthrough in Ebola therapy. This acts by preventing the virus from attaching to the receptors. Another breakthrough was in October 2020 with Inmazeb; the FDA authorized a mixture of three monoclonal antibodies as the first therapy for *Zaire ebolavirus* infection in both adults and children. Following the year 2019, we have 2020, in which multiple breakthroughs and milestones in medical history were reached for the treatment of the viral illness being discussed here. The Food and Drug Administration (FDA) approved a fingerstick test for emergency use in 2018. This functioned by identifying the antigen in relation to the virus. This test is essential for identifying the EBOV in infected individuals with a rapid and precise result. September saw the assistance of statisticians in developing an algorithm to test the hypotheses developed for the treatment of the EBOV. It was approved for anyone over the age of 18. In October 2019, the first quick diagnostic kit for the diagnosis of EBOV was approved. November of the same year witnessed the expansion of biobanks for plasma and other blood products, which were essential for treatment therapies. The FDA then awarded the University of California, Los Angeles, contracts for the expansion of the same. In December 2019, the FDA authorized a vaccine for the prevention of EVD for the first time, marking a significant step forward in the fight against the epidemic.

2 Recent Clinical Trials on Ebola

The features of Ebola, such as its high mortality rate, rapid spread across communities, and grave hazards for exposed health care professionals, exacerbate the anticipated difficulties of conducting clinical trials in asset settings. In the case of a high-fatality, dynamic epidemic, a structured platform is well suited to morally and quickly identifying viable medicines. Academic and research communities must respond as aggressively and quickly as the humanitarian community to enhance medicines in the middle of a rising pandemic (Table 1).

Table 1 Current ongoing clinical trials in the field of Ebola virus therapeutics [23]

Study title	Start	End	Observational model	Objectives	Phase
Prevail III The Partnership for Research on Ebola Virus conducted an Ebola Natural History Study	2 December 2017	April 2022	Cohort	To find out how Ebola impacts the health of survivors and those who live with them	III
In the Democratic Republic of the Congo, a heterologous, two-dose preventive Ebola vaccine was evaluated for effectiveness and safety	14 November 2019	31 January 2024	Ad26 is a single-arm, open-label, nonrandomized interventional experiment that compares the two doses. Ebola preventative vaccination ZEBOV, MVA-BN-Filo		3
Partnership for Research on Ebola Vaccinations (PREVAC)	March 31, 2017	June 2024	Quadruple Intervention, Parallel Assignment (Participant, Care Provider, Investigator, Outcomes Assessor)	The goal of this trial is to see if Ad26.ZEBOV (rHAd26) and rVSVG-ZEBOV-GP are safe and effective in children and adults. The rHAd26 vaccine will be tested with or without an MVA-BN-Filo (MVA) vaccine, and with and without boosting, whereas the rVSV vaccine will be tested with and without boosting	2
Candidates for Ebola vaccines Ad26.ZEBOV and/ or MVA-BN-Filo subjects exposed to the candidate Ebola vaccines Ad26. ZEBOV and/or MVA-BN-Filo in a multi-country prospective clinical safety study	31 May 2016	10 January 2023	Investigational, single group assignment, none (open label)	This is prospective, long-term clinical safety research involving many countries to gather significant adverse events and pregnancy outcomes in people who were engaged in phase I, II, or III clinical studies after getting Ad26.ZEBOV and/or MVA-BN-Filo vaccinations	3

2.1 Potential Therapeutic Targets

Ebolavirus predominantly replicates inside antigen-presenting cells such as dendritic cells and macrophages. Infective dendritic cells and macrophages trigger the uncontrolled secretion of proinflammatory cytokines such as interleukin-1β, tumor necrosis factor α, IL-6, IL10, interleukin-8, monocyte chemoattractant protein-1, macrophage inflammatory protein-1α and growth factor α, reactive oxygen species, and nitric oxide [27].

VP24 is reported as a downregulator of the STAT 1 pathway (IFN signaling pathwayVP24, a protein found in the *Zaire ebolavirus*, inhibits IFN and IFN-induced nuclear accumulation of tyrosine-phosphorylated STAT1 (PY-STAT1), which in turn inhibits IFN and IFN-induced gene expression. VP24 also binds to the PY-STAT1 binding region of the karyopherin 1, 5, or 6, which is involved in molecule transport between the cytoplasm and the nucleus, resulting in the inhibition of PY-STAT1 nuclear transport and thus downregulation of IFN secretion [28].

Ebolavirus and its family counterpart *Marburgvirus* fall into the category of the highest biosafety level agents, that is, biosafety level 4 agents . These deadliest viruses have been reported to have 90% mortality in some outbreaks. EBOV causes hemorrhagic fever, damage to the cardiovascular system, leading to blood leakage from the blood vessels, causing internal bleeding [24]. Several viruses can cause hemorrhagic fevers such as *Dengue virus*, *Yellow fever virus*, Crimean-Congo hemorrhagic fever [25], hantaviruses, Lassa virus, which can range from slight illness to severe disease resulting in death. EVDs progress with nonspecific clinical manifestations such as high fever, headache, fatigue, or gastrointestinal irregularities such as nausea and vomiting. Cardiovascular manifestation bleeding has been reported, but despite the attribution as hemorrhagic fever, not every case was associated with the bleeding. In the first case report of the disease, 75% of the infected individuals manifested bleeding, as low as 30% of the cases reported the bleeding manifestation. Electrolyte disbalance due to dysfunction of the gastrointestinal system causes hypovolemic shock, which may lead to organ failure and is associated with severe disease and mortality [26].

2.2 Possible Therapeutic Targets of Ebola Virus

The RNA viruses have always imposed huge disease burdens on humans, e.g., influenza, HIV, dengue, Ebola, Nipah, severe acute respiratory syndrome , Middle East respiratory syndrome, coronavirus, and the list is getting longer year by year. A major failure in the management of the recent COVID-19 pandemic was the lack of a proper antiviral regime was very vivid and cost many lives. It is unfortunate that despite remarkable progress being achieved in different segments of science and technology, we still struggle to produce efficient antivirals, especially against RNA viruses. Up to now, we have approved antivirals for ten viral pathogens (HIV, hepatitis C virus, hepatitis B virus, herpes, influenza, human cytomegalovirus, varicella zoster virus, respiratory syncytial virus, human papillomavirus, and picornavirus). We certainly live in a precariously balanced environment where the possibilities of new emerging viruses are very high, and it is of utmost necessity to come up with more antiviral strategies. To develop or formulate a new antiviral, one must fundamentally understand the biology of the viral material and its pattern of communication with the host cell. Basic approaches of the antivirals can be of two types, (a) targeting the virus or (b) targeting the host factors associated with the virus infection. Another unfortunate incidence with the antivirals is viral resistance against the drug. Viruses use the host cell types of machinery to replicate, and hence the antiviral target sites are fewer in number; moreover, mutilations leading to specific drug resistance make the scenario more complex and constrict the designing and formulation of new antivirals. At present, only one vaccine with two monoclonal antibody formulations has been approved for EVD [29]. In December 2019, the recombinant vesicular stomatitis virus chimeric construct, 'Ervebo,' manufactured by Merck & Co., Inc., was licensed by the FDA as a vaccine to prevent the spread of *Zaire ebolavirus*. Subsequently, 2020 brought a cocktail of monoclonal antibodies, three different categories (atoltivimab, maftivimab, and odesivimab-ebgn). Inmazeb, which was sanctioned by the FDA in October and 68 days later, another monoclonal antibody, 'Ebanga' (ansuvimab-zykl) was also approved by the agency as prophylaxis against *Zaire ebolavirus* infection [30]. There are currently no licensed antiviral medications for EVD. However, numerous antivirals are in the pipeline, with one reaching phase III studies. It is not necessary to mention that understanding the parameters associated with the life cycle of the virus is essential for designing a new antiviral. The target site, which is the basis of antiviral drug design, is unveiled by an understanding of the host-virus interaction. It should be the primary consideration, as a major part of the drug efficacy is dependent on the selection of the target sites. Ebolavirus consists of an unsegmented negative-strand RNA virus with seven genes that code for nine different viral proteins. Seven of them are structural proteins, whereas the other two are nonstructural proteins) [31]. The basic steps of the viral replication are (1) attachment, (2) entry inside the cytoplasm, (3) uncoating and fusion with endosomal membrane, (4) transcription and replication, (5) assembly and budding. The ten encoded viral proteins execute the replication cycle prominently with the help of a few host cellular proteins. Several vials and host proteins are associated with different steps of the replication cycle. Viral proteins have always been the major target sites for antivirals. Newly marketed formulations and patent trends in the quashing of Ebola virus (drugs and their formulations; Table 2) As mentioned above now, two monoclonal antibodies have been approved by the FDA for prophylaxis, and expected that more will be approved soon looking at the current pace of the developments.

Table 2 List of drugs that are under development and already on the market for Ebola virus therapeutics [32]

Sr. No.	Compound	Target site	Trial phase
1	BCX4430	Viral RNA polymerase L	1st
2	GS-5734	RNA polymerase	2nd
3	Favipiravir	RNA polymerase	3rd
4	Amiodarone	Cationic amphiphilic drug, antiarrhythmic drug, inhibits early stage of replication	2nd
5	Amodiaquine	Cationic amphiphilic drug	Preclinical
6	Brincidofovir	RNA polymerase	2nd
7	TKM-100802/TKM-130803	RNA polymerase	1st and 2nd
8	ZMapp	mAb cocktail	2nd
9	Convalescent whole blood/ plasma (Ebola-Tx)	Targets whole virus or the glycoproteins	2nd/3rd
10	AVI-7537	VP24	1st

3 Theragnostic Approaches

Drug development for EBOV has been progressing for a long time; then again, a new perspective focus on producing treatments and vaccines was actualized right after the 2014 through 2016 outbreak in Western Africa, which raised concerns about the global spread. Various new medicines and concepts are being examined for the therapy and management of the viral Ebola disease, which is supplementary to the supportive care that is provided with the help of electrolyte replacement, hydration therapy, oxygen therapy, as well as by maintaining the acid–base balance of the body [33]. The process of developing a drug comprises many steps, the first one being the identification of credible therapeutic targets that could be a protein, RNA, along with any biotic component that would have a prime role in the pathogen's ability to propagate and infect. Different undertakings have been explored, such as genomics, biochemical, computational, and structural, and are mostly required to identify the specific therapeutic target either from the host or the pathogen. The FDA in recent times has authorized two therapies for EVD [34]. Inmazeb is an amalgamation of three different monoclonal antibodies that was authorized in October 2020. Ebanga is the second medication that was permitted for use in December 2020; the FDA sanctioned a solitary monoclonal antibody. A solitary set of proteins that are developed in a laboratory or any other facility is functionally like that naturally occurring antibody in preventing a pathogen, such as a virus, from reproducing in a host after it has already infected a person. These mAbs attach with the glycoprotein, a part of the EBOV's surface, which inhibits the viral material from entering the body [35].

A. *Antiviral strategies*

(a) *Targeting the entry process*

Vinblastine, vinorelbine, vincristine, colchicine, nocodazole, mebendazole, and albendazole are microtubule inhibitors that are among the most potent anti-EBOV medications [36]. The effect of nocodazole is thought to be linked to microtubule depolymerization, which prevents viral entry. Numerous antibodies target glycoprotein (especially ZMapp, mAb114, and REGN-EB3). Several selective estrogen receptor modulator compounds are much newer and more precise in their inhibitory action on EBOV infection. Clomiphene and toremifene were discovered as successful blockers of EBOV infection subsequently in in vitro and in vivo investigation of selective estrogen receptor modulators [37].

(b) *Targeting viral RNA synthesis machinery*

To pinpoint the viral RNA production, drugs that straightaway target the viral counterparts of the RNA synthesis machinery, as well as agents that target cell contributions to the RNA synthetic chain of the cycle, are utilized. Although our lack of understanding of the physiological components that add to viral RNA generation in EBOV persists in hindering development, both techniques have shown great promise, and the viral replication machinery is a highly appealing target.

B. *Vaccines*

Infections in primary care officials such as physicians and various other health care workers who are in the proximity of infected population need to be avoided. For this very reason, an effective EBOV vaccine is needed, especially in high-risk locations. Three Ebola vaccine candidates have been approved, and three more have finished or are about to complete phase I trials, and two vaccines are now in the phase II of the trials. The final one has conquered phase III of clinical trials. The quest for safe and effective vaccines must continue considering ongoing reports of EVD outbreaks as epidemics in Africa [38]. The delivery of a vaccine, however, is only one of several measures aimed at containing the Ebola spread, according to the WHO. The following section discusses the many vaccines and vaccine platforms being investigated to develop an effective EBOV vaccine.

(a) *Inactivated vaccine*

Even though inactivated vaccines are subject to causing disease in a healthy human being owing to inadequate viral inactivation, numerous plans for emerging innocuous and efficient, nonduplicating vaccine contenders to prevent EBOV infection are being investigated. In a guinea pig model, both heat- and formalin-inactivated EBOV was reported to be protective against EBOV infection. An EBOV researcher's life was saved thanks to the inclusion of an inactivated EBOV E-178 vaccine, as well as interferon (IFN) and immunological plasma [39].

(b) *DNA vaccine*
Plasmids are used in DNA vaccines to produce immunogenic antigens. This vaccine approach is appealing owing to its ease of manufacture and simplicity. Furthermore, DNA vaccines tend to stimulate both humoral as well as cellular immune responses [40]. The development of a low immunity titer for a shorter window would need repeated vaccinations to overcome this problem, even though the first studies utilizing DNA constructs demonstrated acceptable safety profiles. As a result, using a powerful immunization schedule based on DNA vaccine platforms for a broad population does not appear to be rational [41].

C. *Virus-like particles*
EBOV-like particles VLPs are made up of viral transmembrane GP and structural matrix protein in mammalian cells (VP40). They self-assemble and exit from host cells, resembling infectious EBOV particles. Virus-like replicon particles are a better option than live-attenuated vaccinations. The use of this category of particles is to avoid the possibility of live vaccination strains reverting to their original pathogenic state [42].

D. *Recombinant viral vector vaccines*
A fully functional viral vector backbone is built to express an antigen from a foreign transgene in the replication of recombinant vector vaccines. The transgene is not only unneeded, but it may be deleterious to viral propagation [43]. As a result, vaccine revertants that delete or inactivate the transgene could develop to dominate the viral vaccine population during vaccine manufacture and host infection. One source of concern is that the vaccine's antigenicity (the level of immunity) induced against the transgene could be reduced because of its evolution.

Various vaccines have been approved, and some are already in the pipeline for approval. Ervebo, or the VSV-ZEBOV-Ebola Kikwit replication-competent vaccine, is being administered through a single dose, and the vector used is the vesicular Indiana virus. Second to none is the MVA-BN-filo vaccine, which encodes Ebola, Sudan, and Marburg virus GPs as well as Taï Forest virus nucleoprotein utilizing human adenoviral serotype 26 as a vector with a heterological prime-boost regimen. For the Mayinga strain (1976) of EBOV (ChAd3-EBO-Z) with or without MVA-BN-Filo (Ad5-ZEBOV) Makona strain of EBOV (2014), serotype 3 or 5 of adenoviral infection in chimps was taken. The doses administered are homologous prime-boost regimens. Another candidate utilized VSV and Ad5 as vectors for monovalent Zaire (Makona) (GamEvac-Combi and GamEvac-Lyo) (NVX-CoV2373). Nanoparticle recombinant Ebola GP Vaccine and Monovalent Zaire (Makona) contains the fulllength SARSCoV-2 spike protein and MatrixM1 adjuvant. The dosage decided on for administration is two doses with a gap of 21 days. Another double-dose vaccine candidate is plasmid of Ebola epidemic strains from 1976 to 2006 (INO-4201 DNA vaccine) with the same regimen of two doses administered at a gap of 21 days [44].

E. *Repurposed drug*
   Existing medicines can be tested for effectiveness against pathogens through drug repurposing. Because there are no authorized EBOV treatments, the screening of potentially effective medicines indicated that only a few of them might be repurposed for EBOV therapy [45]. Amiodarone, dronedarone, and verapamil, which are used to treat tachycardia, arrhythmias, and high blood pressure or angina, were investigated for their ability to block filoviruses from entering cells in in vitro models and shown to be effective. Statins, angiotensin-converting enzyme inhibitors, and angiotensin receptor blockers have all been recommended as ways to decrease the severity of EBOV infection [46]. Chloroquine and its structural analogs (hydroxychloroquine, pamaquine, primaquine, and plasmoquine) are anti-malarial medicines that operate as osmotrophic agents, inhibiting endosomal/lysosomal acidification and therefore restricting viral contagions.

F. *Gene expression inhibitors*
   For the expression of the viral gene, which is required for virus replication, the host cell machinery is required. BCX4430 (a nucleoside analog) is a viral RNA polymerase inhibitor that protects mice from contracting the fatal EBOV virus [47].

G. *Interferon*
   In vitro experiments using several types of cells have shown that interferons are effective EBOV inhibitors. IFN-1a promotes viral clearance from the bloodstream, resulting in the early remission from illness symptoms [48].

H. *Supportive care*
   This form of treatment is an additive support to the course of treatment, which includes oral rehydration, emphasizes re-establishing the fluid and electrolyte levels. Alongside it, the vital signs as well as the serum biochemistry are monitored, while providing emotional support to family and friends, which could develop a strong support system that would help the patient to recover.

4 Diagnosis

When mucous membranes or abrasions on the skin come into contact with the blood or bodily fluids of infected people, they contract EBOV. EBOV has a broad tropism for several cell types after acquisition, with early replication predominantly occurring in dendritic cells, macrophages, and monocytes. As the infection develops, it spreads to the liver, lymph nodes, and spleen. As the virus spreads, it inhibits the immune system and causes the coagulation cascade to malfunction, resulting in organ failure. All the above-mentioned forms the basis for early diagnosis [24]. The blood test consists of a complete cell count in the blood, a urine analysis, which would clarify the renal function, detecting the presence of the EBOV antigen by PCR testing and ELISA, and also by detecting the presence of antibodies and coagulation profile [49]. The clinical representation and laboratory testing, along with serological testing a low platelet count, show an increase in the enzymes alanine aminotransferase and aspartate aminotransferase in the liver, as well as hemorrhagic coagulation difficulties, which are frequently associated with disseminated intravascular coagulation (DIC) and prolotherapy. OraQuick Ebola Rapid Antigen Test (OraSure Technologies, Inc.), SD Q Line Ebola (SD BioSensor, Inc.), and the DPP Ebola Antigen assays are a few examples of rapid testing sets. With early diagnosis there are chances of differential diagnosis too. This entails differentiating one illness or condition from others with comparable clinical characteristics. Clinicians utilize differential diagnostic methods to identify a patient's specific ailment or, at the very least, to rule out any life-threatening diseases.

ADI-15946 is a monoclonal antibody that was recently found to neutralize EBOV [50]. The current work employed molecular dynamics (MD) simulation and the Poisson–Boltzmann surface area of molecular mechanics in the MMPBSA study to see how the EBOV receptor GPcl interacts with ADI-15946 on a molecular level. The key driving mechanism for ADI-15946 binding on EBOV has been found to be a hydrophobic interaction. The contribution of each amino acid residue to the binding was also evaluated. The advantageous residues for binding were then used to create an affinity binding model (ABM), such as Y107, F108, D109, W110, and R113. Using the ABM of ADI 15946, a biology mimetic design of an EBOV neutralizer was then carried out [21].

The nucleocapsid is crucial to the survival of EBOV. The Ebola nucleocapsid is a single-stranded viral RNA encased in a helical NP structure. Understanding the genetic mechanisms that affect Ebola nucleocapsid stability is critical for anti-EBOV therapy development [51]. Owing to the many degrees of freedom associated with the Ebola nucleocapsid helical assembly, previous modeling investigations were limited to monomers. Ions affect the EBOV nucleocapsids' electrostatic potential and contribute to its stability. In areas of strong electrostatic potential, ions around the formed nucleocapsid were shown to exhibit large occupancies. It describes how ions, NP–NP interactions, and NP–RNA interactions affect the robustness of EBOV nucleocapsids. Viral protein 35 (VP35), the main protein of the *Zaire ebolavirus*, interacts with many human proteins, weakening the human immune system [52]. Despite its significance, the exact structure of EBOV VP35's tetrameric assembly and the mechanisms by which it inhibits autophosphorylation of the region of human protein kinase R are unknown. Across all four chains of the tetramer, the simulations show extremely symmetrical behavior among identical residues. Five symmetrical inter-chain salt bridge networks and two inter-chain disulfide bond networks link neighboring chains and are required for the overall stability of the viral protein assembly [53].

5 Conclusion

Future work should focus on increasing the efficiency and efficacy of clinical trials in general (e.g., by implementing adaptive designs) as well as depending on individual genetic markers or other personalized characteristics. Bringing research drug eligibility down to the individual level has the advantage of proving greater efficacy, but it also has the disadvantage of restricting the available sample size to focus study medication usage more quickly to targets of opportunity.

Filoviruses that are resistant to current vaccinations and conventional virus-directed therapies are unknown treatment approaches, particularly those that rely on antibodies, which could cause future outbreaks. As a result, a greater focus on the development of broad-based active intervention techniques is required. This can be done by looking for antiviral targets in filoviruses, or by targeting host components that are shared by filoviruses in general, or perhaps a broader group of viruses. Many of the challenges associated with the traditional "one bug, one medicine" strategy, such as adopting the latter method, the risks presented by novel viral species, and the limited resources available, can be avoided.

References

1. Baseler L, Chertow DS, Johnson KM, et al. The pathogenesis of Ebola virus disease. Ann Rev Pathol. 2017;12:387–418.
2. Marzi A, Mire CE. Current Ebola virus vaccine progress. BioDrugs. 2019;33(1):9–14.
3. Jadav SS, Kumar A, Ahsan MJ, et al. Ebola virus: current and future perspectives. Infect Disord Drug Targets. 2015;15(1):20–31.
4. Martines RB, Ng DL, Greer PW, et al. Tissue and cellular tropism, pathology and pathogenesis of Ebola and Marburg viruses. J Pathol. 2015;235(2):153–74.
5. Reynolds P, Marzi A. Ebola and Marburg virus vaccines. Virus Genes. 2017;53(4):501–15.
6. Kortepeter MG, Dierberg K, Shenoy ES, et al. Marburg virus disease: a summary for clinicians. Int J Infect Dis. 2020;99:233–42.
7. Martell HJ, Masterson SG, McGreig JE, et al. Is the Bombali virus pathogenic in humans? Bioinformatics. 2019;35(19):3553–8.
8. Muñoz LS, Garcia MA, Gordon-Lipkin E, et al. Emerging viral infections and their impact on the global burden of neurological disease. Semin Neurol. 2018;38(2):163–75.
9. Siddharta A, Pfaender S, Vielle NJ, et al. Virucidal activity of World Health Organization-recommended formulations against enveloped viruses, including Zika, Ebola, and Emerging Coronaviruses. Int J Infect Dis. 2017;215(6):902–6.
10. Thys S, Boelaert M. The origin of Ebola: biomedical approach versus popular interpretations in Macenta. Guinea Sante Publique. 2017;29(4):497–507.
11. Shoman H, Karafillakis E, Rawaf S. The link between the West African Ebola outbreak and health systems in Guinea, Liberia and Sierra Leone: a systematic review. Glob Health. 2017;13(1):1.
12. Formenty P, Boesch C, Wyers M, et al. Ebola virus outbreak among wild chimpanzees living in a rain forest of Côte d'Ivoire. Int J Infect Dis. 1999;179(Suppl 1):S120–6.
13. Holmes EC, Dudas G, Rambaut A, et al. The evolution of Ebola virus: insights from the 2013-2016 epidemic. Nature. 2016;538(7624):193–200.

14. Yu C, Li S, Zhang X, et al. MARCH8 inhibits Ebola virus glycoprotein, human immunodeficiency virus type 1 envelope glycoprotein, and Avian Influenza Virus H5N1 Hemagglutinin maturation. mBio. 2020;11(5):e01882–20.
15. Jacob ST, Crozier I, Fischer WA 2nd, et al. Ebola virus disease. Nat Rev Dis Primers. 2020;6(1):13.
16. Wan W, Kolesnikova L, Clarke M, et al. Structure and assembly of the Ebola virus nucleocapsid. Nature. 2017;551(7680):394–7.
17. Wong G, Zhang Z, He S, et al. Marburg and Ravn virus infections do not cause observable disease in Ferrets. Int J Infect Dis. 2018;218(suppl_5): S471–4.
18. Alazard-Dany N, Ottmann Terrangle M, Volchkov V. Ebola and Marburg viruses: the humans strike back. Med Sci (Paris). 2006;22(4):405–10.
19. Charbe NB, Amnerkar ND, Ramesh B, et al. Small interfering RNA for cancer treatment: overcoming hurdles in delivery. Acta Pharm Sin B. 2020;10(11):2075–109.
20. De Clercq E. Interferon: ten stories in one. A short review of some of the highlights in the history of an almost quinquagenarian. Acta Microbiol Immunol Hung. 2005;52(3–4):273–89.
21. Zhu W, Banadyga L, Emeterio K, et al. The roles of Ebola virus soluble glycoprotein in replication, pathogenesis, and countermeasure development. Viruses. 2019;11(11):999.
22. Dhama K, Karthik K, Khandia R, et al. Advances in designing and developing vaccines, drugs, and therapies to counter Ebola virus. Front Immunol. 2018;9:1803.
23. van Griensven J, De Weiggheleire A, Delamou A, et al. The use of Ebola convalescent plasma to treat Ebola virus disease in resource-constrained settings: a perspective from the field. Clin Infect Dis. 2016;62(1):69–74.
24. Marcinkiewicz J, Bryniarski K, Nazimek K. Ebola haemorrhagic fever virus: pathogenesis, immune responses, potential prevention. Folia Med Cracov. 2014;54(3):39–48.
25. Shayan S, Bokaean M, Shahrivar MR, et al. Crimean-Congo hemorrhagic fever. Lab Med. 2015;46(3):180–9.
26. Cole E, Gillespie S, Vulliamy P, et al. Multiple organ dysfunction after trauma. Br J Surg. 2020;107(4):402–12.
27. Möller MN, Rios N, Trujillo M, et al. Detection and quantification of nitric oxide-derived oxidants in biological systems. J Biol Chem. 2019;294(40):14776–802.
28. Malvy D, McElroy AK, de Clerck H, et al. Ebola virus disease. Lancet. 2019;393(10174):936–48.
29. Furuyama W, Marzi A. Ebola virus: pathogenesis and countermeasure development. Annu Rev Virol. 2019;6(1):435–58.
30. Takada A. Ebola vaccine and treatment. Uirusu. 2015;65(1):61–70.
31. Matz KM, Marzi A, Feldmann H. Ebola vaccine trials: progress in vaccine safety and immunogenicity. Expert Rev Vaccines. 2019;18(12):1229–42.
32. Lu DY, Wu HY, Yarla NS, et al. Ebola therapeutic study and future directions. Infect Disord Drug Targets. 2019;19(1):17–29.
33. Muñoz-Fontela C, McElroy AK. Ebola virus disease in humans: pathophysiology and immunity. Curr Top Microbiol Immunol. 2017;411:141–69.
34. Kang YL, Chou YY, Rothlauf PW, et al. Inhibition of PIKfyve kinase prevents infection by Zaire ebolavirus and SARS-CoV-2. Proc Natl Acad Sci U S A. 2020;117(34):20803–13.
35. King LB, Milligan JC, West BR, et al. Achieving cross-reactivity with pan-ebolavirus antibodies. Curr Opin Virol. 2019;34:140–8.
36. Sharma A, Castellani RJ, Smith MA, et al. 5-Hydroxytryptophan: a precursor of serotonin influences regional blood-brain barrier breakdown, cerebral blood flow, brain edema formation, and neuropathology. Int Rev Neurobiol. 2019;146:1–44.
37. Arao Y, Korach KS. Transactivation Function-1-mediated partial agonist activity of selective estrogen receptor modulator requires homo-dimerization of the estrogen receptor α ligand binding domain. Int J Mol Sci. 2019;20(15):3718.
38. Tambo E, Chengho CF, Ugwu CE, et al. Rebuilding transformation strategies in post-Ebola epidemics in Africa. Infect Dis Poverty. 2017;6(1):71.

39. Clarke DK, Xu R, Matassov D, et al. Safety and immunogenicity of a highly attenuated rVSVN4CT1-EBOVGP1 Ebola virus vaccine: a randomised, double-blind, placebo-controlled, phase 1 clinical trial. Lancet Infect Dis. 2020 Apr;20(4):455–66.
40. O'Donnell K, Marzi A. The Ebola virus glycoprotein and its immune responses across multiple vaccine platforms. Expert Rev Vaccines. 2020;19(3):267–77.
41. Chalkias S, Gorham JM, Mazaika E, et al. ViroFind: a novel target-enrichment deep-sequencing platform reveals a complex JC virus population in the brain of PML patients. PLoS One. 2018;13(1):e0186945.
42. Warfield KL, Posten NA, Swenson DL, et al. Filovirus-like particles produced in insect cells: immunogenicity and protection in rodents. J Infect Dis. 2007;196(Suppl 2):S421–9.
43. Yadav T, Srivastava N, Mishra G, et al. Recombinant vaccines for COVID-19. Hum Vaccin Immunother. 2020;16(12):2905–12.
44. Martins KA, Jahrling PB, Bavari S, Kuhn JH. Ebola virus disease candidate vaccines under evaluation in clinical trials. Expert Rev Vaccines. 2016 Sep 1;15(9):1101–12.
45. De Clercq E. Ebola virus (EBOV) infection: therapeutic strategies. Biochem Pharmacol. 2015;93(1):1–10.
46. Menicucci AR, Jankeel A, Feldmann H, et al. Antiviral innate responses induced by VSV-EBOV vaccination contribute to rapid protection. mBio. 2019;10(3):e00597–19.
47. Taylor R, Kotian P, Warren T, et al. BCX4430 - a broad-spectrum antiviral adenosine nucleoside analog under development for the treatment of Ebola virus disease. J Infect Public Health. 2016;9(3):220–6.
48. Johnson JC, Martinez O, Honko AN, et al. Pyridinyl imidazole inhibitors of p38 MAP kinase impair viral entry and reduce cytokine induction by Zaire ebolavirus in human dendritic cells. Antivir Res. 2014;107:102–9.
49. Kofman A, Linderman S, Su K, et al. Characteristics of Ebola virus disease survivor blood and semen in Liberia: serology and RT-PCR. Clin Infect Dis. 2020:ciaa1331.
50. Hou Q, Zhang L. Biomimetic Design of Peptide Neutralizer of Ebola virus with molecular simulation. Langmuir. 2020;36(7):1813–21.
51. Xu C, Katyal N, Nesterova T, et al. Molecular determinants of Ebola nucleocapsid stability from molecular dynamics simulations. Chem Phys. 2020;153(15):155102.
52. Hume A, Mühlberger E. Marburg virus viral protein 35 inhibits protein Kinase R activation in a cell type-specific manner. J Infect Dis. 2018;218(suppl_5): S403–8.
53. Banerjee A, Mitra P. Ebola virus VP35 protein: modeling of the tetrameric structure and an analysis of its interaction with human PKR. J Proteome Res. 2020;19(11):4533–42.

Drug Delivery Options for Treatment of Ebola Infection

Harshita Krishnatreyya, Hemanga Hazarika, Bhrigu Kumar Das, Neelutpal Gogoi, Abdul Baquee Ahmed, and Kamaruz Zaman

Abstract In West Africa, the Ebola virus disease (EVD) outbreak between 2013 and 2016 is considered as one of the deadliest re-emerging viruses, which began after human contact with infected bats, the host of the Ebola virus (EBOV). WHO has inventoried and provided an updated list of probable drug candidates that demonstrate antiviral efficacy either in vitro or in animal models. However, till date, no anti-EBOV agents have confirmed efficacy on humans. Plant-based drugs mainly comprise of standardized plant extract or plant materials. But regarding the treatment of Ebola, no such products have been developed to date. The success rates for drug discovery process involving target identification, screening of compounds, lead optimization, and preclinical drug candidate selection are within the range of 69% to 85%. We have discussed several therapeutic agents, some having direct anti-EBOV efficacy (favipiravir and BCX4430), interference-based drugs (TKM-Ebola), etc. Nanotechnology is considered one of the most capable technologies because of its effective nature against viral diseases, and there are different categories of nanomaterials which have served as delivery vehicles for antiviral drugs. These include several polymer-based, lipid-based, lipid-polymer-based, inorganic metal–based, carbon-based, stimuli-sensitive materials, etc. and their applications in antiviral therapeutics. Repurposing of antiviral drugs and small molecules to extensively modify their structures and structure-activity relationship (SAR) analyses can help focus on EBOV disease eradication. The capacity to rapidly test these drug com-

H. Krishnatreyya
National Institute of Pharmaceutical Education and Research, Guwahati, SilaKatamur (Halugurisuk), Changsari, Kamrup, Assam, India

H. Hazarika · A. B. Ahmed
Girijananda Chowdhury Institute of Pharmaceutical Science-Tezpur, Dekargaon, Tezpur, Assam, India

B. K. Das
Girijananda Chowdhury Institute of Pharmaceutical Science-Guwahati, Azara, Guwahati, Assam, India

N. Gogoi · K. Zaman (✉)
Department of Pharmaceutical Sciences, Dibrugarh University, Dibrugarh, Assam, India

R. Shegokar, Y. Pathak (eds.), *Viral Drug Delivery Systems*,
https://doi.org/10.1007/978-3-031-20537-8_8

pounds in human trials during epidemics is however, a major concern. Nevertheless, there is an imperative need for active collaboration between industry, academia, and health regulatory organizations to accelerate and share resources that might help develop antivirals against EBOV.

Keywords Ebola · Antiviral · Drug development · Nanoformulations

1 Introduction

The Ebola virus disease (EVD) outbreak in West Africa between 2013 and 2016 appeared to be the deadliest re-emerging virus, probably beginning after human contact with an animal (presumably a bat) host of the Ebola virus (EBOV) [9]. It has received global attention due to its high contagious nature and spread, causing acute hemorrhagic fever, with a case fatality rate (CFR) reaching 90% [97]. The name of the virus originated from the Ebola river in the Democratic Republic of the Congo (previously Zaire), where the first cases of hemorrhagic fever were reported in 1976. Ever since, the virus has infected people regularly, causing episodes of EVD in various African countries. Between 2013 and 2016, an outbreak in West Africa and other affected countries reported 28,652 cases and 11,325 deaths, confirming the significant fatality of this disease [16].

1.1 Ebola Biology

1.1.1 Taxonomy

The Ebola virus is a member of the *Filoviridae* family with a negative-sense single-stranded RNA structure (–ssRNA), from the order *Mononegavirales* and genus *Ebolavirus*. It is composed of five species as tabulated in Table 1 [59]. The family *Filoviridae* derives its name from the Latin word *filum*, which means "thread," because the virion's form mimics a twisted thread when observed through an electron microscope [6]. The *Zaire ebolavirus* (ZEBOV) has the highest mortality rate

Table 1 Taxonomy of Ebola virus (EBOV) [59]

Order	*Mononegavirales*
Family	*Filoviridae*
Genus	*Ebolavirus*
Species	*Zaire ebolavirus* (ZEBOV) *Sudan ebolavirus* (SEBOV) *Tai Forest ebolavirus* (TEBOV; formerly Cote d'Ivoire Ebola virus) *Bundibugyo ebolavirus* (BEBOV) *Reston ebolavirus* (REBOV)

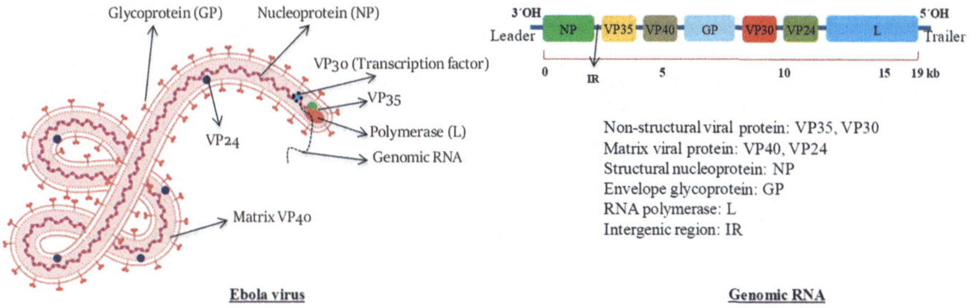

Fig. 1 Ebola virus and genomic negative RNA strand of EBOVs. (Image has been prepared in Biorender.com and Microsoft PowerPoint)

among the EBOV genus (up to 90%). This proportion is substantially lower (approx. 25% and 53%) for the *Bundibugyo ebolavirus* (BEBOV) and *Sudan ebolavirus* (SEBOV). The *Reston ebolavirus* (REBOV) has been identified with the most common respiratory and reproductive illness manifestation in nonhuman primates (NHP) [14, 114].

1.1.2 Genome and Structure

This negative-strand RNA EBOV with a size of roughly 19 kb, a length of 10,000 to 14,000 nm, and a diameter of 50 to 80 nm varies in shape from branch to cylinders or loops. All filoviruses, however, retain a unique thread-like filamentous structure. The genome sequence comprises seven sequentially arranged genes, as shown in Fig. 1. It begins with a 3′OH leader site and ends with a 5′OH trailer site. It encodes with one structural nucleoprotein (NP), non-structural viral protein (VP35 and VP30), viral matrix protein (VP40 and VP24), envelope glycoprotein (GP), and RNA polymerase (L). The virion surface structural glycoprotein (GP) encodes three glycoproteins and is involved in receptor binding and viral entry. Most genes are separated by intergenic regions (IR) of varying lengths [39, 71, 124]. Each of these proteins serves a specific function. The ribonucleoprotein-RNP complex (i.e., NP, VP35, VP30, and RNA polymerase L) is required for viral transcription and replication. The GP, VP40, and VP24, on the other hand, are membrane proteins that promote the formation of filamentous virions [98].

1.1.3 Pathogenesis

Cellular Growth and Viral Invasion

In humans, EBOV disease starts by contacting infected body fluids with skin lesions or mucous membranes, which allow virions to enter through direct contact with the cell surface. Immature dendritic cells (DCs), adrenal cells, macrophages, endothelial cells, monocytes, and Kupffer cells in the liver are all infected by the virions [37]. Several attachment factors/co-receptors of the EBOVs such as glycosylated protein (GP1), C-type lectins, T-cell Ig and mucin 1 (TIM-1), glycosaminoglycans, β_1-Integrin, Axl, folate receptor-α, Tyro3, and Mer have been documented for their ability to bind to a range of host-cell proteins [2, 18, 25, 83, 104]. In addition, macropinocytosis, a non-specific process of endocytosis, facilitates the uptake and processing of the virion by the target cells after attachment, which stimulates the cytoskeleton rearrangement and causes ruffling of the plasma membrane and invasion [1, 19, 76].

Genome: Transcription, Translation, and Replication

The nucleocapsid is released into the cytosol after cellular uptake, where viral RNA is transcribed into mRNA by host cell activity. The development of stem loops at the 5′ end and the assembly of the lengthy non-coding area at the 3′ and 5′ end enhance the stability of EBOV mRNA. The different types of virion proteins such as structural nucleoprotein (NP), non-structural viral protein (VP35 and VP30), viral matrix protein (VP40 and VP24), envelope glycoprotein (GP), and RNA polymerase (L) are translated from the mRNA. This, in turn, allows viral genome replication [77].

Assembling and Releasing

The newly synthesized proteins undergo various post-translational changes and are found in the membrane alongside the viral DNA. Viral VP40 interacts with the host ESCRT machinery (endosomal sorting complexes required for transport), enabling budding and the release of a new virion [129].

1.2 Preclinical Studies: Rodents and Small Animals Mimicking Disease Condition

In the process of designing clinical trials and generating a safe and marketable product, potential therapeutic substances are moved to preclinical studies. The promising therapeutic agents developed for the treatment of EBOVs are evaluated in different animal models mimicking the disease condition. This includes rodents like mice, small animals like guinea pigs, and/or non-human primates (NHPs). But these models lack innate immune population response (in case of guinea pig) and hemorrhage and clotting problems (in case of mice), along with the non-lethality of wild-type EBOVs in immunocompromised mice and/or guinea pigs. Henceforth, the usage of adapted virus from serial passage served the purpose for better response of the therapeutic agents in animals. The limitations linked with the various species influence the selection of an animal model sometimes. Further, the efficiency of these agents in the animals is measured on the basis of weight loss, longevity, and average duration to death as endpoints. The rodent models can also be used to validate the possible mechanisms by directly targeting viruses as one of the therapeutic options. The clinical manifestations of EBOV disease condition in humans, like hemorrhage and coagulation, can be better reproduced in the NHP model [11, 32]. Table 2 summarizes the different preclinical in vivo evidence for EBOV treatment.

Table 2 In vivo preclinical evidence for Ebola virus (EBOV) treatment

Drug/chemical name/ code	Targeting EBOVs/species	Inference	References
AVI-7539, AVI-7537, and AVI-6002	VP24 alone/ Rhesus monkeys	The drugs targeting VP24 protein alone were effective in protecting against deadly EBOV infection on the basis of survival rate	[123]
Convalescent-phase blood	Whole Zaire EBOV/Rhesus macaques	There was no benefit from transfusion of convalescent-phase blood from rhesus macaques susceptible to ZEBOV infection	[48]
Recombinant nematode anticoagulant protein c2 (rNAPc2)	Whole EBOVs/ Rhesus macaques	The post-exposure treatment with the anticoagulant protein rNAPc2 against EBOVs showed a significant survival rate in non-human primates (NHP) ↓ in the activation of fibrinolysis and coagulation ↓ concentrations of IL-6 and MCP-1	[38]
rhAPC (Drotrecogin alfa; activated)	Whole Zaire EBOV/Rhesus macaques	Significant increase in survival/average time to death for the rhAPC-treated macaques when compared to ZEBOVs control	[40]
Recombinant human mannose-binding lectin (rhMBL)	Whole Zaire EBOV/C57B6 mice	Mice with sevenfold higher rhMBL serum concentrations developed immunity to viral rechallenge compared to the controls	[73]
Bepridil, sertraline	Murine Ebola virus/C57BL/6 mice	Prevented a late stage of viral penetration and showed a significant survival benefits in a murine EBOV infection model	[51]
Clomiphene, toremiphene	Murine Ebola virus/C57BL/6 mice	Both clomiphene and toremifene exhibited a statistically significant survival benefit compared to the control animals	[51]

Abbreviations: *VP24* Viral matrix protein; *EBOV* Ebola virus; *ZEBOV* Zaire EBOV; *IL-6* interleukin-6; *MCP-1* monocyte chemoattractant protein-1

2 Pharma Development for the Treatment of Ebola Virus (EBOV)

It is a filovirus, a class of zoonotic pathogens related with severe diseases in humans. They are filamentous, encased with non-segmented, negative-sense RNA genomes. The virus has six species, among which *Zaire ebolavirus*, *Sudan ebolavirus* (SUDV), and *Bundibugyo ebolavirus* (BDBV), which trigger substantial outbreaks in humans, causing substantial illnesses. The largest filovirus epidemic was caused by EBOV in West Africa between 2013 and 2016. This resulted in more than 28,000 infections, more than 11,000 deaths, and the export of infected cases to the United States and Europe [29]. Besides strong pathogenicity, the fatality rates (up to 90%) of EBOV are due to the absence of suitable vaccines and therapeutics to combat the EBOV epidemic. The current approach in designing antivirals is based on detailed knowledge of a given virus's life cycle, genome, proteome, and vital enzymes that can be interrupted or disabled [46]. In the absence of appropriate specific treatment, contemporary medical care chiefly relies on thorough supportive care, particularly with electrolytes and intravenous, oral rehydration to preserve intravascular volume. Management of sepsis and blood transfusion must also be considered at such times [67].

Development and advancement of anti-filovirus therapeutics is demonstrated by studies of drug efficacy in cell cultures, followed by studies in experimental animals like mice, guinea pigs, etc. But non-human primates (NHPs) are considered the "gold standard" for pharmacological efficacy testing of antiviral drugs in animals [93]. Monoclonal antibodies and nucleic acid-based therapeutics have been able to successfully protect NHPs from fatal filovirus challenge. Since September 2014, the WHO has inventoried and provides an updated list of probable drug candidates that demonstrate antiviral efficacy either in vitro or in animal models [67]. However, till date, no anti-EBOV agents have confirmed efficacy on humans.

In the subsequent session, we have discussed several therapeutic agents, some having direct antiviral efficacy (favipiravir and BCX4430), Brincidofovir type I interferons, and RNA interference-based drugs (TKM-Ebola). For each drug candidate, we have discussed very briefly about its structure/composition, mode of action, pharmacokinetic features in humans and/or animals, and existing data on safety and efficacy in non-human primates (NHP) studies [67].

2.1 Drug Candidates

2.1.1 Favipiravir

Toyama Chemical Co Ltd., Japan, developed the broad-spectrum antiviral drug Favipiravir (T-705), which is currently in phase III of clinical development in the USA for treatment of resistant flu [65]. It is a purine nucleic acid analogue, which is phosphorylated intracellularly into its active form, T-705RTP, which then interferes with viral replication by inhibiting the RNA polymerase [35]. Pharmacokinetic study of favipiravir was firstly considered in Japanese healthy volunteers in several doses ranging from 30 to 2400 mg for single administration and 800 to 1200 mg daily for repeated administration. After a single oral dose, favipiravir concentration reaches C_{max} within 2 h and then declines with an elimination rate corresponding to a half-life of 2–5.5 h. Favipiravir is metabolized by aldehyde oxidase, leading to the inactive metabolite T705M1, excreted in hydroxylated forms by the kidney. The portion of metabolites excreted in the urine surges over time to reach 80–100% after 7 days. Reported adverse events of favipiravir (while in use for influenza treatment) include diarrhea, asymptomatic rise of blood transaminases and uric acid, reduction of neutrophil counts, etc. [74]. Favipiravir has shown high efficacious activity against EBOV in vitro. It excellently blocked the creation of infective virus with an EC_{50} of 10 μg/mL in in vitro experiments using wild-type Zaire EBOV Mayinga 1976 strain and Vero E6 cells [85]. Favipiravir recorded positive results during preclinical testing in murine models against EBOV. The sturdy antiviral effect of favipiravir was established in a pharmacokinetic-viral kinetic model that showed efficiency in blocking 99.6% of viral production at steady state. Also, ongoing studies in NHP models are yet to be available [68].

2.1.2 BCX4430

BioCryst Pharmaceuticals, USA, developed BCX4430, which is a broad-spectrum antiviral originally intended for treating hepatitis C virus but was subsequently developed for the treatment of filovirus infections such as EBOV. BCX4430 is metabolized into its active triphosphate form, BCX4430-TP, which reduces viral RNA production by preventing the RNA polymerase activity.

PK: The pharmacokinetics study of BCX4430, in doses ranging from 2 to 50 mg/kg, was evaluated in animal models. BCX4430 concentration decreased rapidly in the plasma of rodents and cynomolgus macaques with a half-life of 5–10 min. The metabolite, BCX4430-TP, however, showed a longer half-life of 6.2 hours when studied in liver of rats. Increased bioavailability and faster absorption via the intramuscular route were observed in experimental animals [122]. Further, in vitro experiments showed no signs of mutagenicity and chromosomal aberrations in human lymphocyte by BCX4430 [128].

Efficacy: The pharmacological efficacy of BCX4430 in countering EBOV infection was evaluated in two different NHP models [44, 67]. With an EC50 of 3.13 μg/mL, BCX4430 depicted a strong in vitro antiviral effect against EBOV infection when studied using HeLa cells and EBOV Kikwit strain [128].

2.1.3 Brincidofovir

The drug brincidofovir exerts its therapeutic activity as an inhibitor of DNA polymerase and RNA polymerase, which are vital for EBOV replication [42]. It is under development by Chimerix, USA, and has already undergone clinical trials to assess its safety and efficacy against several viral diseases. Although EBOV does not belong to the DNA virus type, in vitro tests of brincidofovir exhibited inhibition of EBOV infection as well [34]. The lipid moiety in this drug plays an important role by increasing its cellular uptake, thereby increasing intracellular levels of the drug. Nucleotide analogues comprising phosphonate moieties have long been thought to be utilizable as antiviral agents [22]. Among them, cytosine and cidofovir selectively prevent viral DNA polymerases and has been used for treating viral infections [110]. A clinical trial was scheduled for brincidofovir in late 2014 but was later withdrawn [46].

2.1.4 TKM-Ebola

Arbutus biopharma developed TKM-Ebola, which belongs to a newer therapeutic class based on RNA interference technology. This drug is composed of two minor interfering RNA, siLpol-2 and siVP35-2, sequences which correspond to that of EBOV viral polymerase and VP35 genes separately. siRNA are unstable and hence encapsulated using lipid nanoparticles coated with polyethylene glycol [116]. In TKM-Ebola, the two siRNA constrain mRNA translation and silence the corresponding viral genes and enhance host cell-mediated viral mRNA destruction [15]. The pharmacokinetics of TKM-Ebola was considered in healthy volunteers with doses ranging from 0.075 to 0.5 mg/kg. siLpol-2 and siVP35–2, the two siRNA, showed comparable plasma concentration time profiles, signifying the distribution and metabolism of lipid nanoparticles that affect drug kinetics. Initial data from 24 patients suggested a greater-than-dose-proportional upsurge in C_{max} and in the area under curve (AUC). The efficacy of TKM-Ebola was established in vitro using both Kikwit and Guinea strains on HepG2 cells, with EC_{50} stated between 50 and 250 ng/mL [116]. In the USA, TKM-Ebola was used as compassionate treatment for two adult patients along with supportive care and convalescent plasma, and the patients survived despite certain clinical and biological complications [58]. Lately, there has been news of another small-molecule drug – triazavirin – to show substantial anti-EBOV activity [23]. Considering the severity of the EBOV infection diseases in humans, intense upcoming research in this field are anticipated to be successful in delivering necessary progress.

2.1.5 Inmazeb

On October 14, 2020, the US Food and Drug Administration (FDA) approved Inmazeb, a mixture of three monoclonal antibodies: atoltivimab, maftivimab, and odesivimab-ebgn, as the first FDA-approved treatment for EBOV infection in pediatric and adult patients. Inmazeb targets the surface glycoprotein of Ebola virus and binds to this glycoprotein, blocking the attachment of viral and host cell membranes and thereby preventing the entry of the virus.

Three hundred and eighty-two pediatric and adult patients with confirmed EBOV infection were treated with Inmazeb in the PALM trial conducted in the Democratic Republic of the Congo (DRC) EBOV outbreak during 2018–2019. In this trial, the safety and efficacy of Inmazeb was evaluated in an open-label, multicenter, randomized controlled trial, in which 154 patients received 50 mg of Inmazeb IV as a single infusion and 168 patients received an investigational control. Twenty-eight-day mortality was considered as the endpoint in testing Inmazeb's primary efficacy. Of the 154 patients who received Inmazeb, 33.8% died after 28 days, contrary to 51% death from among 153 patients who received the investigational control. Common symptoms experienced while receiving Inmazeb include chills, fever, tachypnea, tachycardia and vomiting, symptoms that are also common for EBOV infection. Patients receiving Inmazeb may have hypersensitivity reaction, including infusion-related events, in the event of which, the treatment should be discontinued [119].

2.2 *Advances in Drug Delivery*

Having had a brief knowledge of probable candidates for the treatment of EBOV, let us now have a look at the formulation characteristics for delivery of such drugs in viral diseases. EBOV's ability to multiply within the host cells by replicating its own RNA or DNA is the main challenge that affects the development of effective antiviral agents against it. Diagnosing and distinguishing the exact type of viral disease is quite challenging and detrimental for the evolution of its treatment [130]. Difficulties faced in the prevention, detection, or treatment are seen as a red signal by the research community, and newer technologies then have to be discovered to overcome the restrictions of existing therapies. Continuous efforts in the direction of research on antiviral therapies have improved the quality of life of patients suffering from viral infections. However, there is need to discuss primary challenges faced during the development of antiviral treatment therapies: interaction of antiviral drugs with regular prescription medicines [86]. Many of the antiviral drugs have shorter half-life, which means increased frequency of medication; prolonged drug exposure might cause drug resistance in patients; EBOV might spread into remote anatomical regions like the CNS, lymphatic system, synovial fluid, etc., causing reduced therapeutic efficacy of the drug [120]; viral infections may remain in the latent stage, posing as a challenge for their diagnosis and treatment [20].

2.2.1 Nanotechnology-Based Antiviral Formulations

Despite all the shortcomings, there have been sincere efforts by researchers, clinicians, pharmaceutical companies, etc., in recent years, to develop and formulate drugs and medicines that could cure and treat the deadly EBOV. In the current section, we shall briefly converse about the types of drug delivery systems that might be useful and applicable in formulating antiviral drugs and biological vaccines. One should note that there are little formulations available for treatment against EBOV till date. However, discussed in the following section are some drug delivery systems that can be used to formulate antiviral drugs, such as those for EBOV, in general. In the past years, innumerable technologies have been explored for prevention, diagnosis, and treatment of viral infections. Nanotechnology is considered one of the most capable technologies because of its effective nature against viral diseases, bypassing the limitations of out-of-date antiviral medicines. Nanotechnology enhances the drug's biological properties, which upsurges their selectivity and effectiveness to viral cells. There are different categories of nanomaterials which have served as delivery vehicles for the antiviral drugs. These include several polymer-based, lipid-based, lipid-polymer-based, inorganic metal–based, carbon-based, stimuli-sensitive materials, etc. and their applications in antiviral therapeutics [17].

Lipid-Based Nanoformulations

Pharmaceutical formulations, especially nano-based drug delivery systems, comprise of a variety of lipids that serve as carriers for antiviral drug compounds. In contrast to polymers, lipids are advantageous as they are inert, nontoxic, biodegradable, biocompatible, non-immunogenic and cheaper [91]. Furthermore, lipids are smaller sized, possess larger surface area, and have a high drug-loading capacity with enhanced interface interactions, controlled-release properties, etc. [78]. They improve pharmacokinetic profile and drug bioavailability and achieve the desired drug concentration at sites, which are otherwise inaccessible, through various techniques [80]. Triglycerides, lecithins, glyceryl palmitostearate, and fatty acids such as carnauba wax and beeswax are some common lipids used in nanoformulations. They are generally used in combination with co-solvents (up to 10%) and surfactants (up to 30%) to optimize drug solubility [7]. Some such lipid-based nanoformulations include liposomes, solid lipid nanoparticles (SLN), nanoemulsions, nanosuspensions, etc.

Liposomes

Liposomes are small spherical vesicles, 15–1000 nm in size, prepared from phospholipids that entrap an aqueous core in which the drug remains dispersed. Both hydrophilic and hydrophobic drugs can be entrapped and delivered within liposomes. Here, the drug is protected from the gastrointestinal environment by lipid layers in the liposomes, which also help in its sustained release small drug-loading capacities and physical instability during its administration and storage restricts the utility of liposomes despite their promising drug delivery capacities [64].

Solid Lipid Nanoparticles

SLNs are colloidal arrangements consisting of a solid lipid matrix and are in the range of 10–1000 nm in diameter. Solid lipids commonly utilized in SLNs include glycerides, triglycerides, fatty acids, steroidal lipids, etc. SLNs are designed to achieve a controlled-release profile of the drug, improve its stability, increase its loading capacity, and provide targeted drug delivery. SLNs have already been used for the delivery of numerous antiviral drugs like darunavir, ritonavir, maraviroc, zidovudine, efavirenz, and lopinavir, using lipids like Compritol® 888ATO, Gelucire® 44/14, etc. [20].

Nanostructured Lipid Carriers (NLCs)

Nanostructured lipid carriers utilize liquid lipids and, hence, depict better loading capacity, increased stability, and controlled release pattern as compared to SLNs [13]. Furthermore, NLCs can also be surface-modified to attain target specificity. Previously, NLCs formulated for the delivery of nevirapine showed better release kinetics as compared to nevirapine SLNs [60]. Also, NLCs of saquinavir showed increased transport across Caco-2 cell monolayers when compared to pure saquinavir [13].

Lipid Nanoparticles for siRNA Delivery

Recently, the RNA interference (RNAi) approach has been utilized for antiviral treatment. Small interfering RNA (siRNA) is used to target specific genes and cause their short-term silencing, thereby blocking the creation of respective proteins. They are double stranded, short (19–21 nucleotides), premeditated, and synthesized to mark a particular mRNA. Then, they are transfected into cells using cationic lipids or polymers [84]. Lipids are efficient as vehicles for the delivery of siRNA because of their efficient structural features like presence of polyunsaturated chains that destabilize intracellular membranes, ionizable amino groups that help in the mixing of the lipid nanoparticles with the cell membrane, etc. Leung et al. [63] had formulated improved siRNA lipid nanoparticles to treat Ebola, which directly inhibits the production of viral proteins. Their results depicted that for people infected with Ebola virus, RNA interference has potential as a significant post-exposure therapy and that this strategy could also be extended to treat other types of viruses [63].

2.2.2 Polymer-Based Nanoformulations

A variety of natural and synthetic hydrophilic as well as hydrophobic polymers are used in the preparation of nanoformulations. Natural hydrophilic polymers include proteins like albumin, gelatin, pectin, and polysaccharides like alginate, chitosan, etc., while polylactide-co-glycolic acid (PLGA), polylactic acid (PLA), poly(-caprolactone) (PECL), etc. are some synthetic hydrophobic polymers commonly used [53]. Several polymers with surface-modification properties have been used to prevent nonspecific contact with serum proteins and cause favorable alterations in pharmacokinetic parameters of the drug. Also, certain polymeric nanocarrier systems have been premeditated in such a manner that they effect drug release when stimulated by variations in chemical stimuli, environmental pH, etc. This helps in increasing drug bioavailability by preventing its degradation before it reaches its site of absorption [94].

Polymeric Micelles

Polymeric micelles are nanosized systems (10–100 nm) made of amphiphilic block copolymers, each unimer comprising of two segments, one being hydrophobic and the other hydrophilic. Past the critical micellization concentration (CMC), the unimers combine to form within the polymeric micelle, and poorly water-soluble drugs get incorporated into its hydrophobic core with the hydrophilic shell acting as a blockade to shield the drug and lessen the nonspecific interactions with proteins, enzymes, and cells [75].

Polymeric Nanoparticles

Polymeric nanoparticles are the nanocarriers formulated using either natural or synthetic polymers by utilizing a number of preparation techniques like emulsion–diffusion evaporation, solvent evaporation, nanoprecipitation, dialysis, polymer dispersion, coacervation, phase inversion temperature methods, etc. [81]. Polymeric nanoparticles are termed as nanocapsules when the drug is incorporated into its core (<300 nm in size), whereas they are termed as nanospheres when the drug is adsorbed onto its surface or embedded in the matrix (10–200 nm in size). Polymeric nanoparticles are advantageous as they provide controlled release of the drug, improve cellular uptake, prevent drug molecule from degradation site, have low toxicity, etc. [33].

Nanospheres

Nanospheres are even smaller spherical structures ranging from 10 to 200 nm in diameter, with the drug being uniformly dispersed in the polymeric matrix system. It provides site-specific drug delivery, optimized drug release, and rapid drug clearance due to smaller size [108]. Topical treatment of herpes with the drug acyclovir was found effective when the drug was formulated with chitosan nanospheres using the modified nanoemulsion method. The nanospheres of the drug were more effective against herpes simplex viruses 1 and 2, when the drug itself was used alone [27].

Inorganic-Based Nanoformulations

Gold Nanoparticles

Gold nanoparticles (AuNPs) are colloidal compounds incorporated with nanosized particles of gold and depict optical properties in the presence of light. When AuNP comes in contact with light, the electromagnetic field of this light initiates oscillation of the free gold electrons. This electron oscillation on the particle surface causes a dipole oscillation in the electric field of light and increases the radiative characteristics of the nanoparticles. Therefore, AuNPs have a widespread application in the fields of materials science, biological imaging, and electronics [43].

Silver Nanoparticles

Metallic silver naturally possesses antimicrobial activity due to its interaction with the respiratory chain and electron transport enzymes in viral and bacterial DNA. Silver nanoparticles (AgNPs), being smaller sized and having larger surface area, enable rapid dissolution of a quite a number of viruses. They possess unique properties regarding chemical stability, higher conductivity, viral cell catalysis, etc., permitting researchers to employ them in disease therapeutics systems [89].

Zinc Oxide Nanoparticles

Zinc oxide nanoparticles have been studied by Antoine et al. [4], who designed and synthesized zinc oxide tetrapod nanoparticles (ZOTEN) with engineered oxygen vacancies using flame transport synthesis. Female mice were used to test the pharmacological efficacy of ZOTEN. Clinical signs of vaginal infection showed great improvements with the use of this therapy. ZOTEN was able to trap HSV-2 virus and increased the presentation of the virus to mucosal APCs; the virus was then suppressed and destroyed by T cell–mediated and Ab-mediated responses [4].

Polymer-Lipid Hybrid (PLH or LPH) Nanoformulations

Multidrug therapy approach is essential in the treatment of antiviral diseases as it is impossible for a single drug to completely cure any viral infection. Lipid-polymer hybrid nanoparticles become useful when there is a need to deliver more chemotherapeutic drugs with diverse physicochemical properties in a single delivery vehicle. The amphiphilic characteristics of polymers and lipids are utilized in developing such stable nano-delivery systems for the concurrent delivery of several hydrophilic/hydrophobic drugs [26].

Biomimetic Lipid-Polymer Hybrid Nanoformulation

Biomimetic lipid-polymer hybrid nanoparticles, LPH-NPs, are prepared by modifying the nanoparticle surface with ligands that imitate cell surface proteins. These advanced LPH-NPs can offer different advantages like cell-specific targeting, biocompatibility, longer circulation time, and, most importantly, increased efficacy. Two technologies that fall in this category are virosomes and virus-like particles (VLPs). VLPs are formed by integrating the virus-derived envelope proteins into certain naturally occurring proteins like encapsuling, ferritin, lumazine synthase, etc., which depict precise structure and defined surface functionalities [49].

Stimuli-Based Lipid-Polymer Hybrid Nanoformulation

Stimuli-responsive lipid-polymer hybrid nanoparticles, in short SRNPs, regulate the release of the encapsulated drug at the targeted site based on response to certain stimuli, thereby increasing the drug's therapeutic value and reducing its side effects. pH, temperature, and magnetic field can serve as stimuli to obtain such drug responses [52]. Clawson et al. [21] synthesized SRNPs using PLGA core and a lipid-PEG monolayer shell, which disrupts at low pH and releases the drug [21].

Multifunctional Lipid-Polymer Hybrid (LPH) Nanoformulation

Multifunctional LPH-nanoparticles associate different functionalities in a single stable system. A drug particle can be associated to a precise targeting compound, which has the capacity to distinguish the surface characteristics of viral cells. The same particle is also altered with lipids and/or polymers combined to boost drug penetration and modification and to mask the undesirable biological interaction, thereby increasing the drug's therapeutic efficacy. Such nanoparticles function to detect, diagnose, deliver, and destroy the virus [100].

Graphene-Based Nanomaterials

Graphene is one of the most potential carbon-based nanomaterials with antiviral application. They are advantageous because they have greater loading capacity, increased surface area, and greater mechanical strength, which makes them suitable for delivery of antiviral agents. Oxygen-containing functional groups in graphene also act as binding sites for a variety of biological molecules like DNA, RNA, proteins, etc. [28]. Recently, Pokhrel et al. observed interactions between VP40 protein of EBOV and graphene at various interfaces and concluded recommending the disinfectant properties of graphene nanosolutions in preventing the EBOV epidemic [88].

Besides the wide field of nanotechnology, there are additional novel and innovative drug polymer systems that might be prolific in the long fight against viral diseases like Ebola termination.

2.2.3 Polymer Drug Conjugates

Polymer drug conjugates comprise of a polymer and a therapeutic agent bound covalently. The significance of this conjugation is the achievement of greater efficacy by extending plasma stability and safety through targeted delivery. Some polymers show inherent antiviral capacity, and their coupling with antiviral drugs provides viral synergistic action. Conjugation of interferon α2A and α2A with polyethylene glycol (PEG) has been found effective against hepatitis C virus. Polymers like poly(methacrylic acid) (PMMA) and glycosaminoglycans like heparin, heparan sulfate, dermatan sulphate, etc. have revealed effective binding to the HIV virus, thereby shielding viral entry into host cells [62].

2.2.4 Cyclodextrin-Based Delivery Systems

Cyclodextrins (CDs) are cyclic oligosaccharides comprising of six to twelve α-D-glucopyranose monomers joined by α1–4 linkages having a hydrophobic interior and hydrophilic exterior with primary and secondary –OH groups. CDs entrap drugs in an encasing, cage-like cavity, protecting the drug from degradation and increasing its solubility. Due to this, CDs have now evolved as favored delivery system for certain drugs. They have a crystalline structure that offer advantages like formation of inclusion complexes by interacting with a wide range of organic and inorganic lipophilic entities. But their poor aqueous solubility restricts the use of CDs. Currently, structural modifications in CDs are being made to make them more appropriate for pharmaceutical application [107].

2.3 Small-Molecule Therapeutics in Combating EBOV Infection

In terms of severity, the EBOV epidemic is second only to the human immunodeficiency virus (HIV). Besides strong pathogenicity, the fatality rate (up to 90%) of EBOV is due to the absence of suitable vaccines and therapeutics to combat the EBOV epidemic. Major phases of the EBOV life cycle including attachment to specific cell receptors, drug fusion with the virus, its entry, transcription, etc. have already been therapeutically targeted [3]. Consequently, synthesis of nucleotide or nucleoside analogues is considered an effective strategy during the design of antivirals, as they can imitate the building blocks of viral RNA and, therefore, interrupt its regular functioning [46].

Complete protection of rhesus macaques from EBOV challenge was demonstrated by Expanded Access Remdesivir (RDV; GS-5734) through several dosing regimens, including animals that received GS-5734 3 days after infection [121]. GS-5734 was found to be distributed in macaque tissues such as eyes, brain, epididymis, and testes, where EBOV might reside even after recovery from illness. Favipiravir is an additional drug that has broad-spectrum activity against a number of RNA viruses and EBOV as well. Anti-EBOV activity has been demonstrated by nucleoside analogs b-D-N4-hydroxycytidine (NHC), azacytidine, 6-azauridineb, etc., all of which have been identified using antiviral screens with EBOV or minigenome assays [30, 95]. However, enhanced study in this area is essential to determine their efficacy in animal models. Minigenome assays again determined benzoquinoline compounds as RNA synthesis inhibitors in EBOV disease [66]. Heat shock protein 90 (Hsp90) inhibitors reveal EBOV inhibition in cell cultures by the destabilization of EBOV L [24].

Arbidol is another small molecule used clinically in China and Russia to prevent infections of the influenza virus. It has shown broad-spectrum antiviral activity in cell cultures, including anti-EBOV activity, by blocking EBOV entry. This may be through the capacity of Arbidol to bind lipid membranes, but the study needs further exploration [115]. Inhibitors of macropinocytosis are also found to inhibit EBOV entry in human cells. Compound such as ethylisopropylamiloride (EIPA), an inhibitor of the Na+/H+ exchanger; phosphokinase C inhibitor rottlerin; actin polymerization inhibitor latrunculin A, and PI3-kinase inhibitor wortmannin etc. are also reported as promising to combat EBOV infection [79, 115]. Besides, Cathepsins B and L proteolyze EBOV glycoprotein, which prohibits EBOV entry in cells, was studied using pharmacological inhibitors of these proteases such as CA074, CA074 methyl ester etc. [103, 105].

These are just a few examples of the potential of small molecules as drugs used against Ebola and other viral diseases. The fact that such therapeutic small-molecule inhibitors have been efficacious in protecting non-human primates from EBOV challenge is an important milestone, as is the advancement of such therapeutics when reaching the clinical trial stages. There are continued efforts at drug development for filoviruses like EBOV, since no drug is yet clinically approved for human use and also since there is a possibility of emerging viral resistance. Therefore, small-molecule drug development is a priority for treatment of antiviral diseases.

2.4 Novel Antiviral Approaches and Developments for EBOV

Undeniably, several studies acknowledged multiple drugs displaying anti-EBOV activity both in vitro and in vivo [50, 99]. Repurposing of such drugs and small molecules to extensively modify their structures and structure-activity relationship (SAR) analyses can help focus on specific disease eradication. 4-aminoquinoline antimalarial compounds, particularly amodiaquine, were identified as potent EBOV inhibitors during recent screening studies [69]. In humans, amodiaquine gets metabolism by cytochrome CYP2C8 to desethylamodiaquine, which has an extended half-life of 9–18 days. Previous reports exhibited anti-EBOV activity of both amodiaquine and desethylamodiaquine in vitro [131], signifying long-lasting antiviral efficacy in humans. Therefore, Sakurai et al. [99] concluded amodiaquine provides clinical relief for EBOV patients but requires a significant improvement in its potency before being beneficial.

It takes nearly 10–15 years to approve a new chemical entity as a safe and effective drug through the clinical trial stage to registration. Sometimes, it may takeup to 30 years, when the drug development period for essential antivirals is actually much shorter, approval times from Food and Drug Administration (FDA) and other regulatory authorities. In principle, infected humans might be treated with potent antiviral drugs directly to inhibit viral infections. Hence, there have been substantial efforts toward identifying and distinguishing known antiviral drugs that might have anti-EBOV efficacy. Till date, 80 FDA-approved drugs with anti-EBOV activity have been identified, which include calcium channel blockers, antihistamines, antidepressants, and selective estrogen receptor modulators (SERMs) [50, 57]. Specific SERMs like toremifene, tamoxifen, clomiphene, and raloxifene were efficacious in a mouse EBOV infection model. A study stated the effective nature of calcium channel blockers verapamil, amiodarone, and dronedarone against a number of filoviruses by inhibition of filoviral entry [36]. Interestingly, these drugs share a tertiary amine at the same position as aminoquinoline derivatives like chloroquine, amodiaquine, and SERMs like clomiphene and toremifene. Each of them displays anti-EBOV activity in vitro and/or in mice and is believed to act through a similar mechanism [31, 54]. These compounds could aid as the chemi-informatic basis for a common pharmacophore characterization, leading to further drug discovery.

A number of potential drug candidates are being developed, many of which have only just entered clinical phase trials. In case of EBOV, clinical trials offer restricted possibilities as opposed to influenza, whose epidemic occurs annually in developed countries, whereas EBOV occurs in developing countries and is erratic and deadly, with limited medical infrastructure. An alternative approach is the use of 3D cell/tissue culture systems such as organoids. These systems are yet to be experimented on for EBOV, but they have proved efficacious in treating Zika virus [92].

Hopefully, small-molecule EBOV antivirals are in the preclinical development stage in a number of pharmaceutical companies. The capacity to rapidly test these drug compounds in human trials during epidemics is, however, a major concern. Nevertheless, there is an imperative need for improved BSL4 facilities and active collaboration between industry, academia, and health regulatory organizations to accelerate and share resources that might help develop antivirals against EBOV.

2.5 Plant-Based Drugs

Plant-based drugs are mainly comprised of standardized plant extract or plant materials [117]. They are either used in mono-herbal or poly-herbal form. But regarding the treatment of Ebola, no such products have been developed to date.

The development of plant-based vaccine mainly encompasses the incorporation of a transgene into plant cells. The targeted sequence of the specific antigen is incorporated within the vector and then transferred into the expression system by different available techniques [5]. Agrobacterium-mediated gene transfer and transformation, electroporation, agroinfiltration, and sonication are the different techniques to deliver the gene of interest to the target plants for the production of desired vaccines [124]. Several plant-based lead vaccine candidates against viral infection have gone through clinical trials and have been appraised expansively [4]. Ebola outbreak that took place in Western Africa (2013–2016) was challenged and affected due to the availability of limited treatment options for infected patients. Later on, significant resources for developing antibodies/vaccines have been gradually expended to increase the availability and accessibility options of treatment for Ebola.

To fulfil the requirement of new therapies like vaccine for emerging diseases and their outbreak, rapid engineering and scalable invention techniques with alternate hosts are being pursued. Hence, plant-based vaccines/antibodies against Ebola virus are also being considered and under the focus of researchers [117]. In *N. benthamiana*, a geminiviral replicon system was used to develop an immune complex of Ebola to use against the deadly viral disease. The antigen-antibody fusion vaccine with a combination of 6D8 anti-Ebola IgG against GP of Ebola results in 80%t endurance of mice against fatal Ebola infection [12]. The heavy- and light-chain molecules of a protective IgG monoclonal antibody (6D8) have been developed by the process of nuclear transformation at levels of 0.5 mg 6D8 per gram of fresh leaves. In another study, Rios-Huerta et al. (2017) showed that immunogen from tobacco plant directs two counterbalancing Ebola virus epitopes of GP and developed the cost-effective vaccine candidate fusion with *E. coli* heat-labile enterotoxin B subunit against Ebola virus GP1 [96]. The efficacious outcome of the chimeric protein, *E. coli* heat-labile enterotoxin B subunit, in tobacco plant cells with immunogenicity was evaluated in BALB/c mice using subcutaneous and oral routes.

The plant-based monoclonal antibody counter with Ebola virus was introduced after a lethal Ebola virus challenge was successfully passed by non-human primates.

The monoclonal antibody against Ebola GP triple cocktail (13C6, 13F6, 6D8; MB-003) from *N. benthamiana* gave 43–100% survival to the treated rhesus macaques (IV injection), which was dependent on the treatment time after the infection of Ebola virus. The antibody cocktail, ZMapp, obtained from tobacco plant is currently considered as the most favorable antibody-based drug against the Ebola virus infection [82]. Both RNA and DNA viruses were significantly modified to express the heterologous protein required for the plant-based vectors. The efficiency and bioactivity of monoclonal antibody may change due to the glycosylation pattern, comprising its attachment with the antigenic epitope, and therefore these are crucially beneficial for humans. Currently, the ZMapp is a WHO-approved treatment regimen for the infection of Ebola virus disease [47]. Moreover, a report by Phoolcharoen et al. suggests that the GP sequence merged to a monoclonal antibody targeting a GP epitope to form immune complexes of Ebola by self-polymerization process [87].

Plants have been used as bioreactors for antibody production as they offer several potential advantages over other conventional production systems, including using bacteria, yeast, or mammalian cell culture [106]. Plant production facilities are cheaper than equivalent bioreactors and offer rapid protein turnaround time and high scalability. They are not susceptible to contamination with mammalian-tropic pathogens. Post-translational modification (PTM) in plants is controllable and represents an essential advantage over bacteria since many proteins, including most antibody formats, do not fold correctly and have limited functionality when expressed without PTM.

Plants must be transformed with genes encoding antibody proteins to produce antibodies in plants. Typically, the bacterium *Agrobacterium tumefaciens* is used to transfer recombinant regions of DNA encoding for the genes of interest into the plant nucleus through the activity of the vir (virulence) operon. These DNA regions are termed transfer DNAs (T-DNA). T-DNA can integrate into plant chromosomes, generating a stable transgenic cell that can be regenerated into a whole plant. However, a high level of transcriptional activity occurs before integration takes place. This burst of transcription can be utilized to produce large amounts of recombinant protein without the need for time-consuming regeneration steps. Furthermore, the transcription rate can be significantly enhanced through the simultaneous delivery of viral genes encoding proteins directing the replication of RNA or even permitting the cell-to-cell spread of the message.

Crucially, transient expression allows antibodies to be expressed with faithful PTMs at scale and within a concise time frame, without the need for expensive bioreactors or product dedicated production facilities. Transgenic plants require no specialized equipment for growth or antibody production except to control genetically modified organisms and can be grown at an agricultural scale. Downstream processing is similar for both approaches, and protein A or G matrices are commonly used to purify mAbs from plant extracts.

3 Clinical Trials and Observational Studies of Ebola Therapeutics

There are several clinical trial studies going on to evaluate the efficacy of the different treatment strategies for Ebola viral disease. The only random clinical trial evaluated the monoclonal antibody ZMapp, had a low risk of bias, and found a statistically nonsignificant decrease in mortality in Ebola infection [90]. All interventions evaluated in non-randomized studies, including interferon β-1a, convalescent whole blood or plasma, the antimalarial artesunate-amodiaquine, and favipiravir, were associated with moderate-to-serious risk of bias due to confounding and severely limiting inferences regarding treatment effects [45]. The non-randomized evaluations of brincidofovir and TKM-130803 did not provide evidence to continue their evaluation against the Ebola infection. Based on existing data and study designs, several agents with promising preclinical findings or evaluations cannot be evaluated for effectiveness in human trials yet. These include the monoclonal antibody cocktail REGN3470–3471-3479, remdesivir (GS-5734), and the monoclonal antibody MAb114 [125]. These are evaluated by a World Health Organization-convened independent scientific committee in addition to favipiravir and ZMapp to find out possible emergency use of unregistered and investigational interventions against the infection [126].

Evaluating the effect of clinical trial outcomes of patients with Ebola is very challenging because of its uncommon and lethal nature. Few preexisting therapies with strong potential treatment effect have made difficult for clinicians, researchers, regulators, and funders to prioritize new drugs in patients. Till now, Ebola outbreaks have occurred in West Africa and most recently in DR Congo, Central Africa, where the health system is challenged with many factors [10]. Often, in remote areas of these regions, delayed recognition of the onset of outbreak increases mortality [126]. Historically high mortality rate of Ebola virus, including among the healthcare workers, leads to diminished clinical capacity for care and clinical research of Ebola infection. The variable standard of supportive care for Ebola makes estimation of treatment effects against the infection difficult and also contributes to a high mortality rate. This is also leading to selection of patients who are poorly responsive to investigational new agents and an inability to compare therapies across different studies [8, 100]. Uniform adoption of evidence-based supportive care guidelines in future outbreaks may facilitate the evaluation of anti-Ebola virus therapies [61].

During the year 2014–2016, West African nations mostly affected by Ebola but it had no previously experienced Ebola outbreak; hence there was limited pre-existing Ebola specific clinical and research capacity. With overwhelmed national healthcare systems and slow international response, there was little opportunity to evolve interventional research programs parallel with outbreak care. Eventually, early diagnostic and descriptive studies on the disease gave rise to a potential impact of supportive and specific Ebola therapy [61, 100, 111]. Among the 28,616 infected patients in this outbreak, fewer than 5% got any therapies described or evaluated for Ebola infection. From this, only 0.25% patients participated in a random clinical trial to evaluate an investigational therapy. Considering the high mortality rate, sparse treatment options, and high capacity for spread of Ebola, an adequate research capacity in Ebola outbreak-prone regions should be developed under the support of the international community and research organizations. Rigorous prior knowledge in this particular viral infection is critical to plan relevant future research for any kind of therapies.

Currently, the small number of patients exposed to each clinical evaluation and also design-related limitations make it difficult to find out clinical effectiveness of investigational therapeutics. In the future, decision-makers need to be confronted with the impact of Ebola outbreaks to prioritize or avoid system-wide delivery of certain experimental interventions in research and development process. The guideline developers can use the evidence summary to make graded recommendations regarding specific therapies against viral infection. Lastly, insufficiencies of the existing study or strategies must be highlighted so that researchers can design future studies for implementation during an outbreak and to prioritize experimental therapies for future evaluation.

4 Challenges in Drug Discovery and Development

The success rates for drug discovery process are within the range of 69% to 85%, which involves target identification, screening of compounds, lead optimization, and preclinical evaluation [108]. Failure of a drug discovery project accounts for many reasons: (i) unclear mechanism of selected target protein, (ii) lack of lead compounds, (iii) poor potency of the compounds, (iv) lack of efficacy of the compounds, (v) inappropriate drug-like properties of the compounds, and (vi) unexpected high toxicity levels in animal model. However, the success rates vary distinctly for the clinical development of drugs. An experimental drug candidate in phase I clinical trials has only 10% chance to reach the market for clinical use [108]. Inadequate efficacy of drug candidates under investigation accounts for recent phase III clinical trial failures. Half of the clinical trials for phase II and approximately 16% for phase I failed to reach their next phase [108]. Most of the drugs that fail during clinical trials are based on preclinical experiment data performed in in vivo models. As experiments with Ebola virus strains require a safety level of BSL4, it narrows the possibilities and slows the progress in Ebola drug discovery research. The complexities of human biology also pose a significant challenge in the target-based drug discovery [70].

The magnitude of the recent Ebola outbreak has necessitated the need to explore broad-spectrum alternative strategies apart from conventional drug discovery process. Despite these problems, some experimental drug and vaccine development is progressing with some in the earliest stages of the development as product. Repurposing drugs also provide an alternative way to accelerate the process of drug discovery against Ebola. Investigation of FDA-approved drugs from other therapeutic area in new directions open chances for developing a strategy against such challenging diseases. Drug repurposing will bypass phase I of clinical trials, which can accelerate the drug discovery process and will eliminate logistic considerations like manufacturing and distribution. However, any unfavorable data gained from these clinical trials for repurposing drugs may not serve any new purpose against the approved drug, but it might cut down on the development timeline of the drug. Furthermore, identifying the mechanism of action of the repurposed drugs will be difficult. Therefore, new experiments will need to design to identify the mechanism of action of such repurposed drugs for its anti-Ebola potency. As only a small of mutation can drastically change the biological properties of RNA viruses, hence development of drug resistance can also be a major clinical problem for the treatment of infection of Ebola infections [41]. Hence, the statistical significance of preclinical studies is very crucial for the efficiency of clinical trials to avoid unnecessary testing of a large number of drug candidates [110].

It is found that a large proportion of drug candidates proceeding into the clinical trials never showed efficacy in animal model, which leads to a large number of useless drug candidates and waste of resources [72]. Moreover, a sufficient number of participants are required to get statistically significant results from clinical studies [112]. To reflect the patient's situation, Ebola-infected animal models should also be reliable enough. Therefore, the overall aspects of experimental procedures and efficacy should be estimated to a higher level in preclinical study [55]. Additionally, pharmacokinetics properties like plasma half-life are representative of drug efficacy, which need to be estimated precisely with the use of animal models. Although many treatment options for EVD have been proposed, there is no FDA-approved drug approved from the ongoing clinical study.

The genetics and immunological profile vary from one population to another, which poses further challenge for evaluation of drug safety profile in different populations. Identifying hits through in silico study and screening is almost achievable for any target, but hit to lead optimization or development as drug candidate remains difficult. Besides, computational methods are still not completely reliable and need to be optimized to accurately predict binding constants for chemically diverse compounds and large datasets. On the other hand, ADMET properties are difficult to predict for large datasets using in silico technique because it is impossible to simplify them to a single molecular event, which is one of the causes for failure of drug candidates. But currently, more attention has been given to the pharmacokinetic properties during lead optimization, which reduce the failures of clinical trials (mostly in phase I) up to only 10% [56]. With the combined efforts by regulatory institutions, analysis with controlled experimental protocols and performance can deliver safe and unprecedented predictive results for human clinical trials.

5 Conclusion

In recent years, there have been sincere efforts by researchers to develop effective therapy against the deadly EBOV. There are various nanomaterials, viz., polymer-based, lipid-based, lipid-polymer-based, inorganic metal–based, carbon-based, stimuli-sensitive materials, etc., which have served as delivery vehicles for antiviral drugs. Graphene-based nanomaterials, polymer drug conjugates, cyclodextrin-based delivery systems, etc. also play a major role in Ebola drug delivery. Small-molecule drug development is a priority for treatment of antiviral diseases. A number of potential drug candidates are being developed, in recent years, and many of which have only just entered clinical phase trials. Three-dimensional cell/tissue culture systems are yet to be experimented on for EBOV in the future.

References

1. Aleksandrowicz P, Marzi A, Biedenkopf N, Beimforde N, Becker S, Hoenen T, Feldmann H, Schnittler H-J. Ebola virus enters host cells by macropinocytosis and Clathrin-mediated endocytosis. J Infect Dis. 2011;204:S957–67. https://doi.org/10.1093/infdis/jir326.
2. Alvarez CP, Lasala F, Carrillo J, Muñiz O, Corbí AL, Delgado R. C-type lectins DC-SIGN and L-SIGN mediate cellular entry by Ebola virus in cis and in trans. J Virol. 2002;76:6841–4. https://doi.org/10.1128/JVI.76.13.6841-6844.2002.
3. Ansari AA. Clinical features and pathobiology of Ebolavirus infection. J Autoimmun. 2014;55:1–9. https://doi.org/10.1016/j.jaut.2014.09.001.
4. Antoine TE, Hadigal SR, Yakoub AM, Mishra YK, Bhattacharya P, Haddad C, Valyi-Nagy T, Adelung R, Prabhakar BS, Shukla D. Intravaginal zinc oxide tetrapod nanoparticles as novel immunoprotective agents against genital herpes. J Immunol. 2016;196:4566–75. https://doi.org/10.4049/jimmunol.1502373.
5. Arizona State University. Origins of world's first cure for Ebola had roots at ASU [WWW Document]; 2019. https://biodesign.asu.edu/news/origins-world's-first-cure-ebola-had-roots-asu#:~:text=Therootsoftheamazing,producetheEbola therapeuticZMapp.&text=Asof Aug. Accessed 31 Jan 2022.
6. Ascenzi P, Bocedi A, Heptonstall J, Capobianchi MR, Di Caro A, Mastrangelo E, Bolognesi M, Ippolito G. Ebolavirus and Marburgvirus: insight the Filoviridae family. Mol Asp Med. 2008;29:151–85. https://doi.org/10.1016/j.mam.2007.09.005.
7. Attama AA, Momoh MA, Builders PF. Lipid Nanoparticulate drug delivery systems: a revolution in dosage form design and development. In: Recent advances in novel drug carrier systems. InTech.; 2012. https://doi.org/10.5772/50486.
8. Bah EI, Lamah M-C, Fletcher T, Jacob ST, Brett-Major DM, Sall AA, Shindo N, Fischer WA, Lamontagne F, Saliou SM, Bausch DG, Moumié B, Jagatic T, Sprecher A, Lawler JV, Mayet T, Jacquerioz FA, Méndez Baggi MF, Vallenas C, Clement C, Mardel S, Faye O, Faye O, Soropogui B, Magassouba N, Koivogui L, Pinto R, Fowler RA. Clinical presentation of patients with Ebola virus disease in Conakry. Guinea N Engl J Med. 2015;372:40–7. https://doi.org/10.1056/NEJMoa1411249.
9. Baize S, Pannetier D, Oestereich L, Rieger T, Koivogui L, Magassouba N, Soropogui B, Sow MS, Keïta S, De Clerck H, Tiffany A, Dominguez G, Loua M, Traoré A, Kolié M, Malano ER, Heleze E, Bocquin A, Mély S, Raoul H, Caro V, Cadar D, Gabriel M, Pahlmann M, Tappe D, Schmidt-Chanasit J, Impouma B, Diallo AK, Formenty P, Van Herp M, Günther

S. Emergence of Zaire Ebola virus disease in Guinea. N Engl J Med. 2014;371:1418–25. https://doi.org/10.1056/NEJMoa1404505.

10. Barry A, Ahuka-Mundeke S, Ali Ahmed Y, Allarangar Y, Anoko J, Archer BN, Aruna Abedi A, Bagaria J, Belizaire MRD, Bhatia S, Bokenge T, Bruni E, Cori A, Dabire E, Diallo AM, Diallo B, Donnelly CA, Dorigatti I, Dorji TC, Escobar Corado Waeber AR, Fall IS, Ferguson NM, FitzJohn RG, Folefack Tengomo GL, Formenty PBH, Forna A, Fortin A, Garske T, Gaythorpe KA, Gurry C, Hamblion E, Harouna Djingarey M, Haskew C, Hugonnet SAL, Imai N, Impouma B, Kabongo G, Kalenga OI, Kibangou E, Lee TM-H, Lukoya CO, Ly O, Makiala-Mandanda S, Mamba A, Mbala-Kingebeni P, Mboussou FFR, Mlanda T, Mondonge Makuma V, Morgan O, Mujinga Mulumba A, Mukadi Kakoni P, Mukadi-Bamuleka D, Muyembe J-J, Bathé NT, Ndumbi Ngamala P, Ngom R, Ngoy G, Nouvellet P, Nsio J, Ousman KB, Peron E, Polonsky JA, Ryan MJ, Touré A, Towner R, Tshapenda G, Van De Weerdt R, Van Kerkhove M, Wendland A, Yao NKM, Yoti Z, Yuma E, Kalambayi Kabamba G, de Lukwesa Mwati JD, Mbuy G, Lubula L, Mutombo A, Mavila O, Lay Y, Kitenge E. Outbreak of Ebola virus disease in the Democratic Republic of the Congo, April–May, 2018: an epidemiological study. Lancet. 2018;392:213–21. https://doi.org/10.1016/S0140-6736(18)31387-4.
11. Bente D, Gren J, Strong JE, Feldmann H. Disease modeling for Ebola and Marburg viruses. Dis Model Mech. 2009;2:12–7. https://doi.org/10.1242/dmm.000471.
12. Budzianowski J. Tobacco against Ebola virus disease. Przegl Lek. 2015;72:567–71.
13. Bule M, Khan F, Niaz K. Antivirals: past, present and future. In: Recent advances in animal virology. Springer Singapore, Singapore; 2019, pp. 425–446. https://doi.org/10.1007/978-981-13-9073-9_22.
14. Carroll SA, Towner JS, Sealy TK, McMullan LK, Khristova ML, Burt FJ, Swanepoel R, Rollin PE, Nichol ST. Molecular evolution of viruses of the family Filoviridae based on 97 whole-genome sequences. J Virol. 2013;87:2608–16. https://doi.org/10.1128/JVI.03118-12.
15. Castanotto D, Rossi JJ. The promises and pitfalls of RNA-interference-based therapeutics. Nature. 2009;457:426–33. https://doi.org/10.1038/nature07758.
16. Centers for Disease Control and Prevention. What is Ebola Virus Disease? [WWW Document]; 2021. https://www.cdc.gov/vhf/ebola/about.html. Accessed 31 Jan 2022.
17. Chakravarty M, Vora A. Nanotechnology-based antiviral therapeutics. Drug Deliv Transl Res. 2021;11:748–87. https://doi.org/10.1007/s13346-020-00818-0.
18. Chan SY, Empig CJ, Welte FJ, Speck RF, Schmaljohn A, Kreisberg JF, Goldsmith MA. Folate receptor-α is a cofactor for cellular entry by Marburg and Ebola viruses. Cell. 2001;106:117–26. https://doi.org/10.1016/S0092-8674(01)00418-4.
19. Chandran K, Sullivan NJ, Felbor U, Whelan SP, Cunningham JM. Endosomal proteolysis of the Ebola virus glycoprotein is necessary for infection. Science (80-). 2005;308:1643–5. https://doi.org/10.1126/science.1110656.
20. Chaudhuri A. Diagnosis and treatment of viral encephalitis. Postgrad Med J. 2002;78:575–83. https://doi.org/10.1136/pmj.78.924.575.
21. Clawson C, Ton L, Aryal S, Fu V, Esener S, Zhang L. Synthesis and characterization of lipid–polymer hybrid nanoparticles with pH-triggered poly(ethylene glycol) shedding. Langmuir. 2011;27:10556–61. https://doi.org/10.1021/la202123e.
22. De Clercq E, Holý A. Acyclic nucleoside phosphonates: a key class of antiviral drugs. Nat Rev Drug Discov. 2005;4:928–40. https://doi.org/10.1038/nrd1877.
23. ClinicalTrials.gov. Putative investigational therapeutics in the treatment of patients with known Ebola infection [WWW Document]; 2022. https://clinicaltrials.gov/ct2/show/NCT02363322?term=zmapp+ebola&rank=1. Accessed 14 Feb 2022.
24. Connor JH, McKenzie MO, Parks GD, Lyles DS. Antiviral activity and RNA polymerase degradation following Hsp90 inhibition in a range of negative strand viruses. Virology. 2007;362:109–19. https://doi.org/10.1016/j.virol.2006.12.026.
25. Dahlmann F, Biedenkopf N, Babler A, Jahnen-Dechent W, Karsten CB, Gnirß K, Schneider H, Wrensch F, O'Callaghan CA, Bertram S, Herrler G, Becker S, Pöhlmann S, Hofmann-Winkler

H. Analysis of Ebola virus entry into macrophages. J Infect Dis. 2015;212:S247–57. https://doi.org/10.1093/infdis/jiv140.
26. Dave V, Tak K, Sohgaura A, Gupta A, Sadhu V, Reddy KR. Lipid-polymer hybrid nanoparticles: synthesis strategies and biomedical applications. J Microbiol Methods. 2019;160:130–42. https://doi.org/10.1016/j.mimet.2019.03.017.
27. Donalisio M, Leone F, Civra A, Spagnolo R, Ozer O, Lembo D, Cavalli R. Acyclovir-loaded chitosan nanospheres from Nano-emulsion templating for the topical treatment of herpesviruses infections. Pharmaceutics. 2018;10:46. https://doi.org/10.3390/pharmaceutics10020046.
28. Donskyi IS, Azab W, Cuellar-Camacho JL, Guday G, Lippitz A, Unger WES, Osterrieder K, Adeli M, Haag R. Functionalized nanographene sheets with high antiviral activity through synergistic electrostatic and hydrophobic interactions. Nanoscale. 2019;11:15804–9. https://doi.org/10.1039/C9NR05273A.
29. Edwards MR, Basler CF. Current status of small molecule drug development for Ebola virus and other filoviruses. Curr Opin Virol. 2019;35:42–56. https://doi.org/10.1016/j.coviro.2019.03.001.
30. Edwards MR, Pietzsch C, Vausselin T, Shaw ML, Bukreyev A, Basler CF. High-throughput minigenome system for identifying small-molecule inhibitors of Ebola virus replication. ACS Infect Dis. 2015;1:380–7. https://doi.org/10.1021/acsinfecdis.5b00053.
31. Ekins S, Coffee M. FDA approved drugs as potential Ebola treatments. F1000Research; 2015. https://doi.org/10.12688/f1000research.6164.2.
32. Enterlein S, Warfield KL, Swenson DL, Stein DA, Smith JL, Gamble CS, Kroeker AD, Iversen PL, Bavari S, Mühlberger E. VP35 knockdown inhibits Ebola virus amplification and protects against lethal infection in mice. Antimicrob Agents Chemother. 2006;50:984–93. https://doi.org/10.1128/AAC.50.3.984-993.2006.
33. Ferrari R, Sponchioni M, Morbidelli M, Moscatelli D. Polymer nanoparticles for the intravenous delivery of anticancer drugs: the checkpoints on the road from the synthesis to clinical translation. Nanoscale. 2018;10:22701–19. https://doi.org/10.1039/C8NR05933K.
34. Florescu DF, Kalil AC, Hewlett AL, Schuh AJ, Stroher U, Uyeki TM, Smith PW. Administration of brincidofovir and convalescent plasma in a patient with Ebola virus disease. Clin Infect Dis. 2015;61:969–73. https://doi.org/10.1093/cid/civ395.
35. Furuta Y, Gowen BB, Takahashi K, Shiraki K, Smee DF, Barnard DL. Favipiravir (T-705), a novel viral RNA polymerase inhibitor. Antivir Res. 2013;100:446–54. https://doi.org/10.1016/j.antiviral.2013.09.015.
36. Gehring G, Rohrmann K, Atenchong N, Mittler E, Becker S, Dahlmann F, Pöhlmann S, Vondran FWR, David S, Manns MP, Ciesek S, von Hahn T. The clinically approved drugs amiodarone, dronedarone and verapamil inhibit filovirus cell entry. J Antimicrob Chemother. 2014;69:2123–31. https://doi.org/10.1093/jac/dku091.
37. Geisbert TW, Hensley LE, Jahrling PB, Larsen T, Geisbert JB, Paragas J, Young HA, Fredeking TM, Rote WE, Vlasuk GP. Treatment of Ebola virus infection with a recombinant inhibitor of factor VIIa/tissue factor: a study in rhesus monkeys. Lancet. 2003;362:1953–8. https://doi.org/10.1016/S0140-6736(03)15012-X.
38. Geisbert TW, Young HA, Jahrling PB, Davis KJ, Larsen T, Kagan E, Hensley LE. Pathogenesis of Ebola hemorrhagic fever in primate models. Am J Pathol. 2003;163:2371–82. https://doi.org/10.1016/S0002-9440(10)63592-4.
39. Geisbert TW, Jahrling PB. Differentiation of filoviruses by electron microscopy. Virus Res. 1995;39:129–50. https://doi.org/10.1016/0168-1702(95)00080-1.
40. Hensley LE, Stevens EL, Yan SB, Geisbert JB, Macias WL, Larsen T, Daddario-DiCaprio KM, Cassell GH, Jahrling PB, Geisbert TW. Recombinant human activated protein C for the postexposure treatment of Ebola hemorrhagic fever. J Infect Dis. 2007;196:S390–9. https://doi.org/10.1086/520598.
41. Hoenen T, Safronetz D, Groseth A, Wollenberg KR, Koita OA, Diarra B, Fall IS, Haidara FC, Diallo F, Sanogo M, Sarro YS, Kone A, Togo ACG, Traore A, Kodio M, Dosseh A, Rosenke K, de Wit E, Feldmann F, Ebihara H, Munster VJ, Zoon KC, Feldmann H, Sow

S. Mutation rate and genotype variation of Ebola virus from Mali case sequences. Science (80-.). 2015;348:117–9. https://doi.org/10.1126/science.aaa5646.
42. Hostetler KY. Synthesis and early development of Hexadecyloxypropyl-cidofovir: an Oral Antipoxvirus nucleoside phosphonate. Viruses. 2010;2:2213–25. https://doi.org/10.3390/v2102213.
43. Huang X, El-Sayed MA. Gold nanoparticles: optical properties and implementations in cancer diagnosis and photothermal therapy. J Adv Res. 2010;1:13–28. https://doi.org/10.1016/j.jare.2010.02.002.
44. Inc., B.P. BioCryst announces study results for BCX4430 in a Non-Human Primate Model of Ebola Virus Infection. 2014.
45. Ippolito G, Lanini S, Brouqui P, Di Caro A, Vairo F, Fusco FM, Krishna S, Capobianchi MR, Kyobe-Bosa H, Puro V, Wölfel R, Avsic-Zupanc T, Ioannidis JPA, Portella G, Kremsner P, Dar O, Bates M, Zumla A. Non-randomised Ebola trials—lessons for optimal outbreak research. Lancet Infect Dis. 2016;16:407–8. https://doi.org/10.1016/S1473-3099(16)00132-8.
46. Izawa K, Aceña JL, Wang J, Soloshonok VA, Liu H. Small-molecule therapeutics for Ebola virus (EBOV) disease treatment. Eur J Org Chem. 2016;2016:8–16. https://doi.org/10.1002/ejoc.201501158.
47. Jacob ST, Crozier I, Fischer WA, Hewlett A, Kraft CS, de La Vega M-A, Soka MJ, Wahl V, Griffiths A, Bollinger L, Kuhn JH. Ebola virus disease. Nat Rev Dis Prim. 2020;6:13. https://doi.org/10.1038/s41572-020-0147-3.
48. Jahrling PB, Geisbert JB, Swearengen JR, Larsen T, Geisbert TW. Ebola hemorrhagic fever: evaluation of passive immunotherapy in nonhuman primates. J Infect Dis. 2007;196:S400–3. https://doi.org/10.1086/520587.
49. Jin K, Luo Z, Zhang B, Pang Z. Biomimetic nanoparticles for inflammation targeting. Acta Pharm Sin B. 2018;8:23–33. https://doi.org/10.1016/j.apsb.2017.12.002.
50. Johansen LM, Brannan JM, Delos SE, Shoemaker CJ, Stossel A, Lear C, Hoffstrom BG, DeWald LE, Schornberg KL, Scully C, Lehár J, Hensley LE, White JM, Olinger GG. FDA-approved selective estrogen receptor modulators inhibit Ebola virus infection. Sci Transl Med. 2013;5:190ra79. https://doi.org/10.1126/scitranslmed.3005471.
51. Johansen LM, DeWald LE, Shoemaker CJ, Hoffstrom BG, Lear-Rooney CM, Stossel A, Nelson E, Delos SE, Simmons JA, Grenier JM, Pierce LT, Pajouhesh H, Lehár J, Hensley LE, Glass PJ, White JM, Olinger GG. A screen of approved drugs and molecular probes identifies therapeutics with anti–Ebola virus activity. Sci Transl Med. 2015;7:290ra89. https://doi.org/10.1126/scitranslmed.aaa5597.
52. Karimi M, Ghasemi A, Sahandi Zangabad P, Rahighi R, Moosavi Basri SM, Mirshekari H, Amiri M, Shafaei Pishabad Z, Aslani A, Bozorgomid M, Ghosh D, Beyzavi A, Vaseghi A, Aref AR, Haghani L, Bahrami S, Hamblin MR. Smart micro/nanoparticles in stimulus-responsive drug/gene delivery systems. Chem Soc Rev. 2016;45:1457–501. https://doi.org/10.1039/C5CS00798D.
53. Kaushik S. Polymeric and ceramic nanoparticles: possible role in biomedical applications, in: Handbook of polymer and ceramic nanotechnology. Springer International Publishing, Cham; 2020. pp. 1–17. https://doi.org/10.1007/978-3-030-10614-0_39-1.
54. Kazmi F, Hensley T, Pope C, Funk RS, Loewen GJ, Buckley DB, Parkinson A. Lysosomal sequestration (trapping) of lipophilic amine (cationic amphiphilic) drugs in immortalized human hepatocytes (Fa2N-4 cells). Drug Metab Dispos. 2013;41:897–905. https://doi.org/10.1124/dmd.112.050054.
55. Kimmelman J, Mogil JS, Dirnagl U. Distinguishing between exploratory and confirmatory preclinical research will improve translation. PLoS Biol. 2014;12:e1001863. https://doi.org/10.1371/journal.pbio.1001863.
56. Kola I, Landis J. Can the pharmaceutical industry reduce attrition rates? Nat Rev Drug Discov. 2004;3:711–6. https://doi.org/10.1038/nrd1470.
57. Kouznetsova J, Sun W, Martínez-Romero C, Tawa G, Shinn P, Chen CZ, Schimmer A, Sanderson P, McKew JC, Zheng W, García-Sastre A. Identification of 53 compounds that

block Ebola virus-like particle entry via a repurposing screen of approved drugs. Emerg Microbes Infect. 2014;3:1–7. https://doi.org/10.1038/emi.2014.88.

58. Kraft CS, Hewlett AL, Koepsell S, Winkler AM, Kratochvil CJ, Larson L, Varkey JB, Mehta AK, Lyon GM, Friedman-Moraco RJ, Marconi VC, Hill CE, Sullivan JN, Johnson DW, Lisco SJ, Mulligan MJ, Uyeki TM, McElroy AK, Sealy T, Campbell S, Spiropoulou C, Ströher U, Crozier I, Sacra R, Connor MJ, Sueblinvong V, Franch HA, Smith PW, Ribner BS, Nebraska Biocontainment Unit and the Emory Serious Communicable Diseases Unit. The use of TKM-100802 and convalescent plasma in 2 patients with Ebola virus disease in the United States. Clin Infect Dis. 2015;61:496–502. https://doi.org/10.1093/cid/civ334.
59. Kuhn JH, Andersen KG, Bào Y, Bavari S, Becker S, Bennett RS, Bergman NH, Blinkova O, Bradfute S, Brister JR, Bukreyev A, Chandran K, Chepurnov AA, Davey RA, Dietzgen RG, Doggett NA, Dolnik O, Dye JM, Enterlein S, Fenimore PW, Formenty P, Freiberg AN, Garry RF, Garza NL, Gire SK, Gonzalez J-P, Griffiths A, Happi CT, Hensley LE, Herbert AS, Hevey MC, Hoenen T, Honko AN, Ignatyev GM, Jahrling PB, Johnson JC, Johnson KM, Kindrachuk J, Klenk H-D, Kobinger G, Kochel TJ, Lackemeyer MG, Lackner DF, Leroy EM, Lever MS, Mühlberger E, Netesov SV, Olinger GG, Omilabu SA, Palacios G, Panchal RG, Park DJ, Patterson JL, Paweska JT, Peters CJ, Pettitt J, Pitt L, Radoshitzky SR, Ryabchikova EI, Saphire EO, Sabeti PC, Sealfon R, Shestopalov AM, Smither SJ, Sullivan NJ, Swanepoel R, Takada A, Towner JS, van der Groen G, Volchkov VE, Volchkova VA, Wahl-Jensen V, Warren TK, Warfield KL, Weidmann M, Nichol ST. Filovirus RefSeq entries: evaluation and selection of filovirus type variants, type sequences, and names. Viruses. 2014;6:3663–82. https://doi.org/10.3390/v6093663.
60. Kuo Y-C, Chung J-F. Physicochemical properties of nevirapine-loaded solid lipid nanoparticles and nanostructured lipid carriers. Colloids Surf B. 2011;83:299–306. https://doi.org/10.1016/j.colsurfb.2010.11.037.
61. Lamontagne F, Fowler RA, Adhikari NK, Murthy S, Brett-Major DM, Jacobs M, Uyeki TM, Vallenas C, Norris SL, Fischer WA, Fletcher TE, Levine AC, Reed P, Bausch DG, Gove S, Hall A, Shepherd S, Siemieniuk RA, Lamah M-C, Kamara R, Nakyeyune P, Soka MJ, Edwin A, Hazzan AA, Jacob ST, Elkarsany MM, Adachi T, Benhadj L, Clément C, Crozier I, Garcia A, Hoffman SJ, Guyatt GH. Evidence-based guidelines for supportive care of patients with Ebola virus disease. Lancet (London, England). 2018;391:700–8. https://doi.org/10.1016/S0140-6736(17)31795-6.
62. Larson N, Ghandehari H. Polymeric conjugates for drug delivery. Chem Mater. 2012;24:840–53. https://doi.org/10.1021/cm2031569.
63. Leung AKK, Tam YYC, Cullis PR. Lipid nanoparticles for short interfering RNA delivery. Adv Genet. 2014;88:71–100. https://doi.org/10.1016/B978-0-12-800148-6.00004-3.
64. Li T, Cipolla D, Rades T, Boyd BJ. Drug nanocrystallisation within liposomes. J Control Release. 2018;288:96–110. https://doi.org/10.1016/j.jconrel.2018.09.001.
65. Li TCM, Chan MCW, Lee N. Clinical implications of antiviral resistance in influenza. Viruses. 2015;7:4929–44. https://doi.org/10.3390/v7092850.
66. Luthra P, Liang J, Pietzsch CA, Khadka S, Edwards MR, Wei S, De S, Posner B, Bukreyev A, Ready JM, Basler CF. A high throughput screen identifies benzoquinoline compounds as inhibitors of Ebola virus replication. Antivir Res. 2018;150:193–201. https://doi.org/10.1016/j.antiviral.2017.12.019.
67. Madelain V, Nguyen THT, Olivo A, de Lamballerie X, Guedj J, Taburet A-M, Mentré F. Ebola virus infection: review of the pharmacokinetic and Pharmacodynamic properties of drugs considered for testing in human efficacy trials. Clin Pharmacokinet. 2016;55:907–23. https://doi.org/10.1007/s40262-015-0364-1.
68. Madelain V, Oestereich L, Graw F, Nguyen THT, de Lamballerie X, Mentré F, Günther S, Guedj J. Ebola virus dynamics in mice treated with favipiravir. Antivir Res. 2015;123:70–7. https://doi.org/10.1016/j.antiviral.2015.08.015.
69. Madrid PB, Panchal RG, Warren TK, Shurtleff AC, Endsley AN, Green CE, Kolokoltsov A, Davey R, Manger ID, Gilfillan L, Bavari S, Tanga MJ. Evaluation of Ebola virus

inhibitors for drug repurposing. ACS Infect Dis. 2015;1:317–26. https://doi.org/10.1021/acsinfecdis.5b00030.
70. Maggiora GM. The reductionist paradox: are the laws of chemistry and physics sufficient for the discovery of new drugs? J Comput Aided Mol Des. 2011;25:699–708. https://doi.org/10.1007/s10822-011-9447-8.
71. Manicassamy B, Wang J, Jiang H, Rong L. Comprehensive analysis of Ebola virus GP1 in viral entry. J Virol. 2005;79:4793–805. https://doi.org/10.1128/JVI.79.8.4793-4805.2005.
72. Martić-Kehl MI, Schibli R, Schubiger PA. Can animal data predict human outcome? Problems and pitfalls of translational animal research. Eur J Nucl Med Mol Imaging. 2012;39:1492–6. https://doi.org/10.1007/s00259-012-2175-z.
73. Michelow IC, Lear C, Scully C, Prugar LI, Longley CB, Yantosca LM, Ji X, Karpel M, Brudner M, Takahashi K, Spear GT, Ezekowitz RAB, Schmidt EV, Olinger GG. High-dose mannose-binding lectin therapy for Ebola virus infection. J Infect Dis. 2011;203:175–9. https://doi.org/10.1093/infdis/jiq025.
74. Ministry of Health and Labour Welfare. Report on the deliberation results. 2011.
75. Miyata K, Christie RJ, Kataoka K. Polymeric micelles for nano-scale drug delivery. React Funct Polym. 2011;71:227–34. https://doi.org/10.1016/j.reactfunctpolym.2010.10.009.
76. Moller-Tank S, Maury W. Ebola virus entry: a curious and complex series of events. PLoS Pathog. 2015;11:e1004731. https://doi.org/10.1371/journal.ppat.1004731.
77. Mühlberger E. Filovirus replication and transcription. Future Virol. 2007;2:205–15. https://doi.org/10.2217/17460794.2.2.205.
78. Mukherjee S, Ray S, Thakur RS. Solid lipid nanoparticles: a modern formulation approach in drug delivery system. Indian J Pharm Sci. 2009;71:349–58. https://doi.org/10.4103/0250-474X.57282.
79. Mulherkar N, Raaben M, de la Torre JC, Whelan SP, Chandran K. The Ebola virus glycoprotein mediates entry via a non-classical dynamin-dependent macropinocytic pathway. Virology. 2011;419:72–83. https://doi.org/10.1016/j.virol.2011.08.009.
80. Nabi B, Rehman S, Baboota S, Ali J. Insights on oral drug delivery of lipid nanocarriers: a win-win solution for augmenting bioavailability of antiretroviral drugs. AAPS PharmSciTech. 2019;20:60. https://doi.org/10.1208/s12249-018-1284-9.
81. Nagavarma BVN, Yadav HKS, Ayaz A, Vasudha LS, Shivkumar HG. Different techniques for preparation of polymeric nanoparticles-a review. Asian J Pharm Clin Res. 2012;5:16–23.
82. New antibodies best ZMapp in Ebola trial. Nat Biotechnol. 2019;37:1105. https://doi.org/10.1038/s41587-019-0284-y.
83. O'Hearn A, Wang M, Cheng H, Lear-Rooney CM, Koning K, Rumschlag-Booms E, Varhegyi E, Olinger G, Rong L. Role of EXT1 and glycosaminoglycans in the early stage of filovirus entry. J Virol. 2015;89:5441–9. https://doi.org/10.1128/JVI.03689-14.
84. O'Keefe EP. siRNAs and shRNAs: tools for protein knockdown by gene silencing. Mater Methods. 2013;cn3:10.13070/mm.cn.3.197.
85. Oestereich L, Lüdtke A, Wurr S, Rieger T, Muñoz-Fontela C, Günther S. Successful treatment of advanced Ebola virus infection with T-705 (favipiravir) in a small animal model. Antivir Res. 2014;105:17–21. https://doi.org/10.1016/j.antiviral.2014.02.014.
86. Oshikoya KA, Oreagba IA, Ogunleye OO, Lawal S, Senbanjo IO. Clinically significant interactions between antiretroviral and co-prescribed drugs for HIV-infected children: profiling and comparison of two drug databases. Ther Clin Risk Manag. 2013;9:215. https://doi.org/10.2147/TCRM.S44205.
87. Phoolcharoen W, Bhoo SH, Lai H, Ma J, Arntzen CJ, Chen Q, Mason HS. Expression of an immunogenic Ebola immune complex in Nicotiana benthamiana. Plant Biotechnol J. 2011;9:807–16. https://doi.org/10.1111/j.1467-7652.2011.00593.x.
88. Pokhrel R, Sompornpisut P, Chapagain P, Olson B, Gerstman B, Pandey RB. Domain rearrangement and denaturation in Ebola virus protein VP40. AIP Adv. 2018;8:125129. https://doi.org/10.1063/1.5063474.
89. Prabhu S, Poulose EK. Silver nanoparticles: mechanism of antimicrobial action, synthesis, medical applications, and toxicity effects. Int Nano Lett. 2012;2:32. https://doi.org/10.1186/2228-5326-2-32.

90. PREVAIL II Writing Group, Multi-National PREVAIL II Study Team, Davey RT, Dodd L, Proschan MA, Neaton J, Neuhaus Nordwall J, Koopmeiners JS, Beigel J, Tierney J, Lane HC, Fauci AS, Massaquoi MBF, Sahr F, Malvy D. A randomized, controlled trial of ZMapp for Ebola virus infection. N Engl J Med. 2016;375:1448–56. https://doi.org/10.1056/NEJMoa1604330.
91. Puri A, Loomis K, Smith B, Lee J-H, Yavlovich A, Heldman E, Blumenthal R. Lipid-based nanoparticles as pharmaceutical drug carriers: from concepts to clinic. Crit Rev Ther Drug Carrier Syst. 2009;26:523–80. https://doi.org/10.1615/critrevtherdrugcarriersyst.v26.i6.10.
92. Qian X, Nguyen HN, Song MM, Hadiono C, Ogden SC, Hammack C, Yao B, Hamersky GR, Jacob F, Zhong C, Yoon K, Jeang W, Lin L, Li Y, Thakor J, Berg DA, Zhang C, Kang E, Chickering M, Nauen D, Ho C-Y, Wen Z, Christian KM, Shi P-Y, Maher BJ, Wu H, Jin P, Tang H, Song H, Ming G. Brain-region-specific organoids using mini-bioreactors for modeling ZIKV exposure. Cell. 2016;165:1238–54. https://doi.org/10.1016/j.cell.2016.04.032.
93. Qiu X, Wong G, Fernando L, Audet J, Bello A, Strong J, Alimonti JB, Kobinger GP. mAbs and Ad-vectored IFN-α therapy rescue Ebola-infected nonhuman primates when administered after the detection of viremia and symptoms. Sci Transl Med. 2013;5:207ra143. https://doi.org/10.1126/scitranslmed.3006605.
94. Ratemi E. pH-responsive polymers for drug delivery applications. In: Stimuli responsive polymeric Nanocarriers for drug delivery applications, Vol 1. Elsevier; 2018. pp. 121–141. https://doi.org/10.1016/B978-0-08-101997-9.00005-9.
95. Reynard O, Nguyen X-N, Alazard-Dany N, Barateau V, Cimarelli A, Volchkov V. Identification of a new ribonucleoside inhibitor of Ebola virus replication. Viruses. 2015;7:6233–40. https://doi.org/10.3390/v7122934.
96. Ríos-Huerta R, Monreal-Escalante E, Govea-Alonso DO, Angulo C, Rosales-Mendoza S. Expression of an immunogenic LTB-based chimeric protein targeting Zaire ebolavirus epitopes from GP1 in plant cells. Plant Cell Rep. 2017;36:355–65. https://doi.org/10.1007/s00299-016-2088-6.
97. Rivera A, Messaoudi I. Molecular mechanisms of Ebola pathogenesis. J Leukoc Biol. 2016;100:889–904. https://doi.org/10.1189/jlb.4RI0316-099RR.
98. Rojas M, Monsalve DM, Pacheco Y, Acosta-Ampudia Y, Ramírez-Santana C, Ansari AA, Gershwin ME, Anaya J-M. Ebola virus disease: an emerging and re-emerging viral threat. J Autoimmun. 2020;106:102375. https://doi.org/10.1016/j.jaut.2019.102375.
99. Sakurai Y, Sakakibara N, Toyama M, Baba M, Davey RA. Novel amodiaquine derivatives potently inhibit Ebola virus infection. Antivir Res. 2018;160:175–82. https://doi.org/10.1016/j.antiviral.2018.10.025.
100. Sanvicens N, Marco MP. Multifunctional nanoparticles – properties and prospects for their use in human medicine. Trends Biotechnol. 2008;26:425–33. https://doi.org/10.1016/j.tibtech.2008.04.005.
101. Schieffelin JS, Shaffer JG, Goba A, Gbakie M, Gire SK, Colubri A, Sealfon RSG, Kanneh L, Moigboi A, Momoh M, Fullah M, Moses LM, Brown BL, Andersen KG, Winnicki S, Schaffner SF, Park DJ, Yozwiak NL, Jiang P-P, Kargbo D, Jalloh S, Fonnie M, Sinnah V, French I, Kovoma A, Kamara FK, Tucker V, Konuwa E, Sellu J, Mustapha I, Foday M, Yillah M, Kanneh F, Saffa S, Massally JLB, Boisen ML, Branco LM, Vandi MA, Grant DS, Happi C, Gevao SM, Fletcher TE, Fowler RA, Bausch DG, Sabeti PC, Khan SH, Garry RF, KGH Lassa Fever Program, Viral Hemorrhagic Fever Consortium, WHO Clinical Response Team. Clinical illness and outcomes in patients with Ebola in Sierra Leone. N Engl J Med. 2014;371:2092–100. https://doi.org/10.1056/NEJMoa1411680.
102. Schornberg K, Matsuyama S, Kabsch K, Delos S, Bouton A, White J. Role of endosomal Cathepsins in entry mediated by the Ebola virus glycoprotein. J Virol. 2006;80:4174–8. https://doi.org/10.1128/JVI.80.8.4174-4178.2006.
103. Schornberg KL, Shoemaker CJ, Dube D, Abshire MY, Delos SE, Bouton AH, White JM. 5 1-Integrin controls ebolavirus entry by regulating endosomal cathepsins. Proc Natl Acad Sci. 2009;106:8003–8. https://doi.org/10.1073/pnas.0807578106.

104. Shah PP, Wang T, Kaletsky RL, Myers MC, Purvis JE, Jing H, Huryn DM, Greenbaum DC, Smith AB, Bates P, Diamond SL. A small-molecule Oxocarbazate inhibitor of human Cathepsin L blocks severe acute respiratory syndrome and Ebola Pseudotype virus infection into human embryonic kidney 293T cells. Mol Pharmacol. 2010;78:319–24. https://doi.org/10.1124/mol.110.064261.
105. Sharma AK, Sharma MK. Plants as bioreactors: recent developments and emerging opportunities. Biotechnol Adv. 2009;27:811–32. https://doi.org/10.1016/j.biotechadv.2009.06.004.
106. Shelley H, Babu RJ. Role of Cyclodextrins in nanoparticle-based drug delivery systems. J Pharm Sci. 2018;107:1741–53. https://doi.org/10.1016/j.xphs.2018.03.021.
107. Singh A, Garg G, Sharma PK. Nanospheres: a novel approach for targeted drug delivery system. Int J Pharm Sci Rev Res. 2010;5:84–8.
108. Smietana K, Siatkowski M, Møller M. Trends in clinical success rates. Nat Rev Drug Discov. 2016;15:379–80. https://doi.org/10.1038/nrd.2016.85.
109. Snoeck R, Sakuma T, De Clercq E, Rosenberg I, Holy A. (S)-1-(3-hydroxy-2-phosphonylmethoxypropyl)cytosine, a potent and selective inhibitor of human cytomegalovirus replication. Antimicrob Agents Chemother. 1988;32:1839–44. https://doi.org/10.1128/AAC.32.12.1839.
110. Spurney CF, Gordish-Dressman H, Guerron AD, Sali A, Pandey GS, Rawat R, Van Der Meulen JH, Cha H-J, Pistilli EE, Partridge TA, Hoffman EP, Nagaraju K. Preclinical drug trials in the mdx mouse: assessment of reliable and sensitive outcome measures. Muscle Nerve. 2009;39:591–602. https://doi.org/10.1002/mus.21211.
111. Sterne JA, Hernán MA, Reeves BC, Savović J, Berkman ND, Viswanathan M, Henry D, Altman DG, Ansari MT, Boutron I, Carpenter JR, Chan A-W, Churchill R, Deeks JJ, Hróbjartsson A, Kirkham J, Jüni P, Loke YK, Pigott TD, Ramsay CR, Regidor D, Rothstein HR, Sandhu L, Santaguida PL, Schünemann HJ, Shea B, Shrier I, Tugwell P, Turner L, Valentine JC, Waddington H, Waters E, Wells GA, Whiting PF, Higgins JP. ROBINS-I: a tool for assessing risk of bias in non-randomised studies of interventions. BMJ. 2016;355:i4919. https://doi.org/10.1136/bmj.i4919.
112. Suresh K, Chandrashekara S. Sample size estimation and power analysis for clinical research studies. J Human Reprod Sci. 2012;5:7–13. https://doi.org/10.4103/0974-1208.97779.
113. Taniguchi S. Reston ebolavirus antibodies in bats, the Philippines. Emerg Infect Dis. 2011;17:1559. https://doi.org/10.3201/eid1708.101693.
114. Teissier E, Zandomeneghi G, Loquet A, Lavillette D, Lavergne J-P, Montserret R, Cosset F-L, Böckmann A, Meier BH, Penin F, Pécheur E-I. Mechanism of inhibition of enveloped virus membrane fusion by the antiviral drug Arbidol. PLoS One. 2011;6:e15874. https://doi.org/10.1371/journal.pone.0015874.
115. Thi EP, Mire CE, Lee ACH, Geisbert JB, Zhou JZ, Agans KN, Snead NM, Deer DJ, Barnard TR, Fenton KA, MacLachlan I, Geisbert TW. Lipid nanoparticle siRNA treatment of Ebola-virus-Makona-infected nonhuman primates. Nature. 2015;521:362–5. https://doi.org/10.1038/nature14442.
116. Thomford N, Senthebane D, Rowe A, Munro D, Seele P, Maroyi A, Dzobo K. Natural products for drug discovery in the 21st century: innovations for novel drug discovery. Int J Mol Sci. 2018;19:1578. https://doi.org/10.3390/ijms19061578.
117. Tripathy S, Dassarma B, Bhattacharya M, Matsabisa MG. Plant-based vaccine research development against viral diseases with emphasis on Ebola virus disease: a review study. Curr Opin Pharmacol. 2021;60:261–7. https://doi.org/10.1016/j.coph.2021.08.001.
118. USFDA, FDA approves first treatment for Ebola virus [WWW Document]. FDA NEWS RELEASE; 2020. https://www.fda.gov/news-events/press-announcements/fda-approves-first-treatment-ebola-virus
119. Vyas SP, Subhedar R, Jain S. Development and characterization of emulsomes for sustained and targeted delivery of an antiviral agent to liver. J Pharm Pharmacol. 2010;58:321–6. https://doi.org/10.1211/jpp.58.3.0005.

120. Warren TK, Jordan R, Lo MK, Ray AS, Mackman RL, Soloveva V, Siegel D, Perron M, Bannister R, Hui HC, Larson N, Strickley R, Wells J, Stuthman KS, Van Tongeren SA, Garza NL, Donnelly G, Shurtleff AC, Retterer CJ, Gharaibeh D, Zamani R, Kenny T, Eaton BP, Grimes E, Welch LS, Gomba L, Wilhelmsen CL, Nichols DK, Nuss JE, Nagle ER, Kugelman JR, Palacios G, Doerffler E, Neville S, Carra E, Clarke MO, Zhang L, Lew W, Ross B, Wang Q, Chun K, Wolfe L, Babusis D, Park Y, Stray KM, Trancheva I, Feng JY, Barauskas O, Xu Y, Wong P, Braun MR, Flint M, McMullan LK, Chen S-S, Fearns R, Swaminathan S, Mayers DL, Spiropoulou CF, Lee WA, Nichol ST, Cihlar T, Bavari S. Therapeutic efficacy of the small molecule GS-5734 against Ebola virus in rhesus monkeys. Nature. 2016;531:381–5. https://doi.org/10.1038/nature17180.
121. Warren TK, Wells J, Panchal RG, Stuthman KS, Garza NL, Van Tongeren SA, Dong L, Retterer CJ, Eaton BP, Pegoraro G, Honnold S, Bantia S, Kotian P, Chen X, Taubenheim BR, Welch LS, Minning DM, Babu YS, Sheridan WP, Bavari S. Protection against filovirus diseases by a novel broad-spectrum nucleoside analogue BCX4430. Nature. 2014;508:402–5. https://doi.org/10.1038/nature13027.
122. Warren TK, Whitehouse CA, Wells J, Welch L, Heald AE, Charleston JS, Sazani P, Reid SP, Iversen PL, Bavari S. A single phosphorodiamidate morpholino oligomer targeting VP24 protects rhesus monkeys against lethal Ebola virus infection. MBio. 2015;6 https://doi.org/10.1128/mBio.02344-14.
123. Watanabe S, Noda T, Kawaoka Y. Functional mapping of the nucleoprotein of Ebola virus. J Virol. 2006;80:3743–51. https://doi.org/10.1128/JVI.80.8.3743-3751.2006.
124. Watt PC, Sawicki MP, Passaro E. A review of gene transfer techniques. Am J Surg. 1993;165:350–4. https://doi.org/10.1016/S0002-9610(05)80841-4.
125. WHO (World Health Organization). Notes for the record: consultation on monitored emergency use of unregistered and investigational interventions for Ebola Virus Disease (EVD). 2018.
126. WHO (World Health Organization). Ebola virus disease [WWW Document]; 2018. https://www.who.int/en/news-room/fact-sheets/detail/ebola-virus-disease
127. World Health Organization (WHO). Categorization and prioritization of drugs for consideration for testing or use in patients infected with Ebola. 2015.
128. Yu D-S, Weng T-H, Wu X-X, Wang FXC, Lu X-Y, Wu H-B, Wu N-P, Li L-J, Yao H-P. The lifecycle of the Ebola virus in host cells. Oncotarget. 2017;8:55750–9. https://doi.org/10.18632/oncotarget.18498.
129. Zhu J-D, Meng W, Wang X-J, Wang H-CR. Broad-spectrum antiviral agents. Front Microbiol. 2015;6 https://doi.org/10.3389/fmicb.2015.00517.
130. Zilbermintz L, Leonardi W, Jeong S-Y, Sjodt M, McComb R, Ho C-LC, Retterer C, Gharaibeh D, Zamani R, Soloveva V, Bavari S, Levitin A, West J, Bradley KA, Clubb RT, Cohen SN, Gupta V, Martchenko M. Identification of agents effective against multiple toxins and viruses by host-oriented cell targeting. Sci Rep. 2015;5:13476. https://doi.org/10.1038/srep13476.

Polymers for Biosensing Applications in Viral Detection and Diagnosis

Kavyashree Puttananjegowda, Arash Takshi, and Sylvia Thomas

Abstract Technologies for the detection of viruses that are rapid, precise, portable, and on a wide scale are essential for diagnosing the many viral diseases. In order to reduce the transmission of the virus and to contain the illness epidemic, early detection and treatment are crucial. When subjected to extensive and quick testing, the present standard technologies reveal practical constraints and difficulties.

Due to its numerous benefits, including high selectivity and sensitivity, quick detection, low cost, simplicity, flexibility, extended self-life, and ease of use, biosensors, in particular polymer-based biosensors, are seen as prospective alternatives. In order to facilitate the development of point-of-care (POC) systems and home-use biosensors for viral detection, polymer-based biosensors can act as multisensory, mobile biosensors, and wearable biosensors. This book chapter provides various techniques of virus detection using polymers and also provides different strategies for virus biosensing and diagnosis.

Keywords Biosensors · Diagnosis · Detection · Polymers · Virus

1 Introduction

A great number of pathogens are present in the world, and they are primarily responsible for a wide range of diseases in people and animals. Viruses, bacteria, fungus, protozoa, and worms are among the pathogens classified into five categories. They enter the human body and proliferate using the body's resources before departing and infecting a new host, weakening the immune system in the process. Viruses are unique among diseases in that they kill host cells. Pathogens have been a threat to humans for ages, and our capacity to minimize morbidity is evident when we look at our average life duration [1].

K. Puttananjegowda (✉) · A. Takshi · S. Thomas
Department of Electrical Engineering, University of South Florida, Tampa, FL, USA

R. Shegokar, Y. Pathak (eds.), *Viral Drug Delivery Systems*,
https://doi.org/10.1007/978-3-031-20537-8_9

Certain viruses, on the other hand, have sparked renewed interest in biosensor designs in recent years, as their repeated protein subunits arrayed at nanometric spacing can be used to couple useful molecules. They provide efficient immobilization of analyte-specific recognition and detector elements such as antibodies and enzymes at greatest surface densities when utilized as adapters on sensor chip surfaces. When used repeatedly, the display on viral bio-nanoparticles may lead to long-term stability of sensor molecules and has the potential to significantly improve sensor performance when compared to conventional layouts. This has been demonstrated in a variety of biosensor proof-of-concept devices. As a result, widely available plant viral particles that are non-pathogenic to animals or humans could take on new significance if used in the receptor layers of virus biosensing devices.

Infectious diseases spread quickly from contaminated water, food, and bodily fluids, killing humans and animals all over the world. Viruses are among the most dangerous infectious agents, posing a severe threat to public health and the global economy since they are typically difficult to detect and treat. A number of sensors have been described so far, owing to the importance of developing speedy, precise, cost-effective, and in situ approaches for early detection viruses. As a result, early detection and treatment are critical for slowing the virus's transmission and containing the illness outbreak. As a result, new diagnostic techniques and devices for viral identification in clinical samples are required that are faster, more accurate and reliable, easier to use, and less expensive than current ones. Because of its small size, quick response time, label-free functioning without the need for costly and time-consuming labeling steps, the ability to do real-time and multiplexed measurements, and the fact that they are portable and wearable, biosensors are becoming increasingly popular. Antigens, complete particles, antibodies, and nucleic acids such as ribonucleic acids (RNAs) and deoxyribonucleic acids (DNAs) are all used to detect viruses [30–39].

Biosensors, specifically those based on electrochemical, are being investigated as prospective alternatives due to their numerous benefits, including excellent selectivity and sensitivity, rapid detection, low cost, simplicity, adaptability, long self-life, and ease of use. As a result, electrochemical-based biosensors can be used as implantable sensors, mobile biosensors, and wearable biosensors, easing the creation of point-of-care (POC) systems and home-use biosensors for glucose monitoring, virus detection, and heart rate monitoring [2–23]. However, using these biosensors to detect viruses has a number of difficulties, including degradation, limited crystallinity, charge transport characteristics, and weak interaction with biomarkers. To address these issues, this research provides scientific evidence for the possible applications of electrochemical biosensors in virus detection, based on their detection of diverse biomarkers such DNA/RNA, proteins, entire viruses, and antigens. Then, for various types of virus detection, there are promising methodologies for the creation of electrochemical-based biosensors [24, 25].

Many novel materials, including gold nanoparticles, carbon, graphene, graphene oxides, metal oxides, electrically conducting and organic polymers, and carbon nanotubes (CNTs), have been discovered and used to manufacture electrodes for biosensors in recent decades. When compared to other materials, polymers have several unique qualities due to their unique orbital structure and chain conformation modifications, which result in high sensitivity and selectivity for specific biological compounds, as well as quick electrical impulses when used in biosensors. Polymer-based biosensors are also predicted to benefit from the simplicity with which polymer properties can be tailored. It's simple to customize polymer characteristics by functionalizing or connecting polymer monomers with different functional groups, which can result in significant improvements in electronic properties and electrical stability. Furthermore, various critical characteristics, such as size, shape, structure, electrochemical conductivity, and morphology, have been shown to influence the efficacy of polymer-based biosensors. As a result, a range of techniques have been used to improve sensitivity, selectivity, flexibility, stability, reproducibility and a variety of recognized bio-analytes in order to expand the uses of polymers in biosensors. Additionally, apart from functional group grafting, the strategies have primarily focused on the construction of polymer nanostructures such as polymer nanowires, nanotubes, and nanospheres, as well as the association of other functional materials to generate hybrid nanoparticles, composites, and hydrogels. Because of its shown potential and benefits, the polymer-based biosensor is one of the most promising technologies for early virus diagnosis. As a result, the development of novel polymer-based electrochemical biosensors is predicted to garner significant interest among potential viral detection research in the near future. However, in order to use polymer-based biosensor technology in similar applications, it is necessary to consolidate the available scientific data. Furthermore, due to their amorphous structure and low stability, charge transport capabilities, and contact with biomolecular biomarkers, the creation of biosensors based on pure polymers poses considerable hurdles and concerns. Polymers can be used to create flexible, wearable, and implantable biosensors [26–29].

As a result, developing improved technology for early and accurate virus diagnosis is critical. Due to its ability to detect diverse biological analytes such as RNA/DNA, pathogens, viruses, toxins, and disease biomarkers, biosensors, which are analytical instruments containing a transducer, are regarded the next-generation diagnostic tools for combating viruses. Because of its great sensitivity, selectivity, cost-effectiveness, and quick response, polymer-based biosensors have gotten a lot of attention. Furthermore, electrochemical biosensors have been shown to have various advantages over traditional diagnostic procedures for detecting viruses and infections, including the potential for producing portable and wearable sensor devices and commercial products. As a result, the biosensor is regarded as an effective, innovative, and promising tool for early viral detection and prevention [30–45].

In this book chapter, recent developments and significant advancements in the field of detection and diagnostic of viruses with various techniques based on electrochemical [33], impedance spectroscopy [34], electrical-resistance [35], microfluidic [36], reverse transcription-polymerase chain reaction (RT-PCR) [37], cyclic voltammetry [38], differential pulse voltammetry [39], conductometric [40], field-effect-transistor [41], fluorescence [42], colorimetric [43], faradic current [44], and plasmonic [45] are presented.

2 Various Strategies for Virus Biosensing and Diagnosis

There are various strategies that can be used for the detection or the identification of viruses such as (i) sensing of intact virus particle, (ii) sensing of virus nucleic acids, (iii) sensing of viral antibodies, (iv) sensing of viral antigens, and (v) electronic nose for viral diagnosis. The detection and identification component is constantly founded on direct estimations of changes in the electrical properties of viral sensors produced by binding actions. Methods focused on the detection of intact virus particles, particular viral antigens, and nucleic acids are better for diagnosing freshly infected cases, whereas antibody detection techniques are better for determining whether an individual has been infected previously. In addition, a more specific diagnostic method was focused on analyzing patients' breath virus samples using the electronic nose technology. Till date, multiple types of sensors equipped with various recognition elements have been effectively used to identify a variety of harmful viruses [46].

2.1 Sensing of Intact Virus Particles

The viral intensity is determined by the number of days since the commencement of the illness. As a result, detecting intact virus particles can offer clinicians with information about the infection's stage or therapy response. Plaque-based techniques, which include inoculating patient samples into grown cell lines and checking for countable plaques in the confluent cell layer, are frequently used to evaluate viral particle concentration. Direct sensing of intact virus particles can offer doctors or clinicians with information about the infection's stage. Virus particle intensities have traditionally been evaluated by cultivating patient samples on permissive mammalian or insect cell lines and searching for cell death. However, this procedure is slow and ineffective for on-site detection and testing.

As viral particles are often charged throughout a broad pH range, sensors may detect the adsorption or binding of charged intact virus particles onto their gate surface in a label-free electrostatic manner. The sensors surface is frequently functionalized with antibodies against specific surface proteins of the viral particle to achieve selectivity and preferentially capture the entire virus. As far as anyone is concerned, the primary immediate and on-site diagnosis and detection of individual influenza A virus particles was demonstrated in 2004 by utilizing functionalized silicon nanowire (SiNW) biosensors [47].

2.2 Sensing of Virus Nucleic Acids

A viral particle's genome is usually made up of either RNA or DNA. The DNA sensor works by immobilizing a single-stranded oligonucleotide on a transducer surface and using surface hybridization to detect its corresponding DNA sequence. After that, a biosensor converts the hybrid produced on the electrode surface into an analytical signal.

The majority of DNA sensors work by immobilizing single-stranded DNA (ssDNA) capture probes onto the sensing device's surface in order to detect DNA hybridization events. During the DNA hybridization process, a probe that produces a double-stranded DNA (dsDNA) or DNA/RNA helix with two reverse-complementary strands identifies target DNA or RNA within a sample. Hence, the increased charge associated with the hybridization caught target molecule would effectively affect the sensing surface charge, changing the output signal of the virus detecting device, because nucleic acids are negatively charged in near-neutral water solution.

Due to DNA biosensor features such as portability, simplicity, cost-effectiveness, fast response time, high sensitivity, high selectivity, and compatibility with miniaturized detecting technologies, they have gotten a lot of attention. The interaction between DNA-DNA, DNA-RNA, and protein-ligand molecules, as well as the conversion of changes in DNA characteristics and structure into an electrochemical signal, is important to the operation of an electrochemical DNA biosensor. There are also indirect signal amplification strategies that involve electrochemical active DNA intercalators, enzymes, redox mediators, and particles. For detecting various viruses, a number of electrochemical DNA biosensors have been developed.

2.3 Sensing of Viral Antibodies

When the immune system is responding to a specific infection, antibody tests are often employed to determine the presence of virus-specific antibodies in the blood of virus hosts. Antibodies are normally formed between 4 and 10 days following a viral infection, whereas antigen responses occur after 2 weeks of immunization. As a result, antibody detection tests can help determine how many people have been exposed to a virus and have developed a symptomatic or asymptomatic virus infection, as well as aid in the creation of inoculations.

The immobilization of viral capture antigens serving as receptors on the gate surface is used to prepare sensing devices for the detection of specific host antibodies against viruses. Charge shifts caused by affinity binding of host target antibodies to viral antigens are detected by these biosensors. The majority of sensors for detecting antibodies against viruses have only been tested in proof-of-concept trials.

The creation of an electrical signal by the particular interaction between antibody and antigen in the presence of an electrochemical transducer is the operating principle of an electrochemical immunosensor. Covalent or non-covalent bonding is used to attach antibodies to the biosensor surface. Immobilizing antibodies on electrodes with biotin-streptavidin or conductive polymers is a standard practice in these applications. Sandwich-type immunosensors use a complex of immobilized capture antibodies, antigen, and detection antibodies after the capture antibodies are immobilized on a sensor surface. Antibodies can be stabilized by immobilizing them on a solid surface. Due to their high affinity and sensitivity, many immunoassay electrochemical sensing platforms have become accessible [13].

2.4 *Sensing of Viral Antigens*

Another method for detecting viruses is to coat electrodes with antigens and use them to detect circulating antibodies. Antigen-based sensors have many of the same challenges as antibody-based sensors, such as reagent stability and immobilization limitations. Synthetic proteins or peptides are commonly used as antigens, despite the fact that they may not contain the correct structure for which an antibody was produced in the body, necessitating extensive validation of these synthetic peptides and antigens in non-biosensor tests. These elements can make fabricating a useful sensor testing, yet considering them in the advancement cycle will enormously help the chances of an effective assays.

Biosensors for the detection of viral proteins are immunological sensors that detect the affinity binding of viral antigens in a sample to specific antibodies, antibody fragments, or fusion proteins, affibodies, or aptamers fixed on the sensor surface and generate detectable signals. The influenza A, Ebola, dengue, and SARS-CoV-2 viruses were identified using label-free viral antigen detection techniques. Sensing devices were used to identify non-structural proteins of the dengue virus as well as hemagglutinin glycoproteins of the highly pathogenic avian influenza virus H5N1. For example, due to infection, virus secrets surface protein termed hemagglutinin (HA). HA is an antigenic homotrimeric integral membrane glycoprotein. SiNW field-effect-transistors (FETs) are one of the compatible method were used to detect HA1 with ultrahigh sensitivity. Using a top-down SiNW-based technique, this is the first demonstration of electrical detection of the HA1 domain in HA, the viral surface protein of the influenza virus [24].

Biosensors that detect viral antigens, such as virion and non-virion proteins, could be used to quickly identify infected individuals, thereby lowering or eliminating the requirement for costly molecular confirmatory tests for viral nucleic acids. Antigen tests could potentially be one approach to significantly increase testing capacity. Antigen tests, on the other hand, are only accurate if the virus's target viral proteins are present in appropriate amounts in a sample. Antigen material is often in short supply in samples from people who have been infected with respiratory viruses. However, once appropriate antigens have been identified, immunology-based detection is an excellent tool for detecting acute or early infection [25].

2.5 *Electronic Nose for Viral Diagnosis*

While biosensors are used for the direct or indirect detection of the virus in a patient body, coronavirus disease can be diagnosed through symptoms in a patient. In the early months of the pandemic, regular symptoms, such as fever and coughing, were monitored by security officials at airports and public places for screening people. However, due to the commonality of those symptoms with the symptoms in other diseases such as cold and flu, the screening method was not effective. Considering the effect of the coronavirus on the respiratory system, a more specific diagnostic method was focused on analyzing patients' breath samples using the electronic nose (also known as e-nose) technology. Miller et al. relied on the studies of the difference in the concentrations of various gases and volatile organic compounds (VOCs) exhaled from a healthy person and a patient being diagnosed with COVID-19 and developed a breathalyzer using an in-house-made e-nose [48]. Studies show that, specifically, monitoring acetone and NOx in a person's breath is a reliable method for diagnosing respiratory diseases. However, being very contagious, a rigorous disinfection process has to be applied on a breathalyzer before using it for testing another patient. In this regard, it is highly demanded to develop e-nose systems with disposable sensors. This requires shifting from the conventional metal-oxide-based gas sensors to new technologies using polymer-based VOC and gas sensors.

3 Techniques of Virus Detection Using Polymers

Portable, accurate, and point-of-care diagnostic techniques for the detection of viruses are important for regulating the viral infections and diseases. In the event of vast and rapid testing, current conventional methods, such as reverse transcription-polymerase chain reaction (RT-PCR), serological assays, and computed tomography (CT), have practical constraints and obstacles.

Until this point, polymers have been utilized as straightforward and viable devices for scientific examinations due to the essential changes in the physico-chemical qualities which quickly create advancements in charge distribution, orientation or direction, and electronic structures of polymers. The prompt transduction of improvements into reaction empowered polymers to be applied as a fast identification examine that reacted to the intermolecular communication of the objective and comparing biorecognition components, like antibodies, chemicals, and proteins.

In particular, the polymer-based biosensor is one of the most encouraging innovations for the early finding of virus infections because of its established potential and benefits. As a result, in the near future, the development of new polymer-based biosensor technology is predicted to gain substantial interest among possible studies connected to various forms of viral detection. Table 1 summarizes some recent results on several virus detection methods employing polymers that have been

Table 1 Various detection methods of viruses using polymers

Detection method	Virus/biomarkers	Polymer	Ref
Electrochemical	MicroRNAs	Polyaniline (PANI)	[33]
Impedance	Hepatitis E virus	Polyaniline (PANI)	[34]
Electrical resistance	M13 virus	Poly(3,4ethylenedioxythiophene) (PEDOT)	[35]
Microfluidic	DNA	Poly(3,4ethylenedioythiophene)polystyrene sulfonate (PEDOT:PSS)	[36]
RT-qPCR	H_3N_2 viruses	Polydimethylsiloxane (PDMS)	[37]
Cyclic voltammetry	Cowpea mosaic virus	Polypyrrole (PPy)	[38]
Differential pulse voltammetry	Human papilloma virus	Polythionine (PTH)	[39]
Conductometric	Bovine viral diarrhea virus	Polyaniline (PANI)	[40]
Field-effect-transistor	Rotavirus	Polydimethylsiloxane (PDMS)	[41]
Fluorescence	Human cytomegaloviruses	Polystyrene (PS)	[42]
Colorimetric	H_1N_1 virus	Polydiacetylene (PDA)	[43]

published in the literature, including the detection method, virus/biomarkers, and polymer type. This chapter is oriented toward the various detection methods for the viruses [33–45].

3.1 *Electrochemical-Based Virus Detection*

In electrochemical biosensors, new functional electrically conducting materials play a critical role. Because of its established potential and benefits, conducting polymer-based electrochemical biosensor is one of the most encouraging strategies for immediate virus diagnosis. As a result, the development of novel conducting polymer-based electrochemical biosensors is predicted to garner significant interest among potential viral detection research in the near future. However, in order to use electrochemical-based biosensor technology in similar applications, it is necessary to consolidate the available scientific data.

Figure 1 shows how researchers created ultrasensitive and low-fouling microRNA electrochemical biosensors by adding thiol-terminated antifouling compounds such as peptide sequences onto the surface of polyaniline-modified electrodes. The electrochemical biosensor made from peptide-functionalized polyaniline (PANI) has a highly selective property to complement the RNAs when compared to RNA biosensors without surface modification and antifouling capability. It was also established that the presence of antifouling peptides on the PANI surface has no effect on the biosensor's sensitivity. As a result, functionalizing polymers with peptides or other low-fouling compounds could be a promising technique for developing anti-fouling biosensors for detecting viruses [33].

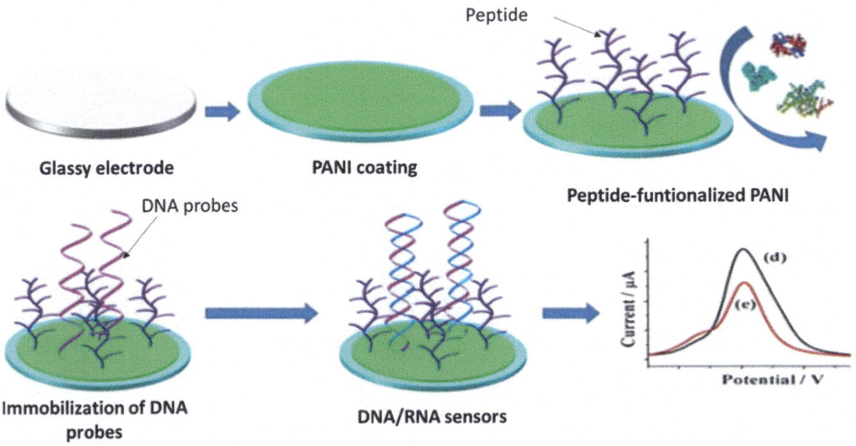

Fig. 1 The preparation of antifouling RNA biosensors employing peptides for surface modification of PANI polymer is depicted schematically. (Reproduced with permission from [33])

Most intriguingly, the experimental results revealed that adding antifouling molecules to the sensing interfaces had no substantial impact on the biosensor's sensitivity. The approach of developing antifouling biosensors using freshly synthesized zwitterionic peptides and conducting polymers has the potential to be used to the construction of various electrochemical sensors and biosensors that do not meet biofouling.

3.2 *Impedance Measurement for Virus Detection*

The hepatitis E virus (HEV) is one of the most common causes of acute viral hepatitis around the world. Graphene quantum dots and gold-embedded polyaniline nanowires were used to make a pulse-triggered ultrasensitive electrochemical sensor, which was made via an interfacial polymerization and subsequent self-assembly method. In comparison to other traditional electrochemical sensors, introducing an external electrical pulse during the viral accumulation process boosts the sensitivity toward HEV due to the increased surface of the virus particle as well as the antibody-conjugated polyaniline chain length.

The nanocomposites were deposited on a finely electropolymerized polyaniline-coated glassy carbon electrode (GCE) to create an antibody conjugated to nitrogen- and sulfur-co-doped graphene quantum dots (Ab-N,SGQDs) with gold-embedded polyaniline nanowire (AuNP-PAni/PAni)-based HEV sensor, as shown in Fig. 2. This viral sensor was used with several external pulses during the virus loading process to achieve great sensitivity at low virus concentrations, which was tuned to achieve the best results at +0.8 V. The suggested biosensor displays its ability to detect HEV in a wide linear range with a low detection limit under these ideal

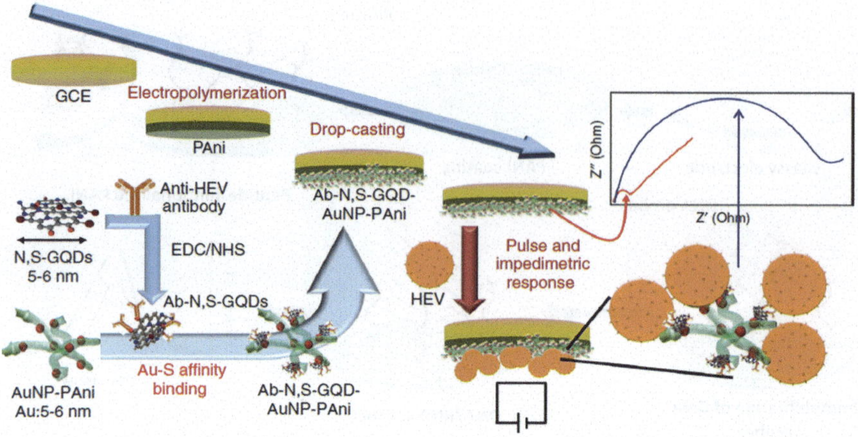

Fig. 2 The electrochemical impedance spectroscopy measurements of the polymeric nanocomposite-based hepatitis E virus sensor electrode are shown schematically [34]

conditions. The suggested biosensor's specificity and sensitivity were next evaluated in HEV-like particles (HEV-LPs) in buffer and then human serum. Furthermore, HEV samples taken from cell culture supernatant and feces of HEV-infected monkeys were utilized to validate its pertinency.

3.3 Electrical Resistance-Based Biosensor for Virus Detection

Biosensing with virus-PEDOT nanowire arrays indicates that fully functional viruses can be successfully included. To make biosensors out of the nanowire arrays, silver contacts and wires were pasted onto the nanowires, exposing the virus-PEDOT nanowires for electrical biosensing measurements; each device exposed 200–300 m lengths of nanowires with hundreds of parallel nanowires. To avoid contact with the liquid, the contacts were painted with an insulating paint, as illustrated in Fig. 3a. The devices were evaluated for resistance values between 30 and 400 k after full drying, indicating that the array had a sufficient number of nanowires for repeatable biosensing. The current across the nanowire array, I, was measured and translated into resistance, $R = E\text{app}/I$, which was recorded in real time during immersion in phosphate-buffered fluoride (PBF) solution using an applied bias, $E\text{app} = 100$ mV. Figure 3b depicts real-time biosensing data with the appropriate injections of negative antibody (n-Ab), positive antibody (p-Ab), or PBF buffer washes. Figure 3c depicts the calibration plot of change in resistance caused by each antibody injection (ΔR), normalized by the electrical resistance (R), measured in a buffer solution PBF. These findings reveal that virus-PEDOT hybrid nanowires are capable of directly electrically transducing an antibody's particular binding to an entrapped virus. The findings show that viruses embedded in PEDOT retain

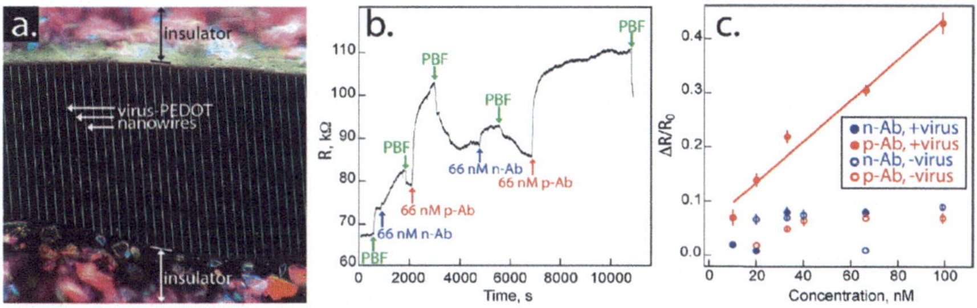

Fig. 3 (**a**) A virus-PEDOT nanowire array device optical microscopic image. (**b**) Real-time biosensing data with the appropriate injections of negative antibody (n-Ab), positive antibody (p-Ab). (**c**) A calibration curve depicting the resistance change upon injection as a function of analyte concentration, compiled from all real-time biosensing data. (Reproduced with permission from [35])

functioning and can change the characteristics of the nanowire by recognizing analytes molecularly. Real-time, reagent-free biosensing could be a valuable tool for detecting and diagnosing diseases early. The disclosed virus-PEDOT nanowires help achieve this goal by allowing for electrical resistance-based sensing in a buffer with physiologically comparable pH, ionicity, and ambient temperature.

3.4 Microfluidic Detection of Virus

The most promising conducting polymer, (CP) poly(3,4-ethylenedioxythiophene) polystyrene sulfonate (PEDOT:PSS), was used as a flexible electrode to fabricate organic electrochemical transistors (OECT)-based biosensors for detecting DNA/RNA, bacteria, glucose, and biomarkers, due to their intrinsic flexibility, tunable conductivity, biocompatibility, and low cost. PEDOT:PSS-based OECT flexible biosensors have been shown to have excellent performance in terms of on-site monitoring, point-of-care detection, and diagnosis. As shown in Fig. 4a, researchers created a DNA biosensor by combining a flexible microfluidic system with an organic electrochemical transistor (OECT) made of a PEDOT:PSS active layer with a gold gate electrode. The OECT transistor is cast and printed on a high-flexibility PET substrate in the proposed system and then integrated with a PDMS-based microfluidic device on top. When both sides of this biosensor system can be bent without causing harm, as shown in Fig. 4b, it demonstrates exceptional flexibility. Figure 4c shows the device's transfer characteristics measured with phosphate-buffered saline solution filled in the microfluidic channel at various bending states, where V_G, V_{DS}, and I_{DS} are the gate voltage, drain voltage, and channel current, respectively. It's worth noting that when the gadget is bent on both sides, the performance is nearly the same, which is critical for real-world applications. Figure 4d depicts the channel current's response to various gate voltages; it reaches steady values after several seconds [32].

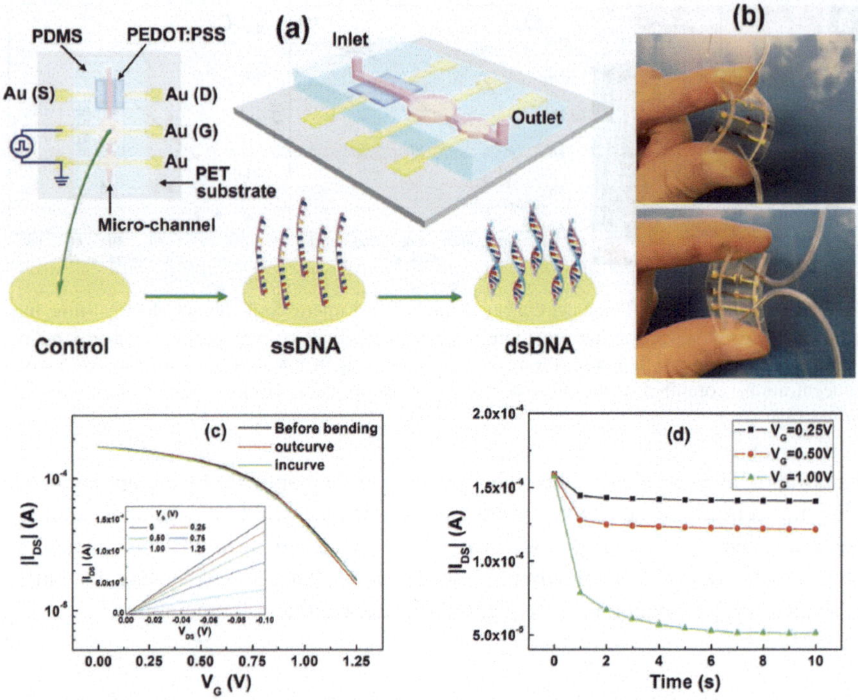

Fig. 4 (**a**) Schematic of an OECT integrated in a flexible microfluidic biosensing device. (**b**) Photographs depict the sensor flexibility. (**c**) An OECT's transfer characteristics and output characteristics (inset) were measured at various bending states. (**d**) At various gate voltages, the time-dependent channel current of an OECT was measured. (Reproduced with permission from [32, 36])

The PEDOT:PSS flexible biosensor device can detect DNA targets at low concentrations of 1 mM in terms of sensing performance. As a result, combining CPs-based OECTs with microfluidic systems could be a promising way to create flexible, highly sensitive, low-cost, and disposable biosensors for DNA and virus detection [32, 36].

3.5 *RT-qPCR-Based Virus Detection*

Influenza epidemics cost the world's economy and people a lot of money every year. Rapid and accurate flu diagnosis procedures, on the other hand, are severely inadequate. Figure 5 shows the direct and selective identification of influenza viruses (H3N2) in diluted exhaled-breath-condensate (EBC) samples obtained from flu patients using the silicon nanowire (SiNW) sensor within minutes. The results shown in Fig. 6 were validated using quantitative reverse transcription-polymerase chain reaction (RT-qPCR) testing, which revealed that stronger PCR signals in EBC

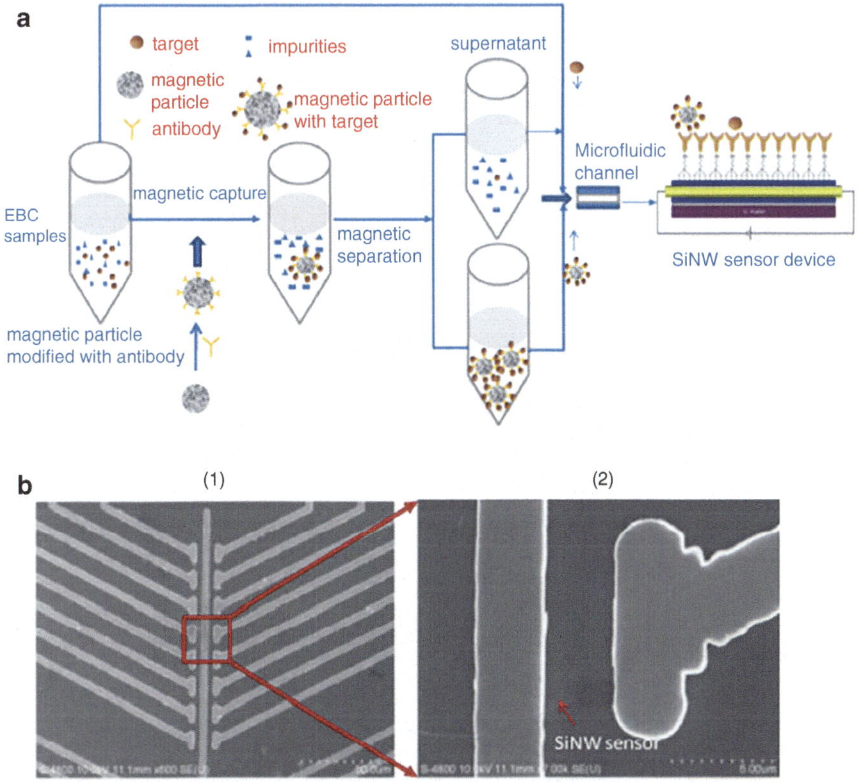

Fig. 5 (**a**) Using a Si nanowire sensor, an experimental approach for viral detection was developed. (**b**) (1) Si nanowire; (2) chip image. (Reproduced with permission from [37])

samples matched to higher SiNW sensor responses. It was also discovered that using magnetic beads improved SiNW sensing for low-level viruses and biomarkers. The research has shown that when calibrated with standards and controls, the SiNW sensor device can be used to quickly diagnose respiratory viral infections in a clinical context [37].

3.6 Cyclic Voltammetry-Based Biosensor for Virus Detection

Researchers are reported that, using the technique that combines colloidal lithography, electrochemical polymerization, and electrostatic adsorption, hierarchical cowpea mosaic virus (CPMV) nanoparticle assemblies were formed on conducting polymer arrays, and the fabrication process is described in-detail as shown in Fig. 7 [38].

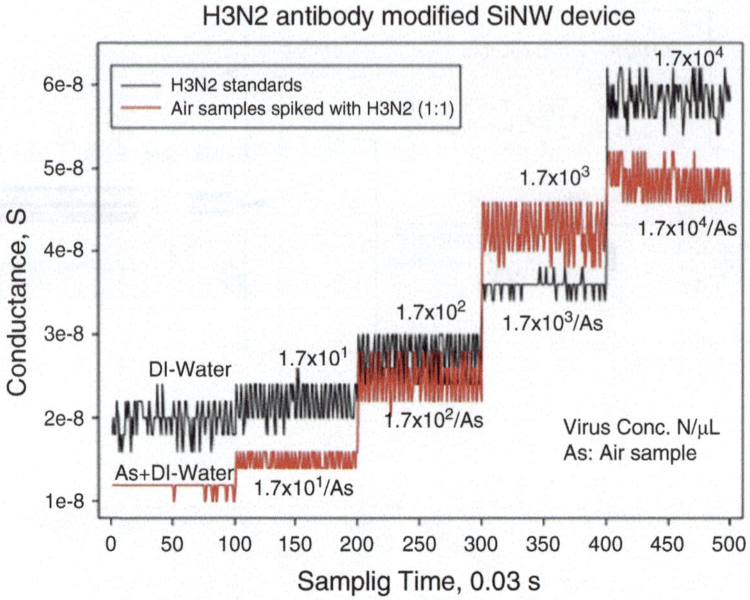

Fig. 6 H3N2 virus detection in air samples utilizing a SiNW device equipped with an H3N2 antibody. (Taken with permission from [37])

In the presence of these ionic probes, Fig. 8 depicts the expected cyclic voltammogram electrochemical behavior of nonpatterned and patterned polymeric arrays, as well as the polymeric CPMV surface. Diffusion of either probe through the electrode is not expected due to the nonpatterned copolymer's inflexible film formation.

3.7 Differential Pulse Voltammetry-Based Biosensor for Virus Detection

In this sensor, the detection of human papillomavirus (HPV) was developed by differential pulse voltammetric (DPV) technique. The fabrication of an enzyme-free electrochemical biosensor based on graphene/Au-nanorod/poly-thionine (G/AuNR/PT) adapted glassy carbon electrode (GCE) for ultra-sensitive detection of HPV DNA is illustrated in Fig. 9. The inclusion of MCH was intended to prevent nonspecific binding sites from forming. Following that, one terminal of TD was hybridized with CP to immobilize it on the electrode, while the other terminal of TD was hybridized with AP1. To hybridize with AP1, AP2 was introduced. The attachment of nanostructured DNA to CP was activated by the released target. Graphene was employed to boost the electrode surface area and electrical conductivity, and AuNRs were used to improve the probe DNA's mobility. Various $[Ru(phen)_3]^{2+}$ were bound

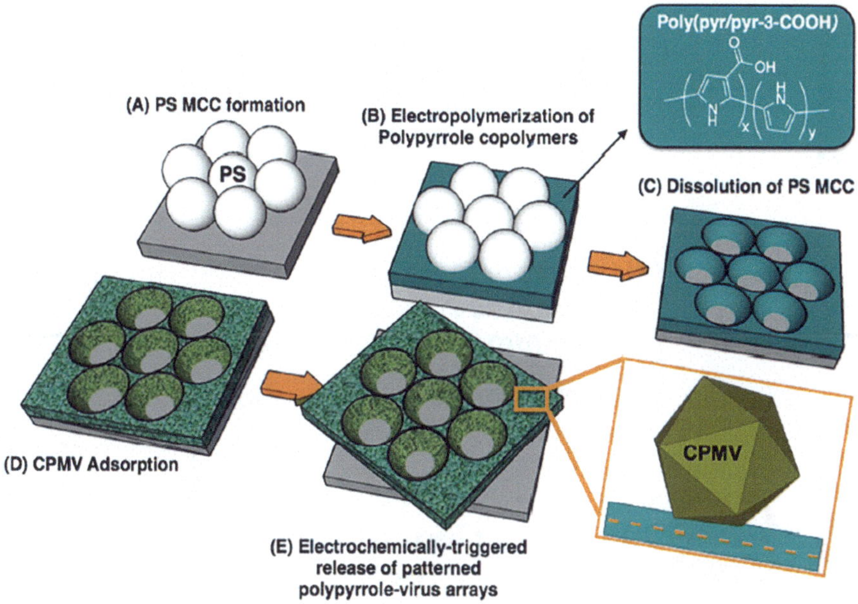

Fig. 7 Stepwise fabrication of free-standing Janus polypyrrole/CPMV arrays: (**a**) Formation of PS monolayer colloidal crystals (MCC), (**b**) electropolymerization of polypyrrole copolymers, (**c**) solvent dissolution of PS MCC, (**d**) adsorption of CPMV on polypyrrole inverse opal, and (**e**) electrochemical lift-off of patterned polypyrrole-virus arrays. (Reproduced with permission from [38])

Fig. 8 Cyclic voltammograms of nonpatterned and patterned CPMV films in an electrolyte solution containing 5 mM $Ru(NH_3)_6^{3+}$. (Reproduced with permission from [38])

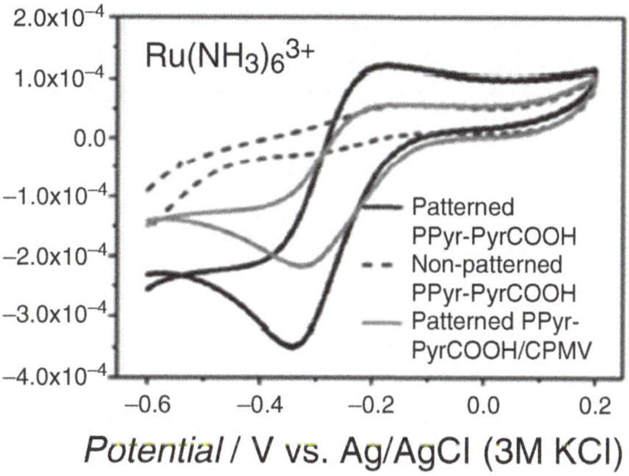

to the DNA nanostructure with negative charges through electrostatic interface and also significantly enhanced electrochemical signals [39].

As demonstrated in Fig. 10, as the concentration of TD was raised, the DPV peak current rose. As can be shown, the current approach for HPVDNA detection has a wider linear range and a lower detection limit than other methods. As a result, the

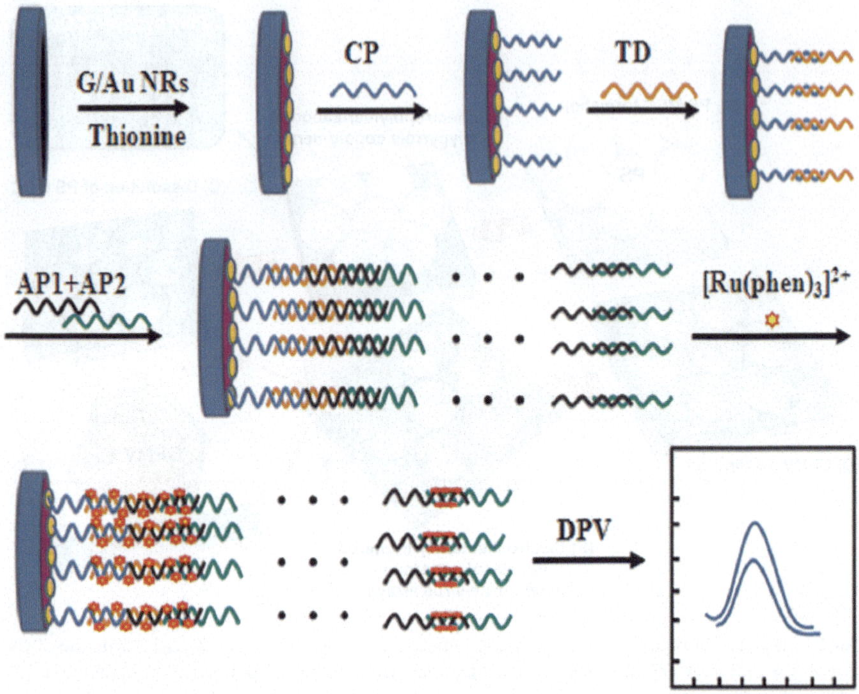

Fig. 9 Schematic representation of the differential pulse voltammetry-based biosensor for virus detection. (Reproduced with permission from [39])

G/AuNR/PTmodified GCE was found to be a reliable platform for electrochemical sensors. There are two possible explanations for the obtained ultrasensitivity. On the one hand, the electrochemical responses were boosted by the large length of the self-assembled DNA nanostructure. Biocompatible and hydrophilic DNA nanostructures, on the other hand, were less prone to non-specific adsorption onto the electrode surface [39].

3.8 Conductometric-Based Biosensor for Virus Detection

Magnetic separation by antibody conductive magnetic nanoparticles is used in the conductometric sensor. Mixing magnetic iron oxide and polyaniline in a phosphate-buffered solution, followed by sonification, is how it's done. For hybridization, monoclonal antibodies in phosphate-buffered solution will be employed, followed by incubation. The pathogen solution is added after serial dilution, and the nanoparticles are vortexed to separate them. Figure 11 depicts a sensor based on this principle, with the resistance change being proportional to the bovine-viral-diarrhea-virus (BVDV) concentrations [40].

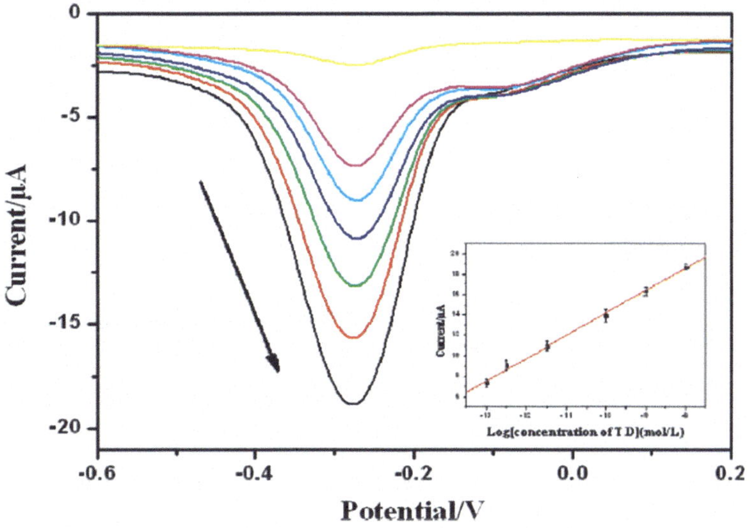

Fig. 10 Differential pulse voltammetry responses at various concentrations of human papillomavirus DNA. (Reproduced with permission from [39])

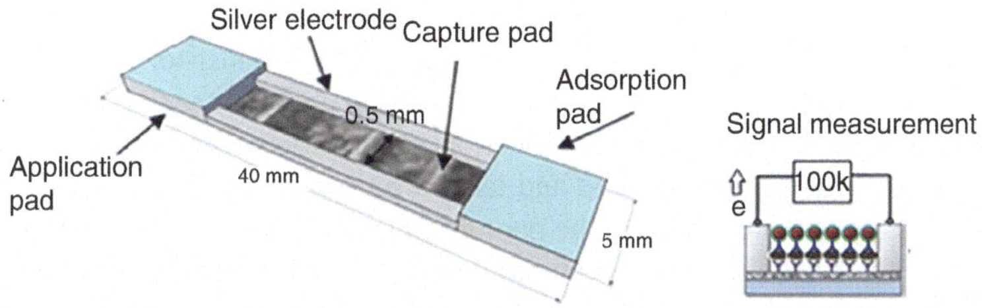

Fig. 11 Conductometric biosensor. (Taken with permission [40])

Experiments with *E. coli* O157:H7 and bovine-viral-diarrhea-virus (BVDV) at various doses were used to verify the sensitivity and efficacy of the electrospun biosensor. Figure 12 shows the sensor conductance signal's time response for varied concentrations of *E. coli* O157:H7 samples. The conductance signal grows with fluctuations during the first 50 s when the antigen-polyaniline combination moves along the capture pad. The immunoreactions bind the polyaniline-coated nanoparticles to the silver electrodes, forming conducting bridges. The sensor conductance gradually reduces once the flow reaches the absorption pad, as the excess reagent is separated and absorbed by capillary action. The conductance signal becomes steady and adequate for sensor reading when absorption reaches equilibrium. The conductivity measurement presented in the figure below confirms that the conducting bridges generated by the captured sandwich complex are proportional to the pathogen concentration [40].

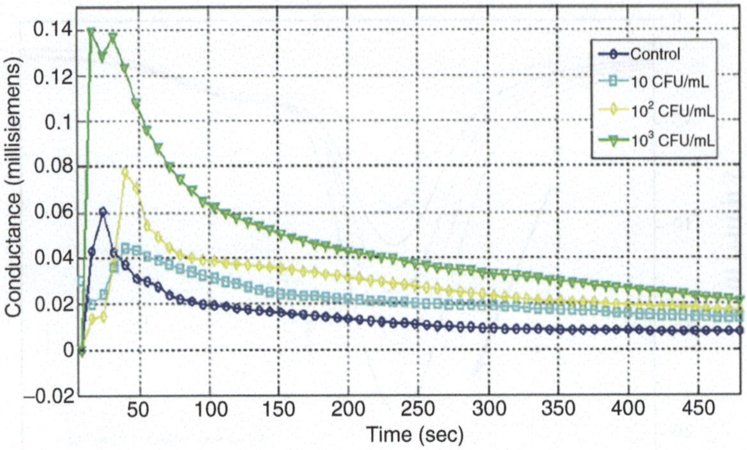

Fig. 12 Biosensor conductance versus test time for *E. coli* O157:H7 bacterium samples with various target concentrations. (Taken with permission [40])

3.9 Field-Effect-Transistor-Based Biosensor for Virus Detection

The micropatterned reduced graphene oxide field-effect transistor (MRGO-FET), which exhibits p-type behavior, was used to develop a real-time, fast, and sensitive biosensor for pathogenic rotavirus detection. A modified Hummers method was used to manufacture single-layered and large-sized graphene-oxide sheets, and an MRGO-FET was created through the photolithography and reduction process. This type of sensor necessitates the traditional way of fabricating FET using optical lithography, followed by modification for virus detection, as shown in Fig. 13. The FET is treated with 1-pyrenebutyric acid N-hydroxysuccinimide (PSE) and then protected with 1-dodecanethiol (DDT), allowing it to be immobilized by antibodies. The virus is immobilized on the antibody in the last phase. In the measurement, the current measured before (Io) and after viral injection (I) is employed. When compared to the traditional ELISA approach, it has the lowest detection limit for rotavirus [41].

3.10 Fluorescence-Based Biosensor for Virus Detection

Fluorescence-based sensing and imaging offers unique advantages such as good sensitivity, high temporal resolution, availability of biocompatible imaging agents, and noninvasive characteristics that make it relevant in research and clinical settings. Fluorescence-based optical biosensors are the single largest group of sensors at present, owing to the commercial availability of numerous fluorescent probes,

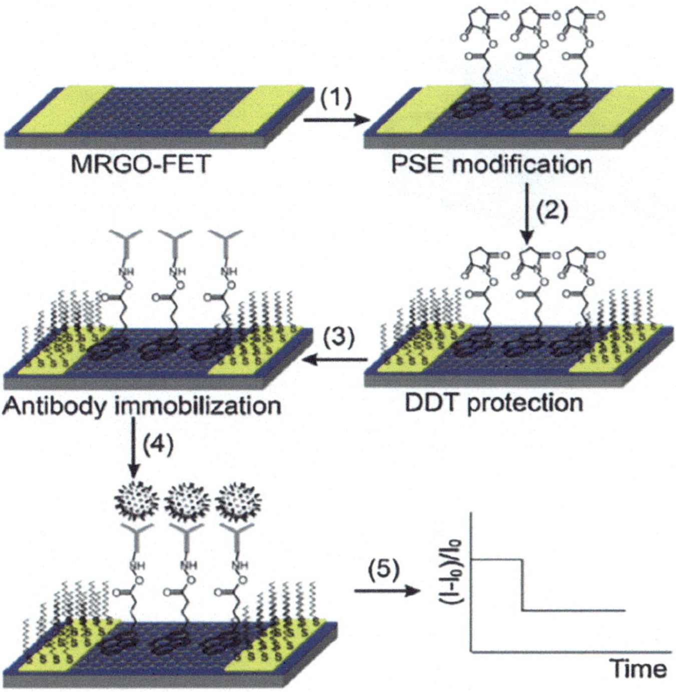

Fig. 13 Fabrication process of MRGO-FET based biosensor for rotavirus detection. (Taken with permission from [41])

high-quality optical fibers, and suitable optical instruments. Fluorescent biosensors have various parameters like intensity, energy transfer, lifetime, and quantum yield that can be exploited for virus detection. One mechanism that is often used in these biosensors to detect close interactions between an analyte and a fluorophore is fluorescence resonance energy transfer (FRET). FRET is the process where incident radiation is absorbed and non-radiatively transferred from a donor to an acceptor by means of long-range dipole-dipole coupling. Recent improvements in FRET research and advancements in optical instrumentation have established FRET microscopy as an effective tool for biological imaging and detection applications [42] (Figs. 14 and 15).

3.11 Colorimetric-Based Biosensor for Virus Detection

Colorimetric biosensors detect the presence of specific substances by causing a color change that can be seen with the bare eye. Colorimetric biosensors are great candidates for POC diagnostics due to their ease of use and the fact that they do not require expensive analytical equipment. Noble metal nanoparticles (NPs), metal

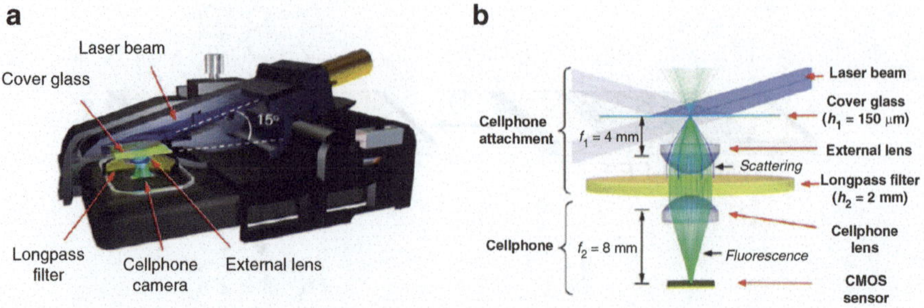

Fig. 14 (**a**) Cell phone schematic illustration. (**b**) Ray-tracing diagram of the cell phone microscope, where scattered beams are indicated with solid blue rays and fluorescent emission is highlighted with solid green rays. (Taken with permission [43])

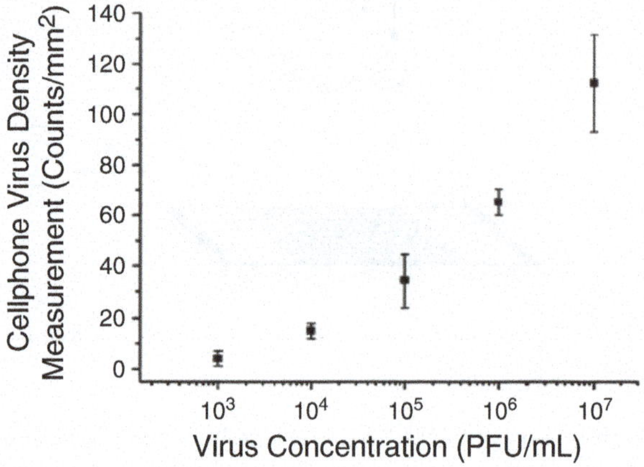

Fig. 15 Fluorescent signal intensity comparison for HCMV particles measured using cell phone-based virus density verses various virus concentrations. (Taken with permission [43])

oxide NPs, carbon nanotubes, and conducting polymers have been used to create smart materials that cause apparent color change (CPs). Metal oxide NPs and carbon nanotubes can change color by stimulating a peroxidase substrate reaction or by having intrinsic peroxidase action.

Researchers created a peptide-functionalized polydiacetylene (PEP-PDA) nano-sensor that allows for naked-eye detection of the pandemic H1N1 virus (pH1N1). PDA nanoparticles were initially generated in the aqueous phase by self-assembly of pentacosadiynoic acid (PCDA) and its derivatives using the nano-precipitation process, as shown in Fig. 16, and then exposed to UV radiation to develop the blue-colored PDA nano-sensor [43].

To identify influenza A (H1N1) virus with the naked eye using the PDA nano-sensors' distinctive chromic properties, a peptide (PEP) capable of specific H1N1 virus binding was added to the PDA nano-sensors. An evident blue-to-red color change and a shift in the absorption spectra, as shown in Fig. 17, validated the colorimetric response of the PEP-PDA nano-sensors in the presence of pH1N1. Because greater pH1N1 concentrations exerted more stress on the PDA backbone, the

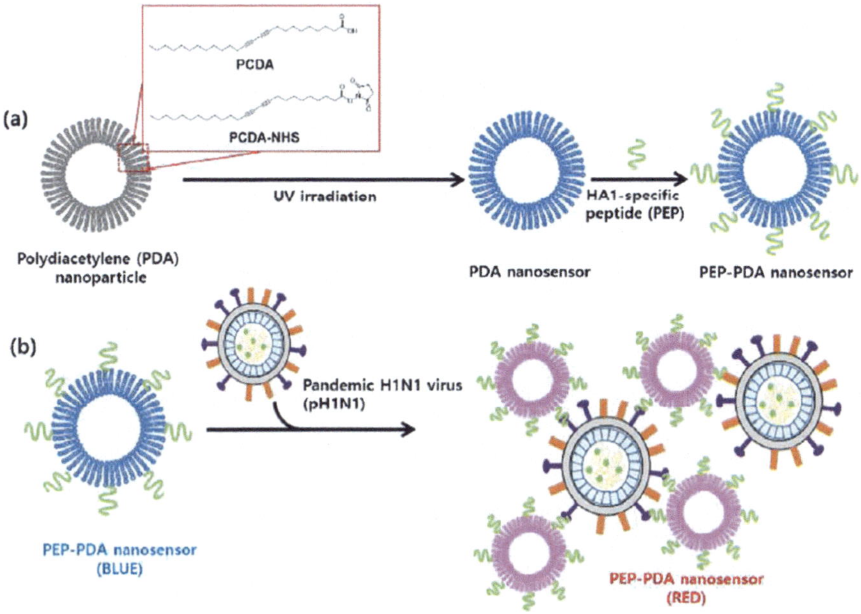

Fig. 16 Schematic representation of colorimetric based H1N1 virus sensor. (Taken with permission [43])

magnitudes of these alterations increased with pH1N1 concentration. In the absence of pH1N1, the PEP-PDA nano-sensors demonstrated maximum absorbance at 628 nm and weaker absorbance at 550 nm, resulting in a blue color, as illustrated in Fig. 17b. When exposed to increasing concentrations of pH1N1, the absorbance at 550 nm increased simultaneously with a reduction in the absorbance at 628 nm, resulting in the color transformation from blue to red.

4 Conclusion

Viruses are one of the most serious risks to the world population's health and survival. To slow the virus's transmission and contain the illness outbreak, early detection and treatment are critical. The virus's impact has brought to light the importance of diagnostic tools in the fight against infectious diseases. As a result, the development of new, rapid, highly sensitive, accurate, and reliable point-of-care diagnostic assays and equipment for viral detection in low-resource settings is critical for medical care.

To detect virus pathogens in laboratories, CT scan, RT-PCR, and ELISA tests, as well as other diagnostic kits, have been developed. For managing rapid detection and diagnosis, sophisticated diagnostic technologies with outstanding ultrasensitivity, specificity, mobility, and wearability are essential. Biosensors have been shown

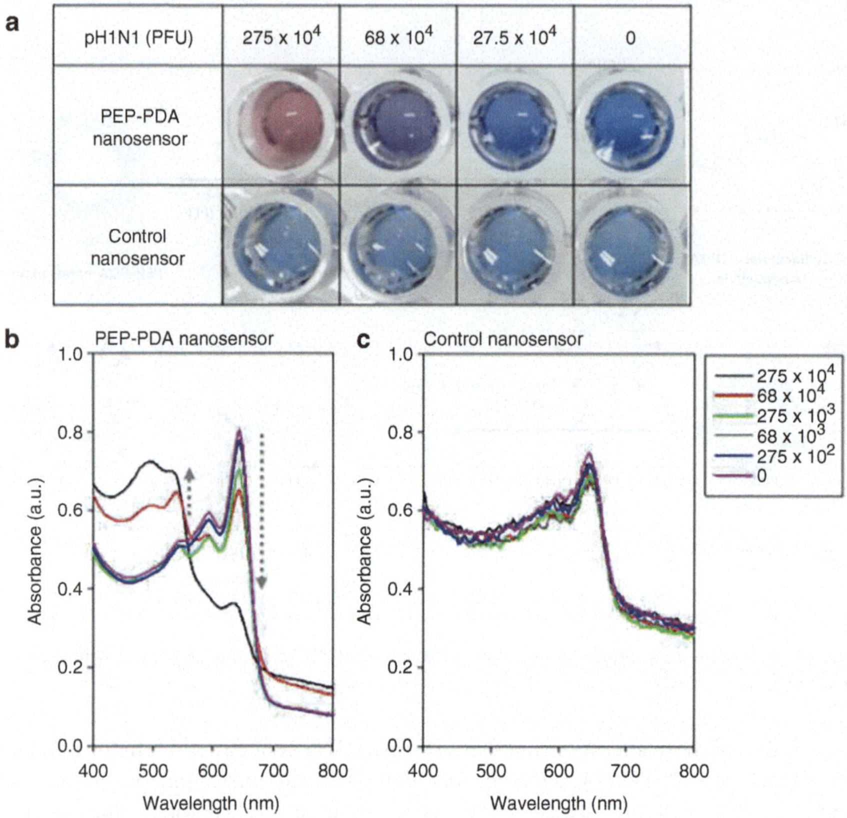

Fig. 17 (**a**) Color transition images after incubation of the PEP-PDA nano-sensor and control nano-sensor with different concentrations of pH1N1. (**b**) Absorbance verses wavelength curve plotted for a PEP-PDA based nano-sensor. (**c**) Absorbance verses wavelength curve plotted for a control-based nano-sensor. (Taken with permission [43])

to be useful tools for virus detection, including early diagnosis and on-site, fast, and ultrasensitive detection. CPs can be used to construct improved biosensors for the detection of viral infections due to their unique features and wide range of commercial advantages. Biosensors based on CP are projected to be developed and widely used in local hospitals, laboratories, doctor's offices, airports, and other high-traffic places, as well as at home, in the future. Finally, connecting CP-based biosensors with the Internet of things and taking into account the simplicity of use for community members would expand the range of viable applications for CP-based biosensors. CP-based biosensors can be utilized for early-stage detection and can help prevent the spread of viruses, thanks to the rapid advancement of CP technology. These sensors also have the potential to improve sensitivity, selectivity, flexibility, electrochemical stability, repeatability, and bioanalyte detecting ability. Recent advances and key developments in the field of polymers for biosensing applications in viral pathogen detection and diagnosis using various types of biosensing strategies are presented in this chapter, which represent one of the most promising techniques for miniaturized biosensors.

References

1. Castillo-Henríquez L, Brenes-Acuña M, Castro-Rojas A, Cordero-Salmerón R, Lopretti-Correa M, Vega-Baudrit JR. Biosensors for the detection of bacterial and viral clinical pathogens. Sensors. 2020;20(23):6926.
2. Puttananjegowda K, Takshi A, Thomas S. Silicon carbide nanoparticles fibrous membrane for implantable glucose sensor integrated with chronoamperometric potentiostat. US Patent App. 17/335,276; Dec 2021.
3. Puttananjegowda K, Takshi A, Thomas S. Silicon carbide nanoparticles electrospun nanofibrous enzymatic glucose sensor. Biosens Bioelectron. 2021;186(11385):1–4.
4. Puttananjegowda K. Electrospun nanofibrous membrane based glucose sensor with integration of potentiostat circuit, Doctoral dissertation. Electrical Engineering, University of South Florida, Tampa, Florida, USA; 2020.
5. Puttananjegowda K, Takshi A, Thomas S. Silicon carbide nanoparticles based Nanofibrous membrane in comparison with thin-film enzymatic glucose sensor. IEEE Trans Nanobioscience. 2021;20(4):577–80.
6. Thomas SW, Khan RR, Puttananjegowda K, Serrano-Garcia W. Conductive polymers and metal oxide polymeric composites for nanostructures and nanodevices. In: Advances in Nanostructured Materials and Nanopatterning Technologies, pp. 243–271; 2020.
7. Puttananjegowda K, Takshi A, Thomas S. Electrospun nanofibrous structures for electrochemical enzymatic glucose biosensing: a perspective. J Electrochem Soc. 2020;167(3):1–6.
8. Puttananjegowda K, Takshi A, Thomas S. Electrospun nanofibrous membrane based electrochemical glucose sensor. IEEE Sens Lett. 2020;4(2):1–4.
9. Puttananjegowda K, Thomas S. Cascode common source transimpedance amplifiers for analyte monitoring systems. US Patent number 11309846; Apr 2022.
10. Ababneh MM, Jasim M, Puttananjegowda K, Perez S, Afroz S, Thomas S. Design of a SiC implantable rectenna for wireless in-vivo biomedical devices. In: IEEE Annual Ubiquitous Computing, Electronics and Mobile Communication Conference, pp. 254–257; 2017.
11. Puttananjegowda K, Ababneh M, Thomas S. The design of ultra low noise CMOS transimpedance amplifier for biosensing applications. In: 2017 IEEE 8th Annual Ubiquitous Computing, Electronics and Mobile Communication Conference (UEMCON); 2017, pp. 16–19. https://doi.org/10.1109/UEMCON.2017.8248971.
12. Puttananjegowda K, Thomas S. A low-power low-noise multi-stage transimpedance amplifier for amperometric based blood glucose monitoring systems. Analog Integr Circuits Signal Process J, Springer. 2020;102:659–66.
13. Puttananjegowda K, Thomas S. A CNTFET based multi-stage transimpedance amplifier for blood glucose monitoring systems. In: IEEE Information Technology, Electronics and Mobile Communication Conference, Vancouver, pp. 383–388; 2018.
14. Ababneh MM, Qaroot A, Puttananjegowda K, Perez S, Thomas S. Optimized power management circuit for implantable rectenna for in-body medical devices. In: IEEE International Conference on Power Electronics and Drive Systems, pp. 30–34; 2017.
15. Puttananjegowda K, Thomas S. A detailed review on physical unclonable function circuits for hardware security. In: 2018 IEEE 9th Annual Information Technology, Electronics and Mobile Communication Conference (IEMCON), pp. 609–612; 2018.
16. Kavyashree P, Yellampalli SS. The design of ultra low power differential CGLNA for IEEE 802.15.4 applications. In: 2015 IEEE International RF and Microwave Conference (RFM), pp. 54–56; 2015.

17. Kavyashree P, Yellampalli S. The design of ultra low power RF CMOS LNA in nanometer technology. In: Sharma M, et al., editors. Design and modeling of low power VLSI systems. IGI Global; 2016. p. 229–51.
18. Kavyashree P, Yellampalli SS. The design of ultra low power CMOS CGLNA in nanometer technology. In: 2014 Fifth International Symposium on Electronic System Design, pp. 15–19; 2014.
19. Kavyashree P, et al. The design of low noise amplifiers in nanometer technology for WiMAX applications. Int J Sci Res Publ. 2013;3(10):1–6.
20. Puttananjegowda K, Cao Y, Green M. DC-DC boost converter for wireless power transfer systems. In: 2021 IEEE 12th Annual Ubiquitous Computing, Electronics & Mobile Communication Conference (UEMCON), pp. 0661–0665; 2021.
21. Thomas S, Seufert B, Serrano-Garcia W, Devisetty M, Khan R, Puttananjegowda K, Alcantar N. Eco-friendly, biodegradable, and biocompatible electrospun nanofiber membranes and applications. In: Sustainable Nanotechnology (eds YV Pathak, G Parayil, JK Patel); 2022. https://doi.org/10.1002/9781119650294.ch11.
22. Puttananjegowda K, Ababneh M, Thomas S. The design of high efficiency voltage mode integrated converter for glucose monitoring systems. In: 2022 IEEE 12th Annual Computing and Communication Workshop and Conference (CCWC), pp. 0649–0653; 2022.
23. Puttananjegowda K, Takshi A, Thomas S (2022) A silicon carbide electrochemical sensor for glucose detection. In Silicon Carbide Technology for Advanced Human Healthcare Applications, Elsevier. https://doi.org/10.1016/B978-0-323-90609-8.00010-7.
24. Critchley P, Dimmock NJ. Binding of an influenza A virus to a neomembrane measured by surface plasmon resonance. Bioorg Med Chem. 2004;12:2773–80.
25. Lee C, Gaston MA, Weiss AA, Zhang P. Colorimetric viral detection based on sialic acid stabilized gold nanoparticles. Biosens Bioelectron. 2013;42:236–41.
26. Skehel JJ, Wiley DC. Receptor binding and membrane fusion in virus entry:the influenza hemagglutinin. Annu Rev Biochem. 2000;69:531–69.
27. Medina RA, Garcia-Sastre A. Influenza A viruses: new research developments. Nat Rev Microbiol. 2011;9:590–603.
28. Sessolo M, Khodagholy D, Rivnay J, Maddalena F, Gleyzes M, Steidl E, Buisson B, Malliaras GG. Easy-to-fabricate conducting polymer microelectrode arrays. Adv Mater. 2013;25:2135–9.
29. Farzin L, Shamsipur M, Samandari L, Sheibani S. HIV biosensors for early diagnosis of infection: the intertwine of nanotechnology with sensing strategies. Talanta. 2020;206:120201.
30. Panghal A, Flora SJ. Chapter 4—Viral agents including threat from emerging viral infections. In: Flora SJ, Pachauri V, editors. Handbook on biological warfare preparedness. Cambridge, MA: Academic Press; 2020. p. 65–81.
31. Martins G, Gogola JL, Caetano FR, Kalinke C, Jorge TR, Duarte CN, Bergamini F, Marcolino L. Quick electrochemical immunoassay for hantavirus detection based on biochar platform. Talanta. 2019;204:163–71.
32. Lin P, et al. Organic electrochemical transistors integrated in flexible microfl uidic systems and used for label-free DNA sensing. Adv Mater. 2011;23:4035–40.
33. Wang D, Wang J, Song Z, et al. Highly selective and antifouling electrochemical biosensors for sensitive MicroRNA assaying based on conducting polymer polyaniline functionalized with zwitterionic peptide. Anal Bioanal Chem. 2021;413:543–53.
34. Chowdhury AD, Takemura K, Li TC, et al. Electrical pulse-induced electrochemical biosensor for hepatitis E virus detection. Nat Commun. 2019;10:3737–42.
35. Arter JA, et al. Virus-PEDOT nanowires for biosensing. Nano Lett. 2010;10(12):4858–62.
36. Vinh VT, et al. Development strategies of conducting polymer-based electrochemical biosensors for virus biomarkers: potential for rapid COVID-19 detection. Biosens Bioelectron. 2021;182:113192.
37. Shen F, et al. Rapid flu diagnosis using silicon nanowire sensor. Nano Lett. 2012;12(7):3722–30.
38. Tiu BD, et al. Free-standing, nanopatterned Janus membranes of conducting polymer-virus nanoparticle arrays. Langmuir. 2016;32(24):6185–93.

39. Huang H, et al. An ultrasensitive electrochemical DNA biosensor based on graphene/ Au nanorod/polythionine for human papillomavirus DNA detection. Biosens Bioelectron. 2015;68:442–6.
40. Luo Y, Nartker S, Miller H, Hochhalter D, Wiederoder M, Wiederoder S, Setterington E, Drzal LT, Alocilja EC. Surface functionalization of electrospun nanofibers for detecting E. coli O157:H7 and BVDV cells in a direct-charge transfer biosensor. Biosens Bioelectron. 2010;26(4):1612–7.
41. Liu F, Kim YH, Cheon DS, Seo TS. Micropatterned reduced graphene oxide based field-effect transistor for real-time virus detection. Sensors Actuators B Chem. 2013;186:252–7.
42. Wei Q, et al. Fluorescent imaging of single nanoparticles and viruses on a smart phone. ACS Nano. 2013;7(10):9147–55.
43. Song S, et al. Colorimetric detection of influenza A (H1N1) virus by a peptide-functionalized polydiacetylene (PEP-PDA) nano sensor. RSC Adv. 2016;6:48566–70.
44. Laila AL, et al. An electrochemical immunosensor for the corona virus associated with the Middle East respiratory syndrome using an array of gold nanoparticle-modified carbon electrodes. Microchim Acta. 2019;186(4):224–34.
45. Shrivastav AM, Cvelbar U, Abdulhalim I. A comprehensive review on plasmonic-based biosensors used in viral diagnostics. Commun Biol. 2021;4:70.
46. Arshak P, et al. Field-effect sensors for virus detection: From Ebola to SARS-CoV-2 and plant viral enhancers. Front Plant Sci. 2020;11:1–14.
47. Patolsky F, Zheng G, Hayden O, Lakadamyali M, Zhuang X, Lieber CM. Electrical detection of single viruses. Proc Natl Acad Sci U S A. 2004;101:14017–22.
48. Miller TC, et al. Electronic nose with detection method for alcohol, acetone, and carbon monoxide in coronavirus disease 2019 breath simulation model. IEEE Sensors J. 21:15935–43.

Nanotechnology: A Stepping Stone Toward Viral Hepatitis Treatment and Prevention in Children and Adults

Kshama Patel, Yash Pagarani, Ranjita Shegokar, and Yashwant Pathak

Abstract Hepatitis in children can occur from viral and nonviral etiologies. Viral hepatitis occurs when an individual contracts the virus from infected people and contaminated food, water, and objects. The implications of obtaining viral hepatitis differ depending on which type of virus is transmitted. Some individuals might experience no symptoms and not even realize they have the virus, and some might experience acute symptoms which can progress to chronic hepatitis. Many hepatitis viruses have been studied with the five main ones including hepatitis A, B, C, D, and E. These viruses can be found all over the world with some countries displaying a higher prevalence of one over the others and vice versa. Vaccinations exist for hepatitis A and hepatitis B and are administered in both children and adults. New research is being conducted to utilize nanotechnology for the development of hepatitis C vaccines along with treatments that offer better efficacy, cheaper costs, and less side effects. Hepatitis not treated appropriately can further cause damage and result in liver disease such as liver cirrhosis, liver failure, or even cancer. Chronic hepatitis B and chronic hepatitis C is also a leading cause of liver cancer in the United States. The World Health Organization (WHO) has created strategic plans to reduce hepatitis infection and death rates by ramping up vaccination in children and adults, increasing access to clean water, and increasing treatment care for infected individuals in countries with low income. This paper will discuss various prevention and treatment methods that are currently offered or that are being researched and will be offered in the future. These include vaccinations with nanotechnology-based delivery systems and genome editing as a possible cure to chronic hepatitis B.

K. Patel
The University of South Florida, Judy Genshaft Honors College, Tampa, FL, USA

Y. Pagarani
The University of South Florida, Tampa, FL, USA

R. Shegokar
CapnoPharm GmbH, Tübingen, Germany

Y. Pathak (✉)
College of Pharmacy, University of South Florida Health, Tampa, FL, USA

R. Shegokar, Y. Pathak (eds.), *Viral Drug Delivery Systems*,
https://doi.org/10.1007/978-3-031-20537-8_10

Keywords Hepatitis · Viral · Nanotechnology · Treatment · Prevention · Children · Vaccinations · Genome editing

1 Introduction

1.1 Viral Hepatitis

Hepatitis in its literal definition means inflammation of the liver. The word stems from the prefix hepato- which means liver and the suffix -itis which means inflammation. The liver is an essential organ of the body whose function is to filter blood, to process nutrients, and to metabolize drugs [1]. Inflammation is a biological response that occurs when tissue reacts to injury or irritants which leads to swelling and pain in the region [2, 3]. Any complications that arise in the liver can cause reduced liver function affecting the digestive system. Nonviral hepatitis can stem from various reasons such as heavy alcohol use, certain medication consumption and abuse, and toxins in the liver. Viral hepatitis, on the other hand, is caused by viruses introduced to the body. Viral hepatitis is serious yet preventable as there are means to reduce the spread in public and eliminate any threats associated with it [4, 5]. There are three main types of viral hepatitis that affect the majority of individuals, and these include hepatitis A, hepatitis B, and hepatitis C. These three viruses differ in genetic components, but all can lead to the same outcome resulting in inflammation of the liver. Hepatitis A is particularly the most common to see in children due to its ease of transmission compared to that of hepatitis B and C [4].

1.1.1 Hepatitis A

Hepatitis A virus (HAV) is one of the several hepatitis viruses that cause inflammation and reduce liver function. Hepatitis A is highly contagious, particularly in children, because of its ease of transmission from one person to the next. The most common form of transmission for hepatitis A is from fecal-oral contact. This can occur via contact with contaminated food from an individual who did not wash their hands, drinking water contaminated with fecal matter, or close contact with an individual or an object infected with the virus [6]. Individuals who do not wash their hands frequently and do not practice proper hygiene are at higher risks for transmission of the virus. Symptoms of hepatitis A include nausea, vomiting, fatigue, abdominal discomfort, dark urine, jaundice, and joint pain. Most individuals experience minor symptoms in which case might not require treatment. It is also likely for individuals who are infected to recover completely and not have permanent liver damage [6]. Risk factors for hepatitis A include traveling to places with high hepatitis A prevalence, living with someone who has the virus, having clotting factor diseases like hemophilia, and having HIV or AIDS [7]. HIV stands for human immunodeficiency virus and is a virus that compromises the immune system. There is no effective cure for HIV, and a progression of this virus can lead to AIDS which stands for acquired immunodeficiency virus [7, 39]. The most common way to prevent hepatitis A infection is to practice good hygiene by washing hands after using the bathroom and before making or consuming food, washing fruits and vegetables before consumption, not eating raw shellfish, and getting the hepatitis A vaccine when appropriate. Children between the ages of 1 and 2 are recommended to get vaccinated in two doses. One dose is given between 12 and 23 months of age, and a second dose is followed six months after the first dose. Individuals not previously vaccinated are recommended to get vaccinated up into adulthood. The vaccine is also highly recommended for individuals who are traveling internationally, men who have sex with men, people with occupational risk, and people experiencing homelessness [8].

1.1.2 Hepatitis B

Hepatitis B is one of the three most common types of viral hepatitis. Hepatitis B is called hepatitis B because it is caused by the hepatitis B virus or HBV. Patients with hepatitis B will experience different types of symptoms from other hepatitis viruses. However, the most common symptoms include but are not limited to abdominal pain, dark urine, fever, joint pain, loss of appetite, nausea, vomiting, weakness, and fatigue. Additionally, many patients will experience jaundice as one of their symptoms in which their skin and the whites of their eyes start to turn into a yellowish hue [2]. Hepatitis B can be acute with symptoms lasting up to a couple weeks or chronic in which case symptoms prolong and progress to a serious illness of chronic hepatitis B [5]. It is most commonly spread when blood or other body fluids from an infected person enters another person's body. This can be via unprotected sexual intercourse and sharing needles, razors, and toothbrushes [9]. Hepatitis B can also be spread if an individual presents no symptoms. It is estimated that two-thirds of individuals do not know they have chronic hepatitis B. It is also estimated that between 900,000 and 1.89 million people are living with HBV in the United States [5]. Hepatitis B is vaccine-preventable and is recommended for all infants and adults if they were not vaccinated before. Risk factors for hepatitis B include having unprotected sex with infected partners, living with individuals who have chronic hepatitis B, jobs with exposure to human blood, and traveling to countries with high prevalence of the virus [10]. Chronic hepatitis B is a serious form of the virus as it can lead to further liver function problems. Chronic HBV can lead to complications such as liver cirrhosis, increased risk for liver cancer and liver failure, and increased risks for kidney disease. Chronic HBV risks are higher for children the younger they are. Nine in ten infants infected with the virus go on to develop chronic hepatitis B, children up to age six have an estimated 33% chance of developing chronic hepatitis B, and children and adults over the age of six have a much lower risk with most not developing chronic HBV [5]. Hepatitis B can be prevented by taking precautionary measures. These measures include knowing the HBV status of potential sexual partners, using contraception such as condoms to reduce risks, and understanding the risks when traveling to international countries with high prevalence of the virus. There is a HBV vaccine available, and, depending on the manufacturer, it is given as either two, three, or four doses within a six month period. Hepatitis B vaccination is recommended for newborns, children, and adults not previously vaccinated, people with HIV, liver disease and kidney disease, and individuals who plan on traveling to areas with high prevalence of the virus [10].

1.1.3 Hepatitis C

Hepatitis C is another one of the three most common viral hepatitis viruses. It is caused by the hepatitis C (HCV) virus. Unlike hepatitis A, hepatitis C is spread via contact with blood from an infected person. The most common way to get infected with HCV is via needle sharing for drug usage. Some people experience short-term illnesses with minor symptoms, but more than half tend to develop chronic hepatitis C [11]. Chronic hepatitis C can lead to serious progressions such as liver cirrhosis and liver cancer. However, many people don't experience symptoms for chronic hepatitis C until advanced liver disease symptoms start to show [11]. The US Preventive Services Task Force recommends adults between ages of 18 and 79 be screened for hepatitis C even if they display no symptoms. When symptoms do show up years down from having the virus, symptoms can include bleeding and bruising easily, fatigue, dark-colored urine, swelling in legs, spider angiomas, and weight loss [12]. There are different genotypes of the HCV virus present globally. Currently, 7 genotypes and over 67 subtypes have been identified with the most common one being type 1 in the United States. Risk factors for HCV include healthcare workers exposed to infected blood, individuals who share needles for drug administration, being born to a woman with hepatitis C, and being born between 1945 and 1965 because of high incidence during this time [12]. The prevalence of HCV is not accurately known due to many individuals not presenting symptoms and getting tested for it; however, the US Department of Health and Human Services (HHS) estimates the number to be between 2.5 million and 4.7 million individuals living with the virus in the United States [13]. There is no vaccination for HCV, and thus prevention lies on individuals to avoid contact with contaminated blood. Steps that should be taken include not sharing needles, not sharing razors, practicing safe sex, and being cautious about tattoos and piercings by having them done in reputable locations with proper safety procedures in place [12].

1.2 Hepatitis in Children

As previously mentioned, hepatitis in children is most commonly caused by the hepatitis virus. However, other viruses can contribute toward causing hepatitis as well. Other viruses include cytomegalovirus, Epstein-Barr virus, herpes simplex virus, varicella zoster virus, adenovirus, and parvovirus. These viruses differ in genetic make-up; however, all can cause onset of liver inflammation. Other complications such as auto-immune liver disease can contribute toward hepatitis as well. This occurs when the body makes antibodies that attack the liver resulting in inflammation leading to hepatitis. Common ways hepatitis A virus is transmitted to children include eating food made by an infected person who didn't wash their hands, drinking water contaminated with fecal matter, and international travel to areas with high prevalence. Hepatitis B can also be transmitted to children if they were born to mothers with the virus, they live in households with an infected individual, or if they have a blood clotting disorder such as hemophilia putting them in a higher risk [14]. Hepatitis C, much like hepatitis B, is passed through infected blood. Mothers can pass it along to their babies if they have the virus during pregnancy. Hepatitis D, another virus that can be transmitted to cause hepatitis, is less prevalent in the United States. Hepatitis D onset can only occur if a child already has hepatitis B. Hepatitis D requires HBV to be present for its replication. This virus is not common in the United States because most children obtain the hepatitis B vaccine and thus are protected from hepatitis D infection [15]. Hepatitis E is another type of virus caused by the hepatitis E virus (HEV). It is similar to hepatitis A in how it is transmitted. Children can obtain it by not practicing proper hygiene and consuming contaminated food or water. HEV is found in the stool of an infected person, and thus not practicing proper hygiene can result in transmission of the virus [16]. Hepatitis E is not common in the United States and is more common in developing countries.

Hepatitis can be diagnosed in children conducting physical examinations followed by blood testing to look at liver enzyme levels, liver function, antibody tests to determine which virus is present, and coagulation tests. Healthcare providers might also take CT scans, ultrasounds, and MRI scans to look at the liver and get images. Further testing might require obtaining a liver biopsy to examine tissue samples for possible scarring or discrepancies. If a child has tested positive for hepatitis, care providers will determine appropriate treatment plans and factor in age, history, and symptoms. Treatment options can include antiviral medications to treat against the virus whether by injection or taken orally and supportive care such as drinking enough fluids, eating a healthy diet, getting enough rest, and monitoring of symptoms [14]. It is important for caregivers to keep track of children and monitor any symptom progressions and follow up with their providers on time to ensure there aren't any complications or further developments (Table 1).

1.3 Nanotechnology

Nanotechnology is a terminology used for science, engineering, and technology studied at the nanoscale. This allows for a systematic approach to designing, producing, and applying materials to science fields like biology, chemistry, physics, and engineering. The concept started when a physicist by the name of Richard Feynman at the California Institute of Technology hosted a talk in 1959 to describe

Table 1 Table summarizing the different types of the hepatitis virus

Virus	Hepatitis A	Hepatitis B	Hepatitis C	Hepatitis D	Hepatitis E
Abbreviation	HAV	HBV	HCV	HDV	HEV
Family	*Picornaviridae*	Hepadnavirus	*Flaviviridae*	*Deltavirus*	Hepeviridae
Transmission	Fecal-oral	Parenteral sex	Parenteral	Parenteral sex	Fecal-oral
Incubation (days)	15–50	30–180	14–180	42–180	21–56
Chronic infection	Rare	Greater than 90% in children and less than 10% in adults; increases risk for hepatocellular carcinoma	75–80%; increased risk for hepatocellular carcinoma	Superinfection: 75%; Coinfection: 5%; increased risk for hepatocellular carcinoma	Most reported in immuno-compromised or posttrans-plant patients
Occurrence	Worldwide; commonly found in the Far East, the Middle East, Africa, and Central and South America	Worldwide; commonly found in Asia, Africa, and South America	Worldwide; commonly found in Europe, Asia, Africa, and South America	Worldwide; commonly found in Eastern Europe, East Asia, the Middle East, and West and Central Africa	Worldwide; commonly found in the Middle East, Asia, Africa, and South America
Vaccine available	Yes	Yes	No	No; however, it is prevented by the HBV vaccines	Yes; however, it is only approved for use in China
Vaccine under research	N/A	N/A	Yes	Yes	Yes
References	[1, 40, 41]	[1, 40, 41]	[1, 40, 41]	[1, 40, 41]	[1, 40, 41]

ways in which scientists would be able to study and manipulate individual atoms and molecules [17]. Over a decade later in 1974, a Japanese scientist by the name of Norio Taniguchi coined the term nanotechnology, and in 1981 it became more popularized when Scanning Tunneling Microscopes were created to look at individual atoms [17]. Nanotechnology has the power to create positive impacts in medicine and engineering and also contribute to a reduction in global greenhouse gas emissions. Nanotechnology in medicine particularly has benefits such as targeting specific areas of interest for nanomedicine utilization. With a greater understanding of nanotechnology, more and more research and studies are being conducted to incorporate nanotechnology for the betterment of medicine, pharmaceuticals, and therapeutic areas [17]. In its relation to hepatitis, nanoparticles are used as carriers for HCV vaccine production, anti-HCV combination, and targeting deliveries to reduce HCV infection in individuals [18].

2 Vaccinations

There are some antiviral medications which have already been approved in treating some viral infections. Most of these include getting vaccinated as a child in order to prevent attracting the hepatitis virus in children. However, the vaccinations can also be taken as an adult because it is never too late to be protected from the virus by simply getting the vaccine [8]. Despite what vaccine it is, the hepatitis vaccinations are a measure of prevention, so individuals do not get infected with the virus rather than a method of treatment.

2.1 Hepatitis A Vaccine

2.1.1 History of Timeline of Hepatitis A Vaccines

As more research has been done, scientists and other individuals began to realize that there are certain populations in which it is crucial to be vaccinated due to the people that the individual is surrounded by, especially when considering the HAV. Early on, in 1996, the Advisory Committee on Immunization Practices (ACIP) recommended that only children who live in communities with a high outbreak rate among teens should be vaccinated. Then, in 1999, the ACIP recommended that children should be vaccinated if the area they live in has a reported annual rate of HAV greater than 20 per 100,000 between the ten years of 1987 to 1997. Additionally, they recommended that children get vaccinated if the rate of HAV was greater than the national average. At the time, the national average was 10 cases per 100,000 people or higher. In 2006, the ACIP proceeded to recommend that all children between the ages of twelve months and twenty-three months should all be vaccinated, despite where they live [19]. However, an emphasis was placed on getting the vaccine if an individual was as follow:

- Traveling to an area where there was an HPV endemic [19]
- If there were men who have sex with other men in the area [19]
- If there were many individuals who use injection and non-injection drugs [19]
- If the individual is in contact with another individual who is in contact with HAV [19]
- If the individual is in contact with an individual who does research with the HAV [19]
- If the individual is receiving clotting factors [19]
- If the individual will be in close contact with an international adoptee [19]
- Individuals with chronic liver disease [19]
- Individuals who are homeless [19]

2.1.2 Types of Hepatitis A Vaccinations

There are many different vaccines which are approved to be used in the United States for the hepatitis A virus. Some of these vaccines include but are not limited to Havrix, Vaqta, and Twinrix. Both Havrix and Vaqta are vaccines that are inactivated single-antigen vaccinations. The efficacy of one of the doses of the 25-unit dose when taking the Vaqta vaccination is 97%. However, this efficacy rate is in children who are as young as two years old to sixteen years old. The second dose of the 25 units for Vaqta is recommended to be taken six to eighteen months after the first dose in individuals who are minors or aged one to eighteen years old. Furthermore, if the individual is nineteen years or older, they should take two doses of the Vaqta vaccine in 50 units, six months after the first dose is administered. While the first dose helps provide immunity against the hepatitis A virus, the second dose is usually administered as a booster shot to help produce lifelong immunity against the virus. Another vaccine which can be used for prevention of hepatitis A includes Havrix. Similar to Vaqta, Havrix is also a two-dose vaccination. The second dose should be taken six to twelve months after the first dose. Hence, the younger an individual gets the hepatitis vaccine, the better it is for the individual [19].

Another vaccine used for hepatitis A is an inactivated virosomal vaccine known as Epaxal that can be used to help with the virus. This vaccine works by imitating the actual virus through an inactive form of the hepatitis A virus in order to build up immunity in the individual. This immunity would allow the child to already know how the virus will react in their bodies which will help protect them in the future since their immune system will already know what to do and how to fight off the virus. Epaxal is a special vaccine because it was one of the first vaccines used that did not use aluminum salts or other vaccine adjuvants in their formula. Due to Epaxal not containing these aluminum salts, individuals were better off because they did not experience some severe symptoms after being vaccinated such as local inflammation and pain to such a severe extent [20].

2.1.3 Side Effects

Children who are under the age of six years old do not usually appear to be symptomatic for HAV, even if they have contracted the virus from someone. This is different from children over the age of six and adults because they will feel the symptoms of HAV, and they will also feel weaker. Common symptoms of HAV in older children and adults appear approximately after two to six weeks. These symptoms include but are not limited to fever, loss of appetite, fatigue, stomach pain, vomiting, dark urine, and jaundice. While these are serious symptoms that can possibly be fatal, if the individual were to take the vaccine for HAV, they would experience similar symptoms but to a lesser extent. For example, symptoms of the vaccine include but are not limited to a sore arm at the vaccination site, headaches, fatigue, fever, and loss of appetite [21].

2.2 *Hepatitis B Vaccine*

2.2.1 History of Timeline of Hepatitis B Vaccines

Hepatitis B is transmitted through the blood. Hence, the first hepatitis B vaccine which was created in the 1980s was created by taking blood samples of patients that were already infected with HBV. The infected HBV blood samples were then used to separate the surface proteins from the HBV virus, and the sample was then purified. However, when performing this purification method, there was a worry among the researchers that there would be cross-contamination. The most common worry was that the vaccine would be contaminated with another virus that is transmitted through blood. At this time in history, their main worry was that HIV would be transmitted through the HBV vaccine. However, nobody ever contracted HIV from the HBV vaccine. There was a reason for this though and it was because of the way the vaccine was created and the amount of purification steps that it went through. The HBV vaccine went through several different chemical treatments that allowed for any possible contaminating viruses inside the blood sample to be inactivated which prevented any viruses from being spread. Furthermore, the vaccine is created in laboratories nowadays which prevents any cross-contamination with another virus [22].

2.2.2 Types of Hepatitis B Vaccinations

There are many different types of HBV vaccines. In the United States, there are variations of vaccines that are given with varying doses. Some HBV vaccines require two shots, whereas others require three or four shots. The most common vaccines usually require three shots for the HBV vaccine [23]. The first HBV vaccine should be administered to newborn babies right at birth or shortly after birth. Then, the second dose should be administered when the child is one to two months old. Lastly, the third dose should be administered when the child is between six to eighteen months of age [24]. Adults can also take the vaccine if it was not administered to them as a child; however, they should speak with their healthcare provider before taking any vaccinations [23].

New Oral Vaccine

A new type of vaccine which is nanotechnology based has been developed and is being tested as a possible vaccine for HBV. This vaccine is interesting because it has an oral delivery system which allows immunity against hepatitis B. This new nanotechnology-based delivery system was developed and studied by researchers in Brazil and Europe. They found that the nanotechnology-based delivery system consists of particles which contain silica and HBsAg, the surface antigen for hepatitis B. This allows the "vaccine" to reach the intestines of the individual completely, and it does not get destroyed by the acidity in the stomach due to the high pH [25].

Some may wonder why this is an advancement in science. Well, there is a simple answer to this question. An oral delivery of the antigen or an "oral vaccine" is non-invasive, and it is safer and less painful for individuals, especially infants, which is typically when the HBV vaccine is received. Not only is it more convenient to administer, but an oral vaccine is also more cost-effective. While it is easier to administer and more cost-effective, creating the technology and the oral vaccine is a very expensive and hard process [25]. However, the advantage of being able to use alternative methods to the traditional intramuscular needle vaccination is that no sterile needles would need to be used and the systemic and mucosal immune responses would be activated, which could potentially initiate more of a significant immune response than the intramuscular injection [26]. The main obstacle that researchers are facing with creating the oral vaccine is how to get past the stomach and the acids in the gastrointestinal tract without damaging or altering the administered vaccine. However, initial tests which were done with nanotechnology showed that there was a comparable immune response produced from the oral vaccine to the intramuscular injection [25]. This seems to be the case because nanotechnology-based delivery systems have a smaller particle size which makes it easier to stimulate and activate the humoral and the cellular immune systems [26].

While there have been significant advancements made in the nanotechnology delivery systems of the hepatitis B vaccine, there is still so much more research to be done before they can be administered at a large scale. While there are many obstacles that have to be overcome, there are also so many questions that will arise such as the price and whether the vaccine will end up being cost-effective. Will insurance cover the costs of the vaccine or will patients have to pay for the vaccine out of pocket since it is more expensive to develop the technology for this so-called oral vaccine? Only after most of those questions are answered will this form of the vaccine be offered to the mass public, and until then individuals will continue to take the injection of the HBV vaccine, as it is still the most effective prevention method to HBV we currently have. However, if researchers can develop this vaccine and get it on the market, it will definitely be more convenient and allow for a less painful option to the currently used injection.

2.2.3 Side Effects

In the United States, approximately twenty-two thousand people are infected with HBV on a yearly basis. Furthermore, approximately two thousand people die annually because they were infected with HBV. HBV is one of the top vaccine-preventable diseases in the United States besides the influenza virus and the COVID-19 virus. Hence, if individuals stick to the recommended vaccine schedule for HBV, many individuals could prevent being infected and possibly dying due to a virus which could easily be prevented or have the side effects lessened by taking a three-dose vaccine. The symptoms of the disease HBV include but are not limited to hepatitis, inflammation of the liver; cirrhosis, severe liver disease; hepatocellular carcinoma, cancer of the liver; or even death. However, the side effects of the vaccine are not as fatal. These include but are not limited to injection site soreness, pain at the injection site, a low-grade fever, or a severe allergic reaction which only occurs in approximately 1 out of every 600,000 doses given. Based on the side effects of the disease itself and the side effects of the vaccine, the side effects of the vaccine are less severe and help protect individuals from the severe side effects which could possibly be fatal to the patient if they have HBV [22, 24].

2.3 Hepatitis C Vaccine

2.3.1 Potentials for HCV Vaccine

Globally, an approximated 170–200 million people have hepatitis C with Egypt having the highest prevalence in the world. There is no approved or effective vaccine available for HCV. Scientists conducted trials with combinations of PEG-interferon and ribavirin as a therapy, but these trials failed due to serious side effects and high costs to manufacture [27]. Several newer forms of medication using direct-acting antiviral medications (DAA) have been approved to treat HCV infection [28]. These medications are used for both acute and chronic hepatitis C. This list includes but is not limited to:

- Daclatasvir (Daklinza) [28]
- Ledipasvir/sofosbuvir (Harvoni) [28]
- Simeprevir (Olysio) [28]
- Elbasvir/grazoprevir (Zepatier) [28]

DAA-based solutions provided positive results; however, studies have shown relapses and reduced antiviral efficacy when given to cirrhotic HCV patients along with other adverse side effects [18]. DAAs also pose another barrier of expenses as these drugs are costly to produce. This creates a socioeconomic barrier for low-income countries, some of which have high prevalence of HCV [29]. New approaches are being taken to utilize novel nanoparticles to act as carriers for DAA and carriers for potential HCV vaccines. A study conducted by Jyoti et al. used poly lactic-co-glycolic acid (PGLA) nanoparticles along with liver targeting peptides to conjoin and encapsulate cyclosporine A, a known HCV inhibitor. This was done by targeting the host factors for cyclosporine A and in turn resulted in high specificity to the liver cells and inhibiting HCV replication [18]. These studies allow for positive correlation between nanotechnology and HCV treatment.

The potential grows further as this same technique is used for HCV vaccine studies. A study conducted by Jiao et al. researched CpG oligodeoxynucleotide together with recombinant HCV NS3 to create a vaccine model in cationic liposomes. Their study found that both cellular and humoral immune responses toward HCV NS3 increased along with a dramatic inducing of Th1 immune response against HCV [28]. More research is needed to determine side effects and efficacy over time, but utilizing nanoparticles as carriers helps reduce the current issues seen with studies for HCV vaccine.

3 Hepatitis B Treatment

Advancements in nanotechnology allow for the development of new methods for detection and screening of HBV. Combinations of nanomaterials including metal and inorganic nanoparticles, carbon nanotubes, along with microfabrication technologies render signals for sensing HBV and other viruses in low volumes of samples. In comparison to traditional serological methods, gold nanoparticles provide a greater amount of surface area to immobilize biomarkers even when there are low amounts of target molecules in the sample [30]. These advancements in turn reduce the time for HBV detection lowering assay costs as well. Nanotechnology enables development of micro-nanofabrication technology to create a biosensing platform consisting of recognition and responding to physiochemical changes on the sensor surface. For the system to be efficient, it needs to have higher sensitivity and selectivity along with label-free detection abilities. New developments are continuously being studied such as incorporation of recombinant antibody-coupled nanomaterials, graphene-based wireless sensors, surface plasmon resonance aptasensor, and microcantilever resonators for diagnosing HBV [30]. It paves the way for a future of better HBV detection reducing discrepancies in results, reducing costs, and timely detection of the virus.

4 Hepatitis C Treatment

Hepatitis C has a number of available treatments; however, some treatments have shown to be less beneficial due to adverse side effects. One treatment including interferon plus ribavirin produces side effects in individuals resulting in anemia, flu-like symptoms, and depression. This treatment is also predominantly exclusive to type 3a and 3b hepatitis C viruses, whereas, in North America and Europe, the dominant viral type is 1a and 2a [31]. Another treatment includes the usage of synthetic and vector-based siRNAs to inhibit replication of the virus. Synthetic-based siRNAs can be designed to target HCV viral type 1a to silence structural genes and have a lower expression in a dose-dependent manner. However, siRNAs have difficulties in their own manner including low cell uptake, rapid nuclease breakdown, and poor blood stability [32].

Nanoparticles created via nanotechnology have been used to address some of the issues associated with siRNAs to limit off-targeting and better the outcomes. A research study conducted by Lakshi Narayanan et al. found a technique that utilized galactose functionalized dendritic nanovector (DG) as a carrier for siRNA against the 5′ untranslated locale of the HCV genome. This allowed for better targeting as the siRNA-DG complex arranged in a way forming a stable shape allowing interaction between its free galactose and asialoglycoprotein receptor [32].

Other nanotechnology-based treatments include using cross-connected polymeric micelles (CCPM) to target HCV in vitro. Micelles are nanosized core structures containing two regions which possess different affinities toward water. They are made up of a hydrophilic fragment shell and a hydrophobic fragment core. This allows for drugs to be encapsulated and enhance their water solubility and stability when administered. The micelles targeted for HCV utilized camptothecin (CPT), an anti-HCV compound. CPT has poor chemical stability and water solubility, and thus utilizing CCPM allowed it to show high loading capacity which in turn kept CPT levels high to reduce HCV toxicity in the test samples [32].

5 A Possible Cure for Chronic Hepatitis B Virus

5.1 Introduction to the Chronic Hepatitis B Virus

Hepatitis B virus tends to cause an array of fatal problems which are discussed earlier in the paper more in depth. There are approximately 350 million people in the world that are carriers of the chronic hepatitis B virus. It may come as a surprise that even children can contract and be infected with chronic hepatitis B (CHB) infection. However, most children that have CHB tend to be asymptomatic, similar to having the virus [33]. Hence, how do they contract it in the first place? The most common way that children will contract HBV is through their mothers or coming in close contact with places that have poor hygiene and sanitation such as school [34]. Acute hepatitis can become chronic hepatitis in children. It is likely that the disease progresses in children if not treated properly, and the child can grow up to have a greater chance of developing liver disease or even liver cancer. This is very likely to occur for the child by the time they are thirty years old [33]. Hence, it is very important to monitor the child and monitor the progression of the disease in the child at all phases in their life.

5.2 Genome Editing

Since CHB is not necessarily curable, it is crucial to monitor the patient and their progression on the disease to keep them from developing fatal diseases such as cirrhosis or cancer [35]. While there are medications that help with the symptoms of CHB, the patient, especially children, will start to become immunotolerant to the drugs, and they will also start to develop a resistance to the medications. This means that the medications will no longer affect the child to the same extent as it would when they first started taking it, and, in the end, it is basically useless to take as it will not be helping with the symptoms of CHB. However, researchers have been looking into a way to bypass this and possibly develop a cure for CHB. One of these new ways that researchers are looking into requires genome editing [36]. Genome editing, also known as gene editing, is when certain genes in individuals are modified or changed to learn more about gene function. From these functions, researchers can then use it to treat genetic diseases and acquired diseases by changing or modifying the genes that are in charge or causing the disease to occur in the first place [37]. Genome editing specifically means that the DNA is altered in some way, shape, or form by correcting the DNA, introducing new DNA, or deleting the existing DNA sequence [37].

In the viral life cycle of HBV, there is a template which is used for transcription to occur and future replication cycles to occur. This template is known as the covalently closed circular DNA or cccDNA [35]. A cure of the CHB can be accomplished by eliminating the viremia which is what the HBV DNA is known as. Additionally, the viral surface antigen, the HBsAg, and the seroconversion to anti-HBsAg antibodies, also known as the anti-HBs, should also be eliminated. If and only if both of these parts of the virus are eliminated can a possible cure be completed. However, a complete cure needs more. For a complete cure, the cccDNA needs to be eliminated from the infected hepatocytes as that is the primary cause of the virus. If the cccDNA can be completely eliminated, then the virus can be eradicated from the individual, and the patient would also be protected from the virus being able to be reactivated [35].

5.2.1 CRISPR-Cas9

Hence, to cure CHB, researchers and scientists will have to find a way to use genome editing to directly target the cccDNA. There have been therapies of other viral infections with the usage of the CRISPR-Cas9 therapy [36]. CRISPR-Cas9 is a genome editing method that is used by bacteria as immune defenses. CRISPR stands for clustered regularly interspaced short palindromic repeats, while the Cas9 is to identify that the associated protein being used is protein 9. CRISPR-Cas9 is used to treat many different viruses because it is a faster and cheaper method than other genome editing methods which makes it a more accurate and more efficient way of getting the job done.

CRISPR-Cas9 creates CRISPR arrays by taking the bacteria and inserting the viruses' DNA inside of the bacteria which creates a specific pattern with different segments. These CRISPR arrays which were created allow the bacteria to remember what the viruses are since they remember the DNA, and, when the virus attacks the individual again, the bacteria will produce RNA. These RNA segments are created from the CRISPR arrays which were previously created. This then allows the RNA segments to recognize the DNA from the virus and attach to specific regions from it. Lastly, the Cas9 enzyme is used by the bacteria which will help the bacteria cut the DNA off from the virus. This damages the virus and prevents it from being active [38]. Now the virus cannot harm the patient.

Most of the studies involving CRISPR-Cas9 were performed in cell culture systems; however, researchers are starting to perform them in vivo. However, the issue with these models and the mice models is that the HBV replication cycle does not allow for the production of the cccDNA. Additionally, it does not allow the hepatocytes to be reinfected with the virus to replicate. Hence, more research has to be done to find an animal model that can be used that replicates the way the HBV virus works in humans to find a complete cure. More importantly, a model that produces the cccDNA is vital as genome editing using CRISPR-Cas9 is done to the cccDNA to completely be able to come up with a cure for CHB [36].

6 Conclusion

Viral hepatitis is a common infectious disease that can be fatal for anybody who has come in contact with it. Most ways that the virus is transmitted is because of a lack of good sanitation and hygiene. However, there are prevention methods that will help the individual have reduced symptoms, even if they get infected with the disease. These include taking the initial dosages for the hepatitis A and hepatitis B vaccine as a child or an adult and taking the booster shots if needed in the future. In the end, the symptoms from the vaccinations are way less severe than the symptoms from the actual virus, as the virus can possibly even be fatal to the individual. If the vaccinations are taken as a child, the risks of contracting the virus are decreased as the body is able to build immunity. Additionally, researchers are always trying to come up with new ways that an individual can take the vaccinations in a more effective and convenient manner which is what the new nanotechnology-based delivery systems are trying to help with when producing oral vaccines. This is also convenient for children as kids tend to panic the moment they hear they are getting a shot. Furthermore, researchers are trying to find cures to chronic hepatitis which includes using nanotechnology such as genome editing, specifically CRISPR-Cas9, to delete the virus from the individual which helps cure the patient by removing the virus from the cell's DNA. However, more research needs to be done in the entire field before any solid conclusions can be made. For the time being, the best prevention method is proper sanitation and hygiene as well as taking the recommended dosages for the vaccinations for viral hepatitis, especially as children.

References

1. What is Viral Hepatitis?. https://www.cdc.gov/hepatitis/abc/index.htm.
2. Hepatitis B. https://www.mayoclinic.org/diseases-conditions/hepatitis-b/symptoms-causes/syc-20366802.
3. Viral Hepatitis. https://my.clevelandclinic.org/health/diseases/4245-hepatitis-viral-hepatitis-a-b%2D%2Dc.
4. Viral Hepatitis. https://www.hhs.gov/hepatitis/index.html.
5. Hepatitis B questions and answers for the public. https://www.cdc.gov/hepatitis/hbv/bfaq.htm#:~:text=Hepatitis%20B%20is%20a%20liver,known%20as%20chronic%20hepatitis%20B.
6. Hepatitis A. https://www.mayoclinic.org/diseases-conditions/hepatitis-a/symptoms-causes/syc-20367007.

7. Hepatitis. https://kidshealth.org/en/kids/hepatitis.html#:~:text=The%20two%20most%20common%20are,hepatitis%20C%20virus%20(HCV).
8. Hep A VIS. https://www.cdc.gov/vaccines/hcp/vis/vis-statements/hep-a.html#:~:text=Children%20need%202%20doses%20of,months%20after%20the%20first%20dose.
9. Hepatitis in Children. https://www.urmc.rochester.edu/encyclopedia/content.aspx?contenttypeid=90&contentid=P02517.
10. Hepatitis B. https://www.niddk.nih.gov/health-information/liver-disease/viral-hepatitis/hepatitis-b.
11. Hepatitis C. https://www.cdc.gov/hepatitis/hcv/index.htm#:~:text=Hepatitis%20C%20is%20a%20liver,to%20prepare%20and%20inject%20drugs.
12. Hepatitis C. https://www.mayoclinic.org/diseases-conditions/hepatitis-c/symptoms-causes/syc-20354278.
13. Viral Hepatitis in the United States: data and trends. https://www.hhs.gov/hepatitis/learn-about-viral-hepatitis/data-and-trends/index.html#:~:text=Millions%20of%20Americans%20from%20all,as%20low%20as%202.5%20million.
14. Hepatitis in children. https://www.stanfordchildrens.org/en/topic/default?id=hepatitis-in-children-90-P02517.
15. Hepatitis D. https://www.who.int/news-room/fact-sheets/detail/hepatitis-d.
16. Hepatitis E. https://www.cdc.gov/hepatitis/hev/index.htm#:~:text=Hepatitis%20E%20is%20a%20liver,virus%20%E2%80%93%20even%20in%20microscopic%20amounts.
17. What is Nanotechnology? https://www.nano.gov/nanotech-101/what/definition.
18. Mostafa E, Noureldien D, Shaker M. Hepatitis C virus management: potential impact of nanotechnology. Virol J. 2017;14:88.
19. John B. Immunization. In: Mandell, Douglas, Bennett, editors. Principles and practice of infectious diseases. 9th ed; 2020.
20. Vandana P, Prajakta D, Ratnesh J. Case studies: Nano-systems in the market. In: Nanoparticulate drug delivery. Woodhead Publishing Series in Biomedicine; 2012. p. 209–20.
21. Hepatitis A. https://www.cdc.gov/vaccines/parents/diseases/hepa.html.
22. A look at each vaccine: Hepatitis B vaccine. https://www.chop.edu/centers-programs/vaccine-education-center/vaccine-details/vaccine-hepatitis-b-vaccine#:~:text=The%20first%20hepatitis%20B%20vaccine,in%20blood%2C%20such%20as%20HIV.
23. Hepatitis B VIS. https://www.cdc.gov/vaccines/hcp/vis/vis-statements/hep-b.html
24. Hepatitis B. https://www.cdc.gov/vaccines/parents/diseases/hepb.html.
25. Nanotechnology used to deliver Hepatitis B vaccine. https://nano-magazine.com/news/2019/7/12/nanotechnology-used-to-deliver-hepatitis-b-vaccine.
26. Priscilla P, Nicolas V, Andrea S, Rodrigo G, Alejandro N, Maria C. Nanotechnology, drug delivery systems and their potential applications in Hepatitis B vaccine. Int J Vaccin Vaccin. 2015;1(2):40–4.
27. Yu Y, Xuesong W, Peilan L, et al. A nanoparticle-based Hepatitis C virus vaccine with enhanced potency. J Infect Dis. 2020;221(8):1304–14.
28. Hepatitis C. https://www.niddk.nih.gov/health-information/liver-disease/viral-hepatitis/hepatitis-c#treatment.
29. Sonia A, Paula M, Irene M, Ricardo M, Veronica B. Nanotechnology: a reality for diagnosis of HCV infectious disease. J Infect. 2020;80(1):8–15.
30. Hakan UY, Faith I, Shuqi W, et al. Recent advances in micro/nanotechnologies for global control of Hepatitis B infection. Biotechnol Adv. 2015;33(1):178–90.
31. Usman A, Muhammad Y, Maida A, Rahat E, Shah J, Obaid U. siRNAs: potential therapeutic agents against Hepatitis C virus. Virol J. 2011;8:276.
32. Kajal C, Shweta P, Devander S. A critical review on nanoscience advancement: in treatment of viral infection. J Drug Deliv Therap. 2021;11(6):225–37.
33. Mona AH, Deirdre K. Chronic Hepatitis B in children and adolescents: epidemiology and management. Paediatr Drugs. 2013;15(4):311–7.

34. Hepatitis in children. https://www.urmc.rochester.edu/encyclopedia/content.aspx?contenttypeid=90&contentid=P02517#:~:text=It's%20spread%20to%20children%20in,a%20problem%20in%20developing%20countries.
35. Elisabetta L, Giovanni V, Fabio C, Mauro B, Pietro A. Chronic Hepatitis B: are we close to a cure? Dig Liver Dis. 2015;47(10):836–41.
36. Daniel S, Kelly RL, Michelle AL, et al. CRISPR-Cas9 gene editing of Hepatitis B virus in chronically infected humanized mice. Mol Ther Methods Clin Dev. 2021;20:258–75.
37. Gene Editing-Digital Media Kit. https://www.nih.gov/news-events/gene-editing-digital-press-kit#:~:text=Genome%20editing%2C%20also%20called%20gene,treat%20genetic%20or%20acquired%20diseases.
38. What are genome editing and CRISPR-Cas9? https://medlineplus.gov/genetics/understanding/genomicresearch/genomeediting/.
39. About HIV. https://www.cdc.gov/hiv/basics/whatishiv.html.
40. Ankur C, Maryann M, Meryl P, Regino PG. Viral Hepatitis in children: a through E. Pediatr Ann. 2016;45(12):420–6.
41. Meryem J, Bisma R, Harunor R, Thao L, Shafquat R. Update on global epidemiology of viral Hepatitis and preventative strategies. World J Clin Cases. 2018;6(13):589–99.

Recent Developments in the Treatment of Influenza

Lachlan Shiver, Caroline Ward, Brian Arciola, Evan Adler, and Charles Preuss

Abstract Influenza infections are associated with elevated levels of mortality and morbidity in young children, the elderly, and the immunocompromised. Unfortunately, the influenza virus's high intrinsic mutation rate makes developing effective therapeutics challenging. Prevention of symptomatic influenza infection focuses on the circulation of the yearly influenza vaccine. This vaccine is based on World Health Organization (WHO) predictions of future circulating strains and varies in effectiveness from 10% to 60%.

Treatment of a primary infection focuses on inhibiting viral replication and symptom management. Evidence-based medicine supports using inhibitors of viral replication, such as neuraminidase inhibitors and viral polymerase inhibitors, only during the first 2 days of symptom onset. Symptomatic management includes using immunomodulating drugs, such as hyperimmune IVIG and sirolimus. In 2018, the last drug to receive FDA approval for influenza infection, baloxavir marboxil, was approved. However, multiple investigational drugs are currently undergoing FDA trials.

Keywords Influenza A · Influenza B · Antiviral therapy · Vaccine development

L. Shiver (✉) · C. Ward · B. Arciola · E. Adler
University of South Florida Morsani College of Medicine, MD Program, Tampa, FL, USA

C. Preuss
University of South Florida Morsani College of Medicine, Department of Molecular Pharmacology & Physiology, Tampa, FL, USA

R. Shegokar, Y. Pathak (eds.), *Viral Drug Delivery Systems*,
https://doi.org/10.1007/978-3-031-20537-8_11

1 Influenza Virus and Disease Progression

1.1 Influenza Strains and Subtypes

Influenza, known colloquially as "the flu," is an acute lower respiratory tract infection caused by multiple strains of the influenza virus. First recognized in the sixteenth century, influenza remains a significant public health burden, with roughly 28,000 flu-related deaths occurring in the United States during the 2018–2019 flu season. Influenza's contagious nature, high mutation rate, and animal reservoir make eradication of this virus unfeasible and development of efficacious therapeutics challenging [1–3].

Influenza viruses (IVs) are single-stranded negative sense RNA viruses within the *Orthomyxoviridae* family. Within this family, three strains (A, B, and C) are known to cause infection in humans. Influenza A, an IV with high rates of infection and mortality, is composed of a lipid bilayer, nine to ten structural proteins, and eight segments of viral RNA. Within medical literature, emphasis is placed on two of these structural proteins due to their polymorphic nature and involvement in epidemic strains: hemagglutinin (HA) and neuraminidase (NA). These two proteins are designated *H & N*, respectively, when naming a strain of influenza A [1, 2]. Two subtypes of influenza A currently circulate in human populations, H1N1 and H3N2 [4].

Influenza B, another IV responsible for high rates of morbidity and mortality, is similar in structure to influenza A. However, influenza B lacks a clear animal reservoir, resulting in less antigenic variability noted in the hemagglutinin and neuraminidase structures when compared to influenza A [5]. Currently, two distinct antigen variants of influenza B circulate in humans: *B/Victoria/2/87 like* and *B/Yamagata/16/88 like* [4].

Influenza C remains outside the scope of this chapter, as it is benign strain known to cause very mild disease [2].

1.2 Influenza Disease Course and Diagnostic Methods

Influenza A and B both present with a similar disease progression. The incubation period is approximately 1–2 days, and onset of symptoms is rapid. An uncomplicated disease course, which presents approximately 98% of the time, is characterized by cough, fevers, myalgias, chills, sweats, and malaise. This persists for approximately 2–8 days. Severe infections, which are characterized by rapid onset of fulminant illness, can present with numerous systemic symptoms. These, as listed in Table 1, include viral pneumonia, acute respiratory distress syndrome (ARDS), conjunctivitis, lacrimation, myocarditis, rhabdomyolysis, aseptic meningitis, and encephalomyelitis [6]. Immunocompromised patients, patients with medical comorbidities, pregnant patients, and patients younger than 2 or older than 65

Table 1 Influenza symptomatology including potential complications from influenza infection

Influenza symptomatology	
Incubation period	1–2 days
Respiratory symptoms	Nonproductive cough Sore throat Nasal discharge Shortness of breath
Systemic symptoms	Fevers Chills Headaches Myalgias Malaise Anorexia
Potential complications	Acute respiratory distress syndrome Bacterial and viral pneumonia Cardiac complications Cytokine storm resulting in multi-organ failure

Adapted from Refs. [2, 6, 7]

Table 2 Various sensitivities and specificities of point of care and NAAT tests for influenza viral strains

	Influenza A		Influenza B	
Virus strain	Sensitivity (%)	Specificity (%)	Sensitivity (%)	Specificity (%)
Traditional rapid influenza diagnostic tests	54.4	99.4	53.2	99.8
Rapid nucleic acid amplification tests	91.6	99.2	95.4	99.4

Data from Ref. [9]

are more likely to undergo a complicated disease course [2, 8]. Mortality rates from influenza remain a point of contention, as over 50% of infections can be asymptomatic and some strains are more virulent than others [6].

Influenza infection often presents with concomitant bacterial pneumonia. This most commonly occurs in immunocompromised patients, as influenza infection can overwhelm the immune response and allows for opportunistic pathogens. *Staphylococcus aureus* and *Streptococcus pneumoniae* are often cultured from the sputum of severely ill influenza patients. These bacterial infections should be treated with proper antibiotic therapy [2].

The diagnosis of influenza can be difficult due to symptoms overlapping with a variety of different respiratory illnesses; however, many different modalities exist for the identification of influenza strains. Point of care tests, as seen in Table 2, are favorable due to their low cost, but they come with a sacrifice in sensitivity and specificity.

1.3 Virus Transmission and Notable Pandemics

Influenza viruses can spread via aerosol, droplet, and contact transmission. The importance of each modality in the spread is debated. Aerosols are small airborne droplets that can remain viable for several hours. Viability on fomites, or surfaces that may carry infectious particles, has been recorded for up to 48 hours. Due to the long viability period and risk of airborne transmission, the WHO recommends the use of N95 masks in situations where aerosols may be generated [2].

All known influenza pandemics are due to antigenic shifts that occur within influenza A. Four such outbreaks have occurred in 1918, 1957, 1968, and 2009. These pandemics were due to the following subtypes, respectively: H1N1, H2N2, H3N2, and H1N1. Today, H3N2 and H1N1 are the two circulating subtypes of the influenza A strain [2, 10].

1.4 Influenza Pathophysiology

Symptoms and disease course of influenza occur due to respiratory epithelial tissue damage caused by both the virus itself and the immune system response. Viral replication occurs in a variety of different cell types, but efficient cleavage of the hemagglutinin cell surface molecule occurs only in respiratory epithelial cells. This cleavage is a prerequisite for influenza cell entry. Multi-organ involvement and systemic symptoms, such as malaise and cardiomyopathies, are believed to be caused by a downstream innate immune response. Of note, IL-1B and IL-18 secretion has been linked to the innate immune response in severe influenza infections, and experimental drugs are in development to target pathways these cytokines are involved in [11]. Drugs targeting influenza virus pathophysiology must both target viral replication and the potentially lethal sequelae of the viral-induced cytokine storm.

1.5 Influenza Replication Cycle

As an obligate intracellular pathogen, IVs require living cells to replicate. Hemagglutinin, one of the membrane glycoproteins found in the influenza virus, is crucial to this process. On initial contact with the host cell, IV uses hemagglutinin to bind to a sialic acid-containing cellular membrane protein. Cell-specific enzymes cleave hemagglutinin, inducing a conformational change which results in the adsorption and endosomal uptake of influenza [1].

After endosomal uptake, the acidic environment of the endosome results in a flow of protons into the viral envelop via the viral M2 protein channel. The viral RNA is then transported rapidly to the nucleus due to nuclear localization signals. Within the nucleus, viral RNA (vRNA) is transcribed into mRNA, which is

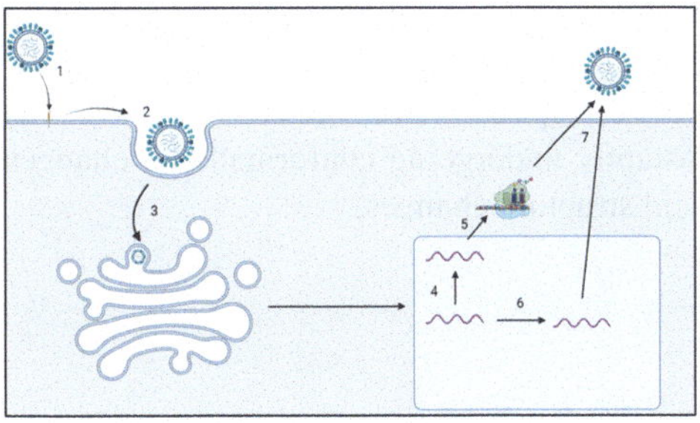

1. Viral attachment via sialic acid (Hemagglutinin-potential drug target)
2. Endocytosis
3. Uncoating and membrane fusion (M2 ion channel – potential drug target)
4. Transcription
5. Translation of viral proteins
6. Replication of vRNA
7. Virion budding and release (Neuraminidase – potential drug target)

Fig. 1 Simplified replication cycle of the influenza A virus. (Created w/BioRender and adapted from Ref. [12])

subsequently exported back into the cytoplasm for assembly of viral structural proteins. At the same time, the vRNA is transcribed into complementary RNA (cRNA), and this cRNA serves as the template for further vRNA synthesis.

Once vRNA synthesis is complete, cellular pathways are activated that allow for nuclear export of the viral nucleotides. These cellular pathways include the Raf/MEK/ERK signal transduction pathway and the activation of Nf-Kb and caspase 3. The viral nucleotides are then encapsulated within the viral structural proteins and integrated into the plasma membrane. The last step of the viral replication cycle is budding, which is the cleavage of virion from cell surface membrane. This occurs via enzymatic activity of viral neuraminidase, resulting in the cleavage of sialic acid moieties on cell surface proteins, allowing for viral exit [1]. The replication cycle of influenza is depicted in Fig. 1. Of note, the M2 ion channel is the hypothesized target of amantadine, which is in the adamantane drug class. However, these drugs were only effective against influenza A and are no longer recommended for clinical use due to drug resistance.

2 Influenza Virus Genomic Variability and Antigenic Shift and Drift

Influenza A's genome consists of eight unique single-stranded viral RNA segments which are bound to viral polymerase and oligomeric nucleoproteins. During stage 7 of the virus replication cycle (Fig. 1), nucleotide packaging sequences found on the vRNA allow for packaging of vRNA into viral particles. The packaging process is a highly selective process, and each vRNA segment has a unique packaging sequence that coordinates RNA-RNA interactions and RNA-amino acid interactions. Despite the highly selective process of vRNA incorporation into viral particles, research suggests this process has numerous redundancies, to prevent a high likelihood that any one mutation will result in deleterious effects. This allows influenza's genome to change gradually throughout time, without significant impacts on the virus's fitness [13].

Developing therapeutics targeted toward influenza virus strains is a challenge due to two evolutionary phenomena, termed antigenic drift and antigenic shift. These phenomena prevent robust immune responses from mammalian hosts, as antigenic targets of antibodies are constantly undergoing conformational changes due to primary and secondary amino acid structure changes.

2.1 Antigenic Drift

Antigen drift is defined as the gradual change of genetic information an influenza strain develops over time. This occurs due to the lack of proof-reading activity of influenza's RNA polymerase complex. During one replication cycle, numerous viral progenies are produced, and this lack of proof-reading produces RNA transcripts that differ from the original copy. While many of these RNA transcripts contain deleterious mutations, the sheer number of progenies produced during one infection result in new strains that have more evolutionary fitness. Thus, over time the genome of an influenza strain will "drift." From an immunological standpoint, this drift is most notable in the viral surface hemagglutinin. Amino acid changes in this hemagglutinin, a high-affinity target for neutralizing antibodies, result in the failure of long-term immunity to influenza after a primary infection [4].

For reasons still unclear, multiple studies have revealed that influenza B undergoes antigenic drift at a slower rate when compared to influenza A. Multiple reasons for this fact have been postulated. One theory focuses on influenza B's de novo mutation rate, which is believed to be half that of influenza A's de novo mutation rate. This is due to experiments that indicate influenza B's RNA polymerase complex is less error prone than influenza A's RNA polymerase complex. Another theory is that a more significant population bottleneck occurs for influenza B when transmission occurs from host to host. This population bottleneck, defined as a significant reduction in viral populations, would eliminate viruses with potentially advantageous mutations and prevent them from spreading [14].

2.2 Antigenic Shift

Antigenic shift is defined as the recombination of viral RNA segments from two distinct viral strains. This phenomenon, referred to as genetic reassortment, is possible due to influenza's segmented genome. Antigenic shift only occurs in influenza A, as influenza B's lack of an animal reservoir and slower mutation rate hinders recombination from two distinct strains.

As previously mentioned, successful antigenic shift results in an influenza virus with a novel combination of hemagglutinin and neuraminidase. These shifts can occur when avian strains recombine with mammalian strains in an avian host, as in

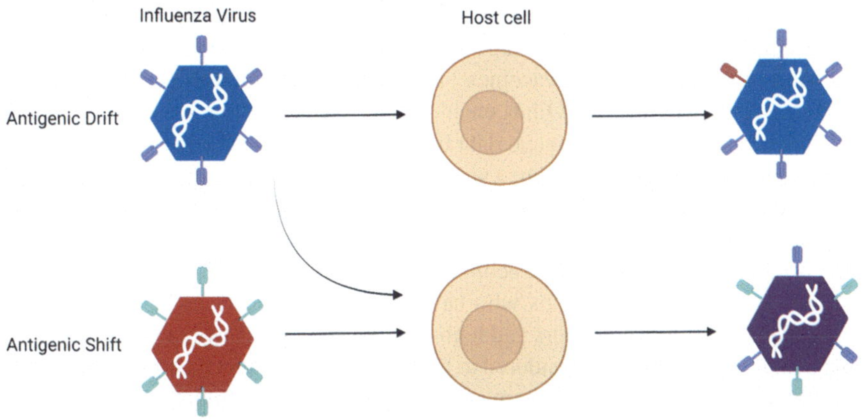

Fig. 2 Representation of antigenic shift and drift. (Created with BioRender)

the 1918 (H1N1) and 1968 (H3N2) pandemic. This recombination can also occur in swine populations, which was observed 2009 H1N1 pandemic.

When two viral strains coinfect the same cell, the eight genomic sequences can recombine in nearly any potential order. This results in 256 theoretical combinations [15]. However, not every theoretical combination will result in a viable virion particle for reasons not completely understood. One in vitro study indicated a lack of compatibility among RNA polymerase subunits from different influenza strains. This lack of compatibility resulted in the inability of the virus to synthesize a functioning ribonucleoprotein complex and thus prevented viral RNA synthesis [16]. Other speculated reasons for limited combination of genomic sequences include preferential packaging of distinct genomic sequences together [17]. However, more research needs to be completed to fully elucidate the nonrandom selection that occurs during antigenic shift. Figure 2 is a representation of antigenic shift and drift.

3 Preventative Medicine: Influenza Vaccine Development

3.1 Influenza Vaccine

The influenza vaccine is a biologically active therapeutic designed to confer acquired immunity to one (monovalent) or several (multivalent) strains of the influenza virus via stimulating antibody production to specific viral proteins, most commonly NA and the head domain of HA [18].

The development of the influenza vaccine began in the 1930s following the isolation of the influenza A virus, which culminated in the creation of a monovalent inactivated vaccine for the influenza A virus [19]. Over the next several decades, numerous advances led to the creation of multivalent influenza vaccines and different vaccination paradigms, including split virion, subunit, live attenuated, virosome, and viral vectored influenza vaccines [19, 20]. Global monitoring of circulating influenza viruses by the WHO has enabled the development of seasonal influenza vaccines based on epidemiological data from previous influenza outbreaks [19, 21].

Vaccination against the influenza virus continues to be the most effective strategy for preventing influenza infection [18, 20, 22]. FDA-approved indications for influenza vaccination include the prevention of influenza A and B in individuals older than 6 months [18]. The CDC recommends annual influenza vaccination for all individuals older than 6 months who have no counterindications [23]. The WHO recommends healthcare workers and high-risk individuals receive the influenza vaccine annually. Individuals considered at high risk for complications following influenza infection include pregnant women, elderly individuals, children aged 6 months to 59 months old, individuals with specific chronic medical conditions, and healthcare workers [24].

The efficacy of the seasonal influenza vaccine varies substantially, ranging from 10% to 60% [18]. This considerable variation in seasonal efficacy is primarily attributed to the antigenic relatedness of the vaccine strains and the presently circulating influenza strains [25]. The lowest levels of efficacy are seen in years where there is a suboptimal match between the vaccine and circulating strains [18]. Other factors affecting seasonal efficacy include the host's age and immune competence [18, 25]. The influenza vaccine is less effective in specific populations, including infants, the elderly, and immunosuppressed individuals; however, vaccination is considered beneficial due to reduced incidence of severe disease and hospitalization [18].

The adverse events associated with influenza vaccination are generally mild and include fever, irritability, myalgia, and injection site reactions. In rare cases, the influenza vaccine has been associated with allergic reactions, urticaria, and anaphylaxis. Contraindications for receiving the influenza vaccine include infants younger than 6 months old, high fever, Guillain-Barre syndrome, or a history of allergy to any vaccine component. Toxicity related to the influenza vaccine is minimal, with no evidence of carcinogenicity or effects on fertility. No dose-dependent toxicity has been noted in the influenza vaccine nor toxicity related to components of the vaccine such as aluminum salt adjuvants [18].

3.2 *Influenza Vaccine Paradigms*

This section will discuss six major categories of experimental and commercially available vaccine strategies.

3.2.1 Whole Inactivated Influenza Vaccine (WIIV)

Whole inactivated influenza vaccines (WIIV) contain the target viruses HA and NA domains but have been rendered unable to replicate by either heat or a chemical reagent. They are developed by initially infecting susceptible cells (usually chicken embryos) with the target virus. The virus is then allowed to replicate in the infected cells, followed by purification, concentration, and inactivation through heat or chemical reagents [20, 22]. A common strategy for developing WIIVs involves RNA virus reassortment, an evolutionary mechanism of segmented RNA viruses that allows for genetic recombination of multiple viruses when coinfecting a cell. This recombination can yield viral progeny with novel genome combinations [22, 26]. RNA virus reassortment can be utilized in WIIV development by coinfecting a susceptible cell with the target virus and a separate high-growth RNA virus. After reassortment, a viral progeny expressing the target virus's NA and HA domains with the high-growth virus's six internal segments (PB2, PB1, PA, NP, M, and NS) is selected by neutralizing antibodies against the HA and NA domains of the high-growth virus. The selection of a viral progeny with the appropriate genetic reassortments is confirmed by genetic sequencing. Finally, the product is purified and inactivated. Adjuvants can be added to the final product to increase immunogenicity. Benefits of this strategy include a high safety profile and a strong humoral immune response. A significant drawback of this type of vaccine is the limited T-cell-mediated response [20, 22].

3.2.2 Split-Virion Influenza Vaccines

This subtype of influenza vaccine is prepared in a similar process to the WIIV. However, an additional step during the final processing of the vaccine requires splitting the viral envelope and further purification of the viral contents via removing large molecular weight proteins and nucleic acids while preserving the HA and NA proteins [20]. This process reduces adverse events associated with vaccine administration compared to whole inactivated influenza vaccines, especially in children [19–21, 25]. Split-virion vaccines stimulate a robust humoral response and have an excellent safety profile. However, as with whole inactivated and subunit influenza vaccines, the T-cell response is minimal [20].

3.2.3 Subunit Influenza Vaccines

This subtype of influenza vaccine was initially developed from whole inactivated influenza vaccines and involved the same process as split-virion vaccines with additional purification steps and adjuvants added [20, 21]. Genetic cloning techniques can also produce this subtype of vaccine by engineering plasmids that contain the genes for antigenic influenza proteins and transferring these plasmids to another cell for production. The antigenic proteins produced from these plasmids can then be purified. The risks and benefits of this type of vaccine are similar to those of split-virion vaccines [20].

3.2.4 Virosome Influenza Vaccines

Virosome influenza vaccines are prepared by solubilization of the influenza virus membrane in a detergent followed by removal of the viral nucleocapsid via centrifugation. Subsequently, the detergent is extracted to allow reconstitution of the viral membrane without the nucleocapsid, which is necessary for viral replication [20]. The reconstituted viral membrane will retain structural surface glycoproteins and the ability to perform receptor-mediated endocytosis [27]. A unique benefit of virosome influenza vaccines is the capacity to deliver adjuvants or other small peptides directly to antigen-presenting cells by receptor-mediated endocytosis; however, despite this capacity, a significant T-cell-mediated response has not been noted. Structural surface glycoproteins will induce a humoral response and antibody production [20].

3.2.5 Live Attenuated Influenza Vaccines (LAIV)

Live attenuated influenza vaccines (LAIV) contain a virus capable of intracellular replication and express the influenza virus's HA and NA domains but that has been selected for low virulence and pathogenicity. This process is done similarly to the WIIV development beginning with the coinfection of a chicken embryo with the target influenza virus and a Master Donor Virus (MDV), followed by the selection of a viral progeny expressing the HA and DA domains of the target influenza virus with internal segments of the Master Donor Virus. The progeny is then repeatedly exposed to primary chicken kidney cells under low temperature and in the presence of antibodies to the MDV's HA and DA domain until the progeny expresses specific mutations that result in a phenotype that is attenuated and has impaired replication at body temperature. LAIV are significantly more immunogenic, producing a T-cell-mediated response, systemic humoral response, and a mucosal humoral response. However, since the virus can actively replicate inside the host's cells, there is a significant safety concern for genetic reversal of the attenuated phenotype, especially in immunocompromised individuals [22].

3.2.6 Viral Vectored Influenza Vaccines

Viral vectored influenza vaccines use genetic engineering to form a recombinant virus expressing the influenza virus antigens (notability HA and NA) within a different virus's genome. Numerous viruses can be used as vectors, including adenoviruses, baculoviruses, and arenaviruses. This type of vaccine will induce both a humoral and a cytotoxic response due to the active production of viral antigens inside the host cells, similar to live attenuated influenza vaccines [20]. A perceived benefit of viral vectored influenza vaccines is the ability to administer the vaccine directly to the mucosa, thereby more effectively mimicking a natural infection [21]. Viral vectored influenza vaccines could also be developed to express antigens from multiple influenza subtypes within the same viral vector facilitating multivalent vaccine development [20].

3.3 Universal Flu Vaccine

The universal influenza vaccine is a vaccine that is meant to provide long-lasting immunity to influenza A and B regardless of antigenic shift or drift. Areas of interest in developing a universal influenza vaccine include highly conserved regions of the influenza virus, such as the stalk domain of HA, specific regions of NA, and the matrix 2 protein. A major barrier to the induction of antibodies to NA and the stalk domain of HA is the immunodominance of the HA head domain. Chimeric HA proteins with modified head regions and recombinant headless stem HA proteins have been created to overcome the head domain's immunodominance. A potential option to improve antibody development to NA could be to increase the concentration of NA proteins inside the vaccine. While the matrix 2 protein has shown minimal potential for mutation, one of the significant limitations of its use as a target in a universal influenza vaccine is decreased levels of antibody production due to low immunogenicity.

Other approaches include T-cell-based vaccines designed to allow recognition of highly conserved epitopes on the nucleoprotein and matrix 1 protein. Yet another significant challenge is protecting high-risk groups. Age-related changes in immune function can significantly alter vaccination effectiveness, which must be accounted for to achieve a universal influenza vaccine. While progress has been made toward developing a universal influenza vaccine, additional research in animal and human systems is necessary [15, 20, 28] (Tables 3 and 4).

Table 3 Selected available influenza vaccines on the market [29]

Drug (manufacturer)	Type	Age indication	Dosage[a]	Route of administration
Afluria Quadrivalent (Seqirus)	IIV	≥6 months	15 μg	IM
Fluarix Quadrivalent (GlaxoSmithKline)	IIV	≥6 months	15 μg	IM
FluLaval Quadrivalent (GlaxoSmithKline	IIV	≥6 months	15 μg	IM
Flublok Quadrivalent (Sanofi Pasteur)	VVIV[b]	≥18 yrs	45 μg	IM
FluMist Quadrivalent (AstraZeneca)	LAIV	2 through 49 yrs	10 fluorescent focus units	NAS

Table on clinical trials if any/for new vaccine candidates

[a]The adult dosage is provided, dosages vary for children and the elderly

[b]This vaccine uses a viral vector to produce the HA and NA antigens which are subsequently purified

Table 4 Selected clinical trials for influenza vaccines [30–33]

Drug (manufacturer)	Type	Ages included	Phase
GSK2321138A (GlaxoSmithKline)	IIV	6–35 months old	3
MVA-NP+M1 (Vaccitech)	Universal	>18 y.o.	2b
VXA-A1.1 (Vaxart)	VVIV	18–49	2
cH8/1 N1 LAIV, AS03-adjuvanted-cH5/1 N1 IIV, and cH5/1 N1 IIV	Universal	19–39	1

4 Neuraminidase Inhibitors Used in Influenza Treatment

4.1 Neuraminidase Inhibitors

Neuraminidase inhibitors (NAI) are a class of drugs that competitively inhibit viral neuraminidase (NA) in influenza A and B. Neuraminidase is a highly conserved surface glycoprotein present in both influenza A and influenza B. It serves the purpose of cleaving the a-ketosidic bond linking a terminal neuraminic acid from an adjacent oligosaccharide moiety. This function is critical for viral penetration of respiratory tract mucus and virion liberation [34]. NA may also play a role in viral entry by facilitating viral movement toward endocytosis regions in the influenza A virus [35, 36]. The NA protein is targeted for developing influenza vaccines due to its abundance, highly conserved nature, and location as a surface glycoprotein [20, 28].

NAIs are considered an effective tool for preventing and treating influenza infections. NAIs have been shown to reduce the duration of the symptoms of influenza infections by 0.5–1.5 days and inhibit viral transmission if given within 2 days of the onset of symptoms [37]. NAIs are also approved for chemoprophylaxis and are the standard of care to treat influenza infection in severely ill hospitalized patients [37, 38]. NAI use may be considered for post-exposure prophylaxis, especially in the case of a nosocomial outbreak. Pre-exposure prophylaxis with NAI treatment may serve as a potential option to reduce the risk of influenza complications in high-risk patients such as organ transplant recipients [37].

4.2 Approved Neuraminidase Inhibitors

Zanamivir was the first developed NAI and is widely licensed to treat and prevent influenza. It has very poor bioavailability and is primarily administered as an inhalant. IV administration of zanamivir is approved by the European Medicines Agency (EMA) for life-threatening influenza in individuals older than 6 months [37]. IV administration of zanamivir is eligible for consideration in emergency use protocol in the United States [39]. Zanamivir is primarily renally excreted in an unchanged form [40, 41]. Zanamivir tends to be well tolerated by individuals with minimal side effects; however, its use is not recommended in individuals with underlying respiratory diseases, including asthma and COPD due to the risk of bronchospasm [41, 42].

Oseltamivir is another widely licensed NAI designed to treat and prevent influenza. It differs from its predecessor zanamivir by increased bioavailability allowing oral administration. Oseltamivir is a prodrug that requires hepatic metabolism and is eliminated unchanged in the urine following its initial activation by hepatic esterase [37, 43]. Oseltamivir is generally well tolerated by individuals; however, there is an increased risk of adverse effects, including nausea, vomiting, psychiatric, and renal events, compared to placebo. The increased toxicity of oseltamivir when compared to zanamivir may be related to oseltamivir's increased bioavailability [44, 45].

Peramivir is an NAI that was FDA and EMA approved for single-dose IV administration in 2014 and 2018 [37, 46]. However, the marketing authorization holder of peramivir withdrew marketing authorization from the EMA in December 2020. Peramivir is indicated for early treatment of uncomplicated influenza in individuals over 2 years old via IV administration [37]. Peramivir is renally excreted in an unchanged form and is generally well tolerated. The effectiveness of peramivir in reducing influenza symptoms is similar to that of oseltamivir [47–49].

Laninamivir octanoate is a long-acting NAI approved for influenza treatment and prevention in Japan only. Laninamivir is a nasally inhaled prodrug hydrolyzed into its active form by respiratory epithelial cells [37]. Laninamivir is well tolerated with a similar side effect profile to other NAIs, and its effectiveness is comparable to oseltamivir and zanamivir [50, 51].

4.3 Neuraminidase Inhibitor Resistance

NAI-resistant strains of influenza pose a substantial risk to public health as NAIs are currently the most widely used antivirals for influenza prophylaxis and treatment [52]. Presently, NAI-resistant influenza strains remain rare with an estimated prevalence of 0.2–0.5% based on WHO susceptibility testing [37]. High-risk groups for contracting an NAI-resistant infection include young children and highly immunosuppressed patients. NAI-resistant infections often emerge following prolonged NAI use in high-risk populations [52].

4.4 Adamantane Antivirals

Adamantanes are a class of antiviral used to treat influenza A infections by inhibiting the M2 ion channel, which subsequently blocks the release of the influenza RNA from intracellular endosomes. This class of antiviral does not affect influenza B viruses due to structural differences in the M2 ion channel. However, this class of antivirals is rarely used clinically due to the emergence and worldwide spread of adamantane-resistant influenza A strains. Almost all human influenza A strains are resistant to adamantane, and its use as an antiviral is not recommended [37].

5 Baloxavir: A First in Class Medication

5.1 Background

Despite the effectiveness of the neuraminidase inhibitors, evidence has shown that influenza strains can develop resistance to this drug class, as discussed in the previous section. The 2009 H1N1 influenza pandemic in Mexico, caused by the neuraminidase-resistant A/H1N1pdm09 strain, was the first global demonstration that neuraminidase-resistant influenza strains represent a notable public health concern, as they do not display a significant fitness disadvantage when compared to nonresistant strains. Neuraminidase-resistant strains, including A/H1N1pdm09, continue to circulate and generate community outbreaks, thus underscoring the need to develop drugs that target distinct regions of the virus [53].

5.2 *Mechanism of Action*

Baloxavir marboxil (trade name Xofluza) was one drug designed to target an area of the influenza virus that had not been targeted before. Baloxavir is a prodrug that is metabolized to baloxavir acid before it can exert its effect as a small molecule inhibitor of the polymerase acidic (PA) subunit of the influenza viral polymerase [54].

To understand the mechanism of action of baloxavir, a brief review of the function of influenza's polymerase is necessary. Influenza's polymerase is a complex consisting of three subunits: PA, polymerase basic protein 1 (PB1), and polymerase basic protein 2 (PB2). Transcription of the influenza viral genome requires the use of a 5′ cap; however, the virus does not carry the machinery to produce this cap by itself. As a result, influenza engages in a phenomenon known as "cap-snatching" in which it cleaves the 5′ cap from host mRNA and uses it as a primer for the transcription of its own viral RNA. The PB2 subunit functions to bind the 5′ cap of host mRNA, and the PA subunit acts as an endonuclease to cleave this cap. The resulting free 5′ cap, with 8–14 nt attached, is then used as a primer for viral transcription by PB1, which is an RNA-dependent RNA polymerase [55]. This process is summarized in Fig. 3.

This process is important because it is conserved across IV strains and plays a critical role in viral replication. Influenza A, B, and C strains use the same PB1 and PB2 proteins; however, the influenza C strain utilizes a unique protein known as P3 instead of the PA protein. It is important to note that the function of P3 is identical to that of PA, which allows for an identical replication process to occur in influenza C strains when compared to A and B strains, despite the difference in this protein [55].

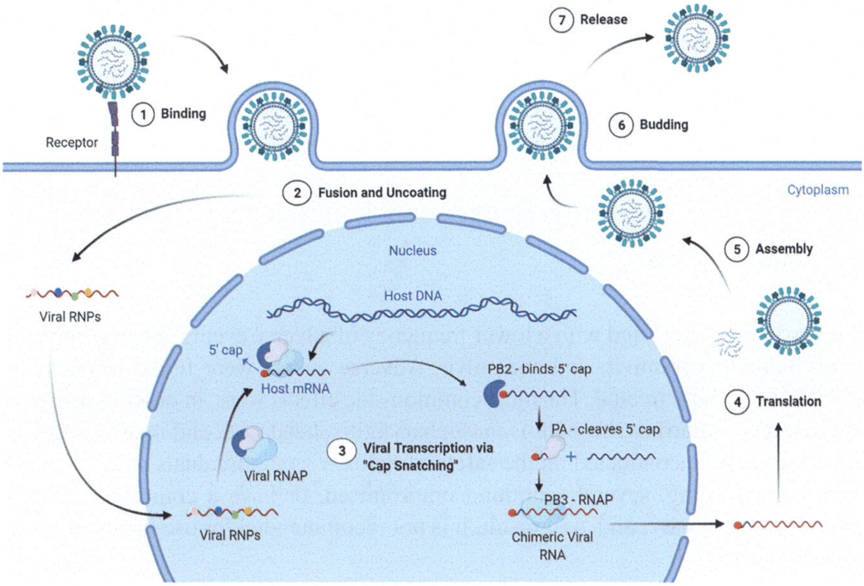

Fig. 3 Simplified diagram demonstrating the influenza viral life cycle with emphasis on the mechanism of viral transcription. RNA, RNA polymerase; RNP, ribonucleoprotein. (Created with BioRender, adapted from Ref. [56])

5.3 Clinical Uses

Baloxavir is currently approved for treatment of acute, uncomplicated flu in individuals 12 years and older within 2 days of symptom onset. It is also approved for use in flu post-exposure prophylaxis in this population. A major clinical benefit of baloxavir is that it is effective in strains of influenza resistant to neuraminidase inhibitors. In addition, its efficacy is similar to that of oseltamivir, with some studies showing a slightly more rapid time to alleviation of symptoms (TTAS) and reduction in viral load with baloxavir [57].

Of note, baloxavir is currently contraindicated in children under 12 years old, which is significant because this population is at an increased risk of developing the flu. This sets baloxavir apart from oseltamivir, which is approved for children after 2 weeks of age. Clinical trials are currently underway to assess the safety and efficacy of baloxavir in this population [58].

5.4 Pharmacology

Baloxavir is given as a single-dose tablet, which may improve adherence when compared to oseltamivir, which is given twice daily for 5 days. Dosing is based upon weight, wherein a 40 mg dose is given to those weighing between 40 and

80 kg, and an 80 mg dose is given to those weighing 80 kg or greater. With respect to elimination, baloxavir is metabolized principally via UGT1A3, with a minor contribution of CYP3A4. Excretion is largely through the feces (80%), with a small amount excreted in urine (15%). Studies have not yet been conducted to assess the safety and effect on baloxavir metabolism of those with liver or kidney failure [58].

5.5 *Side Effects*

Baloxavir was associated with a lower frequency of adverse events, when compared to oseltamivir, peramivir, and zanamivir. Adverse events were found to occur in about 25% of those treated. The most common side effects were, in order of decreasing frequency, diarrhea, bronchitis, nasopharyngitis, headache, and nausea. Studies have not yet been conducted on the safety of baloxavir in individuals who are pregnant, breastfeeding, severely immunocompromised, or have a comorbid complicated chronic disease, and, as a result, it is not recommended for use in any of these populations [57].

5.6 *Resistance*

Mutations in the PA protein were found to confer resistance to baloxavir, resulting in a reported reduction in efficacy varying from 11- to 57-fold. A mutation at I38T/M/F in the PA protein was discovered in 10% of patients treated, with other PA mutations occurring at a rate of 8% [54].

6 Drugs for the Treatment of Hypercytokinemia

6.1 *Definition*

Hypercytokinemia, also called "cytokine storm," refers to an aberrant immune response characterized by the systemic release of cytokines, which can progress to multiple organ failure and death. Causes of cytokine storm vary broadly from iatrogenic causes like chimeric antigen receptor T-cell (CAR-T) therapy to hereditary autoimmune disorders like hemophagocytic lymphohistiocytosis (HLH) and to pathogenic causes like sepsis. The definition of what is considered a cytokine storm, as opposed to a normal immune response, is difficult to determine given that the systemic release of cytokines can represent a feature of a healthy immune response. Only when the cytokine release is considered maladaptive is the term cytokine storm applied. In addition, the effects observed with a cytokine storm vary based on the underlying cause. For example, a cytokine storm observed in the presence of a nonpathogenic cause (e.g., CAR-T therapy) may cause a lower total cytokine release when compared to a cytokine storm observed in the presence of a pathogenic cause (such as influenza), since there is a greater adaptive role for cytokines when a pathogen is present [59].

6.2 *Significance in Influenza*

With respect to influenza, cytokine storms play a critical role in flu-related mortality. Deaths from severe influenza infection occur most frequently by either bacterial superinfection or primary viral pneumonia. In the latter case, the increasing viral load in the lower respiratory tract is often associated with the cytokine storm phenomenon, which further exacerbates the patient's condition and contributes to a greater observed mortality [60].

During the infamous 1918 flu pandemic, there was an unusually high number of flu-related deaths in previously healthy young adults (ages 15–30). This phenomenon has been hypothesized to have been the result of influenza-induced dysregulation of the immune response and subsequent cytokine storm production in these individuals. Similarly, the H1N1 virus from the 2009 pandemic and the avian H5N1 virus have been shown to produce a similar dysfunctional immune response in animal models. In a retrospective analysis of the 2009 influenza pandemic, an association was discovered between patients with known HLH mutations causing increased susceptibility to cytokine storm production and flu-related mortality secondary to coagulopathies. This association suggests that genetic susceptibility may also play a role in a subset of cases [61].

6.3 *Clinical Features*

The first sign of cytokine storm is often fever, which can be followed by numerous other systemic symptoms. These include tachypnea, tachycardia, headache, myalgias, fatigue, rash, diarrhea, and delirium. The patient's condition may then deteriorate quickly into disseminated intravascular coagulation (DIC), hypotensive shock, ARDS, and possibly death. With respect to lab values, the CRP is frequently elevated, which serves as a nonspecific marker of inflammation. Blood counts typically show abnormalities such as leukocytosis/leukopenia, anemia, and thrombocytopenia. Additionally, increased D-dimer and ferritin levels may be observed.

6.4 *Pathogenesis*

The pathogenesis of cytokine storm is related to aberrant immune system activation. In general, it is mediated by an imbalance between pro-inflammatory and anti-inflammatory signals, resulting in the favoring of pro-inflammatory functions and the stimulation of a runaway process without negative feedback. This type of runaway positive feedback may be due to a number of factors, including the antigenicity of the virus itself, as well as host characteristics [59].

6.5 Pharmacology

Pharmacological agents for the treatment of cytokine storm in influenza are currently under investigation. As of now, clinical guidelines do not recommend the use of adjuvant immunomodulatory therapy for severe influenza infection. Listed below are several of the most promising treatments that are currently being studied [62, 63].

1. *Hyperimmune Intravenous Immunoglobulin (IVIG)*: IVIG is an antibody-blood product derived from roughly 10,000 pooled donor serum samples. Hyperimmune IVIG is the same product but derived specifically from patients who have been exposed to a pathogen of interest, ensuring that antibodies to that pathogen are present. In addition to providing passive immunization, IVIG has demonstrated immunomodulatory effects that have made it a useful treatment for inflammatory conditions. These effects function to dampen the immune system through the inhibition of macrophage activation, inhibition of neutrophils and eosinophils, and enhanced regulatory T-cell (T-reg) activity. Most recently, one randomized controlled trial (RCT) across five hospitals has shown decreased viral load and mortality in severe A(H1N1)pdm09 infections with the use of hyperimmune IVIG [64].
2. *Oseltamivir, Clarithromycin, and Naproxen Combination Therapy*: Oseltamivir is a neuraminidase inhibitor as discussed in a previous section, clarithromycin is a macrolide antibiotic, and naproxen is an NSAID. It has been shown that macrolide antibiotics, in addition to inhibiting the bacterial 50S ribosomal subunit, also exhibit anti-inflammatory effects. Naproxen has been shown to demonstrate various anti-inflammatory effects via its reversible inhibition of COX-2. A RCT conducted in 2015 demonstrated that this triple combination therapy reduced mortality at 30 and 90 days and produced a more rapid reduction in viral load in patients who were hospitalized with A(H3N2) infection [63].
3. *Sirolimus*: Sirolimus is an mTOR inhibitor traditionally used as an immunosuppressant. However, mTOR mediates several pathways, which allows sirolimus to exhibit its effects within these various immune pathways. One of which includes the indirect inhibition of PI3K by sirolimus, which has been shown to decrease viral replication. A 2014 RCT studied patients placed on ventilators due to complications of A(H1N1)pdm09 infection and found that use of adjuvant sirolimus reduced ventilator time and resulted in more rapid viral clearance [65].

In addition to the three treatments listed above, several other therapies have been investigated with mixed or inconclusive results. These include peroxisome proliferator-activated receptor (PPAR) agonists, statins, mycophenolate mofetil, and the anti-C5a antibody.

It is important to note that corticosteroids have been shown to have deleterious effects when used in influenza cases, as they can potentially increase mortality and risk of nosocomial pneumonia. Corticosteroids have also been shown to increase the risk of opportunistic infections like aspergillosis, as well as the risk of fracture and venous thromboembolism (VTE). Unfortunately, corticosteroids are widely prescribed in severe influenza infections under the assumption that their immunosuppressive properties function to counteract the cytokine storm associated with infection.

Table 5 summarizes the available pharmacological therapies under investigation for cytokine storm along with their associated recommendations.

Table 5 Summary of different immunomodulatory drugs under investigation for use in severe influenza infection

Recommendation	Drug	Proposed mechanism	Findings
May improve patient outcome	Hyperimmune IVIG	Passive immunization, anti-inflammatory	Lower mortality
May improve patient outcome	Oseltamivir, clarithromycin, and naproxen combination therapy	Naproxen inhibits viral replication, clarithromycin is anti-inflammatory, and oseltamivir blocks viral neuraminidase	Lower mortality, more rapid viral clearance
May improve patient outcome	Sirolimus	Inhibition of mTOR, inhibition of viral replication	Shorter duration of IMV
Unclear	PPAR agonists	Inhibition of lipid and inflammatory CK production	Reduced inflammation and lower mortality in mouse model
Unclear	Statins	Inhibition of HMG-CoA, inhibition of T cell activation	Epidemiological data showing lower mortality following long-term use
Unclear	Mycophenolate mofetil	Immunosuppressant, antiviral activity via inhibition IMPDH	Reduced viral titer in mouse model
Unclear	Anti-C5a antibody	Sequestration of C5a, which initiates lung remodeling in ARDS	Reduced ALI and inflammation in monkey model
Likely harmful	Systemic CCS	Immunosuppression via blockage of LTs and PGs, inhibition of inflammatory CKs, and inhibition of WBC migration	Increased mortality, risk of nosocomial pneumonia, and risk of opportunistic infections

CCS corticosteroids, *LT* leukotriene, *PG* prostaglandin, *CK* cytokine, *WBC* white blood cell, *IMV* invasive medical ventilation, *ALI* acute lung injury, *IMPDH* inosine monophosphate dehydrogenase

Adapted from Ref. [63]

7 Drugs for Treatment of Flu Symptoms

In treating influenza, several drugs have been developed with the goal of reducing the presence of symptoms associated with infection. Examples of these drugs include neuraminidase inhibitors (oseltamivir, zanamivir, peramivir, and laninamivir) and cap-dependent endonuclease inhibitors (baloxavir marboxil) [37]. These drugs attempt to reduce the time necessary for symptom alleviation, the viral load, and the number of laboratory-confirmed influenza cases, especially among those who may be high risk for complications of influenza [37].

7.1 Neuraminidase Inhibitors

Among the neuraminidase inhibitors (NAI), oseltamivir, zanamivir, and peramivir are three WHO-approved first-generation antiviral drugs that have been used extensively in the reduction of influenza symptoms. Oseltamivir is available as an oral capsule or powder and is highly recommended for both the prophylaxis and treatment of adults, children, and full-term infants due to its high oral bioavailability [37, 66]. Dobson et al. examined the association between oseltamivir use and resolution of symptoms and found that, when compared to placebo, oseltamivir reduced the symptom duration by 25.2 hours or 1.05 days [67]. In addition, oseltamivir administration led to fewer lower respiratory complications, including pneumonia and bronchitis, and reduced subsequent antibiotic use in patients 48 hours after randomization [67]. An additional study conducted by Hayden et al. compared oseltamivir to placebo based on similar outcomes as Dobson et al.'s study. This study found that the oseltamivir group had a 1.75-day reduction in symptoms when compared to placebo, in addition to a reduction in upper respiratory complications and abnormalities in middle ear pressure [68].

Peramivir is an intravenous NAI that has also been shown to alleviate influenza-associated symptoms when compared to placebo. In Kohno et al., individuals hospitalized with influenza were randomized into groups receiving 300 mg IV peramivir, 600 mg IV peramivir, or placebo [69]. Among the peramivir groups, the study found that the time to symptomatic alleviation was 59.1 hours (300 mg) and 59.9 hours (600 mg) [69]. In the placebo group, time to symptom alleviation was 82 hours, suggesting that peramivir's use helped reduce the overall symptomatic duration by approximately 22 hours [69]. Ison et al. evaluated two different peramivir regimens among hospitalized patients, 300 mg twice daily versus 600 mg once daily, and found no significant difference in symptomatic or virologic end points between the two groups, concluding that either regimen is satisfactory in treating inpatient influenza cases [70]. Potential adverse effects associated with peramivir's use were documented in Komeda et al.'s post-marketing analyses and included nausea, vomiting, and diarrhea; however, these adverse events were mild and were not reported as serious among the study population [71, 72].

Zanamivir is approved for use in adults and children over the age of 5 and is available via intravenous administration [37, 73]. Hedrick et al. sought to evaluate zanamivir's efficacy in reducing the number of days of symptomatic infection and found that among cases of confirmed influenza infection in children ages 5–12, zanamivir's use was associated with a reduction of 1.25 symptomatic days when compared to placebo [73]. This study also found that patients tolerated the zanamivir well and required fewer additional relief medications than those in the placebo group [73]. A similar study conducted by Hayden et al. examined the relationship between the time of symptom onset and zanamivir administration in patients under 13 years old [74]. These patients had been experiencing symptoms for less than 48 hours, including fever, cough, myalgias, and sore throat [74]. The study found that zanamivir's use resulted in reduction of 0.7 symptomatic days among patients who were afebrile upon presentation and 1.4 days among febrile patients [74]. The study also found that if patients were treated within 30 hours of symptom onset, there was an associated reduction of 1.9 symptomatic days compared to placebo [74]. Other studies have evaluated the reduction of symptomatic days following zanamivir's use and have found values ranging from 1.25 to 4.5 days when compared to the placebo group [75, 76]. It is important to note that zanamivir has been found to increase the risk of bronchospasm among patients with COPD and other restrictive airway diseases and is thus contraindicated in these individuals [41, 42].

Despite the success of oseltamivir and zanamivir in reducing symptoms among influenza patients, there has been an emerging resistance to neuraminidase inhibitors among certain strains of influenza, most frequently against oseltamivir. This resistance can be seen in community-based clusters of influenza; however, resistance was observed on a global scale during the 2008–2009 H1N1 influenza A outbreak [53]. This pattern of resistance prompted the development of additional antiviral agents, such as the cap-dependent endonuclease inhibitors, to provide adequate management of influenza patients and their associated symptoms.

7.2 *Cap-Dependent Endonuclease Inhibitors*

One commonly used cap-dependent endonuclease inhibitor is baloxavir marboxil, which selectively targets the PA protein subunit within influenza's polymerase complex [77–79]. The polymerase complex consists of 3 protein subunits: PB1, PB2, and PA [77–79]. The PB2 protein binds to the cap on host pre-mRNA, which allows the PA subunit to subsequently cleave this cap, providing influenza with the RNA primers necessary for its replication [80]. Baloxavir marboxil inhibits PA, ultimately preventing influenza from completing a necessary step in its replication process. The endonuclease inhibitor also provides a broad spectrum of coverage and provides an alternative treatment for influenza strains exhibiting resistance against neuraminidase inhibitors [81].

Hayden et al.'s study attempted to evaluate baloxavir marboxil's efficacy in reducing influenza-associated symptoms and time to recovery [54]. To achieve this goal, the study compared baloxavir marboxil and a placebo group in both children and adults and included populations from both Japan and the United States [54]. The study found a 26.5-hour reduction in symptom duration among the adult baloxavir group and a 38.6-hour reduction in the child baloxavir group [54]. In addition, the study found a larger difference in the time needed for symptom resolution when patients began baloxavir treatment within 24 hours after symptom onset (32.8 fewer hours needed for symptom resolution with baloxavir) versus starting treatment later (13.2 fewer hours with baloxavir), when compared to the placebo group [54]. Adverse events associated with baloxavir marboxil most commonly include diarrhea but may also include headaches and elevated liver transaminases [82].

7.3 Treatment Indications

In treating influenza among healthy patients, there has been no proven benefit to combining NAIs or cap-dependent endonuclease inhibitors or administering higher than recommended dosages; however, among patients with comorbidities, a prolonged course of treatment may be indicated [62, 83]. Among patients that develop lower respiratory disease as a complication of influenza infection, treatment resistance should be suspected in addition to natural disease progression [83].

8 Drugs in Development and Future Research Directives

There have been no new FDA-approved pharmaceuticals approved in the past 2–3 years for influenza. The newest was approved in Baloxavir®, in 2018. However, Several new anti-influenza therapeutics have recently been developed and are currently being studied to determine efficacy and potential resistance mechanisms. Examples of the recent developments include cyanovirin-N, conjugated sialidase, thiazolides, and small interfering RNA (siRNA).

8.1 Cyanovirin-N

Cyanovirin-N, initially developed as an anti-HIV therapeutic agent, is currently being evaluated for potential antiviral effects exhibited against influenza A and B strains [84]. Cyanovirin-N demonstrates its antiviral effects by binding in a concentration-dependent manner to oligosaccharides with high-mannose content on cell surface glycoproteins, specifically the HA glycoprotein on influenza, and prevents viral entry into cells [84, 85]. Cyanovirin-N is limited in its use due to its dependence on the glycosylation of hemagglutinin, as influenza strains lacking this glycosylation (e.g., 2009 H1N1 pandemic) can display resistance to this treatment

[86]. Among animals infected with strains containing glycosylated HA proteins, however, cyanovirin-N was associated with a reduction in viral titers among lung and nasal samples [87].

8.2 Conjugated Sialidase (DAS181)

In initiating infection among respiratory epithelial cells, influenza must first bind to sialic acid residues present on epithelial cells [85]. DAS181 is a conjugated sialidase that functions to remove sialic acids present on host epithelial cells, thus rendering influenza incapable of entering and infecting respiratory epithelial cells [85]. DAS181 was created as a conjugate protein consisting of a sialidase enzyme from *Actinomyces viscosus* and a respiratory epithelial anchoring domain and is primarily administered in an inhaled form [88]. The conjugated sialidase has been shown to reduce viral titers in lung and nasal washes of mice and ferrets, respectively, and has also been shown to demonstrate inhibitory effects on the H5N1 influenza A strain, as well as the H1N1 strain of the 2009 pandemic [89, 90]. In addition, DAS181 has also shown beneficial effects among oseltamivir-resistant strains, suggesting a potential use as a prophylactic and therapeutic agent for treating influenza [91].

8.3 Thiazolides

Thiazolides, specifically nitazoxanide, comprise a drug class that was initially developed to treat parasitic infections; however, drugs within this class have been shown to exhibit antiviral activity against DNA and RNA viruses [92, 93]. Nitazoxanide interferes with the terminal glycosylation of the HA glycoprotein, which prevents its translocation from the endoplasmic reticulum to the cell surface [94]. Studies are currently being conducted regarding nitazoxanide's use in vivo; however, in vitro studies against influenza, hepatitis B, and hepatitis C have demonstrated beneficial antiviral effects with limited resistance potential [95].

8.4 Small Interfering RNA (siRNA)

The final drug class under development for influenza treatment includes the small interfering RNAs (siRNAs). siRNAs are short sequences of single-stranded RNAs that target double-stranded influenza mRNA [96]. In in vitro and animal models, siRNAs have shown inhibitory effects on influenza's replication, and in vivo effects have been elicited through inhaled and IV administration [97–100]. However, studies have shown a potential resistance mechanism to siRNAs through influenza's high mutation rate, so the addition of host targets may be beneficial in decreasing the resistance potential [85, 101–103].

9 Conclusion

The influenza virus is a negative sense single-stranded RNA virus within the *Orthomyxoviridae* family. Influenza A and B are both responsible for high rates of mortality and morbidity, and effective treatments for these viruses remain elusive due to the intrinsically high mutation rate of these strains [1]. This high mutation rate has prevented the development of a universal influenza vaccine and has resulted in the development of yearly influenza vaccines with efficacies ranging from 10% to 60% [18]. In addition, influenza's genetic instability has also led to resistance against adamantanes, resulting in the discontinuation of this drug class.

Treatment of influenza can focus on direct inhibition of viral replication and/or symptom management. Drugs inhibiting viral replication include neuraminidase inhibitors and the viral polymerase inhibitor, baloxavir marboxil. Evidence-based medicine indicates that these drugs should be administered within 2 days of symptom onset, when viral loads are still elevated [37]. Other interventions that lower mortality and/or ventilation time include hyperimmune IVIG, sirolimus, and a combination therapy of oseltamivir, clarithromycin, and naproxen [63–65]. Of note, the use of systemic corticosteroids is associated with an increased risk of mortality and should be avoided [63]. In diagnosing a patient with acute influenza pneumonia, clinicians should use an evidence-based medicine approach that emphasizes antiviral drugs and specific immunosuppressants [37].

References

1. Pleschka S. Overview of influenza viruses. Curr Top Microbiol Immunol. 2013;370:1–20. https://doi.org/10.1007/82_2012_272. PubMed PMID: 23124938
2. Paules C, Subbarao K. Influenza. Lancet. 2017;390(10095):697–708. Epub 20170313 PubMed PMID: 28302313. https://doi.org/10.1016/S0140-6736(17)30129-0.
3. Xu X, Blanton L, Elal AIA, Alabi N, Barnes J, Biggerstaff M, et al. Update: influenza activity in the United States during the 2018–19 season and composition of the 2019–20 influenza vaccine. MMWR Morb Mortal Wkly Rep. 2019;68(24):544–51. Epub 20190621. PubMed PMID: 31220057; PubMed Central PMCID: PMC6586370. https://doi.org/10.15585/mmwr.mm6824a3.
4. Kim H, Webster RG, Webby RJ. Influenza virus: dealing with a drifting and shifting pathogen. Viral Immunol. 2018;31(2):174–83. Epub 20180126. PubMed PMID: 29373086. https://doi.org/10.1089/vim.2017.0141.
5. Sharma L, Rebaza A, Dela Cruz CS. When "B" becomes "A": the emerging threat of influenza B virus. Eur Respir J. 2019;54(2). Epub 20190815 PubMed PMID: 31416813 https://doi.org/10.1183/13993003.01325-2019.
6. Dharmapalan D. Influenza. Indian J Pediatr. 2020;87(10):828–32. Epub 20200211. PubMed PMID: 32048225; PubMed Central PMCID: PMC7091034. https://doi.org/10.1007/s12098-020-03214-1.
7. Peteranderl C, Herold S, Schmoldt C. Human Influenza virus infections. Semin Respir Crit Care Med. 2016;37(4):487–500. Epub 20160803. PubMed PMID: 27486731; PubMed Central PMCID: PMC7174870. https://doi.org/10.1055/s-0036-1584801.

8. Kondrich J, Rosenthal M. Influenza in children. Curr Opin Pediatr. 2017;29(3):297–302. Epub 2017/03/28. PubMed PMID: 28346272. https://doi.org/10.1097/MOP.0000000000000495.
9. Merckx J, Wali R, Schiller I, Caya C, Gore GC, Chartrand C, et al. Diagnostic accuracy of novel and traditional rapid tests for Influenza infection compared with reverse transcriptase polymerase chain reaction: a systematic review and meta-analysis. Ann Intern Med. 2017;167(6):394–409. Epub 20170905. PubMed PMID: 28869986. https://doi.org/10.7326/M17-0848.
10. Kilbourne ED. Influenza pandemics of the 20th century. Emerg Infect Dis. 2006;12(1):9–14. https://doi.org/10.3201/eid1201.051254. PubMed PMID: 16494710; PubMed Central PMCID: PMC3291411
11. Kalil AC, Thomas PG. Influenza virus-related critical illness: pathophysiology and epidemiology. Crit Care. 2019;23(1):258. Epub 20190719. PubMed PMID: 31324202; PubMed Central PMCID: PMC6642581. https://doi.org/10.1186/s13054-019-2539-x.
12. Hutchinson EC. Influenza virus. Trends Microbiol. 2018;26(9):809–10. Epub 20180613. PubMed PMID: 29909041. https://doi.org/10.1016/j.tim.2018.05.013.
13. Bolte H, Rosu ME, Hagelauer E, Garcia-Sastre A, Schwemmle M. Packaging of the Influenza virus genome is governed by a plastic network of RNA- and nucleoprotein-mediated interactions. J Virol. 2019;93(4). Epub 20190205. PubMed PMID: 30463968; PubMed Central PMCID: PMC6363987 https://doi.org/10.1128/JVI.01861-18.
14. Valesano AL, Fitzsimmons WJ, McCrone JT, Petrie JG, Monto AS, Martin ET, et al. Influenza B viruses exhibit lower within-host diversity than Influenza A viruses in human hosts. J Virol. 2020;94(5). Epub 20200214. PubMed PMID: 31801858; PubMed Central PMCID: PMC7022338 https://doi.org/10.1128/JVI.01710-19.
15. Pica N, Palese P. Toward a universal influenza virus vaccine: prospects and challenges. Annu Rev Med. 2013;64:189–202. https://doi.org/10.1146/annurev-med-120611-145115. PubMed PMID: 23327522
16. Li C, Hatta M, Watanabe S, Neumann G, Kawaoka Y. Compatibility among polymerase subunit proteins is a restricting factor in reassortment between equine H7N7 and human H3N2 influenza viruses. J Virol. 2008;82(23):11880–8. Epub 20080924. PubMed PMID: 18815312; PubMed Central PMCID: PMC2583690. https://doi.org/10.1128/JVI.01445-08.
17. Varich NL, Gitelman AK, Shilov AA, Smirnov YA, Kaverin NV. Deviation from the random distribution pattern of influenza A virus gene segments in reassortants produced under non-selective conditions. Arch Virol. 2008;153(6):1149–54. Epub 20080415. PubMed PMID: 18414973. https://doi.org/10.1007/s00705-008-0070-5.
18. Kalarikkal SM, Jaishankar GB. Influenza vaccine. Treasure Island: StatPearls; 2022.
19. Barberis I, Myles P, Ault SK, Bragazzi NL, Martini M. History and evolution of influenza control through vaccination: from the first monovalent vaccine to universal vaccines. J Prev Med Hyg. 2016;57(3):E115–E20. PubMed PMID: 27980374; PubMed Central PMCID: PMC5139605
20. Chen J, Wang J, Zhang J, Ly H. Advances in development and application of influenza vaccines. Front Immunol. 2021;12:711997. Epub 20210713. doi: 10.3389/fimmu.2021.711997. PubMed PMID: 34326849; PubMed Central PMCID: PMC8313855
21. Wong SS, Webby RJ. Traditional and new influenza vaccines. Clin Microbiol Rev. 2013;26(3):476–92. https://doi.org/10.1128/CMR.00097-12. PubMed PMID: 23824369; PubMed Central PMCID: PMC3719499
22. Blanco-Lobo P, Nogales A, Rodriguez L, Martinez-Sobrido L. Novel approaches for the development of live attenuated influenza vaccines. Viruses. 2019;11(2) https://doi.org/10.3390/v11020190. Epub 20190222. PubMed PMID: 30813325; PubMed Central PMCID: PMC6409754
23. Rizzo C, Rezza G, Ricciardi W. Strategies in recommending influenza vaccination in Europe and US. Hum Vaccin Immunother. 2018;14(3):693–8. https://doi.org/10.1080/21645515.2017.1367463. Epub 20180109. PubMed PMID: 28922083; PubMed Central PMCID: PMC5861797

24. Influenza: Vaccination [cited 2022 March]. Available from: https://www.euro.who.int/en/health-topics/communicable-diseases/influenza/vaccination#:~:text=WHO%20recommends%20that%20health%20care,months%20with%20certain%20chronic%20diseases.
25. Fiore AE, Bridges CB, Cox NJ. Seasonal influenza vaccines. Curr Top Microbiol Immunol. 2009;333:43–82. https://doi.org/10.1007/978-3-540-92165-3_3. PubMed PMID: 19768400
26. McDonald SM, Nelson MI, Turner PE, Patton JT. Reassortment in segmented RNA viruses: mechanisms and outcomes. Nat Rev Microbiol. 2016;14(7):448–60. https://doi.org/10.1038/nrmicro.2016.46. Epub 20160523. PubMed PMID: 27211789; PubMed Central PMCID: PMC5119462
27. Wilschut J. Influenza vaccines: the virosome concept. Immunol Lett. 2009;122(2):118–21. https://doi.org/10.1016/j.imlet.2008.11.006. Epub 20081225. PubMed PMID: 19100779
28. Estrada LD, Schultz-Cherry S. Development of a universal influenza vaccine. J Immunol. 2019;202(2):392–8. https://doi.org/10.4049/jimmunol.1801054. PubMed PMID: 30617121; PubMed Central PMCID: PMC6327971
29. Grohskopf LA, Alyanak E, Ferdinands JM, Broder KR, Blanton LH, Talbot HK, et al. Prevention and control of seasonal influenza with vaccines: recommendations of the advisory committee on immunization practices, United States, 2021–22 influenza season. MMWR Recomm Rep. 2021;70(5):1.
30. Dbaibo G, Amanullah A, Claeys C, Izu A, Jain VK, Kosalaraksa P, et al. Quadrivalent influenza vaccine prevents illness and reduces healthcare utilization across diverse geographic regions during five influenza seasons: a randomized clinical trial. Pediatr Infect Dis J. 2020;39(1):e1.
31. Evans TG, Bussey L, Eagling-Vose E, Rutkowski K, Ellis C, Argent C, et al. Efficacy and safety of a universal influenza a vaccine (MVA-NP+ M1) in adults when given after seasonal quadrivalent influenza vaccine immunisation (FLU009): a phase 2b, randomised, double-blind trial. Lancet Infect Dis. 2022;22:857.
32. Liebowitz D, Gottlieb K, Kolhatkar NS, Garg SJ, Asher JM, Nazareno J, et al. Efficacy, immunogenicity, and safety of an oral influenza vaccine: a placebo-controlled and active-controlled phase 2 human challenge study. Lancet Infect Dis. 2020;20(4):435–44.
33. Nachbagauer R, Feser J, Naficy A, Bernstein DI, Guptill J, Walter EB, et al. A chimeric hemagglutinin-based universal influenza virus vaccine approach induces broad and long-lasting immunity in a randomized, placebo-controlled phase I trial. Nat Med. 2021;27(1):106–14.
34. Gubareva LV, Kaiser L, Hayden FG. Influenza virus neuraminidase inhibitors. Lancet. 2000;355(9206):827–35. https://doi.org/10.1016/S0140-6736(99)11433-8. PubMed PMID: 10711940
35. Dou D, Revol R, Ostbye H, Wang H, Daniels R. Influenza A virus cell entry, replication, virion assembly and movement. Front Immunol. 2018;9:1581. https://doi.org/10.3389/fimmu.2018.01581. Epub 20180720. PubMed PMID: 30079062; PubMed Central PMCID: PMC6062596
36. Sakai T, Nishimura SI, Naito T, Saito M. Influenza A virus hemagglutinin and neuraminidase act as novel motile machinery. Sci Rep. 2017;7:45043. https://doi.org/10.1038/srep45043. Epub 20170327. PubMed PMID: 28344335; PubMed Central PMCID: PMC5366856
37. Duwe SC, Schmidt B, Gartner BC, Timm J, Adams O, Fickenscher H, et al. Prophylaxis and treatment of influenza: options, antiviral susceptibility, and existing recommendations. GMS Infect Dis. 2021;9:Doc02. https://doi.org/10.3205/id000071. Epub 20210430. PubMed PMID: 34113534; PubMed Central PMCID: PMC8165743
38. Hurt AC, Kelly H. Debate regarding oseltamivir use for seasonal and pandemic influenza. Emerg Infect Dis. 2016;22(6):949–55. https://doi.org/10.3201/eid2206.151037. PubMed PMID: 27191818; PubMed Central PMCID: PMC4880079
39. Slain D. Intravenous zanamivir: a viable option for critically ill patients with influenza. Ann Pharmacother. 2021;55(6):760–71. Epub 20201005. PubMed PMID: 33016090. https://doi.org/10.1177/1060028020963616.

40. Cass LM, Efthymiopoulos C, Bye A. Pharmacokinetics of zanamivir after intravenous, oral, inhaled or intranasal administration to healthy volunteers. Clin Pharmacokinet. 1999;36(Suppl 1):1–11. https://doi.org/10.2165/00003088-199936001-00001. PubMed PMID: 10429835
41. RELENZA (zanamivir for inhalation) U.S. Food and Drug Administration; [cited 2022 March]. Available from: https://www.accessdata.fda.gov/drugsatfda_docs/label/1999/21036lbl.pdf.
42. Heneghan CJ, Onakpoya I, Thompson M, Spencer EA, Jones M, Jefferson T. Zanamivir for influenza in adults and children: systematic review of clinical study reports and summary of regulatory comments. BMJ. 2014;348:g2547. https://doi.org/10.1136/bmj.g2547. Epub 20140409. PubMed PMID: 24811412; PubMed Central PMCID: PMC3981976
43. TAMIFLU (oseltamivir phosphate): U.S. Food and Drug Administration; [cited 2022 March]. Available from: https://www.accessdata.fda.gov/drugsatfda_docs/label/2016/021087s068,021246s051lbl.pdf.
44. Jefferson T, Jones M, Doshi P, Spencer EA, Onakpoya I, Heneghan CJ. Oseltamivir for influenza in adults and children: systematic review of clinical study reports and summary of regulatory comments. BMJ. 2014;348:g2545. https://doi.org/10.1136/bmj.g2545. Epub 20140409. PubMed PMID: 24811411; PubMed Central PMCID: PMC3981975
45. Jefferson T, Jones MA, Doshi P, Del Mar CB, Hama R, Thompson MJ, et al. Neuraminidase inhibitors for preventing and treating influenza in adults and children. Cochrane Database Syst Rev. 2014;(4):CD008965. https://doi.org/10.1002/14651858.CD008965.pub4. Epub 20140410. PubMed PMID: 24718923; PubMed Central PMCID: PMC6464969
46. RAPIVAB (peramivir injection): U.S. Food and Drug Administration [cited 2022 March]. Available from: https://www.accessdata.fda.gov/drugsatfda_docs/label/2014/206426lbl.pdf.
47. Alame MM, Massaad E, Zaraket H. Peramivir: a novel intravenous neuraminidase inhibitor for treatment of acute influenza infections. Front Microbiol. 2016;7:450. https://doi.org/10.3389/fmicb.2016.00450. Epub 20160331. PubMed PMID: 27065996; PubMed Central PMCID: PMC4815007
48. Scott LJ. Peramivir: a review in uncomplicated influenza. Drugs. 2018;78(13):1363–70. https://doi.org/10.1007/s40265-018-0981-8. PubMed PMID: 30196350
49. Ison MG, Hui DS, Clezy K, O'Neil BJ, Flynt A, Collis PJ, et al. A clinical trial of intravenous peramivir compared with oral oseltamivir for the treatment of seasonal influenza in hospitalized adults. Antivir Ther. 2013;18(5):651–61. Epub 20121030. PubMed PMID: 23111657. https://doi.org/10.3851/IMP2442.
50. Higashiguchi M, Matsumoto T, Fujii T. A meta-analysis of laninamivir octanoate for treatment and prophylaxis of influenza. Antivir Ther. 2018;23(2):157–65. https://doi.org/10.3851/IMP3189. PubMed PMID: 28869418
51. Sugaya N, Ohashi Y. Long-acting neuraminidase inhibitor laninamivir octanoate (CS-8958) versus oseltamivir as treatment for children with influenza virus infection. Antimicrob Agents Chemother. 2010;54(6):2575–82. https://doi.org/10.1128/AAC.01755-09. Epub 20100405. PubMed PMID: 20368393; PubMed Central PMCID: PMC2876358
52. Lee N, Hurt AC. Neuraminidase inhibitor resistance in influenza: a clinical perspective. Curr Opin Infect Dis. 2018;31(6):520–6. https://doi.org/10.1097/QCO.0000000000000498. PubMed PMID: 30299356
53. Li TC, Chan MC, Lee N. Clinical implications of antiviral resistance in influenza. Viruses. 2015;7(9):4929–44. https://doi.org/10.3390/v7092850. Epub 20150914. PubMed PMID: 26389935; PubMed Central PMCID: PMC4584294
54. Hayden FG, Sugaya N, Hirotsu N, Lee N, de Jong MD, Hurt AC, et al. Baloxavir marboxil for uncomplicated influenza in adults and adolescents. N Engl J Med. 2018;379(10):913–23. https://doi.org/10.1056/NEJMoa1716197. PubMed PMID: 30184455
55. te Velthuis AJW, Fodor E. Influenza virus RNA polymerase: insights into the mechanisms of viral RNA synthesis. Nat Rev Microbiol. 2016;14(8):479–93. https://doi.org/10.1038/nrmicro.2016.87.
56. Walker AP, Fodor E. Interplay between Influenza Virus and the Host RNA Polymerase II Transcriptional Machinery. Trends Microbiol. 2019;27(5):398–407. https://doi.org/10.1016/j.

tim.2018.12.013. Epub 20190111. PubMed PMID: 30642766; PubMed Central PMCID: PMC6467242
57. Liu JW, Lin SH, Wang LC, Chiu HY, Lee JA. Comparison of antiviral agents for seasonal influenza outcomes in healthy adults and children: a systematic review and network meta-analysis. JAMA Netw Open. 2021;4(8):e2119151. https://doi.org/10.1001/jamanetworkopen.2021.19151. Epub 20210802. PubMed PMID: 34387680; PubMed Central PMCID: PMC8363918
58. Ng KE. Xofluza (Baloxavir Marboxil) for the treatment of acute uncomplicated influenza. Pharm Ther. 2019;44(1):9–11. PubMed PMID: 30675086; PubMed Central PMCID: PMC6336199
59. Fajgenbaum DC, June CH. Cytokine Storm. N Engl J Med. 2020;383(23):2255–73. https://doi.org/10.1056/NEJMra2026131. PubMed PMID: 33264547; PubMed Central PMCID: PMC7727315
60. Krammer F, Smith GJD, Fouchier RAM, Peiris M, Kedzierska K, Doherty PC, et al. Influenza. Nat Rev Dis Primers. 2018;4(1):3. https://doi.org/10.1038/s41572-018-0002-y.
61. Short KR, Kedzierska K, van de Sandt CE. Back to the future: lessons learned from the 1918 influenza pandemic. Front Cell Infect Microbiol. 2018;8 https://doi.org/10.3389/fcimb.2018.00343.
62. Uyeki TM, Bernstein HH, Bradley JS, Englund JA, File TM, Fry AM, et al. Clinical practice guidelines by the Infectious Diseases Society of America: 2018 update on diagnosis, treatment, chemoprophylaxis, and institutional outbreak management of seasonal influenzaa. Clin Infect Dis. 2019;68(6):895–902. https://doi.org/10.1093/cid/ciy874. PubMed PMID: 30834445; PubMed Central PMCID: PMC6769232
63. Hui DS, Lee N, Chan PK, Beigel JH. The role of adjuvant immunomodulatory agents for treatment of severe influenza. Antiviral Res. 2018;150:202–16. https://doi.org/10.1016/j.antiviral.2018.01.002. Epub 20180108. PubMed PMID: 29325970; PubMed Central PMCID: PMC5801167
64. Arumugham VB, Rayi A. Intravenous immunoglobulin (IVIG). Treasure Island: StatPearls; 2022.
65. Wang CH, Chung FT, Lin SM, Huang SY, Chou CL, Lee KY, et al. Adjuvant treatment with a mammalian target of rapamycin inhibitor, sirolimus, and steroids improves outcomes in patients with severe H1N1 pneumonia and acute respiratory failure. Crit Care Med. 2014;42(2):313–21. https://doi.org/10.1097/CCM.0b013e3182a2727d. PubMed PMID: 24105455
66. Amarelle L, Lecuona E, Sznajder JI. Anti-influenza treatment: drugs currently used and under development. Arch Bronconeumol. 2017;53(1):19–26. https://doi.org/10.1016/j.arbres.2016.07.004. Epub 20160809. PubMed PMID: 27519544; PubMed Central PMCID: PMC6889083
67. Dobson J, Whitley RJ, Pocock S, Monto AS. Oseltamivir treatment for influenza in adults: a meta-analysis of randomised controlled trials. Lancet. 2015;385(9979):1729–37. Epub 20150130. PubMed PMID: 25640810. https://doi.org/10.1016/S0140-6736(14)62449-1.
68. Hayden FG, Treanor JJ, Fritz RS, Lobo M, Betts RF, Miller M, et al. Use of the oral neuraminidase inhibitor oseltamivir in experimental human influenza: randomized controlled trials for prevention and treatment. JAMA. 1999;282(13):1240–6. https://doi.org/10.1001/jama.282.13.1240. PubMed PMID: 10517426
69. Kohno S, Kida H, Mizuguchi M, Shimada J, Group SCS. Efficacy and safety of intravenous peramivir for treatment of seasonal influenza virus infection. Antimicrob Agents Chemother. 2010;54(11):4568–74. https://doi.org/10.1128/AAC.00474-10. Epub 20100816. PubMed PMID: 20713668; PubMed Central PMCID: PMC2976170
70. Ison MG, Fraiz J, Heller B, Jauregui L, Mills G, O'Riordan W, et al. Intravenous peramivir for treatment of influenza in hospitalized patients. Antivir Ther. 2014;19(4):349–61. Epub 20130828. PubMed PMID: 23985625. https://doi.org/10.3851/IMP2680.

71. Komeda T, Ishii S, Itoh Y, Ariyasu Y, Sanekata M, Yoshikawa T, et al. Post-marketing safety and effectiveness evaluation of the intravenous anti-influenza neuraminidase inhibitor peramivir (I): a drug use investigation. J Infect Chemother. 2014;20(11):689–95. Epub 20140811. PubMed PMID: 25131292. https://doi.org/10.1016/j.jiac.2014.07.006.
72. Komeda T, Ishii S, Itoh Y, Ariyasu Y, Sanekata M, Yoshikawa T, et al. Post-marketing safety and effectiveness evaluation of the intravenous anti-influenza neuraminidase inhibitor peramivir. II: a pediatric drug use investigation. J Infect Chemother. 2015;21(3):194–201. Epub 20141126. PubMed PMID: 25523716. https://doi.org/10.1016/j.jiac.2014.11.009.
73. Hedrick JA, Barzilai A, Behre U, Henderson FW, Hammond J, Reilly L, et al. Zanamivir for treatment of symptomatic influenza a and B infection in children five to twelve years of age: a randomized controlled trial. Pediatr Infect Dis J. 2000;19(5):410–7. https://doi.org/10.1097/00006454-200005000-00005. PubMed PMID: 10819336
74. Hayden FG, Osterhaus AD, Treanor JJ, Fleming DM, Aoki FY, Nicholson KG, et al. Efficacy and safety of the neuraminidase inhibitor zanamivir in the treatment of influenzavirus infections. GG167 influenza study group. N Engl J Med. 1997;337(13):874–80. https://doi.org/10.1056/NEJM199709253371302. PubMed PMID: 9302301
75. Boivin G, Goyette N, Hardy I, Aoki F, Wagner A, Trottier S. Rapid antiviral effect of inhaled zanamivir in the treatment of naturally occurring influenza in otherwise healthy adults. J Infect Dis. 2000;181(4):1471–4. Epub 20000413. PubMed PMID: 10762579. https://doi.org/10.1086/315392.
76. The MIST. Randomised trial of efficacy and safety of inhaled zanamivir in treatment of influenza A and B virus infections. (management of influenza in the Southern Hemisphere Trialists) study group. Lancet. 1998;352(9144):1877–81. PubMed PMID: 9863784
77. Stevaert A, Naesens L. The influenza virus polymerase complex: an update on its structure, functions, and significance for antiviral drug design. Med Res Rev. 2016;36(6):1127–73. https://doi.org/10.1002/med.21401. Epub 20160829. PubMed PMID: 27569399; PubMed Central PMCID: PMC5108440
78. Pflug A, Guilligay D, Reich S, Cusack S. Structure of influenza A polymerase bound to the viral RNA promoter. Nature. 2014;516(7531):355–60. Epub 20141119. PubMed PMID: 25409142. https://doi.org/10.1038/nature14008.
79. Reich S, Guilligay D, Pflug A, Malet H, Berger I, Crepin T, et al. Structural insight into cap-snatching and RNA synthesis by influenza polymerase. Nature. 2014;516(7531):361–6. Epub 20141119. PubMed PMID: 25409151. https://doi.org/10.1038/nature14009.
80. Yuan P, Bartlam M, Lou Z, Chen S, Zhou J, He X, et al. Crystal structure of an avian influenza polymerase PA(N) reveals an endonuclease active site. Nature. 2009;458(7240):909–13. Epub 20090204. PubMed PMID: 19194458. https://doi.org/10.1038/nature07720.
81. Noshi T, Kitano M, Taniguchi K, Yamamoto A, Omoto S, Baba K, et al. In vitro characterization of baloxavir acid, a first-in-class cap-dependent endonuclease inhibitor of the influenza virus polymerase PA subunit. Antiviral Res. 2018;160:109–17. Epub 20181011. PubMed PMID: 30316915. https://doi.org/10.1016/j.antiviral.2018.10.008.
82. Heo YA. Baloxavir: first global approval. Drugs. 2018;78(6):693–7. https://doi.org/10.1007/s40265-018-0899-1. PubMed PMID: 29623652
83. Influenza antiviral medications: summary for clinicians: centers for disease control and prevention [updated 4 Feb 2022; cited 2022 April]. Available from: https://www.cdc.gov/flu/professionals/antivirals/summary-clinicians.htm.
84. O'Keefe BR, Smee DF, Turpin JA, Saucedo CJ, Gustafson KR, Mori T, et al. Potent anti-influenza activity of cyanovirin-N and interactions with viral hemagglutinin. Antimicrob Agents Chemother. 2003;47(8):2518–25. https://doi.org/10.1128/AAC.47.8.2518-2525.2003. PubMed PMID: 12878514; PubMed Central PMCID: PMC166092
85. Boltz DA, Aldridge JR Jr, Webster RG, Govorkova EA. Drugs in development for influenza. Drugs. 2010;70(11):1349–62. https://doi.org/10.2165/11537960-000000000-00000. PubMed PMID: 20614944; PubMed Central PMCID: PMC5558450

86. Igarashi M, Ito K, Yoshida R, Tomabechi D, Kida H, Takada A. Predicting the antigenic structure of the pandemic (H1N1) 2009 influenza virus hemagglutinin. PLoS One. 2010;5(1):e8553. https://doi.org/10.1371/journal.pone.0008553. Epub 20100101. PubMed PMID: 20049332; PubMed Central PMCID: PMC2797400
87. Smee DF, Bailey KW, Wong MH, O'Keefe BR, Gustafson KR, Mishin VP, et al. Treatment of influenza A (H1N1) virus infections in mice and ferrets with cyanovirin-N. Antiviral Res. 2008;80(3):266–71. https://doi.org/10.1016/j.antiviral.2008.06.003. Epub 20080702. PubMed PMID: 18601954; PubMed Central PMCID: PMC2740641
88. Malakhov MP, Aschenbrenner LM, Smee DF, Wandersee MK, Sidwell RW, Gubareva LV, et al. Sialidase fusion protein as a novel broad-spectrum inhibitor of influenza virus infection. Antimicrob Agents Chemother. 2006;50(4):1470–9. https://doi.org/10.1128/AAC.50.4.1470-1479.2006. PubMed PMID: 16569867; PubMed Central PMCID: PMC1426979
89. Chan RW, Chan MC, Wong AC, Karamanska R, Dell A, Haslam SM, et al. DAS181 inhibits H5N1 influenza virus infection of human lung tissues. Antimicrob Agents Chemother. 2009;53(9):3935–41. https://doi.org/10.1128/AAC.00389-09. Epub 20090713. PubMed PMID: 19596886; PubMed Central PMCID: PMC2737896
90. Triana-Baltzer GB, Gubareva LV, Nicholls JM, Pearce MB, Mishin VP, Belser JA, et al. Novel pandemic influenza A(H1N1) viruses are potently inhibited by DAS181, a sialidase fusion protein. PLoS One. 2009;4(11):e7788. https://doi.org/10.1371/journal.pone.0007788. Epub 20091106. PubMed PMID: 19893747; PubMed Central PMCID: PMC2770640
91. Triana-Baltzer GB, Gubareva LV, Klimov AI, Wurtman DF, Moss RB, Hedlund M, et al. Inhibition of neuraminidase inhibitor-resistant influenza virus by DAS181, a novel sialidase fusion protein. PLoS One. 2009;4(11):e7838. https://doi.org/10.1371/journal.pone.0007838. Epub 20091106. PubMed PMID: 19893749; PubMed Central PMCID: PMC2770896
92. Korba BE, Montero AB, Farrar K, Gaye K, Mukerjee S, Ayers MS, et al. Nitazoxanide, tizoxanide and other thiazolides are potent inhibitors of hepatitis B virus and hepatitis C virus replication. Antiviral Res. 2008;77(1):56–63. Epub 20070904. PubMed PMID: 17888524. https://doi.org/10.1016/j.antiviral.2007.08.005.
93. Rossignol JF, Keeffe EB. Thiazolides: a new class of drugs for the treatment of chronic hepatitis B and C. Future Microbiol. 2008;3(5):539–45. https://doi.org/10.2217/17460913.3.5.539. PubMed PMID: 18811238
94. Rossignol JF, La Frazia S, Chiappa L, Ciucci A, Santoro MG. Thiazolides, a new class of anti-influenza molecules targeting viral hemagglutinin at the post-translational level. J Biol Chem. 2009;284(43):29798–808. https://doi.org/10.1074/jbc.M109.029470. Epub 20090728. PubMed PMID: 19638339; PubMed Central PMCID: PMC2785610
95. Korba BE, Elazar M, Lui P, Rossignol JF, Glenn JS. Potential for hepatitis C virus resistance to nitazoxanide or tizoxanide. Antimicrob Agents Chemother. 2008;52(11):4069–71. https://doi.org/10.1128/AAC.00078-08. Epub 20080818. PubMed PMID: 18710916; PubMed Central PMCID: PMC2573111
96. Hannon GJ. RNA interference. Nature. 2002;418(6894):244–51. https://doi.org/10.1038/418244a. PubMed PMID: 12110901
97. Ge Q, McManus MT, Nguyen T, Shen CH, Sharp PA, Eisen HN, et al. RNA interference of influenza virus production by directly targeting mRNA for degradation and indirectly inhibiting all viral RNA transcription. Proc Natl Acad Sci U S A. 2003;100(5):2718–23. https://doi.org/10.1073/pnas.0437841100. Epub 20030219. PubMed PMID: 12594334; PubMed Central PMCID: PMC151407
98. Tompkins SM, Lo CY, Tumpey TM, Epstein SL. Protection against lethal influenza virus challenge by RNA interference in vivo. Proc Natl Acad Sci U S A. 2004;101(23):8682–6. https://doi.org/10.1073/pnas.0402630101. Epub 20040601. PubMed PMID: 15173583; PubMed Central PMCID: PMC423255
99. Zhou H, Jin M, Yu Z, Xu X, Peng Y, Wu H, et al. Effective small interfering RNAs targeting matrix and nucleocapsid protein gene inhibit influenza A virus replication in cells and mice.

Antiviral Res. 2007;76(2):186–93. Epub 20070810. PubMed PMID: 17719657. https://doi.org/10.1016/j.antiviral.2007.07.002.

100. Zhang W, Wang CY, Yang ST, Qin C, Hu JL, Xia XZ. Inhibition of highly pathogenic avian influenza virus H5N1 replication by the small interfering RNA targeting polymerase A gene. Biochem Biophys Res Commun. 2009;390(3):421–6. Epub 20090913. PubMed PMID: 19755113. https://doi.org/10.1016/j.bbrc.2009.09.039.
101. Karlas A, Machuy N, Shin Y, Pleissner KP, Artarini A, Heuer D, et al. Genome-wide RNAi screen identifies human host factors crucial for influenza virus replication. Nature. 2010;463(7282):818–22. Epub 20100117. PubMed PMID: 20081832. https://doi.org/10.1038/nature08760.
102. Konig R, Stertz S, Zhou Y, Inoue A, Hoffmann HH, Bhattacharyya S, et al. Human host factors required for influenza virus replication. Nature. 2010;463(7282):813–7. https://doi.org/10.1038/nature08699. PubMed PMID: 20027183; PubMed Central PMCID: PMC2862546
103. Hao L, Sakurai A, Watanabe T, Sorensen E, Nidom CA, Newton MA, et al. Drosophila RNAi screen identifies host genes important for influenza virus replication. Nature. 2008;454(7206):890–3. https://doi.org/10.1038/nature07151. Epub 20080709. PubMed PMID: 18615016; PubMed Central PMCID: PMC2574945.

COVID-19: Molecular Pathogenesis and Prospective Therapeutic Interventions

Priya Shrivastava and Suresh P. Vyas

Abstract COVID-19 is a catastrophe taking a massive toll on humanity across the world. It is a highly transmissible airborne infection. The pathogen responsible for causing COVID-19 is severe acute respiratory syndrome coronavirus 2 (SARS-CoV-2). Coronaviruses (CoV) are named owing to the existence/appearance of the crown-like spike proteins on their surface. In the twenty-first century, they have emerged as severely devastating, transmissible, and pathogenic coronavirus to human community after the SARS-CoV (severe acute respiratory syndrome coronavirus) and the MERS-CoV (Middle East respiratory syndrome coronavirus). The symptoms of COVID-19 are typically similar to viral pneumonia. It is spreading persistently to almost all countries, with more than 457 million infected people resulting in more than 6 million deaths reported globally. The COVID-19 outbreak has been labeled a global pandemic by the world health organization (WHO). The pathogen (SARS-CoV-2) targets the epithelial cells of the respiratory system, resulting in diffuse alveolar damage. Here, in this chapter, we have comprehensively reviewed the information regarding SARS-CoV-2 infections, its molecular basis of pathogenesis, and prospective therapeutic interventions for the treatment of COVID-19 infection.

Keywords COVID-19 · SARS-CoV-2 · Molecular pathogenesis · Nanoparticles · Vaccines

1 COVID-19: A Global Emergency

History shows that infectious diseases are a recurrent problem resulting in sheer devastation, of human lives, and also the economic front [1]. The world is aware of the Spanish flu of 1918 resulted in the death of nearly 50 million people worldwide

P. Shrivastava · S. P. Vyas (✉)
Drug Delivery Research Laboratory, Department of Pharmaceutical Sciences, Dr. Harisingh Gour Vishwavidyalaya, Sagar, MP, India

R. Shegokar, Y. Pathak (eds.), *Viral Drug Delivery Systems*,
https://doi.org/10.1007/978-3-031-20537-8_12

and a severe economic depression [2]. Even during the influenza pandemics of 1957 and 1968, the development of vaccines was delayed, failing to provide effective protection during the severe phases of the outbreaks [3]. The pandemic has led to the spread of infection on a large scale. As a result, morbidity and mortality have increased greatly over a wide geographic area. This causes significant economic and social concerns [4].

Today, the world is facing a severe existential challenge as novel coronavirus disease has turned out to be a psychophysical and social crisis of unprecedented magnitude [5]. The novel coronavirus like any other virus is not a complete living being and depends on the cell machinery of the host for its survival, yet it has gripped the whole planet barring a few isolated islands [6]. World Health Organization (WHO) declared this as a global pandemic on March 12, 2020, and with the rising toll of precious human lives, the United Nations Secretary-General called this humanity's worst crisis since World War II. This is a new version of world war not among the nations but along the nations of the world.

The coronavirus disease 2019 (COVID-19) is a highly transmittable airborne infection. Pathogenic SARS-CoV-2 is the causative organism. The emergence of highly pathogenic coronaviruses in 2003 SARS-CoV, in 2012 MERS CoV, now in 2019 pathogenic SARS-CoV-2, is associated with a global "pandemic" situation. Patients with COVID-19 show clinical manifestations including fever, fatigue, myalgia, nonproductive cough, dyspnea, decreased leucocyte counts, and radiographic evidence of pneumonia, which are similar to the symptoms of MERS-CoV and SARS-CoV infections. In humans, the effects of these viruses are correlated with viral pneumonia and severe respiratory tract infections [7].

The entry of the viral particle relies on the fine interplay between the virion and the host cell [8]. The initiation of the infection depends on the cross talk between viral spike (S) glycoprotein and angiotensin-converting enzyme 2 receptor (ACE2, cell surface receptor) [9]. ACE2 receptor is excessively expressed in airway epithelial cells (AECs). This initial binding with the receptor mediates the viral entry into airway epithelial cells (AECs) and establishes host tropism [10]. After their binding with the receptor, enveloped viruses undergo fusion with the host cell membrane to deliver their nucleocapsid to the target cell [11]. The viral spike glycoprotein plays a pivotal role in entry into the host cell by facilitating receptor binding and membrane fusion. The process of fusion entails the large conformational alterations of the viral spike glycoprotein. Apart from the viral receptor interaction, the proteolytic cleavability of S glycoprotein has also been considered as an important determinant of disease severity [12].

A plethora of literature reported the occurrence of impaired host immune response and production of an exaggerated inflammatory response particularly cytokines against viral infection. The term cytokine storm is now closely associated with the development and progression of SARS-CoV-2 infection. The cytokine storm can lead to excessive production of inflammatory mediators resulting in damage to the host cells. The pro-inflammatory cytokine IL-6, in particular, is considered one of the pivotal mediators during SARS-CoV-2 infection and in an early phase of virus-receptor interaction [13]. The understanding of the high virulence capacity of SARS-CoV-2 and the molecular and cellular factors involved in immune dysregulation will help in the development of potential targeted therapy against it. The three key principles in the management of the COVID-19 pandemic are (i) prevention, (ii) early detection, and (iii) targeted treatment [14]. "Drug repurposing" and "molecular docking analysis" are also considered attractive and emerging alternative strategies in analyzing suitable therapeutic candidates to address COVID-19 infection. Vaccine-based prevention has already been proven to be an effective way to stop infectious diseases and is currently instrumental in controlling the pandemic. Early detection of infected patients is crucial to identify the infection hotspots and accelerate treatment. Finally, targeted treatment might destroy the virus while minimizing damage to healthy tissue. Hence, novel technologies to facilitate prevention and early detection followed by targeted treatment of COVID-19 are of paramount importance for global health. The purpose of this chapter is to focus on the clinical features, pathogenesis, diagnosis, and options and opportunities in the treatment of COVID-19.

2 Molecular Basis of Pathogenesis of COVID-19 Infection

2.1 Virology of SARS-CoV-2

SARS-CoV-2 are positive-sense, single-stranded enveloped RNA viruses from the *Coronaviridae* family. The genome of the coronavirus encodes two major groups of proteins, namely, (A) structural proteins, such as Spike (S) glycoproteins marking all coronaviruses, nucleocapsid (N) that protects the genetic information of virus, matrix (M), and envelope (E), and (B) replicase complex encoding nonstructural proteins, such as proteases (nsp3 and nsp5) and RdRp (nsp 12) [15]. This enzyme is an ideal target as it helps the virus to replicate, meaning that the viral replication and virulence depend on protease. Recent studies revealed that coronavirus, in particular SARS-CoV-2, shares a similar type of genomic organization with other beta coronaviruses. Like others, it produces ~800 kDa polypeptide upon transcription of the genome. The proteolytic processing is facilitated by papain-like protease (PLpro) and 3-chymotrypsin-like protease (3CLpro). It leads to the production of various nonstructural proteins required for the replication of viral particles. These enzymes could serve as a potential target site for the inhibitors of COVID-19 infection [16].

The SARS-CoV-2 exploits host machinery to produce its lipoprotein envelope, which is comprised of several S proteins that give the virus a crown-like appearance. The M glycoprotein is also called the viral matrix protein. These proteins are associated with the envelope, linking the capsid to the viral glycoproteins inserted into the lipid bilayers. The N protein is pivotal for the morphogenesis phase of the life of a virus. The viral transmembrane spike (S) glycoprotein belongs to a trimeric class I fusion protein. It facilitates the entry of SARS-CoV-2 into the cell. It plays an essential role in facilitating cell attachment and subsequently its fusion with the host cell membrane. It comprises of two distinct major domains, the RBD (receptor binding domain) at the amino-terminus of the S1 subunit and the carboxy-terminus (CTD) of the S2 subunit, which facilitates the fusion of the virus with the host cell membrane [17]. SARS-CoV and SARS-CoV-2 share a highly conserved S protein RBD with 79.5% genome sequence identity [18].

Previous research has demonstrated that the activity of spike glycoprotein depends upon its cleavage into the S1 and S2 subunits by host proteolysis enzyme. The coupling of the viral particle is promoted by the S1 subunit, whereas the S2 subunit aids in the fusion of the virus with the host cell membrane. Furthermore, the N-terminal domain (NTD) and the C-terminal domain (CTD) of the S1 subunit can function as receptor-binding entities [19]. Both SARS-CoV and MERS-CoV use the CTD region of the S1 subunit to recognize the RBD; however, the region responsible for SARS-CoV-2 S-glycoprotein-hACE2 receptor interaction is not yet known. Each monomer of S-glycoprotein has revealed that it has 22 N-linked glycosylation sites and 4 predicted O-linked glycosylation sites, according to a comprehensive genomic analysis. Cryo-electron microscopy (Cryo-EM) studies have demonstrated that 14–16 N-glycans exist on 22 potential sites in S-glycoprotein [20]. These are accountable for proper protein folding and priming of the host proteases. It is worth mentioning that the glycosylation pattern of the S-glycoprotein is one of the essential features that serve as a potential site for mutation. Moreover, it also allows SARS-CoV-2 to circumvent both innate as well as adaptive immune responses. According to a recent study, the prognosis of spike glycoprotein of SARS-CoV-2 has a significant impact on the understanding of mechanism of cell entry as well as viral camouflage [21]. Additionally, it may promote the design and development of new bioactive(s), antibodies, vaccines, and screening of the human host targets.

2.2 Role of ACE2 Receptor in SARS-CoV-2 Infection

Angiotensin-converting enzyme 2 (ACE2) receptor protein is the common binding site for both SARS-CoV-2 and SARS-CoV to gain an entry into the host cell, whereas MERS-CoV bind to dipeptidyl peptidase 4 (DPP4 (a host receptor)) through the RBD (CTD) region [22]. ACE2 exists on the surface of the membrane of the host cells. After viral particle-ACE2 receptor binding, proteolytic activity is pivotal for the fusion of the S protein; its exposure allows the fusion of the virus with the host cell membrane. Subsequently, the S protein is proteolytically cleaved into S1 and S2 subunits. The proteolytic cleavage is achieved by the activity of the transmembrane protease serine 2 (TMPRSS2) and human airway trypsin-like protease (HAT). The S2 subunit of the S protein consists of a fusion peptide (FP) and heptad repeat domains HR1 and HR2. After insertion of the hydrophobic FP into the membrane of the cell, the three HR1 regions arrange into a coiled coils trimer. The three HR2 regions bind to the hydrophobic grooves of the HR1 trimer in an antiparallel manner. They interact and undergo irreversible conformational changes. As a result, the formed assembly that contains both HR1 and HR2 domains is known as a fusion core or six-helix bundle. The conformational rearrangement brings the viral particle and host cell membrane in close proximity to initiate the fusion events [23].

Several newly reported studies revealed that the three factors are the possible contributors to the virulence of SARS-CoV-2 infection. They could impact the pathogenicity of the infection. They include the following: (1) variations/differences in the RBD of the S protein, (2) variations/differences in characteristics of accessory proteins, and (3) the insertion of a polybasic cleavage site in the S protein. It is found that five of the six critical amino acids in the RBD of the S protein are pivotal for the S protein-ACE2 receptor cross talk/interaction. They are different in the case of SARS-CoV-2 and SARS-CoV. The S protein of the SARS-CoV-2 shows a stronger binding affinity for the ACE2 receptor as compared to the S protein of the SARS-CoV [24]. This could have reportedly rendered higher transmissibility and contagiousness to SARS-CoV-2 during the current pandemic.

2.3 *Pathogenesis and Immunopathology of COVID-19*

Once the viral particle enters the cells, its antigens will be presented to the antigen presentation cells (APC) in association with MHC 1, which is a crucial part of the body's antiviral immunity. In humans, antigenic peptides are presented by major histocompatibility complex (MHC) or human leukocyte antigen (HLA). They are recognized by virus-specific cytotoxic T lymphocytes (CTLs). The antigen presentation of SARS-CoV primarily relies on MHC I molecules. However, MHC II also contributes to its presentation.

The S protein present in the membrane of SARS-CoV-2 is considered as the most potent virus entry determinant into the host cell through the ACE2 receptor. According to study reported by Belouzard and coworkers, a significant cleavage event occurs by proteolytic enzymes (TMPRSS2 protease) at position S20 of S protein of the SARS-CoV-2. This results in membrane fusion as well as viral infectivity followed by the release of viral RNA [25]. Other studies reported that the entry of viral RNA into the host cell is dependent not only on membrane fusion but also on the clathrin-dependent and/or clathrin-independent endocytosis [26]. When the viral RNA genome is released into the cytoplasm, it is translated into two structural proteins and poly-proteins which help in viral replication. Upon infection with the SARS-CoV-2 genome, the host cell is activated. A rapid and well-coordinated immune response, i.e., an innate and adaptive immune response, occurs. It represents the defense mechanism against viral infection. In endosome, membrane-specific pattern recognition receptors (PRRs), like Toll-like receptors (TLR3, TLR8, TLR7, and TLR9), secretary type PRRs like Mannose-binding lectin (MBL), and C-reactive protein (CRP) or the cytosolic RNA sensor, RIG/MDA5 can recognize viral RNA as pathogen-associated molecular patterns (PAMPs) [27].

Interferon (IFN) type 1 induces a potent innate immune response against viral infection and also elicits an effective adaptive immune response. This recognition initiates a complex signaling cascade by recruiting adaptor proteins like mitochondrial antiviral-signaling protein (MAVS), IFN-β (TRIF), and stimulator of interferon genes protein (STING) and activates downstream cascade molecules including adaptor molecule MyD88. This interaction activates transcription factors such as nuclear factor-κB (NF-κB) and interferon regulatory factor 3 (IRF3) and thus helps in nuclear translocation. In the nuclei, these transcription factors induce the production of type I interferons (IFN-α/β) and a plethora of pro-inflammatory cytokines especially IL-6 [28]. Thus, interactions/cross talk between the host cell and virus build a complex set of first line of defense against the virus at the entry site. Type I IFN-mediated activation of the JAK-STAT pathway initiates the transcription of IFN-stimulated genes (ISGs) under the control of the IFN-stimulated response element (ISRE). Accumulation of type I IFN can suppress viral replication and serves as an immune modulator that promotes phagocytosis of antigens by macrophage, as well as NK cell-mediated restriction of infected cells. Thus, blocking the production of IFNs or disorder of the JAK-STAT signaling pathway or altered expression of macrophages has a direct effect on the survival of the virus within the host cell [15].

Generally, Th1-mediated immune response plays a predominant role in adaptive immunity against viral infections. The responses of the T cell primarily rely upon the presence of the APC (antigen-presenting cells)-mediated cytokine microenvironment. CD8+ cytotoxic T cells (CTLs) are essential for the eradication of virus-infected cells. They carried out their killing function by secreting a cluster of molecules such as perforin, IFN-γ, and granzymes. Nevertheless, CD4+ helper T cells are the most important cells in adaptive immunity. They facilitate the overall adaptive immune response by assisting cytotoxic T cells. Furthermore, B-cell-mediated humoral immune response plays a protective role by producing the neutralizing antibody and also impedes reinfection [29].

Some recent reports suggest that, in SARS-CoV-2-infected patients, an elevated level of chemokines and plasma cytokines, like interleukins (IL-1, IL-2, IL-4, IL-7, IL-10, 1L-12, 1L-13, and IL-17), IP-10, hepatocyte growth factor (HGF), macrophage colony-stimulating factor (MCSF), MCP-1, GCSF, TNF-α, IFN-γ, MIP-1α, etc., are associated with disease severity [30]. Like in SARS and MERS, the existence of "cytokine storm" and "lymphopenia" may have a significant or important role in the pathogenesis of COVID-19. Additionally, the persistence of cytokine storm might stimulate necrosis or apoptosis of T cells and may lead to their exhaustion like in cancer and other chronic infections. This "cytokine storm" is accountable for commencing viral sepsis followed by lung injury induced due to inflammation, which may manifest in other complications like acute respiratory distress syndrome (ARDS), pneumonitis, respiratory failure, sepsis shock, organ failure, and potentially death. Patients associated with severe COVID-19 infection showed a marked decrease in the number of circulating B cells, CD8+ cells, CD4+ cells, and natural killers (NK) cells, as well as a decrease in eosinophils, monocytes, and basophils [31].

Table 1 Variants of SARS-CoV-2 that are currently designated as variants of concern

Variant name	Lineage	Status
Alpha (α)	B.1.1.7	It was first identified in the United Kingdom in December, and it is estimated to be 50% more contagious. Now it is the dominant variant in the United States
Beta (β)	B.1.351	In December, it appeared in South Africa. Against this variant, the effectiveness of some vaccines is reported to be limited
Gamma (γ)	P.1	In late 2020, it appeared in Brazil. The variants are comparable to those found in B.1.351
Delta	B.1.617.2	It is very prevalent in India. Among all other variants, only this variant carries the L452R spike mutation
Omicron	B.1.529	Detected in Botswana, South Africa, and Hong Kong in November 2021, it has emerged as a new SARS-CoV-2 variant of concern
Omicron XE	BA.1; BA.2, BA.3	Omicron has since continued to evolve to have multiple different lineages or genetically related subvariants. This includes the original Omicron BA.1 and also BA.2 and BA.3. These are the recombinant variants of omicron. A "recombinant" variant has emerged "Omicron XE," which is the result of two omicron strains merging together in a single host and then going on to infect others. BA.2 is more infectious than BA.1 and has now taken over or outcompeted BA.1 to become the new dominant form of the SARS-CoV-2 virus worldwide

2.4 Coronavirus Variants and Mutations

Invariably coronavirus possess almost 30,000 RNA letters. In the course of infection, this genetic information is used by the virus to replicate within infected cells. Mutations arise when the virus amplifies in the host cell. If the developed mutation doesn't induce excessive variations, lineages are formed. A new strain is formed if diverse mutations arise in a lineage and may significantly change the viral epidemiology. COVID-19 is caused by a new coronavirus, which belongs to the SARS-CoV-2 strain. Scientists are concerned about various SARS-CoV-2 variants produced throughout the outbreak because they can culminate into prolonged pandemic and reduced the effectiveness of vaccines. Many variants of this novel virus have been identified, but scientists are more concerned about four more dangerous and five designated variants of interest [32–34]. Table 1 presents these four variants that are currently designated as variants of concerns.

3 Diagnosis of COVID-19

The current standard molecular technique that is used to detect SARS-CoV-2 infection (COVID-19) is the real-time reverse transcription-polymerase chain reaction (rRT-PCR), which detects RNA from SARS-CoV-2 and provides accurate results [35]. However, the technique has some limitations of requiring sophisticated equipment and a laboratory with biosafety level 2 or above. Furthermore, the time required to obtain the results is up to 3 days. The method is time-consuming. It is not exceptionally suitable for public health emergencies such as the one we are presently experiencing after the release of SARS-CoV-2 [36]. Several alternative testing techniques targeting different parts of the SARS-CoV-2 genetic profile have been developed worldwide, e.g., the rapid antigen test (RAT), enzyme-linked immune sorbent assay (ELISA), and chemiluminescence immunoassay [37]. However, critical issues related to the low accuracy and reliability of some of these test kits, especially at the beginning of the outbreak, were reported, affecting health care systems globally. Additionally, the huge amount of reagents required for testing has also become a bottleneck.

In the present scenario, the development of point-of-care (PoC), low cost, and efficient devices, capable of providing robust, fast, and reliable responses, is urgently needed. The tools and methods developed for this purpose should be simple enough to be used on-site and in the field. These should not necessarily require trained specialists to operate them.

4 Treatment Strategies for COVID-19

The new coronavirus (SARS-CoV-2) has claimed lives of more than a million people and is affecting thousands daily. Clinical trials are testing different drugs to find a potential therapeutic cure for COVID-19. They are being carried out in which potential antiviral treatment targets are being explored, such as restraining viral proteins identified for genome replication or blocking viral sections into human cells. There are numerous pharmacological strategies that may possibly fight against the SARS-CoV-2 virus including drug-based therapies or drug repurposing, stem cell therapies, plasma therapies, monoclonal antibodies, immunization/vaccines, and nanotechnology-based targeted therapies. Effective solutions against coronavirus can be based on their active component: those which act on viral proteins and proteins involved in RNA replication and union and those which act on the viral auxiliary proteins, restraining self-assembly or blocking the infection. The S protein could be a prime target for vaccine development. The possible treatment strategies for SARS-CoV-2 infection including drug therapy, plasma therapy, vaccine methodology, and nanotechnology-based targeted therapy are discussed in this chapter.

4.1 Repurposed Drug-Based Therapy

Vaccines are essentially protective; however, they are not entirely effective in preventing serious illness or the spread of the virus, especially in the case of newer variants. Some of the drugs that are being applied as investigational medicines against SARS-CoV-2 infection are repurposed medications primarily designed to cure other viral diseases. Chloroquine (CQ) is a drug that is widely used against malaria, and in 2006 it was observed that it has an antiviral potential too [38]. The functional mechanism of CQ is that it reportedly increases the endosomal pH and thus arrests the endosomal maturation, and as a result cytosolic entry of the virus is inhibited. This also interferes with the glycoproteins of cellular receptors of SARS-CoV-2 which bind to their targets. Consequently, by interfering with the binding of viral particles to the host cell surface receptors, CQ can disrupt the pre-entry step of the viral cycle [39, 40]. A study demonstrated the efficiency of chloroquine against SARS-CoV-2 infection. In vitro infection studies were carried out on Vero E6 cells at a multiplicity of infection (MOI) of 0.05. Chloroquine was found significantly effective in reducing viral replication. It can block SARS-CoV-2 infection at a low concentration (half-maximal effective concentration (EC_{50}) of 1.13 μM and a half-cytotoxic concentration (CC_{50}) larger than 100 μM [41].

Another study reported that the in vitro replication of HCoV-229E in epithelial lung cell cultures was inhibited upon treatment with chloroquine. It is also effective in vitro against the Middle East respiratory syndrome coronavirus (MERS-CoV) [42]. Hydroxychloroquine (HCQ) is a derivative of chloroquine. It is one of the antirheumatic bioactive(s). It is used for healing many rheumatic diseases. It demonstrates a strong immunomodulatory capacity which prevents inflammation flare-ups and organ damage. Both chloroquine and hydroxychloroquine can raise the intracellular pH and inhibit the fusion process between viruses and endosomes. They can also inhibit nucleic acid replication, glycosylation of viral proteins, virus assembly, new virus particle transport, virus release, and other processes to achieve antiviral outcomes [39]. The pharmacological activity of chloroquine and hydroxychloroquine was checked out in one study. The Vero cells were infected with SARS-CoV-2. In vitro studies revealed that hydroxychloroquine (EC_{50} = 0.72 μM) was more potent than chloroquine (EC_{50} = 5.47 μM) with relatively low EC_{50} value [43, 44].

Successful and effective pharmacotherapeutic strategies against COVID-19 infection can be explored either using specialized therapies to block viral spike connections like peptide fusion inhibitors or utilization of the broad-spectrum antiviral bioactives like protease inhibitors that cleave the bonds between long strings of protein subunits. These subunits are used by the SARS-CoV-2 to make the proteins needed for its replication within human cells. Without them, the virus cannot multiply [45].

Anti-SARS-CoV-2 neutralizing monoclonal antibodies and anti-ACE2 monoclonal antibodies are potential pharmacotherapeutic alternatives. Clinical trial-based reports indicate that a few antiviral drugs, such as favipiravir, remdesivir, and molnupiravir, are helpful in some cases. Molnupiravir is a new antiviral bioactive which demonstrated activity against SARS-CoV-2. It has been proven that the drug is safe and has antiviral activity in humans. It is under phase 3 clinical trial. Favipiravir is another drug that has been used for treating SARS-CoV-2 infection that may reduce hospitalization time and may exclude the need for mechanical ventilation [46]. Favipiravir inhibits RNA-dependent RNA polymerase. Besides its action against the influenza virus, this antiviral drug can inhibit the replication of other RNA viruses. Once it enters the cells, it gets transformed into an active form by becoming phosphoribosylated (favipiravir-RTP). It inhibits viral RNA polymerase by recognizing its activity. Thus, favipiravir could be the potential drug candidate for SARS-CoV-2 infection. Previous studies have demonstrated that the drug is independently demonstrated its role in fast and active viral clearance. It showed significant prognosis in chest imaging. Moreover, the drug has a positive impact on patients with COVID-19 positive tests. However, it should be noted that patients with mild-to-moderate COVID-19 have shown a better prognosis than patients with severe diseases. Favipiravir probably has no beneficial effect on mortality in the general group of patients with mild-to-moderate symptoms of COVID-19 [47, 48].

In October 2020, FDA approved the antiviral bioactive remdesivir for the treatment of COVID-19. It is an adenosine nucleoside analog. Vero E6 cells were used for testing remdesivir's antiviral activity. In vitro infection studies demonstrated that the EC_{50} value of remdesivir against SARS-CoV-2 was 0.77 μm, and the EC_{90} was 1.76 μm. SARS-CoV-2-infected patients in Washington, USA, demonstrated clinical improvement after treatment with remdesivir. No noticeable side effects were observed. The result of real-time RT-PCR analysis from the oropharyngeal swab was negative for SARS-CoV-2 after 13 days of treatment with remdesivir. Clinical trials showed that, in these patients, remdesivir can moderately accelerate/improve recovery time [49, 50]. Ribavirin is another bioactive that is easily available at low cost; thus, this may support its potential for positive in the treatment of COVID-19 infections. Ribavirin is a nucleotides derivative that competes with physiological nucleotide RNA-dependent RNA polymerases active site [51]. One of the studies reported that the EC_{50} against SARS-CoV-2 for ribavirin was found to be 109.5 μM. The combination of IFN and ribavirin reduces the viral replication and severity of the SARS-CoV-2 infection in animal models [52]. Similarly, nitazoxanide has been found to inhibit both cell entry and viral particle assembly. Thus, it could improve viral clearance in patients with symptomatic COVID-19 [53].

Dexamethasone is a glucocorticoid drug preferably used to treat rheumatic problems, some skin conditions, severe allergies, asthma, chronic obstructive pulmonary disease, brain swelling, and eye pain after eye surgery and as an adjunct therapy for tuberculosis. Remdesivir and dexamethasone are said to be alternative therapies that are effective against SARS-CoV-2-related diseases. Dexamethasone has been found to reduce the incidence of death in ventilated patients [54].

Angiotensin-converting enzyme 2 (ACE2) is an essential severe acute respiratory syndrome coronavirus 2 (SARS-CoV-2) receptor. It protects multiple tissues, including the lung, from injury as a regulator of the rennin-angiotensin system. Thus, ACE2 has become the target for drug design and development for the treatment of COVID-19 infection [55]. Human recombinant soluble ACE2 (hrsACE2) was developed by APN01 [Apeiron Biologics Vienna, Austria] in one of the newly developed drug moieties [56]. It has two modes of action that seem be of benefit in SARS-CoV-2 infection. The first includes binding or linking the viral spike protein to the receptor and thereby neutralizing SARS-CoV-2, and the second it minimizes injury to multiple organs, including the lungs, kidneys, and heart because of unabated rennin-angiotensin system hyperactivation and increased angiotensin II concentrations. hrsACE2 has been tested in 89 patients, with an acceptable safety profile including healthy volunteers in phase I studies and in patients with acute respiratory distress syndrome (ARDS) in phase 2 clinical studies. Moreover, hrsACE2 could reduce SARS-CoV-2 load by a factor of 1000–5000 when tested in in vitro cell-culture experiments and engineered organoids. This demonstrates that the ACE2 can effectively neutralize SARS-CoV-2. PF-07321332 is a completely different drug that prevents the virus from multiplying within the host. Pfizer developed this drug in the early 2000s as a potential treatment for SARS caused by the SARS-CoV corona virus. At the beginning of the COVID-19 epidemic, they retrained it against SARS-CoV-2, which has similar biology. When the drug was given orally to mice, it could effectively block the coronavirus [57–60]. Tables 2 and 3 present the drug-dosage form regimens for COVID-19 in adults and children.

4.2 Plasma-Based Therapy

Convalescent plasma therapy, often referred to as COVID plasma therapy, is an experimental procedure to cure SARS-CoV-2 infection. The treatment involves the collection of plasma from a person who has recovered from the COVID-19 infection and injected into a person who is suffering with the disease [72]. The plasma contains the antibodies that help a patient in fighting the pathogen and recovering from the infection. The USFDA has given emergency authorization for convalescent plasma therapy with high antibody levels to treat SARS-CoV-2 infection. The therapy may be given to people with COVID- 19 who are in the hospital and are an early stage of their illness or have a weak immune system. This may help people recover from COVID-19. It may alleviate the severity or shorten the length of the disease. However, this therapy is also associated with certain drawbacks. These include allergic reactions, lung damage and difficulty in breathing, infections such as hepatitis B and C and, HIV, etc. It's not yet known if this therapy could be an effective treatment for SARS-CoV-2 infection. However, the therapy is yet to be explored for its use in COVID-19 infection [73].

4.3 Vaccine-Based Therapy

In January 2020, the genetic sequence of the coronavirus (SARS-CoV-2) that causes COVID-19 was identified. The genetic sequencing motivated global initiations for making an efficient vaccine for this disease. The impact of the COVID-19 outbreak on humanity and the economy globally necessitated the development and evaluation of next-generation vaccine technology. The first COVID-19 vaccine candidate entered a human clinical trial for testing with unprecedented quickness was on March 16, 2020 [74]. According to the WHO report, there were 320 vaccine candidates in progress, as of October 8, 2021, with 194 in the preclinical development stage while 126 in clinical progressions.

Table 2 Overview table: Drug-dosage form regimens of treatment for adults for SARS-COV-2 infection

Bioactive(s) (monoclonal antibodies, repurposed drugs, etc.)	Targets	Mechanism of action/description	Dose/administration	References
Drug-dosage form regimens of treatment for adults				
Repurposed USFDA-approved bioactive(s)				
Favipiravir	Viral RNA polymerase	Blocks purine analog in order to synthesize viral RNA	Doses vary based on indication. Available as 200-mg tablet. Administration: Tablets can be crushed or mixed with liquid, bioavailability>95%	[61]
Remdesivir			200 mg × 1100 mg every 24 h IV infusion. Available as 5-mg/ml vial. Administration: 30-min IV (intravenous) infusion	[62]
Ribavirin		It is a guanosine (ribonucleic) analog used to block viral mRNA capping and ultimately viral synthesis	500 mg IV BID (twice a day) or TID (thrice a day)	[63]
Umifenovir/Arbidol	SARS-CoV-2 spike glycoproteins	It impedes trimerization of SARS-CoV-2 spike glycoproteins and inhibits host cell adhesion. It is being trialed for prophylactic action against SARS-CoV-2 infection	200 mg every 8 h by mouth 7–14 days (d). Available as 50 mg and 100 mg tablets, capsules, and granules.	[64, 65]

(continued)

Table 2 (continued)

Bioactive(s) (monoclonal antibodies, repurposed drugs, etc.)	Targets	Mechanism of action/description	Dose/administration	References
Chloroquine/hydroxychloroquine	Heme polymerase and ACE2	It increases endosomal pH and terminal glycosylation of ACE2 in order to inhibit SARS-CoV-2 entry	*Chloroquine* 500 mg by mouth 12–24 h × 5–10 d. Available as 250-mg tablets (salt), 500-mg tablets (salt), and 500-mg tablets of chloroquine phosphate (salt) = 300-mg chloroquine base *Hydroxychloroquine* 400 mg by mouth every 12 h × 1 d and then 200 mg by mouth every 12 h × 4 d. Alternative dosing: 400 mg by mouth daily×5 d or 200 mg by mouth 3times/d for 10 d	[66]
Lopinavir	Viral protease	It inhibits PLpro and 3CLpro, thereby disrupting the process of viral replication in SARS-CoV-2 infection	400 mg/100 mg by mouth every 12 h for up to 14 d. Available as lopinavir/ritonavir, 200 mg/50 mg tablets	[67]
Nitazoxanide	Glutathione-S-transferase	It alters pH and inhibits viral maturation. Previously used in protozoan infection, helminthic infection, and tuberculosis	500–600 mg twice daily for 5 d	[61]
Ritonavir	Cytochrome P450 3A	It binds to the protease active site and inhibits the enzyme's activity	400 mg/100 mg by mouth every 12 h for up to 14 d. Available as lopinavir/ritonavir, 200 mg/50 mg tablets	[68]
Monoclonal antibody therapy				
Tocilizumab	IL-6 receptor	Obstructs IL-6-mediated signal transduction	400 mg IV or 8 mg/kg × 1–2 doses. Second dose 8–12 h after first dose if inadequate response. Available as IV infusion injections (20 mg/ml)	[69–71]

Table 3 Overview table: Drug-dosage form regimens of treatment for children for SARS-COV-2 infection

Bioactive(s)	Targets	Age	Drug-dosage form regimens of COVID-19 in children
IFN-α	Enhances interferon-stimulated gene expression (ISG) via JAK/STAT signaling. They hinder viral multiplication and shedding	Nebulization: using with caution in neonates and infants younger than 2 months	Nebulization: 200,000–400,000 IU/kg or 2–4 μg/kg in 2 ml sterile water, twice daily for 5–7 d. Spray: 1–2 sprays on each nostril 8–10 sprays on the oropharynx, once every 1–2 h, 8–10 sprays/d for 5–7 d
Remdesivir	Viral RNA polymerase	≥3 to <40 kg	5 mg/kg IV dose on day 1, followed by 2.5 mg/kg IV every 24 hours
Arbidol	SARS-CoV-2 spike glycoproteins	≥2 years for influenza in Russia	No recommendation
Lopinavir/ritonavir	Lopinavir: viral protease Ritonavir: cytochrome P450 3A	China: oral solutions ≥6 months, tablets ≥2 years. USA: oral solutions ≥14 days, tablets ≥6 months	Body weight (kg) 7–15: 12 mg/3 mg/kg/time, twice daily for 1–2 weeks
Chloroquine diphosphate	Heme polymerase and ACE2	Using with caution	No recommendation
Ribavirin	Viral RNA polymerase	China: oral dosage forms ≥6 years. United States and Europe: oral dosage forms ≥3 years	Intravenous infusion at a dose of 10 mg/kg every time (500 mg every time), 2–3 times daily

The vaccine candidates that recently received clinical approval by international and national agencies/authorities differ in their design. They include subunit vaccine, deactivated vaccine, nucleic acid vaccine, live-attenuated vaccine, and viral vector vaccine [75]. Inactivated or killed viral vaccines are referred to a type of vaccine that contains pathogens (virus/bacteria) which have been killed applying a physical or chemical process. They are capable of replicating in human or animal bodies [76]. The physicochemical agents that are used in the process include heat treatment, radiation (e.g., gamma radiation), pH, formaldehyde, glutaraldehyde, etc. The live-attenuated vaccines consist of the weakened forms of the virus that can poorly replicate in the cell but can't cause a severe disease. These vaccines have been used by billions of people and usually offer immunity for decades [77]. Another type of vaccine is subunit vaccines which are based on viral particles. These vaccines typically contain adjuvants or molecules that increase the magnitude with moderate immune response. The conjugated vaccines such as polysaccharide-based antigens are carried by protein carriers. This type of vaccine can be classified as a subclass of subunit vaccines [78]. Another class of vaccine is viral-vector-based vaccine. They can enhance immunogenicity without an adjuvant. They elicit robust cytotoxic T lymphocyte (CTL) response to eliminate the cells infected with viruses. The typical/conventional viral vectors are adenoviruses, poxviruses, recombinant bacteria, plasmid DNA and RNA, etc. RNA- or DNA-based vaccines are expressed in the host cell instead of direct injection of the antigen or complete virus particles [79].

Vaccines are potentially designed to make numerous diverse antibodies to identify special parts of the virus. So, if one particle of the virus mutates, the antibodies could figure out another part of the virus [80]. In view of the ongoing COVID-19 outbreak, the vaccine companies are making a new class of vaccines that ought to work against the latest strains of SARS-CoV-2.

4.3.1 Mechanisms of Vaccines Against SARS-CoV-2

US FDA has approved various types of vaccines that are being used globally. The functional mechanism of each vaccine in the body is different and unique. In this section, we briefly introduce the major mechanisms of vaccines against novel coronavirus (SARS-CoV-2).

mRNA-Based Vaccines

The dynamic element that encodes for viral spike glycoprotein (s) of SARS-CoV-2 is the nucleoside-modified messenger mRNA (modRNA). This mRNA acts as a template for producing the specific protein that initiates the host's immune response against the virus. The mRNA in mRNA vaccines is wrapped in lipid nanoparticles (LNPs), which aids in the RNA transport and protects it from degradation with salts that function as a buffer. Another consideration is sucrose, which serves as a cryoprotectant [81]. The LNP-mRNA cargos enter the muscle cells via endocytosis shortly after infusion, and then the mRNA is transcribed. Subsequent to, invading antigen-presenting cells (APCs) can be recruited by a network of blood arteries next to the muscles [82]. The vaccine molecules then collide with the membrane and penetrate the cells. Afterward, the mRNA is released and then translates inside the host to form the SARS-CoV-2 S glycoprotein [83]. Spike proteins are then assembled to form spikes that travel to the cell's surface and stick out their tips from the membrane. Apart from this, the vaccinated cells start fragmentation of a few spike proteins so that the immune system can recognize them. When a vaccinated cell dies, it leaves a lot of spike proteins and protein fragments behind as debris. They are then picked up by specialized immune cells called APCs. The APCs display the pieces of the uptaken spike proteins on their surface. Subsequent to MHC-associated presentation, helper T cells, another type of immune cell, detect these pieces and send out the signals for the B cells which offer assistance to fight the infection by synthesizing antibodies. When B cells come in contact with coronavirus spikes on the surface of vaccinated cells or free-floating spike protein, they get attached to it, draw it inside, and show spike protein fragments on its surface. If these B cells are activated by helper T cells at that point, they will multiply and produce antibodies against the spike proteins from joining alternative cells. APCs can also activate a

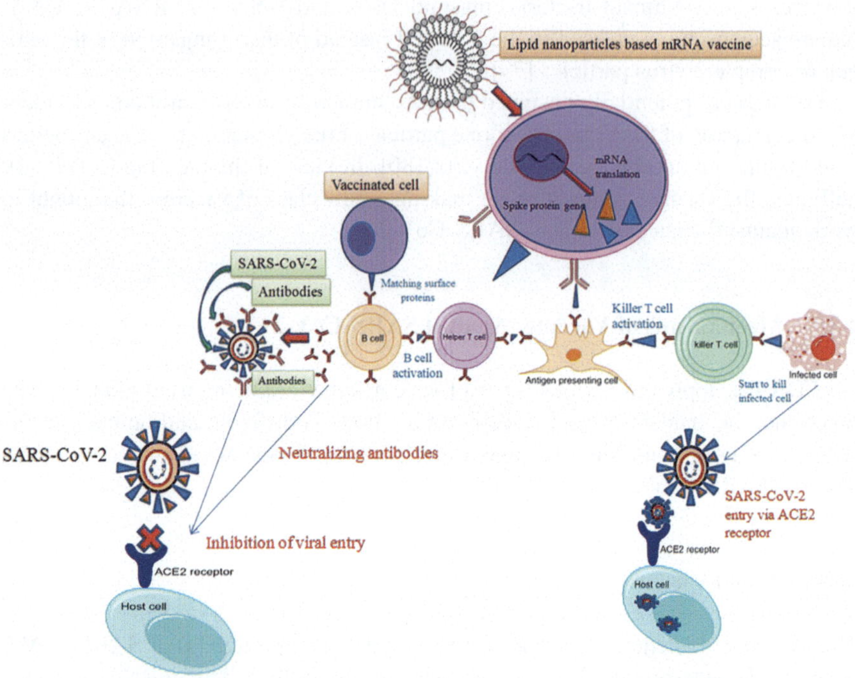

Fig. 1 Schematic representation of immunological mechanism of lipid nanoparticle-based mRNA vaccines against SARS-CoV-2

type of immune cell known as killer T cells with spike protein fragments on their surfaces [84]. The mechanism is shown in Fig. 1.

Adenovirus-Based Vaccines

This vaccine has been produced by adding the coronavirus spike gene into different adenoviruses, such as Ad26 and Ad5. These adenoviruses can invade cells but can't replicate [85]. In the Gamaleya vaccine, both of these adenoviruses have been used. In the CanSino vaccine, the coronavirus spike protein gene was converted to AD5, and, in Janssen, AD26 was utilized. However, the basis of Oxford-AstraZeneca has been ChAdOx1, a modified variant of a chimpanzee adenovirus. After vaccination, these adenoviruses strike the cells. Subsequently, they plug into their surface proteins. The virus is then endocytosed by the cell. Afterward, the viral RNA migrates to the nucleus, and the cell begins to produce spike proteins. Some spike proteins get fragmented and travel to the surface of the cells where they stick out through their tips. As said above, the antigens are recognized by specialized immune cells (APCs, T helper cells, and B cells), and antibodies are produced. If the virus invasion happens again, antibodies can lock onto neutralizing coronavirus spikes, label

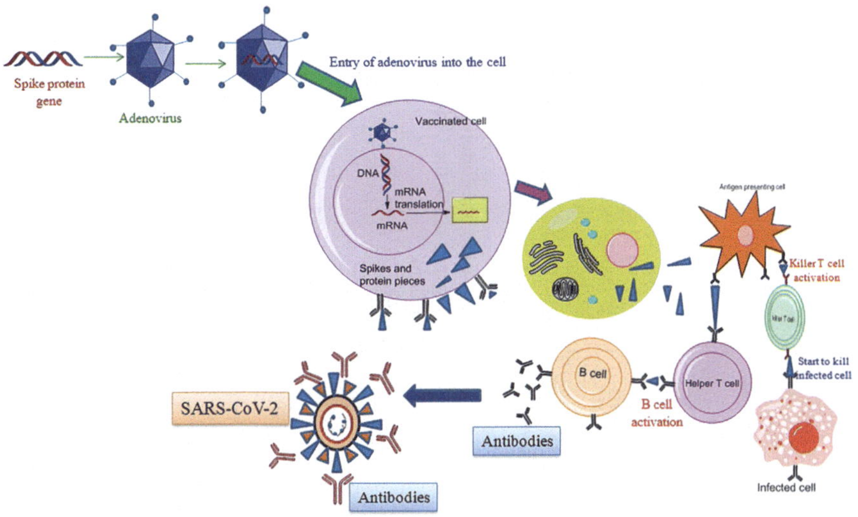

Fig. 2 Diagrammatic representation of immunological mechanism of adenovirus vector-based vaccines against SARS-CoV-2

the virus for destruction, and inhibit the infection [86, 87]. The mechanism is shown in Fig. 2.

The major problem that researchers are concerned about is that the immune system could respond to an adenovirus vaccine by producing antibodies, causing the second dose ineffective. The Russian vaccine, i.e., the Sputnik V vaccine, has solved this problem [88]. The vaccine is two doses vaccine consisting of two components. It utilizes two adenoviruses. Recombinant Ad26 is used in the first injection and Ad5 for the second dose to boost the vaccine's effect. The adenovirus-based vaccines against COVID-19 are more durable than Pfizer and Moderna's mRNA vaccines, because, one, the adenovirus's strong protein shield helps safeguard the genetic information (DNA) within and, two, RNA is much more sensitive and delicate than DNA. Therefore, adenovirus-based vaccines should be stored in the refrigerator. Nevertheless, the vaccine does not require minimum storage temperatures (−80 °C). A list of different vaccines developed so far against COVID-19 is discussed in Table 4.

5 Nanotechnological Interventions in COVID-19 Vaccine Development

Nanotechnology-based approaches could help in providing potential solutions to combat SARS-CoV-2 infection. Nanodiagnostics, for example, depend on the binding of nanoparticles to the targeting moiety of interest to induce a quantifiable/

Table 4 List of vaccines developed so far against SARS-CoV-2 infection

Vaccine manufacturer	Major mechanism/ technology	Route of administration	Efficacy	Dosing	Storage conditions	Status
Pfizer	mRNA	Intramuscular	95%	2 doses (28 days interval)	−80 °C to −60 °C or −25 °C to −15 °C (2 weeks)	Approved in the United States and other countries. Emergency use in EU and other countries
Moderna	mRNA	Intramuscular	94.10%	2 doses (28 days interval)	−25 °C to −15 °C or 2–8 °C (30 days)	Approved in Switzerland. Emergency use the United States, EU, and other countries
Covishield (Oxford/ AstraZeneca formulations)	Recombinant adenovirus	Intramuscular	76%	2 doses (84–112 days interval)	2–8 °C (6 months)	Approved in non-European countries. Emergency use in the United Kingdom
Sputnik V (Russian vaccine)	Recombinant adenovirus	Intramuscular	91.6%	2 doses (21–30 days interval)	2–8 °C	Emergency use in Russia and other countries
Covaxin (Bharat Biotech/ICMR)	Inactivated virus vaccine	Intramuscular	78%	2 doses (28 days interval)	2–8 °C	India, Brazil, Philippines, Iran, Mexico etc.
Janssen (Johnson & Johnson)	Recombinant adenovirus	Intramuscular	77%	1 dose	2–8 °C	Emergency use in the United States, EU, and other countries
CanSinoBIO	Recombinant viral vector vaccine	Intramuscular	65.70%	1 dose	2–8 °C	Approved in China. Emergency use in other countries
Novavax	Protein subunit vaccine	Intramuscular	95.60%	2 doses (21–28 days interval)	2–8 °C	Approved in EU, the United Kingdom, and New Zealand

measurable signal, allowing the recognition or identification of biomarkers or pathogens [89]. This approach enables the modulation and assessment of single molecules. It relies on the use of smaller devices, and platforms to exploit nanoscale attributes resulted as a consequence of interactions or cross talk between surfaces and biomolecules. With innovative nanotechnological developments, the diagnosis is being carried out on a molecular level. This leads to the utilization of handheld devices that are easy to use, simple, and marketable. Additionally, they are stable, highly sensitive, and accurate. Rapid testing and the detection of diseases or infections at early stages could be possible by using this technology. Besides, in case of nanodiagnostics (especially biosensors), the amount of chemical reagents needed for the analysis is significantly reduced. In addition to this, the established methodologies do not necessitate specialized apparatus or instrumentation. This could pave the way to attain simple-comprehensive responses in the biosensors.

In regard to vaccination, the most remarkable strategy for the development of nanosized vaccines relies on virus-like particles (VLPs). They are macromolecular complexes formed by structural viral proteins. They are capable of mimicking the virus. However, they lack the genome, and therefore VLPs do not replicate and are noninfective. The potential benefits of vaccines based on VLPs include increased safety (in comparison to whole virus-based vaccines), high immunogenicity owing to their complexity (in comparison to soluble antigens), and the preservation of the native antigenic determinants. Consequently, robust and protective immune responses could be achieved [90].

The vaccines' development against COVID-19 is under progress, and several clinical trials are underway; most of them are based on nucleic acids given their simpler production when compared to protein-based vaccines. The most advanced candidates are RNA vaccines expressing the spike (S) protein. They are considered an attractive target to lead to virus neutralization by antibodies elicited by vaccination [91]. An efficient packing mechanism is essential to avoid RNA degradation before delivery to the host translational machinery. Cationic lipid nanoparticles, initially designed for nucleic acid delivery in cancer immunotherapy and other vaccine applications, can help meet this need. Using cationic lipids, mRNA is efficiently condensed into solid-lipid nanoparticles which are taken up by the host APCs through endocytic or phagosomal routes. On the other hand, full-length spike protein or viral components can be encapsulated or self-assembled into nanoparticles as an option for using nanoparticles for vaccine development against SARS-CoV-2 [92]. Nanoparticles not only protect the native structure of the antigen but also enhance antigen delivery and presentation to APCs. Many biological systems, including viruses (particularly SARS-CoV-2) and proteins, are nanosized. Therefore, vaccination via nanocarriers provides some advantages. Nanoparticles can be administered via subcutaneous, oral, intranasal, and intramuscular routes. They are being developed to overcome the tissue barriers such as epithelial barriers (airway, nasal, gastrointestinal, etc.) and target critical sites such as lymph nodes and mucosal sites. Another application of nanotechnological strategies in vaccine development is VANs (vaccine adjuvant nanoparticles), which are considered to increase an overall effectiveness and safety of the elicited immune response. Vaccine adjuvants are essential, especially in the case of the SARS-CoV-2 infection, for decreasing the needed antigen dosage (dose-sparing), allowing for the manufacture of more units, and making it available to a broader population [93, 94]. Various other emerging vaccinology trends are entering the scene to produce innovative candidates that are safe, accessible, and easy to administer.

6 COVID-19 Global Vaccination Campaign

The current COVID-19 vaccination program seeks to attain worldwide vaccination coverage, which will be great support in pandemic control. As a result, individuals who refuse to get vaccinated or neglect COVID-19 vaccination may slow overall vaccination coverage. This leads to lower vaccination rates [95]. Moreover, this may impede worldwide efforts to control the spread of SARS-CoV-2. The unvaccinated individuals may act as SARS-CoV-2 reservoirs, producing more outbreaks. According to the WHO, vaccine hesitancy is one of the most serious threats to global health. The importance of social attitudes in developing herd immunity cannot be overstated. It is important to note that, in order to achieve a successful and effective vaccination program, governments and social media platforms should motivate individuals to work together or collaborate to address the public health problems [96]. They should also provide information to alleviate their fears and anxieties. Concerns about the distribution of vaccinations and how they are delivered are also significant issues for COVID-19 immunization. People are sorted into groups based on their vaccination priority. First, this is critical for ensuring a fair and equitable distribution of vaccinations worldwide to respond effectively to the SARS-CoV-2 outbreak [97]. A proper vaccination distribution, according to the centers for disease control and prevention (CDC), will contain the following: phase 1a involves vaccination of healthcare personnel and long-term care residents, phase 1b involves essential frontline employees and persons over the age of 75, and phase 1c involves people aged 65–74 or 16–64 with underlying medical problems and other essential workers. Following that, when vaccination doses become available, more categories will be included. The affluent nations have already secured 60% of the entire COVID-19 vaccine supply for their populations. Vaccine dosages adequate to vaccinate their people several times have been preordered in some of these nations. The worldwide vaccination coverage will be attained only by guaranteeing equitable access to COVID-19 vaccines. COVAX is a worldwide movement by WHO, the coalition of Epidemic Preparedness Innovations, and Gavi, the Vaccine Alliance, that guarantees equitable access to COVID-19 vaccines [98, 99].

7 Conclusion and Future Prognosis

The ongoing global COVID-19 outbreak is an emerging global pandemic threat. Global attention is urgently required to prevent more disasters, including significant social, psychological, and economic dilemmas. It can cause severe permanent respiratory syndromes and other organ malfunctionality and in some cases death. It has become a potential clinical concern to healthcare workers and the general public worldwide. The knowledge and information about this novel coronavirus are still limited, and we are still learning more and more about it. Therefore, the abovementioned challenges urgently require the development and availability of fast, cheap, reliable, and accurate diagnostic tools and techniques and effective therapeutic strategies for the prevention and cure of SARS-CoV-2 infection. More research and studies are needed to further investigate and explore the transmission and pathogenicity mechanism of the emerging novel coronaviruses. Antiviral therapy and vaccination are actively being evaluated and developed as viable and effective options. However, there is a lack of specific or effective antiviral bioactive(s) for many SARS-CoV-2 strains. Several clinical trials are going on, and lots of potential herbal/chemical therapies are being used. However, among all mentioned therapies, vaccination seems to be the most promising approach to reduce mortality and permanent side effects of this disease. The development of nanodiagnostic tools and nanoscale drug delivery systems for the detection or inactivation of the virus is the need of the hour. They are expected to play a paramount role in the success of prophylactic and therapeutic strategies. SARS-CoV-2 infection is not only a critical threat for the population with risk factors but also generates a dramatic economic impact and drift in terms of morbidity and the overall interruption of economic activities. What we can do now is to explicitly implement measures to control the infection and to prevent the spread of the SARS-COV-2 via human-to-human transmission. Perspectives on how vaccines and antiviral nanosystems can be implemented to fight against COVID-19 infection should be well reviewed and envisioned. The approaches that can be implemented in the short term and those that deserve long-term research to cope with respiratory viruses-related pandemics in the future should be identified. Public health officials or authorities may recommend additional actions. They should keep monitoring the situation, as the more we learn about this novel SARS-CoV-2 virus and its associated outbreaks, the better we can respond and prepare to square up the problems.

References

1. Baker RE, Mahmud AS, Miller IF, Rajeev M, Rasambainarivo F, Rice BL, Takahashi S, Tatem AJ, Wagner CE, Wang LF, Wesolowski A. Infectious disease in an era of global change. Nat Rev Microbiol. 2021;13:1–3.
2. Nickol ME, Kindrachuk J. A year of terror and a century of reflection: perspectives on the great influenza pandemic of 1918–1919. BMC Infect Dis. 2019;19(1):1.

3. Petersen E, Koopmans M, Go U, Hamer DH, Petrosillo N, Castelli F, Storgaard M, Al Khalili S, Simonsen L. Comparing SARS-CoV-2 with SARS-CoV and influenza pandemics. Lancet Infect Dis. 2020;20(9):e238–44.
4. Meloni S, Perra N, Arenas A, Gómez S, Moreno Y, Vespignani A. Modeling human mobility responses to the large-scale spreading of infectious diseases. Sci Rep. 2011;1(1):1–7.
5. Sohrabi C, Mathew G, Franchi T, Kerwan A, Griffin M, Del Mundo JS, Ali SA, Agha M, Agha R. Impact of the coronavirus (COVID-19) pandemic on scientific research and implications for clinical academic training–a review. Int J Surg. 2021;86:57–63.
6. Noorimotlagh Z, Jaafarzadeh N, Martínez SS, Mirzaee SA. A systematic review of possible airborne transmission of the COVID-19 virus (SARS-CoV-2) in the indoor air environment. Environ Res. 2021;193:110612.
7. Hu B, Guo H, Zhou P, Shi ZL. Characteristics of SARS-CoV-2 and COVID-19. Nat Rev Microbiol. 2021;19(3):141–54.
8. Perlman S, Netland J. Coronaviruses post-SARS: update on replication and pathogenesis. Nat Rev Microbiol. 2009;7(6):439–50.
9. Kuppusamy M, Wankhar W, Gurugubelli KR, Mahadevappa VH, Lepcha L, Kumar Choudhary A. Angiotensin-converting enzyme 2 (ACE2): COVID 19 gate way to multiple organ failure syndromes. Respir Physiol Neurobiol. 2021;283:103548.
10. Enjuanes L, Almazán F, Sola I, Zuñiga S. Biochemical aspects of coronavirus replication and virus-host interaction. Annu Rev Microbiol. 2006;13(60):211–30.
11. White JM, Delos SE, Brecher M, Schornberg K. Structures and mechanisms of viral membrane fusion proteins: multiple variations on a common theme. Crit Rev Biochem Mol Biol. 2008;43(3):189–219.
12. Supekar VM, Bruckmann C, Ingallinella P, Bianchi E, Pessi A, Carfí A. Structure of a proteolytically resistant core from the severe acute respiratory syndrome coronavirus S2 fusion protein. Proc Natl Acad Sci. 2004;101(52):17958–63.
13. Rabaan AA, Al-Ahmed SH, Garout MA, Al-Qaaneh AM, Sule AA, Tirupathi R, Mutair AA, Alhumaid S, Hasan A, Dhawan M, Tiwari R. Diverse immunological factors influencing pathogenesis in patients with COVID-19: a review on viral dissemination, immunotherapeutic options to counter cytokine storm and inflammatory responses. Pathogens. 2021;10(5):565.
14. Skwarek A, Gąsecka A, Jaguszewski M, Szarpak Ł, Dzieciątkowski T, Filipiak K. Nanoparticles: breakthrough in COVID-19 prevention, diagnosis and treatment. Arch Med Sci. 2021. https://doi.org/10.5114/aoms/142103
15. Zhou Z, Ren L, Zhang L, Zhong J, Xiao Y, Jia Z, Guo L, Yang J, Wang C, Jiang S, Yang D. Heightened innate immune responses in the respiratory tract of COVID-19 patients. Cell Host Microbe. 2020;27(6):883–90.
16. Yang Y, Liu C, Du L, Jiang S, Shi Z, Baric RS, Li F. Two mutations were critical for bat-to-human transmission of Middle East respiratory syndrome coronavirus. J Virol. 2015;89(17):9119–23.
17. Siu YL, Teoh KT, Lo J, Chan CM, Kien F, Escriou N, Tsao SW, Nicholls JM, Altmeyer R, Peiris JS, Bruzzone R. The M, E, and N structural proteins of the severe acute respiratory syndrome coronavirus are required for efficient assembly, trafficking, and release of virus-like particles. J Virol. 2008;82(22):11318–30.
18. Guan WJ, Ni ZY, Hu Y, Liang WH, Ou CQ, He JX, Liu L, Shan H, Lei CL, Hui DS, Du B. Clinical characteristics of coronavirus disease 2019 in China. N Engl J Med. 2020;382(18):1708–20.
19. Yuan M, Wu NC, Zhu X, Lee CC, So RT, Lv H, Mok CK, Wilson IA. A highly conserved cryptic epitope in the receptor binding domains of SARS-CoV-2 and SARS-CoV. Science. 2020;368(6491):630–3.
20. Song W, Gui M, Wang X, Xiang Y. Cryo-EM structure of the SARS coronavirus spike glycoprotein in complex with its host cell receptor ACE2. PLoS Pathog. 2018;14:e1007236.

21. Millet JK, Kien F, Cheung CY, Siu YL, Chan WL, Li H, Leung HL, Jaume M, Bruzzone R, Malik Peiris JS, Altmeyer RM. Ezrin interacts with the SARS coronavirus Spike protein and restrains infection at the entry stage. PLoS One. 2012;7(11):e49566.
22. Wang Q, Zhang Y, Wu L, Niu S, Song C, Zhang Z, Lu G, Qiao C, Hu Y, Yuen KY, Wang Q. Structural and functional basis of SARS-CoV-2 entry by using human ACE2. Cell. 2020;181(4):894–904.
23. Millet JK, Whittaker GR. Host cell proteases: critical determinants of coronavirus tropism and pathogenesis. Virus Res. 2015;16(202):120–34.
24. Chen J. Pathogenicity and transmissibility of 2019-nCoV—a quick overview and comparison with other emerging viruses. Microbes Infect. 2020;22(2):69–71.
25. Belouzard S, Chu VC, Whittaker GR. Activation of the SARS coronavirus spike protein via sequential proteolytic cleavage at two distinct sites. Proc Natl Acad Sci. 2009;106(14):5871–6.
26. Wang H, Yang P, Liu K, Guo F, Zhang Y, Zhang G, Jiang C. SARS coronavirus entry into host cells through a novel clathrin-and caveolae-independent endocytic pathway. Cell Res. 2008;18(2):290–301.
27. Perlman S, Netland J. Coronaviruses post-SARS: update on Coronaviruses post-SARS: update on. Nat Rev Microbiol. 2009;7(6):439–50.
28. Li H, Zhou Y, Zhang M, Wang H, Zhao Q, Liu J. Updated approaches against SARS-CoV-2. Antimicrob Agents Chemother. 2020;64(6):e00483–20.
29. Li X, Geng M, Peng Y, Meng L, Lu S. Molecular immune pathogenesis and diagnosis of COVID-19. J Pharm Anal. 2020;10(2):102–8.
30. Moore JB, June CH. Cytokine release syndrome in severe COVID-19. Science. 2020;368(6490):473–4.
31. Catanzaro M, Fagiani F, Racchi M, Corsini E, Govoni S, Lanni C. Immune response in COVID-19: addressing a pharmacological challenge by targeting pathways triggered by SARS-CoV-2. Signal Transduct Target Ther. 2020;5(1):1.
32. Barati F, Pouresmaieli M, Ekrami E, Asghari S, Ziarani FR, Mamoudifard M. Potential drugs and remedies for the treatment of COVID-19: a critical review. Biol Proced Online. 2020;22(1):1–7.
33. Singh J, Rahman SA, Ehtesham NZ, Hira S, Hasnain SE. SARS-CoV-2 variants of concern are emerging in India. Nat Med. 2021;27(7):1131–3.
34. Altmann DM, Boyton RJ, Beale R. Immunity to SARS-CoV-2 variants of concern. Science. 2021;371(6534):1103–4.
35. Nguyen T, Duong Bang D, Wolff A. 2019 novel coronavirus disease (COVID-19): paving the road for rapid detection and point-of-care diagnostics. Micromachines. 2020;11(3):306.
36. Corman VM, Landt O, Kaiser M, Molenkamp R, Meijer A, Chu DK, Bleicker T, Brünink S, Schneider J, Schmidt ML, Mulders DG. Detection of 2019 novel coronavirus (2019-nCoV) by real-time RT-PCR. Eur Secur. 2020;25(3):2000045.
37. Mak GC, Lau SS, Wong KK, Chow NL, Lau CS, Lam ET, Chan RC, Tsang DN. Evaluation of rapid antigen detection kit from the WHO emergency use list for detecting SARS-CoV-2. J Clin Virol. 2021;134:104712.
38. Colson P, Rolain JM, Lagier JC, Brouqui P, Raoult D. Chloroquine and hydroxychloroquine as available weapons to fight COVID-19. Int J Antimicrob Agents. 2020;55(4):105932.
39. Vincent MJ, Bergeron E, Benjannet S, Erickson BR, Rollin PE, Ksiazek TG. Chloroquine is a potent inhibitor of SARS coronavirus infection and spread. Virol Chloroquine is a potent inhibitor of SARS coronavirus infection and spread J. 2005;2:69.
40. Kwiek JJ, Haystead TA, Rudolph J. Kinetic mechanism of quinone oxidoreductase 2 and its inhibition by the antimalarial quinolines. Biochemistry. 2004;43(15):4538–47.
41. Zhengli S, Team of 10 researchers at the WIV. Remdesivir and chloroquine effectively inhibit the recently emerged novel coronavirus (2019-nCoV) in vitro. Cell Res. 2020;30(3):269–71.
42. Blau DM, Holmes KV. Human coronavirus HCoV-229E enters susceptible cells via the endocytic pathway. In: The Nidoviruses. Boston, MA: Springer; 2001. p. 193–8.

43. Fox RI. Mechanism of action of hydroxychloroquine as an antirheumatic drug. In: Seminars in arthritis and rheumatism 1993 Oct 1 (Vol. 23, No. 2). WB Saunders; 1993. p. 82–91.
44. Yao X, Ye F, Zhang M, Cui C, Huang B, Niu P, Liu X, Zhao L, Dong E, Song C, Zhan S. In vitro antiviral activity and projection of optimized dosing design of hydroxychloroquine for the treatment of severe acute respiratory syndrome coronavirus 2 (SARS-CoV-2). Clin Infect Dis. 2020;71(15):732–9.
45. Saadat YR, Khatibi SM, Vahed SZ, Ardalan M. Host serine proteases: a potential targeted therapy for COVID-19 and influenza. Front Mol Biosci. 2021;8:725528.
46. Delang L, Abdelnabi R, Neyts J. Favipiravir as a potential countermeasure against neglected and emerging RNA viruses. Antivir Res. 2018;153:85–94.
47. Furuta Y, Komeno T, Nakamura T. Favipiravir (T-705), a broad spectrum inhibitor of viral RNA polymerase. Proc Jpn Acad Ser B. 2017;93(7):449–63.
48. Cai Q, Yang M, Liu D, Chen J, Shu D, Xia J, Liao X, Gu Y, Cai Q, Yang Y, Shen C. Experimental treatment with favipiravir for COVID-19: an open-label control study. Engineering. 2020;6(10):1192–8.
49. Sheahan TP, Sims AC, Graham RL, Menachery VD, Gralinski LE, Case JB, Leist SR, Pyrc K, Feng JY, Trantcheva I, Bannister R. Broad-spectrum antiviral GS-5734 inhibits both epidemic and zoonotic coronaviruses. Sci Transl Med. 2017;9(396):eaal3653.
50. Ko WC, Rolain JM, Lee NY, Chen PL, Huang CT, Lee PI, Hsueh PR. Arguments in favour of remdesivir for treating SARS-CoV-2 infections. Int J Antimicrob Agents. 2020;55(4):105933.
51. Elfiky AA, Ismail A. Molecular dynamics and docking reveal the potency of novel GTP derivatives against RNA dependent RNA polymerase of genotype 4a HCV. Life Sci. 2019;238:116958.
52. Markland W, McQuaid TJ, Jain J, Kwong AD. Broad-spectrum antiviral activity of the IMP dehydrogenase inhibitor VX-497: a comparison with ribavirin and demonstration of antiviral additivity with alpha interferon. Antimicrob Agents Chemother. 2000;44(4):859–66.
53. Dos Santos WG. Natural history of COVID-19 and current knowledge on treatment therapeutic options. Biomed Pharmacother. 2020;129:110493.
54. Zhang Y, Hu S, Wang J, Xue Z, Wang C, Wang N. Dexamethasone inhibits SARS-CoV-2 spike pseudotyped virus viropexis by binding to ACE2. Virology. 2021;554:83–8.
55. Bourgonje AR, Abdulle AE, Timens W, Hillebrands JL, Navis GJ, Gordijn SJ, Bolling MC, Dijkstra G, Voors AA, Osterhaus AD, van Der Voort PH. Angiotensin-converting enzyme 2 (ACE2), SARS-CoV-2 and the pathophysiology of coronavirus disease 2019 (COVID-19). J Pathol. 2020;251(3):228–48.
56. Zhang H, Penninger JM, Li Y, Zhong N, Slutsky AS. Angiotensin-converting enzyme 2 (ACE2) as a SARS-CoV-2 receptor: molecular mechanisms and potential therapeutic target. Intensive Care Med. 2020;46(4):586–90.
57. Monteil V, Kwon H, Prado P, Hagelkrüys A, Wimmer RA, Stahl M, Leopoldi A, Garreta E, Del Pozo CH, Prosper F, Romero JP. Inhibition of SARS-CoV-2 infections in engineered human tissues using clinical-grade soluble human ACE2. Cell. 2020;181(4):905–13.
58. Kuba K, Imai Y, Rao S, Gao H, Guo F, Guan B, Huan Y, Yang P, Zhang Y, Deng W, Bao L. A crucial role of angiotensin converting enzyme 2 (ACE2) in SARS coronavirus–induced lung injury. Nat Med. 2005;11(8):875–9.
59. Crackower MA, Sarao R, Oudit GY, Yagil C, Kozieradzki I, Scanga SE, Oliveira-dos-Santos AJ, da Costa J, Zhang L, Pei Y, Scholey J. Angiotensin-converting enzyme 2 is an essential regulator of heart function. Nature. 2002;417(6891):822–8.
60. Wong DW, Oudit GY, Reich H, Kassiri Z, Zhou J, Liu QC, Backx PH, Penninger JM, Herzenberg AM, Scholey JW. Loss of angiotensin-converting enzyme-2 (Ace2) accelerates diabetic kidney injury. Am J Pathol. 2007;171(2):438–51.
61. Guo D. Old weapon for new enemy: drug repurposing for treatment of newly emerging viral diseases. Virol Sin. 2020;35(3):253–5.

62. Wang M, Cao R, Zhang L, Yang X, Liu J, Xu M, Shi Z, Hu Z, Zhong W, Xiao G. Remdesivir and chloroquine effectively inhibit the recently emerged novel coronavirus (2019-nCoV) in vitro. Cell Res. 2020;30(3):269–71.
63. Omrani AS, Saad MM, Baig K, Bahloul A, Abdul-Matin M, Alaidaroos AY, Almakhlafi GA, Albarrak MM, Memish ZA, Albarrak AM. Ribavirin and interferon alfa-2a for severe Middle East respiratory syndrome coronavirus infection: a retrospective cohort study. Lancet Infect Dis. 2014;14(11):1090–5.
64. Warren TK, Wells J, Panchal RG, Stuthman KS, Garza NL, Van Tongeren SA, Dong L, Retterer CJ, Eaton BP, Pegoraro G, Honnold S. Protection against filovirus diseases by a novel broad-spectrum nucleoside analogue BCX4430. Nature. 2014;508(7496):402–5.
65. Konstantinidou SK, Papanastasiou IP. Repurposing current therapeutic regimens against SARS-CoV-2. Exp Ther Med. 2020;20(3):1845–55.
66. Schrezenmeier E, Dörner T. Mechanisms of action of hydroxychloroquine and chloroquine: implications for rheumatology. Nat Rev Rheumatol. 2020;16(3):155–66.
67. Sheahan TP, Sims AC, Leist SR, Schäfer A, Won J, Brown AJ, Montgomery SA, Hogg A, Babusis D, Clarke MO, Spahn JE. Comparative therapeutic efficacy of remdesivir and combination lopinavir, ritonavir, and interferon beta against MERS-CoV. Nat Commun. 2020;11(1):1–4.
68. Cao B, Wang Y, Wen D, Liu W, Wang J, Fan G, Ruan L, Song B, Cai Y, Wei M, Li X. A trial of Lopinavir–Ritonavir in adults hospitalized with severe Covid-19. N Engl J Med. 2020;382:1787.
69. Chi Z, Zhao W, Jia-Wen L, Hong Z, Gui-Qiang W. The Cytokine Release Syndrome (CRS) of severe COVID-19 and Interleukin-6 Receptor (IL-6R) antagonist Tocilizumab man be the key to reduce the mortality. Int J Antimicrob Agents. 2020;55:105954.
70. Temesgen Z, Assi M, Shweta FN, Vergidis P, Rizza SA, Bauer PR, Pickering BW, Razonable RR, Libertin CR, Burger CD, Orenstein R. GM-CSF neutralization with lenzilumab in severe COVID-19 pneumonia: a case-cohort study. In: Mayo Clinic Proceedings 2020 Nov 1 (Vol. 95, No. 11). Elsevier; 2020. p. 2382–94.
71. Fakharian A, Barati S, Mirenayat M, Rezaei M, Haseli S, Torkaman P, Yousefian S, Dastan A, Dastan F. Evaluation of adalimumab effects in managing severe cases of COVID-19: a randomized controlled trial. Int Immunopharmacol. 2021;99:107961.
72. Focosi D, Anderson AO, Tang JW, Tuccori M. Convalescent plasma therapy for COVID-19: state of the art. Clin Microbiol Rev. 2020;33(4):e00072–20.
73. Nagoba B, Gavkare A, Jamadar N, Mumbre S, Selkar S. Positive aspects, negative aspects and limitations of plasma therapy with special reference to COVID-19. J Infect Public Health. 2020;13(12):1818–22.
74. Le TT, Andreadakis Z, Kumar A, Román RG, Tollefsen S, Saville M, Mayhew S. The COVID-19 vaccine development landscape. Nat Rev Drug Discov. 2020;19(5):305–6.
75. Chung YH, Beiss V, Fiering SN, Steinmetz NF. COVID-19 vaccine frontrunners and their nanotechnology design. ACS Nano. 2020;14(10):12522–37.
76. Delrue I, Verzele D, Madder A, Nauwynck HJ. Inactivated virus vaccines from chemistry to prophylaxis: merits, risks and challenges. Expert Rev Vaccines. 2012;11(6):695–719.
77. Pulendran B, Ahmed R. Immunological mechanisms of vaccination. Nat Immunol. 2011;12(6):509–17.
78. Vartak A, Sucheck SJ. Recent advances in subunit vaccine carriers. Vaccine. 2016;4(2):12.
79. Schleef M, Blaesen M, Schmeer M, Baier R, Marie C, Dickson G, Scherman D. Production of non viral DNA vectors. Curr Gene Ther. 2010;10(6):487–507.
80. Livingston EH. Necessity of 2 doses of the Pfizer and Moderna COVID-19 vaccines. JAMA. 2021;325(9):898.
81. Noor R. Developmental status of the potential vaccines for the mitigation of the COVID-19 pandemic and a focus on the effectiveness of the Pfizer-BioNTech and Moderna mRNA vaccines. Curr Clin Microbiol Reports. 2021;8(3):178–85.

82. Park JW, Lagniton PN, Liu Y, Xu RH. mRNA vaccines for COVID-19: what, why and how. Int J Biol Sci. 2021;17(6):1446.
83. Ewer KJ, Barrett JR, Belij-Rammerstorfer S, Sharpe H, Makinson R, Morter R, Flaxman A, Wright D, Bellamy D, Bittaye M, Dold C. T cell and antibody responses induced by a single dose of ChAdOx1 nCoV-19 (AZD1222) vaccine in a phase 1/2 clinical trial. Nat Med. 2021;27(2):270–8.
84. US Food and Drug Administration. Vaccines and related biological products Advisory Committee Meeting: EUA Amendment Request for Pfizer-BioNTech COVID-19 vaccine for use in children 5 through 11 years of age. In: United States. Food and Drug Administration. United States: Food and Drug Administration; 2021.
85. Kisby T, Yilmazer A, Kostarelos K. Reasons for success and lessons learnt from nanoscale vaccines against COVID-19. Nat Nanotechnol. 2021;16(8):843–50.
86. Lanini S, Capone S, Antinori A, Milleri S, Nicastri E, Camerini R, Agrati C, Castilletti C, Mori F, Sacchi A, Matusali G. GRAd-COV2, a gorilla adenovirus-based candidate vaccine against COVID-19, is safe and immunogenic in younger and older adults. Sci Transl Med. 2021;14(627):eabj1996.
87. Jones I, Roy P. Sputnik V COVID-19 vaccine candidate appears safe and effective. Lancet. 2021;397(10275):642–3.
88. Burki TK. The Russian vaccine for COVID-19. Lancet Respir Med. 2020;8(11):e85–6.
89. Jackson TC, Patani BO, Ekpa DE. Nanotechnology in diagnosis: a review. Adv Nanoparticles. 2017;6(3):93–102.
90. Qian C, Liu X, Xu Q, Wang Z, Chen J, Li T, Zheng Q, Yu H, Gu Y, Li S, Xia N. Recent progress on the versatility of virus-like particles. Vaccine. 2020;8(1):139.
91. Nel AE, Miller JF. Nano-enabled COVID-19 vaccines: meeting the challenges of durable antibody plus cellular immunity and immune escape. ACS Nano. 2021;15(4):5793–818.
92. Chauhan G, Madou MJ, Kalra S, Chopra V, Ghosh D, Martinez-Chapa SO. Nanotechnology for COVID-19: therapeutics and vaccine research. ACS Nano. 2020;14(7):7760–82.
93. Ballester M, Nembrini C, Dhar N, De Titta A, De Piano C, Pasquier M, Simeoni E, Van der Vlies AJ, McKinney JD, Hubbell JA, Swartz MA. Nanoparticle conjugation and pulmonary delivery enhance the protective efficacy of Ag85B and CpG against tuberculosis. Vaccine. 2011;29(40):6959–66.
94. Slütter B, Bal S, Keijzer C, Mallants R, Hagenaars N, Que I, Kaijzel E, van Eden W, Augustijns P, Löwik C, Bouwstra J. Nasal vaccination with N-trimethyl chitosan and PLGA based nanoparticles: nanoparticle characteristics determine quality and strength of the antibody response in mice against the encapsulated antigen. Vaccine. 2010;28(38):6282–91.
95. Dhama K, Sharun K, Tiwari R, Dhawan M, Emran TB, Rabaan AA, Alhumaid S. COVID-19 vaccine hesitancy–reasons and solutions to achieve a successful global vaccination campaign to tackle the ongoing pandemic. Hum Vaccin Immunother. 2021;17(10):3495–9.
96. Griffith J, Marani H, Monkman H. COVID-19 vaccine hesitancy in Canada: content analysis of tweets using the theoretical domains framework. J Med Internet Res. 2021;23(4):e26874.
97. García-Montero C, Fraile-Martínez O, Bravo C, Torres-Carranza D, Sanchez-Trujillo L, Gómez-Lahoz AM, Guijarro LG, García-Honduvilla N, Asúnsolo A, Bujan J, Monserrat J. An updated review of SARS-CoV-2 vaccines and the importance of effective vaccination programs in pandemic times. Vaccine. 2021;9(5):433.
98. Mahase E. Covid-19: Vaccine roll out could take a year and will require difficult prioritisation decisions. BMJ. 2020;371:m3846.
99. Sharun K, Dhama K. COVID-19 vaccine diplomacy and equitable access to vaccines amid ongoing pandemic. Arch Med Res. 2021;52(7):761–3.

Nano-Drug Delivery Systems for COVID-19 Drug Delivery

Komal Parmar and Jayvadan Patel

Abstract Currently, SARS-CoV-2 (COVID-19) is the dominant disease in the globe. It has resulted into major economic disruption and social disturbance all around the world. Many antivirals are under trial for the treatment of the infection. However, efficacy of drugs is often limited due to certain factors associated with the drug. Such problems can be overcome by using novel drug delivery systems employing nanotechnology. These delivery systems can be manipulated for the desired site of action and performance. This chapter outlines the detail about SARS-CoV-2, novel nano-drug delivery systems specific to nanoparticles that can be utilized for the delivery of antiviral drugs.

Keywords SARS-CoV-2 · COVID-19 · Nanotechnology · Viral treatment · Nano-drug delivery

1 Introduction

SARS-CoV-2 (severe acute respiratory syndrome-coronavirus-2) has become a serious pandemic, contributing to the current increase in morbidity and fatality rates around the world. Coronavirus gets its name from the crown-like spikes on the virus's outer surface [77]. SARS-COV-2 has been categorized as the seventh member of the corona virus family linked with human infection by the International Committee on Taxonomy of Viruses [16], after the World Health Organization's

K. Parmar (✉)
ROFEL, Shri G.M. Bilakhia College of Pharmacy, Vapi, Gujarat, India

J. Patel
Nootan Pharmacy College, Faculty of Pharmacy, Sankalchand Patel University, Visnagar, Gujarat, India

R. Shegokar, Y. Pathak (eds.), *Viral Drug Delivery Systems*,
https://doi.org/10.1007/978-3-031-20537-8_13

(WHO) classification of SARS-COV-2 as a worldwide epidemic [50]. The seafood wholesale market in Wuhan, China, in 2019 was linked to one of the first instances of SARS-CoV-2 infection. In just a few months, the outbreak swept through China and then the rest of the world [33].

Coronaviruses are enclosed, single-stranded, positive-sense m-ribonucleic acid (RNA) viruses that have a distinctive look that resembles a sun corona due to the protrusion of their characteristic crown-like spikes (Fig. 1) and has a very unusual replication method that causes mutations in some strains and is lethal to humans [24]. There are four classes of coronavirus, namely, alpha, beta, gamma, and delta. SARS-CoV-2 (COVID-19) virus belongs to the *Coronaviridae* family, the *Nidovirales* order, and the coronavirus genus *Beta* class [64, 95]. *Beta* class coronavirus was also responsible for earlier outbreaks, viz., severe acute respiratory syndrome (SARS) in 2002 and the Middle East respiratory syndrome (MERS) in 2012 [80]. The virus is quite small, measuring 65 to 125 nanometers in diameter. It has a single-stranded RNA nucleic material with a length of 26 to 32 kbs [59]. SARS-CoV-2 shares 80% of its genome with SARS-CoV and 96% with BatCoV RaTG13 [95]. Four proteins keep the SARS-CoV particle intact: (i) the S protein (spike glycoprotein) that promotes SARS-CoV entrance into host cells by allowing the virus to bind to host cells followed by membrane fusion; (ii) the abundant M protein (membrane) that maintains the membrane integrity of the viral particle; (iii) the E protein (envelope) which is the smallest protein and has a structural role in assembly and budding; and (iv) the N protein (nucleocapsid) which predominantly binds to the SARS-CoV [48, 55]. COVID-19 infection has a broad clinical spectrum, ranging from asymptomatic through hospitalization and respiratory assistance [92]. COVID-19 causes a human infection with clinical manifestations such as fever, loss

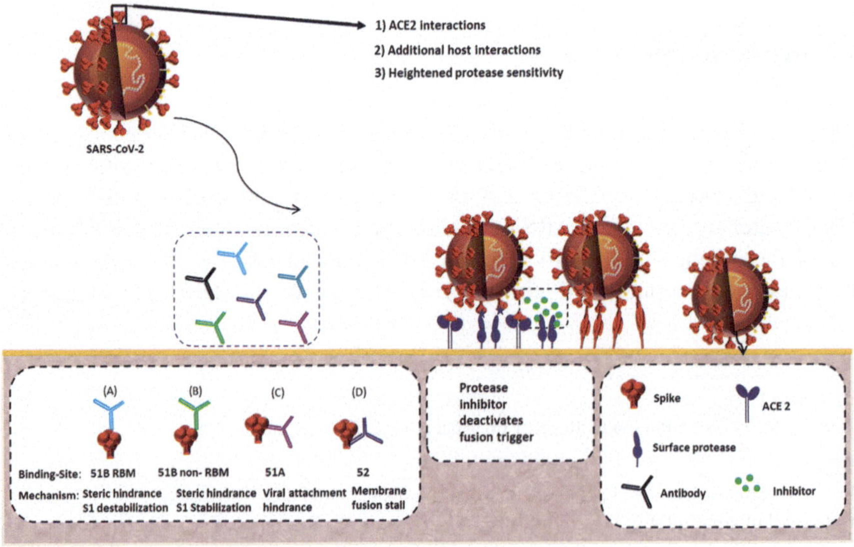

Fig. 1 Mechanism of SARS-COV-2 infecting the cell

of smell and taste, sore throat, cough, weariness, gastric problems, breathlessness, and lymphopenia [10, 17, 90]. Human infections can sometimes induce complications, such as pneumonia, organ dysfunction, and even death [22, 53, 93]. The transmission rate of SARS-CoV-2 (COVID-19) has been shown to be high due to a genetic recombination event at S-protein in the receptor-binding domain region, which may have boosted the virus's ability to transmit [78]. There are many variants of coronavirus reported till date. Among them, delta strain (emergence in India) has remained the deadliest, responsible for increased hospitalization or deaths [49]. However, recently, the new variant of concern is Omicron. It is reported to be more transmissible than all other variants of SARS-CoV-2. It was emerged from South Africa and is now spreading around the world [27]. SARS-CoV-2 is diagnosed by reverse transcription polymerase chain reaction testing, albeit false-negative test findings can occur in patients, depending on the quality and timing of the tests [6, 42, 44]. According to pathological findings, SARS-CoV-2 penetrates the alveolar epithelial cells of the lower lung by passage through the respiratory system, starting with mucous membranes in the nasopharynx [26]. SARS-CoV-2 utilizes the angiotensin-converting enzyme 2 (ACE2) receptor as its primary receptor, same like SARS-CoV [32, 47, 70]. It is largely expressed in vascular endothelium, respiratory epithelium, alveolar monocytes, and macrophages [66, 85]. COVID-19 is found to be associated with coagulation disorder in patients which is characterized by elevations in fibrinogen and D-dimer levels and mild prolongation of PT/aPTT. This is found to be attributed to activation of coagulation system by release of cytokines like IL-6 and in-turn suppression of fibrinolytic system [3]. Furthermore, a defective coagulation system boosts the aggressive immune response which can result into autoimmune and inflammatory diseases such as respiratory failure, immune thrombocytopenic purpura, paralysis, kidney and liver failure, deep vein thrombosis, pulmonary embolism, and stroke [14, 21, 94].

COVID-19 infection is split into three phases based on its clinical course: viremia/presymptomatic phase, acute respiratory phase (pneumonia phase), and severe (multisystemic clinical syndrome with impaired/disproportionate and/or defective immunity) or recovery phase [54, 84]. Patients with good immune function and no risk factors like comorbidities were able to withstand the infection and suppress the virus in the first or the second phase with immune hyper-reaction. COVID-19 individuals with immunological malfunction, on the other hand, may have a higher probability of losing the initial phase and becoming a critical type with a higher mortality rate. As a result, COVID-19 therapy should be chosen by the patients' stage. COVID-19 infection is linked to an increase in the number of people infected every day, and its management is found to be exceedingly difficult due to the virus's high infectiousness in some situations, a lack of effective antivirals and vaccines, and the possibility of a large asymptomatic population [11, 58]. There are currently few particular therapeutic methods for the treatment of SARS-CoV-2 [13, 15]. Considering the huge economic loss and rising infection rate, studies to aid in the development of specialized therapeutics to manage the virus is important for reducing the mortality due to COVID-19 pandemic or any such outbreaks in the future.

2 Role of Novel Drug Delivery in COVID-19

The term "novel drug delivery" refers to a modern technique that combines innovative preparations, better technology, and unique methodology for effectively delivering bioactive molecules in the body as required to produce their desired pharmacological effects. The purpose of the novel drug delivery is to enhance efficacy and safety of drug, to provide site specificity with an optimum dose, to decrease toxicity, and to fulfil the patient's needs [4]. Till date, various novel drug delivery approaches are reported with antiviral agents for the improved performance against viral infections [67, 76]. Clinical evidence of drug resistance has been found in antiviral drugs [81]. Drug resistance is primarily caused by nonspecific cell targeting and inadequate concentration of the drug at the target location. In addition, unsatisfactory pharmacodynamic physicochemical and biological characteristics including poor aqueous solubility, poor permeability, strong affinity for plasma proteins, short biological half-lives, and rapid elimination from the blood circulation are the major governing factors that determine subpar drug concentration at the specific site, which may lead to drug resistance development. The efficacy of several drugs and medications in treating SARS-CoV-2 is now being studied in clinical trials. Various medications include antiviral drugs that inhibit virus replication, e.g., remdesivir, favipiravir; immunomodulators, e.g., tocilizumab; human antibodies including immune globulin (IVIG) and monoclonal bevacizumab; corticosteroids such as methyl-prednisolone and dexamethasone; and antibiotics including azithromycin and ivermectin. Table 1 tabulates the possible medications used for the COVID-19 management. The COVID-19 situation necessitates a thorough examination of all existing nanotechnology-based approaches.

3 Nanostructured Novel Drug Delivery Systems (NNDDS)

Various antiviral-based NNDDS have been reported in recent years as a treatment for HIV, herpes zoster, viral C hepatitis, and other viral illnesses. They are typically constituted of lipids, polymers, metals, and inorganic nanoparticles [18]. Such nanostructured systems can be prepared using synthetic and natural materials. Small particle sizes and large specific surface areas characterize nanoparticle delivery systems, which can be advantageous for various applications due to their quick digestion, penetration, and absorption. Furthermore, nanoparticle composition, structure, and interfacial properties can be tailored to increase the dispersibility and stability of bioactive substances enclosed within them. For SARS-CoV-2, NNDDS can be used to target viral proteases (3CLpro and PLpro) and RNA polymerase (RdRp) or interact with viral S protein in general which will lead to inhibition of virus replication or virus entry into the cell. The antiviral activity of NNDDS made up of lipids, such as liposomes, solid lipid nanoparticles, nanostructured lipid carriers, and nanoemulsions, and polymers, such as nanoparticles, cyclodextrins, and dendrimers, has

Table 1 Therapy medications for COVID-19 disease

Medication	Mechanism of action	References
Remdesivir	It is an intravenous nucleotide prodrug of an adenosine analog. It inhibits RdRp (RNA-dependent RNA polymerase)	[25]
Favipiravir	A purine nucleic acid analog and broad spectrum RdRp inhibitor	[40]
Lopinavir/ritonavir combination	Lopinavir inhibits enzyme 3-chymotrypsin-like protease, thereby disrupting the process of viral replication	[57]
Umifenovir	Causes the viral glycoprotein to become structurally stiff, limiting the conformational changes that occur during membrane fusion and virus penetration	[61]
Camostat mesylate	Inhibits human transmembrane surface protease, TMPRSS2	[9]
Baricitinib	Prevents viral endocytosis and inhibition of the cytokine release blocking JAK1/2	[65]
Saquinavir	Inhibits dimeric SARS-CoV2 main proteinase	[7]
Tocilizumab	Reduces inflammation by blocking the interleukin-6 receptor	[12]
Sotrovimab	Prevents entry of virus into the cell by binding the spike protein of SARS-CoV-2	[30]
Bamlanivimab and etesevimab combination	Both types of antibody target the surface spike protein of SARS-CoV-2	[19]
Hydroxychloroquine and azithromycin	Inhibition of replication of SARS-CoV-2	[31]

been described. Figure 2 illustrates few nanoparticles for the antiviral drug delivery. The potential of nanoparticles to encapsulate and transport antivirals has been established by a number of earlier studies (Table 2). Further, the review focuses on the mentioned NNDDS as possible drug delivery for SARS-CoV-2.

3.1 Liposomes

Liposomes have been widely employed to encapsulate antiviral drugs for various applications. Liposomes are made up of a bilayer of phospholipid molecules with nonpolar tails on the inner side and hydrophilic head groups on the outer side. Antivirals can be encapsulated by adding them to phospholipids prior to the creation of liposomes or by introducing them into the liposomes after they have been formed. Furthermore, liposomes' surfaces can be functionalized by integrating specific ligands, improving their capacity to deliver antivirals to specific targets. The ability of PEGylated liposomes was investigated for targeted delivery of interferon alpha-2b (IFN α-2b) to inhibit human papilloma virus. The study showed that surface-functionalized liposomes could penetrate the epithelium more easily than the free

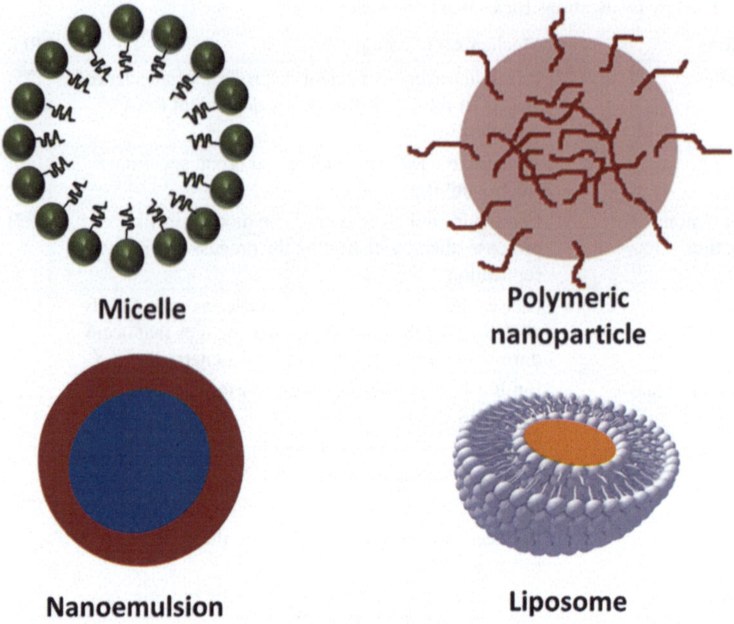

Fig. 2 Various nanoparticles for antiviral drug delivery

Table 2 Antiviral medicines delivered via various nanoparticle delivery systems

Nanoparticle	Antiviral drug	References
Lipid-coated mesoporous silica nanoparticles	ML336, chemical inhibitor of Venezuelan equine encephalitis virus	[45]
Nanoemulsion	Polyprenols from ginkgo leaves	[87]
Nanoemulsion	Curcumin	[56]
Niosomes	Nevirapine	[89]
Solid lipid nanoparticles	Zanamivir	[79]
Micelle	Cidofovir	[52]
Solid lipid nanoparticles	Ritonavir	[37]
Surfactant-based nanoparticles	Ritonavir	[35]
Polymeric nanoparticles	Lopinavir	[39, 68]
Polymeric nanoparticles	Efavirenz	[74, 83]
Nanoemulsion	Efavirenz	[41, 73]
Solid lipid nanoparticles	Lopinavir	[69]
Proliposome	Lopinavir	[63]
Nanoemulsion	Lopinavir	[29, 62]
Nanoemulsion	Darunavir	[28, 34]

form of IFN α-2b [38]. Antibody-coated nanoliposomes were produced in another investigation to improve the efficacy of dapivirine, an antiviral drug. The nanoconstruct demonstrated efficient binding affinity to HIV-1 envelope glycoprotein gp120 and neutralizing ability for HIV virus [88].

3.2 Nanoemulsions

Nanoemulsions are type of nanostructured colloidal systems with relatively small droplet size, i.e., <200 nm, and have a great potential to encapsulate and deliver the antiviral drugs [36]. They are typically prepared from oil, water, emulsifier, and co-surfactants as stabilizer. Nanoemulsions can be utilized to deliver antiviral medicines through any of the major routes of administration. In a study, w/o/w nanoemulsion encapsulating acyclovir was reported for dermal delivery against herpes [72]. In another similar type of research, self-nanoemulsifying drug delivery was prepared for dermal permeation for herpes simplex virus infection. Nanoemulsion system was prepared using garlic oil, mixture of tween 20 and span 20 as surfactant, and propylene glycol as co-surfactant. Ex vivo skin permeation study demonstrated a 2.3-fold augmented permeation of acyclovir [5].

3.3 Micelle

Micelles are surfactant-based nanoparticles with a diameter of 5–100 nm that can be used to create thermodynamically stable antiviral and other active drug delivery systems. This colloidal dispersion is normally made from the bottom up by combining the right types and amounts of surfactant, oil, and water in the right environment (e.g., temperature, pH, and ionic strength). The surfactants are arranged in such a way that their hydrophilic head groups get into interface with water, while their hydrophobic tails form a nonpolar core within the particle. In one study, polymeric prodrug micelles for delivery of acyclovir were synthesized using hydrophilic methoxy poly(ethylene glycol)/chitosan [71]. Stearic acid-g-chitosan oligosaccharide micelles encapsulating lamivudine stearate were developed. The prepared micelle showed high entrapment efficiency and drug loading. In addition, micelles presented significant drug release and high cellular drug uptake [46].

3.4 Lipid Nanoparticles

Solid lipid nanoparticles (SLN) are made up of lipid phase which is completely solidified into a highly regular crystalline structure, whereas nanostructured lipid carriers (NLC) are oil in water type of emulsion with oil phase as solid lipid. Hydrophobic bioactive components are encapsulated within the solidified lipid matrix of SLN/NLC [20]. The presence of a solid lipid phase can impede molecular mobility, preventing bioactive components from being degraded or released. Both SLN and NLC are non-biotoxic as the lipids employed are biodegradable. SLNs and NLCs have been utilized to encapsulate and deliver antiviral agents. Recently, acyclovir solid lipid nanoparticles were investigated for the efficient therapy against herpes simplex infection using a mouse model. Sustained drug release was obtained with single dose of drug-loaded SLN in comparison to 400 mg of three times a day dose for 5 days [43]. In another study, acyclovir-loaded solid lipid nanoparticles (ACV-SLNs) were formulated to target the brain. Chitosan and Tween 80-coated SLN were prepared and studied for in vivo pharmacokinetic parameters. The values of AUC0-∞ and MRT of coated ACV-SLNs were higher than free drug by about twofold, 233.36 ± 41.56 μg.h/mL, and 1.81 ± 0.36 h, respectively. Results indicated the effectiveness of ACV-SLNs for brain targeting [23]. Nanostructured lipid carrier encapsulating adefovir dipivoxil were prepared via solvent emulsification diffusion technique employing Geleol, Gelucire 50/13, Precirol ATO-5, Gelucire 44/14, and Compritol 888 as lipid polymer. Results suggested efficient liver targeting with improved oral bioavailability [1]. In another study, chitosan-coated nanostructured lipid carriers were developed for the ocular delivery of acyclovir. NLCs were fabricated by the hot microemulsion technique. Corneal bioavailability of NLC-loaded acyclovir studied in albino rabbits was enhanced by 4.5-fold as compared to commercially available preparation [75].

3.5 *Polymeric Nanoparticles*

Polymeric nanoparticles are small particles with a diameter ranging from 1 to 1000 nm that can encapsulate drug or be surface-adsorbed onto the polymeric core. Biodegradable/synthetic polymers are widely employed for fabrication of nanoparticles [2]. Synthetic polymers like poly(N-isopropylacrylamide), poly(N-isopropylacrylamide-co-acrylic acid), and poly(ethylene glycol)-b-poly (methacrylic acid), whereas biodegradable polymers like poly(lactic-co-glycolic acid) (PLGA), polylactic acid, chitosan, and albumin, are used as potent vectors for drug delivery [8]. Recently, antiviral activity of curcumin-loaded PLGA nanoparticles was evaluated against Zika virus. Results showed reduction in virus yield as well as viral RNA synthesis and protein expression demonstrating efficiency of formulation in comparison to free drug [60]. In another study, antiviral activity of chitosan nanoparticles encapsulating curcumin was evaluated against hepatitis C virus genotype 4a (HCV-4a) in human hepatoma cell line. Curcumin-chitosan nanocomposite inhibited HCV-4a entry and replication compared to curcumin alone [51].

Antivirals can also be encapsulated and delivered with the help of protein-based nanoparticles. Suwannoi and coresearchers produced acyclovir-loaded bovine serum albumin (BSA) nanoparticles to treat ocular herpes viral infections. Results demonstrated improved transport of acyclovir through the multilayers of corneal epithelial cells [82]. In another study, bovine serum albumin-coated tellurium nanoparticles showed enhanced antiviral activity against porcine reproductive and respiratory syndrome virus, a model of arterivirus, and porcine epidemic diarrhea virus, a model of coronavirus [96].

4 Conclusion

Currently, remdesivir is the antiviral drug approved by US FDA. Many researchers have investigated various nano-systems for the efficient delivery of remdesivir [86, 91]. However, in vivo studies are needed for the pharmacokinetics and pharmacodynamics information. Furthermore, other antivirals are under trials for the positive outcomes, and further research should be carried out for the application of such novel drug delivery systems in order to improve the therapeutic efficacy of the drugs used in COVID-19 management.

References

1. Abd El-Halim SM, Abdelbary GA, Amin MM, Zakaria MY, Shamsel-Din HA, Ibrahim AB. Stabilized oral nanostructured lipid carriers of Adefovir Dipivoxil as a potential liver targeting: estimation of liver function panel and uptake following intravenous injection of radioiodinated indicator. Daru. 2020;28(2):517–32. https://doi.org/10.1007/s40199-020-00355-8.
2. Adhikari C. Polymer nanoparticles-preparations, applications and future insights: a concise review. Polym Plast Technol Mater. 2021;60(18):1996–2024. https://doi.org/10.1080/25740881.2021.1939715.
3. Aggarwal M, Dass J, Mahapatra M. Hemostatic abnormalities in COVID-19: an update. Indian J Hematol Blood Transfus. 2020;36(4):1–11. https://doi.org/10.1007/s12288-020-01328-2.
4. Agrahari V. Novel drug delivery systems, devices, and fabrication methods. Drug Deliv Transl Res. 2018;8:303–6. https://doi.org/10.1007/s13346-017-0459-3.
5. Almehmady AM, Ali SA. Transdermal film loaded with garlic oil-acyclovir Nanoemulsion to overcome barriers for its use in alleviating cold sore conditions. Pharmaceutics. 2021;13(5):669. https://doi.org/10.3390/pharmaceutics13050669.
6. Arevalo-Rodriguez I, Buitrago-Garcia D, Simancas-Racines D, Zambrano-Achig P, Del Campo R, Ciapponi A, Sued O, Martinez-García L, Rutjes AW, Low N, Bossuyt PM, Perez-Molina JA, Zamora J. False-negative results of initial RT-PCR assays for COVID-19: a systematic review. PLoS One. 2020;15(12):e0242958. https://doi.org/10.1371/journal.pone.0242958.
7. Bello M, Martínez-Muñoz A, Balbuena-Rebolledo I. Identification of saquinavir as a potent inhibitor of dimeric SARS-CoV2 main protease through MM/GBSA. J Mol Model. 2020;26(12):340. https://doi.org/10.1007/s00894-020-04600-4.
8. Bolhassani A, Javanzad S, Saleh T, Hashemi M, Aghasadeghi MR, Sadat SM. Polymeric nanoparticles: potent vectors for vaccine delivery targeting cancer and infectious diseases. Hum Vaccin Immunother. 2014;10(2):321–32. https://doi.org/10.4161/hv.26796.
9. Breining P, Frølund AL, Højen JF, Gunst JD, Staerke NB, Saedder E, Cases-Thomas M, Little P, Nielsen LP, Søgaard OS, Kjolby M. Camostat mesylate against SARS-CoV-2 and COVID-19-Rationale, dosing and safety. Basic Clin Pharmacol Toxicol. 2021;128(2):204–12. https://doi.org/10.1111/bcpt.13533.
10. Byambasuren O, Dobler CC, Bell K, Rojas DP, Clark J, McLaws ML, Glasziou P. Comparison of seroprevalence of SARS-CoV-2 infections with cumulative and imputed COVID-19 cases: systematic review. PLoS One. 2021;16(4):e0248946. https://doi.org/10.1371/journal.pone.0248946.
11. Cao W, Li T. COVID-19: towards understanding of pathogenesis. Cell Res. 2020;30(5):367–9. https://doi.org/10.1038/s41422-020-0327-4.

12. Salama C, Han J, Yau L, Reiss WG, Kramer B, Neidhart JD, Criner GJ, Kaplan-Lewis E, Baden R, Pandit L, Cameron ML, Garcia-Diaz J, Chávez V, Mekebeb-Reuter M, de Menezes FL, Shah R, González-Lara MF, Assman B, Freedman J, Mohan SV. Tocilizumab in patients hospitalized with Covid-19 pneumonia. N Engl J Med. 2021;384:20–30. https://doi.org/10.1056/NEJMoa2030340.
13. Chen PL, Lee NY, Cia CT, Ko WC, Hsueh PR. A review of treatment of coronavirus disease 2019 (COVID-19): therapeutic repurposing and unmet clinical needs. Front Pharmacol. 2020;11:584956. https://doi.org/10.3389/fphar.2020.584956.
14. Cheng Y, Luo R, Wang K, Zhang M, Wang Z, Dong L, Li J, Yao Y, Ge S, Xu G. Kidney disease is associated with in-hospital death of patients with COVID-19. Kidney Int. 2020;97(5):829–38. https://doi.org/10.1016/j.kint.2020.03.005.
15. Coppock D, Baram M, Chang AM, Henwood P, Kubey A, Summer R, Zurlo J, Li M, Hess B. COVID-19 treatment combinations and associations with mortality in a large multisite healthcare system. PLoS One. 2021;16(6):e0252591. https://doi.org/10.1371/journal.pone.0252591.
16. Coronaviridae Study Group of the International Committee on Taxonomy of Viruses. The species Severe acute respiratory syndrome-related coronavirus: classifying 2019-nCoV and naming it SARS-CoV-2. Nat Microbiol. 2020;5:536–44. https://doi.org/10.1038/s41564-020-0695-z.
17. da Rosa MR, Francelino Silva Junior LC, Santos Santana FM, Farias de Oliveira T, Campos Alcântara R, Monteiro Arnozo G, da Silva R, Filho E, Galdino Dos Santos AG, Oliveira da Cunha EJ, Salgueiro de Aquino SH, Freire de Souza CD. Clinical manifestations of COVID-19 in the general population: systematic review. Wien Klin Wochenschr. 2021;133(7–8):377–82. https://doi.org/10.1007/s00508-020-01760-4.
18. Delshadi R, Bahrami A, McClements DJ, Moore MD, Williams L. Development of nanoparticle-delivery systems for antiviral agents: a review. J Control Release. 2021;10;331:30–44. https://doi.org/10.1016/j.jconrel.2021.01.017.
19. Dougan M, Nirula A, Azizad M, Mocherla B, Gottlieb RL, Chen P, Hebert C, Perry R, Boscia J, Heller B, Morris J, Crystal C, Igbinadolor A, Huhn G, Cardona J, Shawa I, Kumar P, Adams AC, Van Naarden J, Custer KL, Durante M, Oakley G, Schade AE, Holzer TR, Ebert PJ, Higgs RE, Kallewaard NL, Sabo J, Patel DR, Dabora MC, Klekotka P, Shen L, Skovronsky DM. Bamlanivimab plus Etesevimab in mild or moderate Covid-19. N Engl J Med. 2021;385:1382–92.
20. Duan Y, Dhar A, Patel C, Khimani M, Neogi S, Sharma P, Kumar NS, Vekariya RL. A brief review on solid lipid nanoparticles: part and parcel of contemporary drug delivery systems. RSC Adv. 2020;10:26777–91.
21. Ehrenfeld M, Tincani A, Andreoli L, Cattalini M, Greenbaum A, Kanduc D, Alijotas-Reig J, Zinserling V, Semenova N, Amital H, Shoenfeld Y. Covid-19 and autoimmunity. Autoimmun Rev. 2020;19(8):102597. https://doi.org/10.1016/j.autrev.2020.102597.
22. Elezkurtaj S, Greuel S, Ihlow J, Michaelis EG, Bischoff P, Kunze CA, Sinn BV, Gerhold M, Hauptmann K, Ingold-Heppner B, Miller F, Herbst H, Corman VM, Martin H, Radbruch H, Heppner FL, Horst D. Causes of death and comorbidities in hospitalized patients with COVID-19. Sci Rep. 2021;11(1):4263. https://doi.org/10.1038/s41598-021-82862-5.
23. El-Gizawy SA, El-Maghraby GM, Hedaya AA. Formulation of acyclovir-loaded solid lipid nanoparticles: 2. Brain targeting and pharmacokinetic study. Pharm Dev Technol. 2019;24(10):1299–307. https://doi.org/10.1080/10837450.2019.1667386.
24. Fehr AR, Perlman S. Coronaviruses: an overview of their replication and pathogenesis. Methods Mol Biol. 2015;1282:1–23. https://doi.org/10.1007/978-1-4939-2438-7_1.
25. Frediansyah A, Nainu F, Dhama K, Mudatsir M, Harapan H. Remdesivir and its antiviral activity against COVID-19: a systematic review. Clin Epidemiol Glob Health. 2021;9:123–7. https://doi.org/10.1016/j.cegh.2020.07.011.
26. Gallo O, Locatello LG, Mazzoni A, Novelli L, Annunziato F. The central role of the nasal microenvironment in the transmission, modulation, and clinical progression of SARS-CoV-2 infection. Mucosal Immunol. 2021;14(2):305–16. https://doi.org/10.1038/s41385-020-00359-2.

27. Gao SJ, Guo H, Luo G. Omicron variant (B.1.1.529) of SARS-CoV-2, a global urgent public health alert! J Med Virol. 2021; https://doi.org/10.1002/jmv.27491.
28. Garg B, Beg S, Kaur R, Kumar R, Katare OP, Singh B. Long-chain triglycerides-based self-nanoemulsifying oily formulations (SNEOFs) of darunavir with improved lymphatic targeting potential. J Drug Target. 2018;26(3):252–66. https://doi.org/10.1080/1061186X.2017.1365875.
29. Garg B, Katare OP, Beg S, Lohan S, Singh B. Systematic development of solid self-nanoemulsifying oily formulations (S-SNEOFs) for enhancing the oral bioavailability and intestinal lymphatic uptake of lopinavir. Colloids Surf B Biointerfaces. 2016;141:611–22. https://doi.org/10.1016/j.colsurfb.2016.02.012.
30. Gupta A, Gonzalez-Rojas Y, Juarez E, Crespo Casal M, Moya J, Falci DR, Sarkis E, Solis J, Zheng H, Scott N, Cathcart AL, Hebner CM, Sager J, Mogalian E, Tipple C, Peppercorn A, Alexander E, Pang PS, Free A, Brinson C, Aldinger M, Shapiro AE, COMET-ICE Investigators. Early treatment for Covid-19 with SARS-CoV-2 neutralizing antibody Sotrovimab. N Engl J Med. 2021;385(21):1941–50. https://doi.org/10.1056/NEJMoa2107934.
31. Hache G, Rolain JM, Gautret P, Deharo JC, Brouqui P, Raoult D, Honoré S. Combination of Hydroxychloroquine plus Azithromycin as potential treatment for COVID-19 patients: safety profile, drug interactions, and management of toxicity. Microb Drug Resist. 2021;27(3):281–90. https://doi.org/10.1089/mdr.2020.0232.
32. Hoffmann M, Kleine-Weber H, Schroeder S, Krüger N, Herrler T, Erichsen S, Schiergens TS, Herrler G, Wu NH, Nitsche A, Müller MA, Drosten C, Pöhlmann S. SARS-CoV-2 cell entry depends on ACE2 and TMPRSS2 and is blocked by a clinically proven protease inhibitor. Cell. 2020;181(2):271–280.e8. https://doi.org/10.1016/j.cell.2020.02.052.
33. Hui DS, Azhar IE, Madani TA, Ntoumi F, Kock R, Dar O, Ippolito G, Mchugh TD, Memish ZA, Drosten C, Zumla A, Petersen E. The continuing 2019-nCoV epidemic threat of novel coronaviruses to global health - the latest 2019 novel coronavirus outbreak in Wuhan, China. Int J Infect Dis. 2020;91:264–6. https://doi.org/10.1016/j.ijid.2020.01.009.
34. Inugala S, Eedara BB, Sunkavalli S, Dhurke R, Kandadi P, Jukanti R, Bandari S. Solid self-nanoemulsifying drug delivery system (S-SNEDDS) of darunavir for improved dissolution and oral bioavailability: in vitro and in vivo evaluation. Eur J Pharm Sci. 2015;74:1–10. https://doi.org/10.1016/j.ejps.2015.03.024.
35. Jain S, Sharma JM, Jain AK, Mahajan RR. Surface-stabilized lopinavir nanoparticles enhance oral bioavailability without coadministration of ritonavir. Nanomedicine (Lond). 2013;8(10):1639–55. https://doi.org/10.2217/nnm.12.181.
36. Jaiswal M, Dudhe R, Sharma PK. Nanoemulsion: an advanced mode of drug delivery system. 3. Biotech. 2015;5(2):123–7. https://doi.org/10.1007/s13205-014-0214-0.
37. Javan F, Vatanara A, Azadmanesh K, Nabi-Meibodi M, Shakouri M. Encapsulation of ritonavir in solid lipid nanoparticles: in-vitro anti-HIV-1 activity using lentiviral particles. J Pharm Pharmacol. 2017;69(8):1002–9. https://doi.org/10.1111/jphp.12737.
38. Jøraholmen MW, Basnet P, Acharya G, Škalko-Basnet N. PEGylated liposomes for topical vaginal therapy improve delivery of interferon alpha. Eur J Pharm Biopharm. 2017;113:132–9. https://doi.org/10.1016/j.ejpb.2016.12.029.
39. Joshi G, Kumar A, Sawant K. Bioavailability enhancement, Caco-2 cells uptake and intestinal transport of orally administered lopinavir-loaded PLGA nanoparticles. Drug Deliv. 2016;23(9):3492–504. https://doi.org/10.1080/10717544.2016.1199605.
40. Joshi S, Parkar J, Ansari A, Vora A, Talwar D, Tiwaskar M, Patil S, Barkate H. Role of favipiravir in the treatment of COVID-19. Int J Infect Dis. 2021;102:501–8. https://doi.org/10.1016/j.ijid.2020.10.069.
41. Kamble RN, Mehta PP, Kumar A. Efavirenz self-Nano-emulsifying drug delivery system: in vitro and in vivo evaluation. AAPS PharmSciTech. 2016;17(5):1240–7. https://doi.org/10.1208/s12249-015-0446-2.
42. Kanji JN, Zelyas N, MacDonald C, Pabbaraju K, Khan MN, Prasad A, Hu J, Diggle M, Berenger BM, Tipples G. False negative rate of COVID-19 PCR testing: a discordant testing analysis. Virol J. 2021;18:13. https://doi.org/10.1186/s12985-021-01489-0.

43. Kondel R, Shafiq N, Kaur IP, Singh MP, Pandey AK, Ratho RK, Malhotra S. Effect of Acyclovir solid lipid nanoparticles for the treatment of herpes simplex virus (HSV) infection in an animal model of HSV-1 infection. Pharm Nanotechnol. 2019;7(5):389–403. https://doi.org/10.2174/2211738507666190829161737.
44. Kucirka LM, Lauer SA, Laeyendecker O, Boon D, Lessler J. Variation in false-negative rate of reverse transcriptase polymerase chain reaction-based SARS-CoV-2 tests by time since exposure. Ann Intern Med. 2020;173(4):262–7. https://doi.org/10.7326/M20-1495.
45. LaBauve AE, Rinker TE, Noureddine A, Serda RE, Howe JY, Sherman MB, Rasley A, Brinker JC, Sasaki DY, Negrete OA. Lipid-coated mesoporous silica nanoparticles for the delivery of the ML336 antiviral to inhibit encephalitic alphavirus infection. Sci Rep. 2018;8:13990. https://doi.org/10.1038/s41598-018-32033-w.
46. Li Q, Du YZ, Yuan H, Zhang XG, Miao J, Cui FD, Hu FQ. Synthesis of lamivudine stearate and antiviral activity of stearic acid-g-chitosan oligosaccharide polymeric micelles delivery system. Eur J Pharm Sci. 2010;41(3–4):498–507. https://doi.org/10.1016/j.ejps.2010.08.004.
47. Li W, Moore M, Vasilieva N, Sui J, Wong SK, Berne MA, Somasundaran M, Sullivan JL, Luzuriaga K, Greenough TC, Choe H, Farzan M. Angiotensin-converting enzyme 2 is a functional receptor for the SARS coronavirus. Nature. 2003;426:450–4. https://doi.org/10.1038/nature02145.
48. Liu DX, Fung TS, Chong KK, Shukla A, Hilgenfeld R. Accessory proteins of SARS-CoV and other coronaviruses. Antivir Res. 2014;109:97–109. https://doi.org/10.1016/j.antiviral.2014.06.013.
49. Liu Y, Rocklöv J. The reproductive number of the Delta variant of SARS-CoV-2 is far higher compared to the ancestral SARS-CoV-2 virus. J Travel Med. 2021;28(7):taab124. https://doi.org/10.1093/jtm/taab124.
50. Liu YC, Kuo RL, Shih SR. COVID-19: the first documented coronavirus pandemic in history. Biom J. 2020;43(4):328–33. https://doi.org/10.1016/j.bj.2020.04.007.
51. Loutfy SA, Elberry MH, Farroh KY, Mohamed HT, Mohamed AA, Mohamed EB, Faraag AHI, Mousa SA. Antiviral activity of chitosan nanoparticles encapsulating curcumin against Hepatitis C virus genotype 4a in human Hepatoma cell lines. Int J Nanomedicine. 2020;15:2699–715. https://doi.org/10.2147/IJN.S241702. Erratum in: Int J Nanomedicine. 2021;16:1927
52. Ma F, Nan K, Lee S, Beadle JR, Hou H, Freeman WR, Hostetler KY, Cheng L. Micelle formulation of hexadecyloxypropyl-cidofovir (HDP-CDV) as an intravitreal long-lasting delivery system. Eur J Pharm Biopharm. 2015;89:271–9. https://doi.org/10.1016/j.ejpb.2014.12.010.
53. Mahendra M, Nuchin A, Kumar R, Shreedhar S, Mahesh PA. Predictors of mortality in patients with severe COVID-19 pneumonia - a retrospective study. Adv Respir Med. 2021;89(2):135–44. https://doi.org/10.5603/ARM.a2021.0036.
54. Mason RJ. Pathogenesis of COVID-19 from a cell biology perspective. Eur Respir J. 2020;55(4):2000607. https://doi.org/10.1183/13993003.00607-2020.
55. Masters PS. The molecular biology of coronaviruses. Adv Virus Res. 2006;66:193–292. https://doi.org/10.1016/S0065-3527(06)66005-3.
56. Nabila N, Suada NK, Denis D, Yohan B, Adi AC, Veterini AS, Anindya AL, Sasmono RT, Rachmawati H. Antiviral action of curcumin encapsulated in Nanoemulsion against four serotypes of Dengue virus. Pharm Nanotechnol. 2020;8(1):54–62. https://doi.org/10.2174/2211738507666191210163408.
57. Nutho B, Mahalapbutr P, Hengphasatporn K, Pattaranggoon NC, Simanon N, Shigeta Y, Hannongbua S, Rungrotmongkol T. Why are Lopinavir and Ritonavir effective against the newly emerged Coronavirus 2019? Atomistic insights into the inhibitory mechanisms. Biochemistry. 2020;59(18):1769–79. https://doi.org/10.1021/acs.biochem.0c00160.
58. Oran DP, Topol EJ. Prevalence of asymptomatic SARS-CoV-2 infection: A narrative review. Ann Intern Med. 2020;173(5):362–7. https://doi.org/10.7326/M20-3012.

59. Ouassou H, Kharchoufa L, Bouhrim M, Daoudi NE, Imtara H, Bencheikh N, Elbouzidi A, Bnouham M. The pathogenesis of Coronavirus disease 2019 (COVID-19): evaluation and prevention. J Immunol Res. 2020:1357983. https://doi.org/10.1155/2020/1357983.
60. Pacho MN, Pugni EN, Sierra JBD, Morell ML, Sepúlveda CS, Damonte EB, García CC, D'Accorso NB. Antiviral activity against Zika virus of a new formulation of curcumin in poly lactic-co-glycolic acid nanoparticles. J Pharm Pharmacol. 2021;73(3):357–65. https://doi.org/10.1093/jpp/rgaa045.
61. Padhi AK, Seal A, Khan JM, Ahamed M, Tripathi T. Unraveling the mechanism of arbidol binding and inhibition of SARS-CoV-2: insights from atomistic simulations. Eur J Pharmacol. 2021;894:173836. https://doi.org/10.1016/j.ejphar.2020.173836.
62. Patel G, Shelat P, Lalwani A. Statistical modeling, optimization and characterization of solid self-nanoemulsifying drug delivery system of lopinavir using design of experiment. Drug Deliv. 2016;23(8):3027–42. https://doi.org/10.3109/10717544.2016.1141260.
63. Patel GM, Shelat PK, Lalwani AN. QbD based development of proliposome of lopinavir for improved oral bioavailability. Eur J Pharm Sci. 2017;108:50–61. https://doi.org/10.1016/j.ejps.2016.08.057.
64. Paules CI, Marston HD, Fauci AS. Coronavirus infections-more than just the common cold. JAMA. 2020;323(8):707–8. https://doi.org/10.1001/jama.2020.0757.
65. Petrone L, Petruccioli E, Alonzi T, Vanini V, Cuzzi G, Najafi Fard S, Castilletti C, Palmieri F, Gualano G, Vittozzi P, Nicastri E, Lepore L, Grifoni A, Antinori A, Vergori A, Ippolito G, Cantini F, Goletti D. In-vitro evaluation of the immunomodulatory effects of Baricitinib: implication for COVID-19 therapy. J Infect. 2021;82(4):58–66. https://doi.org/10.1016/j.jinf.2021.02.023.
66. Pons S, Fodil S, Azoulay E, Zafrani L. The vascular endothelium: the cornerstone of organ dysfunction in severe SARS-CoV-2 infection. Crit Care. 2020;24:353.
67. Pradhan D, Biswasroy P, Goyal A, Ghosh G, Rath G. Recent advancement in nanotechnology-based drug delivery system against viral infections. AAPS PharmSciTech. 2021;22(1):47. https://doi.org/10.1208/s12249-020-01908-5.
68. Ravi PR, Vats R, Dalal V, Gadekar N, N A. Design, optimization and evaluation of poly-ε-caprolactone (PCL) based polymeric nanoparticles for oral delivery of lopinavir. Drug Dev Ind Pharm. 2015;41(1):131–40. https://doi.org/10.3109/03639045.2013.850710.
69. Ravi PR, Vats R. Comparative pharmacokinetic evaluation of lopinavir and lopinavir-loaded solid lipid nanoparticles in hepatic impaired rat model. J Pharm Pharmacol. 2017;69(7):823–33. https://doi.org/10.1111/jphp.12716.
70. Samavati L, Uhal BD. ACE2, much more than just a receptor for SARS-COV-2. Front Cell Infect Microbiol. 2020;10:317. https://doi.org/10.3389/fcimb.2020.00317.
71. Sawdon AJ, Peng CA. Polymeric micelles for acyclovir drug delivery. Colloids Surf B Biointerfaces. 2014;122:738–45. https://doi.org/10.1016/j.colsurfb.2014.08.011.
72. Schwarz JC, Klang V, Karall S, Mahrhauser D, Resch GP, Valenta C. Optimisation of multiple W/O/W nanoemulsions for dermal delivery of aciclovir. Int J Pharm. 2012;435(1):69–75. https://doi.org/10.1016/j.ijpharm.2011.11.038.
73. Senapati PC, Sahoo SK, Sahu AN. Mixed surfactant based (SNEDDS) self-nanoemulsifying drug delivery system presenting efavirenz for enhancement of oral bioavailability. Biomed Pharmacother. 2016;80:42–51. https://doi.org/10.1016/j.biopha.2016.02.039.
74. Seremeta KP, Chiappetta DA, Sosnik A. Poly(ε-caprolactone), Eudragit® RS 100 and poly(ε-caprolactone)/Eudragit® RS 100 blend submicron particles for the sustained release of the antiretroviral efavirenz. Colloids Surf B Biointerfaces. 2013;102:441–9. https://doi.org/10.1016/j.colsurfb.2012.06.038.
75. Seyfoddin A, Sherwin T, Patel DV, McGhee CN, Rupenthal ID, Taylor JA, Al-Kassas R. Ex vivo and in vivo evaluation of Chitosan coated nanostructured lipid carriers for ocular delivery of Acyclovir. Curr Drug Deliv. 2016;13(6):923–34. https://doi.org/10.2174/1567201813666151116142752.

76. Sharma P, Chawla A, Arora S, Pawar P. Novel drug delivery approaches on antiviral and antiretroviral agents. J Adv Pharm Technol Res. 2012;3(3):147–59. https://doi.org/10.4103/2231-4040.101007.
77. Shen K, Yang Y, Wang T, Zhao D, Jiang Y, Jin R, Zheng Y, Xu B, Xie Z, Lin L, Shang Y, Lu X, Shu S, Bai Y, Deng J, Lu M, Ye L, Wang X, Wang Y, Gao L, China National Clinical Research Center for Respiratory Diseases; National Center for Children's Health, Beijing, China; Group of Respirology, Chinese Pediatric Society, Chinese Medical Association; Chinese Medical Doctor Association Committee on Respirology Pediatrics; China Medicine Education Association Committee on Pediatrics; Chinese Research Hospital Association Committee on Pediatrics; Chinese Non-government Medical Institutions Association Committee on Pediatrics; China Association of Traditional Chinese Medicine, Committee on Children's Health and Medicine Research; China News of Drug Information Association, Committee on Children's Safety Medication; Global Pediatric Pulmonology Alliance. Diagnosis, treatment, and prevention of 2019 novel Coronavirus infection in children: experts' consensus statement. World J Pediatr. 2020;16(3):223–31. https://doi.org/10.1007/s12519-020-00343-7.
78. Shereen MA, Khan S, Kazmi A, Bashir N, Siddique R. COVID-19 infection: origin, transmission, and characteristics of human coronaviruses. J Adv Res. 2020;24:91–8. https://doi.org/10.1016/j.jare.2020.03.005.
79. Shi LL, Cao Y, Zhu XY, Cui JH, Cao QR. Optimization of process variables of zanamivir-loaded solid lipid nanoparticles and the prediction of their cellular transport in Caco-2 cell model. Int J Pharm. 2015;478(1):60–9. https://doi.org/10.1016/j.ijpharm.2014.11.017.
80. Song Z, Xu Y, Bao L, Zhang L, Yu P, Qu Y, Zhu H, Zhao W, Han Y, Qin C. From SARS to MERS, thrusting Coronaviruses into the spotlight. Viruses. 2019;11(1):59. https://doi.org/10.3390/v11010059.
81. Strasfeld L, Chou S. Antiviral drug resistance: mechanisms and clinical implications. Infect Dis Clin North Am 2010;24(2):413–37 doi: 101016/jidc201001001 Corrected and republished in: Infect Dis Clin North Am. 2010;24(3):809–33.
82. Suwannoi P, Chomnawang M, Sarisuta N, Reichl S, Müller-Goymann CC. Development of Acyclovir-loaded albumin nanoparticles and improvement of Acyclovir permeation across human corneal epithelial T cells. J Ocul Pharmacol Ther. 2017;33(10):743–52. https://doi.org/10.1089/jop.2017.0057.
83. Tshweu L, Katata L, Kalombo L, Chiappetta DA, Hocht C, Sosnik A, Swai H. Enhanced oral bioavailability of the antiretroviral efavirenz encapsulated in poly(epsilon-caprolactone) nanoparticles by a spray-drying method. Nanomedicine (Lond). 2014;9(12):1821–33. https://doi.org/10.2217/nnm.13.167.
84. Turk C, Turk S, Malkan UY, Haznedaroglu IC. Three critical clinicobiological phases of the human SARS-associated coronavirus infections. Eur Rev Med Pharmacol Sci. 2020;24(16):8606–20. https://doi.org/10.26355/eurrev_202008_22660.
85. Valyaeva AA, Zharikova AA, Kasianov AS, Vassetzky YS, Sheval EV. Expression of SARS-CoV-2 entry factors in lung epithelial stem cells and its potential implications for COVID-19. Sci Rep. 2020;10:17772. https://doi.org/10.1038/s41598-020-74598-5.
86. Vartak R, Patil SM, Saraswat A, Patki M, Kunda NK, Patel K. Aerosolized nanoliposomal carrier of remdesivir: an effective alternative for COVID-19 treatment in vitro. Nanomedicine. 2021;16(14):1187–202. https://doi.org/10.2217/nnm-2020-0475.
87. Wang CZ, Li WJ, Tao R, Ye JZ, Zhang HY. Antiviral activity of a nanoemulsion of polyprenols from ginkgo leaves against influenza A H3N2 and hepatitis B virus in vitro. Molecules. 2015;20(3):5137–51. https://doi.org/10.3390/molecules20035137.
88. Wang SX, Michiels J, Ariën KK, New R, Vanham G, Roitt I. Inhibition of HIV virus by neutralizing Vhh attached to dual functional liposomes encapsulating Dapivirine. Nanoscale Res Lett. 2016;11(1):350. https://doi.org/10.1186/s11671-016-1558-7.
89. Witika BA, Walker RB. Development, manufacture and characterization of niosomes for the delivery for nevirapine. Pharmazie. 2019;74(2):91–6. https://doi.org/10.1691/ph.2019.8168.

90. Wong SH, Lui RN, Sung JJ. Covid-19 and the digestive system. J Gastroenterol Hepatol. 2020;35(5):744–8. https://doi.org/10.1111/jgh.15047.
91. Wu J, Wang H, Li B. Structure-aided ACEI-capped remdesivir-loaded novel PLGA nanoparticles: toward a computational simulation design for anti-SARS-CoV-2 therapy. Phys Chem Chem Phys. 2020;22(48):28434–9. https://doi.org/10.1039/d0cp04389c.
92. Yang P, Wang X. COVID-19: a new challenge for human beings. Cell Mol Immunol. 2020;17:555–7. https://doi.org/10.1038/s41423-020-0407-x.
93. Zaim S, Chong JH, Sankaranarayanan V, Harky A. COVID-19 and multiorgan response. Curr Probl Cardiol. 2020;45(8):100618. https://doi.org/10.1016/j.cpcardiol.2020.100618.
94. Zhang C, Shi L, Wang FS. Liver injury in COVID-19: management and challenges. Lancet Gastroenterol Hepatol. 2020;5(5):428–30. https://doi.org/10.1016/S2468-1253(20)30057-1.
95. Zhou P, Yang XL, Wang XG, Hu B, Zhang L, Zhang W, Si HR, Zhu Y, Li B, Huang CL, Chen HD, Chen J, Luo Y, Guo H, Jiang RD, Liu MQ, Chen Y, Shen XR, Wang X, Zheng XS, Zhao K, Chen QJ, Deng F, Liu LL, Yan B, Zhan FX, Wang YY, Xiao GF, Shi ZL. A pneumonia outbreak associated with a new coronavirus of probable bat origin. Nature. 2020;579(7798):270–3. https://doi.org/10.1038/s41586-020-2012-7.
96. Zhou Y, Jiang X, Tong T, Fang L, Wu Y, Liang J, Xiao S. High antiviral activity of mercaptoethane sulfonate functionalized Te/BSA nanostars against arterivirus and coronavirus. SC Adv. 2020;10(24):14161–9.

Exploring the Link Between Malaria and COVID-19

Orhan E. Arslan

Abstract Several observations, some supported with statistical and clinical data, have been put forth regarding the prevalence of COVID-19 in *malaria*-endemic countries. There is a remarkable difference between the industrialized nations and developing countries, particularly *malaria*-endemic African countries, in the incidence and mortality rate of COVID-19. Many factors enter into the equation when these observations are analyzed including cultural norms, testing availability, preparedness tools, and outreach programs. The restriction placed on close contacts in COVID-19-stricken areas rendered the *malaria* control measures, such as indoor-spraying of insecticide, use of insecticide-treated nets (ITNs), and chemoprevention, less effective. Population demographics and age structure may play a role in widening the gap in morbidity and mortality rate of COVID-19 between economically advanced countries and countries with limited per capita income.

As the discussion in this article will show, the difference in the prevalence of COVID-19 among populations is correlated uneven propagation of the ACEI/D and the ACE2 (C1173T substitution) polymorphisms. The similarity of presentations between *malaria* and COVID-19 and the possibility of simultaneous dual infections by both pathogens can pose a diagnostic challenge to health practitioners. This study intended to examine the link between *malaria* endemicity and reported reduction in COVID-19 mortality.

Keywords *Malaria* · COVID-19 · Endemic · Syndemic · Pandemic · SARS-CoV-2 · Hydroxychloroquine · Chloroquine · ARDS

O. E. Arslan (✉)
Department of Cellular Biology &Pharmacology, Florida International University Herbert Wertheim College of Medicine, Miami, FL, USA

R. Shegokar, Y. Pathak (eds.), *Viral Drug Delivery Systems*,
https://doi.org/10.1007/978-3-031-20537-8_14

1 Background

The life-threatening condition that results from *malaria* infection is caused by the parasitic infection of the genus *Plasmodium*. The rapid transmission of the disease by *Anopheles* mosquitoes accounts for the persistent global health threat it poses. With the identification of a new coronavirus in 2019 (COVID-19), and the epidemic proportion of respiratory infection associated with this virus that caused global health emergency, the virus is termed as "severe acute respiratory syndrome coronavirus 2 (SARS-CoV-2)." Later the infection became a worldwide pandemic.

The commonly shared symptoms between *malaria* and COVID-19 have created challenges to physicians and other health practitioners in the diagnosis of *malaria* versus COVID-19. The appearance of COVID-19 and the limitations imposed on the population and health delivery system logistically hindered prevention and therapeutic methodologies used to counter malarial infection, including treatment by insecticide of bed nets and chemoprevention of the disease.

Also, minimizing risks to healthcare workers during the COVID-19 pandemic while addressing public health issues associated with *malaria* endemics must be achieved to maintain key healthcare delivery services and essential public health functions. To preserve continuity of health services, an active social support by the community is needed including prevention, diagnosis, and treatment of *malaria*. This will have additional critical role in the response to COVID-19 pandemics, but the advantage-disadvantage analysis must be carefully considered when determining the optimum delivery methods for healthcare.

2 Morbidity and Mortality of COVID-19

COVID-19 is a contagious viral disease caused by a SARS-CoV-2. The outbreak began in China in 2019, and it was proclaimed a pandemic by the World Health Organization (WHO). COVID-19 is transmitted through air by infected droplets from breathing, sneezing, coughing, and talking of infected individuals. Patients' manifestations may vary from fever or chills, headaches, cough, fatigue, dyspnea, ageusia, anosmia smell, back pain, and sore throat. It has been strongly recommended that the public adhere to certain standards, such as use of face masks, social distancing, isolation, frequent hand washing, and travel restrictions.

When the morbidity and mortality rates associated with COVID-19 infection are examined, a clear demarcation appears between industrial nations and that of developing countries. Reports indicate that the population in Western countries with higher per capita income and social standards experiences higher rates of COVID-19-related mortality. This is in contrast to low-income population in developing countries where the rate is significantly lower. The latter finding is associated with age and life expectancy. The markedly large elderly population in developed nations, which is prone to developing malignancy, hypertension, and diabetes, poses risk of death due to COVID-19 infection [1].

The variation in the COVID-19 morbidity and mortality seems in direct correlation with the age of the affected population. This fact becomes more evident when median youngest age of European inhabitants is compared to that of Africa. This comparison demonstrates a remarkable age difference of two decades in favor of African communities, an interesting finding that may shed the light on the perceived or actual gap between these diverse groups of individuals [2].

Age may not be the only factor in this regard considering other elements, such as limited space, overcrowding, work environment, issue of literacy, unusual lifestyle, culture, and belief systems, as well as socioeconomic factors. Recent reports revealed that the age-related mortality was fourfold more in European countries and North American continent when compared to the age group in Africa. Similarly, age-related deaths in Africa, the second largest and most populated continent, appears to be far less, nearly twofold, when compared to similar age populations in South America and also Asia.

To overcome shortcomings of targeted examination and inspection, accurately assess the prevalence of SARS-CoV-2 infections in urban, suburban, and rural settings within a region, a well-prepared study must be put into action to identify asymptomatic or mildly symptomatic COVID-19 patients.

Studies have shown that rate of spread of COVID-19 is slower in developing and *malaria*-endemic societies. This low prevalence led to the assumption that hydroxychloroquine, used in the treatment of *malaria*, may have a protective immune effect on the spread of COVID-19 and its manifestations. Several research studies dismissed the possible positive impact of this drug; others maintained that low rate of COVID mortality is associated directly or indirectly with *malaria* endemics, a fact supported by the disparity in mortality rates between European and African nations [3].

Several disease control and prevention organizations in developing countries embarked on serious effort to ascertain the validity of argument that drugs used in treatment and prevention of *malaria* endemic, such as hydroxychloroquine (HCQ) and chloroquine (CQ), antibodies, and angiotensin-converting enzyme 2 (ACE2,) have a role in reducing COVID-19 in *malaria*-endemic populations. These efforts fell short of achieving lasting changes due to the instability that COVID-19 caused in basic public health delivery system and the ability to respond to crisis in a timely manner, and thus the low mortality rate as envisioned in the past may not be feasible in the future [4].

3 Malaria vs. COVID-19 Presentations

As indicated above, patients with COVID-19 may be asymptomatic and mildly symptomatic or exhibit severe manifestations. They are commonly febrile, dyspneic, and fatigued with dry or productive cough. Several other presentations may be exhibited by the patients including myalgia, headache, joint pain, vomiting, and nausea [5].

It is conceivable to have, as studies have shown, cross immunity, where one *malaria* pathogen provides immunity against another pathogen [6, 7].

Some of these signs and symptoms are also shared by *malaria* patients, such as fever, headache, fatigue, joint paint, and vomiting, but in addition to the above they also present chills, sweating, and diarrhea [8].

Because of the shared presentations between these two clinical conditions, a greater chance of misdiagnosis can occur. Additionally, multi-organ failure, acute respiratory distress syndrome (ARDS), and septic shock can develop as complications associated with both COVID-19 and *malaria*. Although symptom-based screening of COVID-19 patients seems a valid approach, it fails to identify one half of COVID-19 cases even in areas with robust health systems [9].

The inclination to evaluate a patient with fever is skewed toward COVID-19 rather than *malaria*. Thus, a febrile patient may be assessed for COVID-19 but not *malaria*, opening the door for a deadly *malaria* infection.

4 Syndemic of *Malaria* and COVID-19

This phenomenon refers to a combination of two or more concomitant or sequential disease clusters in a population with biological interfaces that leads to worsening disease processes and prognosis. The possibility that a patient may have coinfection with both COVID-19 and *malaria* (syndemic) will add enormous complications to treatment methodologies and limit the extent of success in eradicating these diseases.

Malaria or COVID-19 may coexist with one or more infections that produce far more complications than either infection can cause. HIV *Plasmodium falciparum* syndemic presents with combined increase in HIV viral load and acute manifestations of *malaria* that are more frequent and severe. *Malaria* may induce B-lymphocyte replication and increase in the viral load of Epstein–Barr virus (EBV) when coinfection occurs in the same patient.

When it comes to COVID-19, the onset of presentation may vary between 4 and 14 days [10]. However, an acute deterioration of patient's condition (cytokine storm) may occur within a week through a surge of cytokines and other mediators [11]. This surge, which involves a drastic increase in the levels of tumor necrosis factor (TNF) and interleukins (IL), is responsible for acute respiratory distress syndrome (ARDS) [12].

Reports indicate that manifestations may be critical to advance to multiple organ failure and septic shock [13].

5 Pathogenesis of *Malaria* and COVID-19

In *malaria*, the merozoites, which are uninuclear cells that originate from ruptured multinucleate Schizonts of the *Plasmodium*, gain access to the blood and invade the erythrocytes through a coordinated approach and eventually mature into schizonts. The latter disintegrates, freeing new generations of merozoites into the systemic circulation, accompanied by the release of TNF, interleukin, and interferon-gamma, which are responsible for the common manifestations observed in *malaria* patients [14, 15]. Studies have shown that interferons produced by lymphocytes in response to *malaria* pathogen appear to be effective against COVID-19 in vitro and in vivo [16, 17].

Thus, a crucial balance between pro-inflammatory Th1 (Type 1 helper cell) response, which includes tumor necrosis factor (TNF), interferon-gamma, interleukin-6 and interleukin-12, and that of the anti-inflammatory Th2 (Type 2 Helper cell) response, which primarily encompasses interleukin-4 and interleukin-10, is achieved. When this critical balance shifts in favor of the Th2 response, severe manifestations of *malaria* will be experienced by the patient leading to poor prognosis [18].

Similar events have been proposed for COVID-19 and that a surge in pro-inflammatory responses is responsible for the acute presentations. Respiratory manifestations associated with *P. falciparum malaria* can be mild or progress to acute respiratory distress syndrome (ARDS) in adults, particularly in pregnant women, but rarely in young children [19]. Research data identify several key factors in the development of acute respiratory distress syndrome (ARDS) experienced by children and adult populations with severe form of *P. falciparum malaria* including metabolic disorder, adhesion of infected erythrocytes to the lumen of the pulmonary vessels, presence of pneumonia pathogens, and hemopoietic disorder, such as anemia [20].

Alveolar damage in ARDS is related to the cytokines-induced endothelial injury and subsequent increase in the permeability of capillaries. These pathologic changes are seen individually in both SARS-CoV-2 and *Plasmodium* species infections; however, the worsening of patient's condition with serious complications and sub-optimal prognosis are more evident when coinfection occurs. A differentiating factor here is that ARDS manifestations continue well beyond treatment regiments [20].

Both COVID-19 and *malaria* infections cause a procoagulant state. SARS-CoV-2 produces a procoagulant state by impeding the function of the endothelium, activating the Toll-like (pattern recognition) receptors, and increasing von Willebrand factor levels. Other findings are also detected, such as elevated levels of D-dimer, fibrin degradation, and lengthening of prothrombin time, and the presence of latter indicators is usually correlated with a poor prognosis [21].

Poor prognosis is also associated with the hypercoagulable state induced by COVID-19, in which venous and arterial thrombosis including pulmonary embolism may occur [22, 23].

Overactivation of coagulation cascades in COVID-19 causes platelet lysis, and the development of thrombocytopenia [24] further increases the likelihood for developing disseminated intravascular coagulation (DIC) particularly in severe cases leading to high mortality [25, 26]. In *malaria*, the extent of activation of coagulation cascade is proportionate to disease progression and activation of IL-6 and TNF [27] and usually produces microthrombi and less frequently macrothrombi in the pulmonary trunk and cerebral veins [28, 29].

Coagulopathy becomes more pronounced, and disease manifestations intensify and assume more severe form when SARS-CoV-2-*Plasmodium* spp. coinfection occurs. Thus, steps must be taken by healthcare workers, particularly in *malaria*-endemic countries, to timely explore and assess the potential of coinfections by COVID-19/*malaria* and conduct test/screening accordingly to lessen the impact of these diseases.

The angiotensin-converting enzyme-2 (ACE-2), a type I transmembrane aminopeptidase, has been identified as the receptor for the SARS-CoV-2 viral entry into the host cells. Thus, it became the center of renewed attention in the development of antiviral medications against SARS-CoV-2. ACE2 was first discovered as an ACE homologue sharing homology with angiotensin-converting enzyme 1 (ACE1) and expressed on the apical surfaces of several pulmonary structures, such as type II alveolar cells as well as extrapulmonary cell types, including cardiac, vascular, renal, gastrointestinal, and endothelial cells [1]. When the latter cells are infected by SARS-COV-2, systemic vasculitis, disseminated intravascular coagulation (DIC), and thromboembolism may ensue causing "cytokine storm." Corollary to this fact, blockers of angiotensin II receptors (ARBs) are widely utilized in the treatment of hypertension and related cardiovascular diseases.

In addition to the membrane-bound form, ACE-2 also exhibit soluble forms in the plasma and urine. ACE-2 receptors produce the lung-protective Ang- [1–7] from angiotensin II (ANG II) and transform angiotensin I to angiotensin- [1–9].

ANG II—the substrate for ACE2—begins to accumulate when the activity of ACE-2 receptor undergoes downregulation. Accumulated ANG II produces neutrophilia that increases vascular permeability leading to pulmonary edema and eventually ARDS [30]. Studies have demonstrated that ANG II causes a reduction in the accumulated sporozoites in the salivary glands of mosquitoes by directly disrupting the membrane of the parasite [31]. However, other reports have demonstrated a possible protective role for ANG II in *malaria*.

Further, the frequency of intron variants rs72717040 and rs17400517 of FCGR2A is associated with the prevalence of COVID-19. FCGR2A encodes a member of immunoglobulin family located on the surface of macrophages and neutrophils and involved in the process of phagocytosis and clearing of immune complexes and has also been associated with severe *malaria* that occurred in Gambia and in Kenya [32, 33].

The genetic deletion/insertion (D/I) polymorphism in intron 16 of ACE-1 enzyme is variable [34] and directly correlated with the levels of freely circulating and tissue-bound forms of ACE. As such, diminished expression of ACE2 receptor occurs when the D is the dominant allele. The decreased expression of ACE2 receptor may play a protective function against COVID-19 [34]. It has also been reported that the mild form of *malaria* is associated with an increase in the level angiotensin II due to the presence of D-allele of ACEI/D polymorphism.

Research data indicate that the log-transformed prevalence of COVID-19 infection was seen to be inversely proportional to the ACE D allele frequency. Data indicate that individuals of African ancestry show ACEI/D polymorphism with a remarkable increase in the plasma levels of ANG II. Studies report that geographical variation of D/I polymorphism may account for the difference in COVID-19 infection prevalence [35], and COVID-19 related mortality. This is evident by the low frequency of D allele in countries where the spread of COVID-19 reached a peak level, such as China and Korea.

Another phenomenon that contributes to the variable spread of COVID-19 is the ACE-2 polymorphism (C1173T substitution). This type of polymorphism boosts the level of ANG II by lowering the ACE2 receptor expression in the presence of T allele leading to an increased in the concentration of ACE2.

A concerning polymorphism is the ACE2 polymorphism (C1173T substitution), which reduces ACE2 receptor expression in the presence of the T allele and consequently increases ANG II. Since ACE2 receptors are primarily responsible for transforming angiotensin II to Ang- [1–7], a reduced expression in ACE2 would produce an increase in angiotensin II. Despite the lack of a clear-cut answer, this type of polymorphism may serve as an additional element that potentially explains the enigma surrounding the difference in the rate of COVID-19 spread among countries.

In addition to the above, polymorphisms have been linked to the natural selection process [36]. Case in point, hypertension presents an advantage from an evolutionary point of view protecting persistently subjected populations to *malaria* infections. Epidemiological findings demonstrate that people with African ancestry have a higher incidence of hypertension contrasted to whites from *malaria*-free areas [37, 38]. Therefore, hypertension could serve as a safeguard for people from developing nervous system *malaria* [39].

6 Control Mechanisms of COVID-19 Pandemic in *Malaria* Endemic

Since exposure to *malaria* does not confer life-long immunity, control of *malaria* endemic is accomplished through public health measures and facilities, mass education, as well as effective control programs. The equitable coverage of *malaria*-stricken population requires unhindered coordination, an interrupted access and communication, and the ability to conduct examination and offer individual treatment. Unfortunately, the outbreak of COVID-19 and the many restrictions brought with it have adversely impacted the campaign against *malaria* leading to inadequate treatment and subsequent rise in morbidity and mortality rates of this condition. To alleviate the strain on core health services in *malaria* endemic, several measures may have to be implemented through massive campaign and individualized medical care. These measures include indoor insecticide spraying, establishment of insecticides-treated net, seasonal *malaria* chemoprevention, preventive treatment directed specifically toward children and infant, as well as pregnant women. The implementation of the programs must consider the goal of safeguarding the communities and health delivery personnel as well as reducing mortality. At the same time, measures must be enforced to protect individuals and community from COVID-19 [40].

7 Interferons Efficacy in *Malaria* and COVID-19 Outbreak

Research data demonstrate that the lymphocyte-generated interferons as an immune response to *malaria* infection have in vivo and in vitro neutralizing effect on coronaviruses including COVID-19 [16, 17, 41]. Plasmodium-specific antigens induce the development of IgG antibodies in *malaria*-infected patients that aim at the glycosylphosphatidylinositol (GPI) molecules, which in turn activate the leukocytes initiating the release of pro-inflammatory cytokines. This chain of events continues with the upregulation of the adhesion molecules via Toll-like receptors 2 and 4. Thus, anti-GPI antibodies may be utilized to render the toxic effects of plasmodium GPI inactive. Membrane GPs, spike GPs, and GPs that possess acetyl esterase and hemagglutination characteristics in the SARS-CoV-2 enable anti-GPI antibodies to recognize and inactivate them.

It must also be noted that suppression of GPI after initial exposure to *malaria* parasitic infection does not provide full protection against the disease, as evidenced by the frequency of infections confronted by individuals in *malaria*-endemic regions, yet the intensity, duration, and severity of the symptoms remain less than in nonimmune patients [42].

Furthermore, SARS-CoV-2 has various glycoproteins (GPs): spike and membrane GPs, as well as GPs that have acetyl esterase and hemagglutination features. These GPs could be identified and neutralized by the anti-GPI antibodies producing some degree of immunity or rendering the disease progression less severe [43].

8 Therapeutic and Preventive Roles of Hydroxychloroquine and Chloroquine

A vigorous worldwide research are being conducted on the efficacy of vaccines against the spread of COVID-19. At present, some antiviral medications have been adopted to treat COVID-19, but development of these specific drugs is a time-consuming endeavor. Hence, the general view of scientific community is to repurpose available drugs, such as antimalarial therapeutics, as they offer safety, cost-effectiveness, and accessibility, have an antiviral property for off-label use [44], and provide a quick method of treatment. Repurposing of drugs is promising for treating and reducing the symptoms of the disease, and it a fast, easy, and safe method to address the crisis, because of their previously known applications. For instance, antimalarial drugs have been repurposed because of the positive results in vitro and in vivo [45–47].

In order to develop new effective antiviral drugs that curtail the spread of COVID-19, large funding resource and years of effort are required, and the chance of success remains [48]. Further development of drugs comprises: discovery, in vitro and in vivo research studies, FDA review, and post-approval monitoring. In contrast, repurposing of drugs necessitates identification, acquisition, and clinical research, followed by FDA market monitoring [49].

In vitro and in vivo studies are still defined, effective treatments and their recovery approaches are urgently needed to overcome this pandemic disease. Moreover, developing a new effective drug takes time to examine its biotherapeutics, and this is unacceptable especially for the current difficult situation. So, repurposing of medication is needed because of its known side effects, pharmacokinetics, safety, and exact dosage; as a result, their usage will be faster with low cost of clinical trials [45–47]. However, although many antiviral and anti-inflammatory drugs have been used, no drug has been confirmed as an effective treatment (553). Besides, the development process of the new effective drug needs more than a billion dollars taking from 10 to 15 years to be produced with only a 2.01% of success rate [47].

Drug development includes five steps: discovering, in vitro studies, in vivo studies, FDA reviewing, and finally FDA post-market monitoring of the safety. However, drug repurposing follows less complicated process: compound identification, compound procurement, clinical research, and FDA safety post-market monitoring [48]. There is global pressure to discover a cheaper and faster solution for this pandemic disease. Also, there is a global agreement to use repurposed drugs as a faster solution [3, 16, 17, 21–51]. Thus, re-tasking, i.e., repurposing, enables the use of preapproved or preapproved but discontinued or investigational compounds to contagious diseases [52–58].

Through repurposing, a single drug will be able to act on several targets including new targets, and these targets correlate with a disease pathogenesis that furthers the development of new application avenues [47]. Thus, repurposing is a cost-effective, efficient, and fast treatment methodology [52, 59].

Upon analysis of the repurposing process, three orderly steps have been determined, (1) recognition of the contender drug for the identified purpose, (2) early on determination of the effectiveness of the drug in preclinical phase, and (3) efficacy in phase II clinical trials, in order to achieve the desired results in recognition of target interactions, molecular docking (models the interaction between a molecule and a target protein at a binding site), signature matching (analysis of gene expression whereby an expression profile used to identify previously systematically arranged and listed biological conditions most related to their profile), genetic association method, data source screening, and pathway mapping. Similarly, experimental approaches may be employed, such as phenotypic screening and binding essay [60].

A number of challenges face the repurposing process; some are technical while others relate to the dosage and preclinical records that render the process prohibitively costly and cumbersome. Presence of a gap between the speed with which biomedical data are collected and the ability to interpret, analyze, and integrate the data also poses an issue in data entry and incorporation. The dissimilarity and heterogeneity in the collected data, the selection of the optimal doses when the indications are variable, and the unsuitability of the drug for the new indication because of a history of toxicologic and preclinical data pose additional challenges [60]. The choice of antimalarial drugs for repurposing against SARS-CoV-2 is based on the immunomodulatory and anti-inflammatory properties, wide usage across age groups, and the record of effectiveness in therapy and prophylaxis [61, 62].

9 Chloroquine (CQ)

This drug is a quinine derivative used commonly in the treatment of *malaria*. It was recommended for the treatment of COVID-19 in China [10, 12]. Studies in vitro conducted in 1960 revealed the antiviral properties of this drug and particularly against coronavirus [63, 64]. It reduced hospitalization period and has been proven to be more efficacious in patients previously treated with *lopinavir* and *ritonavir.* This was evident in the improvement of lung detected in radiologic examination [65]. CQ evaluated on animal models and in vitro exhibited favorable results [34]. COVID-19 *patients treated with* CQ exhibited positive outcomes [66].

The rate of COVID-19 spread and associated high mortality created extraordinary public health predicaments. Absence of vaccines or antiviral medications and urgent need for medical intervention forced medical practitioners to repurpose approved drugs currently in the market to counter this disease. Potential drugs that received the US Food and Drug Administration (FDA) certification were chloroquine (CQ) and hydroxychloroquine (HCQ). In 2020, an Emergency Use Authorization was declared for HCQ in COVID-19 treatment which was later rescinded by the FDA. Clinical trials of chloroquine, hydroxychloroquine, remdesivir, lopinavir, and ritonavir were conducted by the World Health Organization (WHO) but soon discontinued. Here, we will attempt to present a critical assessment of the clinical trials and assess the efficacy of repurposed drugs. Immense efforts have been directed into the development of antiviral drugs and vaccines and that target SARS-CoV-2 [67].

Due to lack of FDA-approved drugs for COVID-19 treatment, control measures, such as isolation, social distancing, and enforcement of travel restrictions to limit the spread of the disease, affected patients received supportive medical care such as intravenous fluid infusion, extracorporeal oxygen supplementation, and mechanical ventilatory assistance [68].

Based on early clinical findings of COVID-19 therapy regiments of patients, several FDA-approved drugs have been repurposed, including *chloroquine (CQ), hydroxychloroquine (HCQ)* [69], *lopinavir* [70], *ritonavir* [71], *remdesivir* [72], *ribavirin* [73], *griffithsin* [74], *tocilizumab* [75], *sarilumab* [76], *interferon* [77], *immunoglobulins* [78], *and corticosteroids* [79], *to hinder lung injury by decreasing viral load.* Below is a discussion of the logic behind repurposing CQ and HCQ, their in vitro and in vivo antiviral impacts on coronaviruses, an insight to clinical trials on COVID-19 patients, and analysis of the process for FDA and WHO approval and disapproval of drugs for use in therapy of specific diseases.

CQ, an affordable and most widely prescribed medicine has been utilized and considered as the most widely prescribed medicine for the treatment of *malaria* throughout the past seven decades [80]. It is chemically represented as N4-(7-chloroquinolin-4-yl)-N1,N1-diethylpentane-1,4-diamine, and its pharmacological effect is not restricted to antimalarial activity but encompasses a wide range of actions as anti-inflammatory, immunomodulatory, and antiviral activities [81].

CQ's antiviral activity was discovered in vitro for the first time in the late 60s and in the subsequent years several published works on this specific property of CQ [82, 83]. The anti-COVID-19 activities of both CQ and HCQ represented in growth inhibition of coronavirus in cell culture was reported by Vincent et al. [84]. CQ also produced a significant reduction in the expression of pro-inflammatory cytokines, interferons (IFN-β and IFN-g), tumor necrosis factor (TNF-α), and interleukins (IL-6 and IL-12) [85].

It has been reported that use of 50 mg/ml of CQ caused a significant reduction in virus production in dengue (DENV-2)-infected U937 cells. However, the same dose did not cause any toxic or harmful effects on normal cells [86]. Kevarerts et al. reported the in vivo antiviral effect of CQ against human coronavirus OC43 [87]. Other researchers documented CQ in vivo action against enterovirus EV-A71 [88], Zika [89], and influenza A H5N1 viruses [90].

Despite the efficacy of CQ in vitro antiviral activity on numerous viruses, the in vivo effectiveness on CHIKV-infected primates was unsatisfactory. In fact, CQ worsens the disease progression by exacerbating the acute fever and delaying the cellular immune response to an incomplete CHIKV viral clearance [85]. In view of the above, CQ exhibited efficacy in in vitro antiviral activities against a number of viruses; however, the compiled data has yet to be assessed in preclinical studies.

The ability of CQ, with an effective concentration (EC_{50}) of 0.306 ± 0.0091 μm and a lethal concentration (LD_{50}) of 419 ± 192.5 μm, to inhibit the replication of a human coronavirus (HCoV-OC43) in HRT-18 cells has been documented. Further, the wide safety margin of CQ is evident in the selectivity index of SI. 1369 [91].

Further, the highest survival rate of 98.6% has been achieved in an in vivo study conducted by the administration of CQ to a newborn C57BL/6 mice infected with a lethal HCoV-OC43. Overall, the data support the efficacy of CQ against HCoV-OC43 in in vitro and in vivo antiviral effects [91]. CQ's antiviral effect on other viruses, such as SARS-CoV-2, has shown similar efficacy in in vitro studies which led to the notion that CQ may serve a potential therapy for patients infected with SARS-CoV-2 and thus become part of the treatment protocols of COVID-19 patients [72].

10 Hydroxychloroquine (HCQ)

HCQ is a hydroxylated derivative of CQ and proven to be an effective antimalarial agent [92]. HCQ utilization is not restricted to *malaria* treatment but includes the treatment of autoimmune diseases, such as rheumatoid arthritis and systemic lupus erythematosus [93]. The fact that HCQ maintains a wider safety profile than CQ makes it more acceptable than CQ in COVID-19 patients [94]. Control of the cytokine storm in severe forms of SARS-CoV-2 infection may be accomplished by the immunomodulatory activity of HCQ [95].

11 In Vitro and In Vivo Comparison of Antiviral Activity of CQ and HCQ

In vitro studies conducted by Yao et al. have evaluated the antiviral activity and prophylactic effect of CQ and HCQ. To foresee drug concentrations under different treatment regimens, the authors established physiologically based pharmacokinetic models (PBPKs) for CQ and HCQ [95].

SARS-CoV-2-infected Vero cells were used in vitro to assess the antiviral activity of HCQ and CQ. The EC_{50} values of CQ in Vero cells were estimated at 23.90 and 5.47 μm at 24 and 48 h, respectively. It was also determined that HCQ ($EC_{50} = 0.72$ μm) remained more active than CQ ($EC_{50} = 5.47$ μm) [95]. Unlike CQ, the EC_{50} values for HCQ ranged between 6.14 and 0.72 μm at 24 and 48 h, respectively. Research data indicate that EC_{50} values for CQ in the drug pretreatment method were >100 and 18.01 μm at 24 and 48 h, respectively, while the values for HCQ ranged between 6.25 and 5.85 μm at 24 and 48 h, respectively. CQ's inhibition rate remained around 50% irrespective of the concentration assessed.

Inhibition of viral replication appears concentration-dependent function for both CQ and HCQ. These data indicate that HCQ is superior in vitro inhibition of SARS-CoV-2 compared to CQ [95]. In vitro and in vivo studies show that HCQ maintains clear edge over CQ as a potent antiviral drug against coronaviruses [95]. This characteristic is also attributed to the immunomodulatory function of HCQ as well as clinical safety within suitable doses which lend support to the inclusion of HCQ in the treatment of COVID-19 [96].

Yet, other data add caution to the short-term use of HCQ in COVID-19 treatment as it causes cardiac arrhythmias, hypoglycemic episodes, seizures, and dermatologic complications and thus creates concerns over its use [97].

Water solubility and less toxicity of HCQ than CQ make it most appropriate for repurposing. In addition, the safety margin combined with antiviral effect of HCQ against SARS-CoV-2 renders it as an ideal candidate for COVID-19 treatment. Despite the distinct differences, both CQ and HCQ have been utilized in the treatment of *malaria* and inflammatory disease.

12 Clinical Trials of Chloroquine and Hydroxychloroquine

The experimental application of CQ and HCQ in the treatment of severe forms of COVID-19 during the pandemic was the result of intolerable and rapid high mortality and enormous challenge faced by public health delivery systems to bring about an effective cure for this lethal disease [98].

In an effort to obtain credible data on the therapeutic efficacy and safety margin of effective cure for COVID-19, a large number of clinical trials were conducted across the world on the repurposed drugs [99]. By the end of 2020, hundreds of clinical trial databases worldwide were recorded for CQ and HCQ individually or collectively and in combination with specific drugs in the treatment of COVID-19 [100].

In a news briefing by a Chinese agency in February 2020, Gao et al. announced the outcomes of the initial clinical trials of CQ using over one hundred COVID-19-infected patients. The research group asserted the findings that CQ phosphate shortened the disease course, enhanced a virus-negative conversion, curtailed worsening signs of pneumonia, and improved disease-related pulmonary radiographic changes with no adverse effects [64].

The findings are a compilation of data from different clinical trials at various hospitals across China and clinical remain unsupported within the scientific community. In a retrospective triple therapy study of low-dose HCQ, azithromycin and zinc were assessed on 141 COVID-19 in outpatient clinics. This trial study which lasted 5 days showed four of the patients were hospitalized and only one death in the treatment group compared to 13 patients in the untreated cohort group. There were no recorded cardiac side effects in this study [101].

The clinical benefits of HCQ in the treatment of COVID-19 patients were evaluated through randomized clinical trial in a pilot study conducted on 30 confirmed COVID-19 patients. In this trial, patients were randomized 1:1 to both HCQ group and control group and were given an HCQ plus conventional therapy or conventional treatment alone. On day seven of the clinical trial, 13 patients were given HCQ, and in the control group 14 patients were negative for COVID-19 nucleic acid in throat swabs. Improvements are also observed in both groups with regard to clinical manifestations such as fever considering the time needed to attain normal temperature and in pneumonia. In the treatment group, there were limited adverse effects to HCQ as it occurred in one patient who was administered HCQ and developed severe pulmonary complications. The limited participants in this open-label clinical trial did not yield convincing data on the therapeutic and preventive value of HCQ on COVID-19 patients [102].

Another clinical trial conducted within a university setting in France included hospitalized COVID-19 patients treated with HCQ alone and also a combination of HCQ plus azithromycin. In this trial that lasted 2 weeks, a single-arm practice (a design in which a group of patients with a targeted condition subjected to an experimental therapy and observed over specific time) was implemented in which participants administered 600 mg of HCQ daily with viral load testing on a daily basis through nasopharyngeal swabs. The regular HCQ treatment was supplemented with azithromycin on the clinical presentation of COVID-19 patients [103].

On day six of the clinical trial, the endpoint was determined by the presence or absence of SARS-CoV-2 virus in nasopharyngeal swabs. A significant reduction of the viral load has been observed in 20 of the COVID-19 patients treated in this study compared to control groups. The combination of azithromycin and HCQ was more effective in virus elimination.

Given the circumstances, the study concluded that combination drug of HCQ and azithromycin was effective in eliminating 100% of viral load as compared to a group that received HCQ alone, which represented 57.1%. The recovery rate in the control group was limited to 12.5%. The most important outcome of this clinical trial was that all the patients treated with a combination of HCQ and azithromycin evaluated negative for COVID-19 on day 6 and that the synergistic effect of azithromycin boosted the pharmacologic effect of HCQ in the treatment of SARS-CoV-2 infection [103]. Despite the findings, scientific community's endorsement of the results was tenuous [104].

The possible therapeutic and preventive role of hydroxychloroquine (HCQ), chloroquine (CQ), and other antimalarial medications on COVID-19 came under scrutiny when the rate of COVID infection in *malaria*-endemic population appears to lag far behind the *malaria*-free populations. Among the reasons attributed to this inverse phenomenon is the widely utilized hydroxychloroquine (HCQ), chloroquine (CQ) in *malaria*-endemic areas [104].

The efficacy of these drugs as a therapy of corona virus diseases has been the topic of scientific research in the past [83, 105]. Some scientists suggested that specific therapy plan of HCQ be implemented for certain period to counter the adverse effects of corona virus infection [94].

The use of HCQ in the treatment of SARS-Cov-2 has also been proposed at twice daily doses of 400 mg in the first day with subsequent 200 mg daily for 4 days [106].

An in vitro research study conducted by Vincent et al. revealed a prophylactic role for CQ through its inhibitory action against SARS-CoV in healthy and diseased cells [105]. Although yet to be supported by in vivo and in vitro studies, there is an indication that both CQ and HCQ share both the molecular mechanism and the effect on prevention and progression of the disease processes [107].

Later, an alarming finding was made through clinical investigations that the risks associated with HCQ or CQ administration in the treatment of COVID-19 outweigh the benefits. This finding and associated data were later challenged, and the published paper consequently was withdrawn [104, 108].

Despite the lack of evidence relative to the efficacy of CQ and its derivative, the drug resistance that patients develop in *malaria*-endemic areas and the WHO recommendations against it utilization, the use of CQ remains popular [104]. Some of the above factors may have solidified the belief that the consumption of antimalarial drugs, which are readily available, serve as unintended chemoprophylaxis against SARS-CoV-2 and that it could slow down coronavirus infection rate among healthcare practitioners [109, 110]. The possible role of HCQ in protecting personnel of the health maintenance system and COVID patients placed on isolation has been the topic of investigation by clinicians [111]. No noted difference in COVID prevention have been documented between groups of asymptomatic yet high-risk patients administered placebo and those who received HCQ in the context of double-blind studies [112].

To determine the impact of HCQ treatment on the rate of death and hospital admission in outpatient clinics, the National Institute of Health (NIH) undertook a phase 2b placebo-based clinical trial. The results of which were supported by the findings of a randomized clinical trial study in Europe in which patients are divided into two groups, with one group of patients receiving HCQ, while the second group, which represented a large population, administered a routine medical care for COVID-19. The variance in the treatment outcome between HCQ group and standard care group appeared negligible. The table below shows selective clinical trial studies on the efficacy of chloroquine and hydroxychloroquine in COVID-19 patients and the in vitro investigations on SARS-CoV-2.

Research reference	Population (*n* Patients)	Therapeutic methodology	Key outcomes
[112]	In vitro research study with SARS_CoV2-infected Vero cells	Infected Vero cells were treated with CQ or HCQ at 0.032, 0.16, 0.80, 4, 20, or 100 μM for 24 or 49 h	CQ and HCQ reduced viral replication in a concentration-dependent manner EC_{50} values for CQ were 23.90 and 5.47 μM at 24 and 48 h, respectively EC_{50} values for HCQ were 6.14 and 0.72 μM at 24 and 48 h, respectively
[113]	In vitro study with Vero cells	Vero cells were pretreated with CQ or HCQ at 0.032, 0.16, 0.80, 4, 20, or 100 μM for 2 h and were then infected with SARS-CoV-2 and incubated for 24 or 48 h	HCQ showed a higher in vitro antiviral impact in comparison with CQ EC_{50} values for CQ were greater than 100 and 18.01 μM at 24 and 48 h, respectively EC_{50} values for HCQ were 6.25 and 5.85 μM at 24 and 48 h, respectively

(continued)

Research reference	Population (n Patients)	Therapeutic methodology	Key outcomes
[114]	In vitro study with African green monkey kidney VeroE6 cells	SARS-CoV-2-infected cells at four different multiplicities of infection (MOI) and treated with CQ or HCQ up to 50 μM for 48 h	CC50 values of CQ and HCQ were 273 and 250 μM, respectively, indicating lack of significant difference At all MOI (0.01, 0.02, 0.2, and 0.8), EC_{50} for HCQ (4.51, 4.06, 17.31, and 12.96 μM) was higher than that of CQ (2.71, 3.81, 7.14, and 7.36 μM Statistical values of the variations in EC_{50} were significant at MOI of 0.01 ($P < 0.01$)
[115]	In vitro study with Vero E6 cells	Cells were infected with SARS-CoV-2 at MOI (multiplicity of infection) of 0.05 in the presence of different concentrations of CQ, penciclovir, ribavirin, nafamostat, nitazoxanide, remdesivir, favipiravir, and chloroquine	EC_{50}, SI index, and CC_{50} values for CQ were 1.13 μM, > 100 μM, and 88.5 These values were higher for ribavirin (EC_{50} = 110 μM, CC_{50} > 400 μM, and SI > 3.65), penciclovir (EC_{50} = 96.0 μM, CC_{50} > 400 μM, SI > 4.17), and favipiravir (EC_{50} = 61.9 μM, CC_{50} > 400 μM, SI > 6.46), nafamostat (EC_{50} = 22.50 μM, CC_{50} > 100 μM, SI > 4.44) and were comparable to nitazoxanide (EC_{50} = 2.12 μM; CC_{50} > 35.53 μM; SI > 16.76) and remdesivir (EC_{50} = 0.77 μM; CC_{50} > 100 μM; SI > 129.87) for EC_{50}
[116]	Age >12 years and positive for SARS-CoV-2. Patients with HCQ or CQ allergy were excluded or had another recognized contraindication for treatment with the drug. Pregnant and breastfeeding patients were excluded	Oral HCQ 200 mg TD × 10 days (n = 20) Symptomatic treatment and AZT (n = 6; 500 mg/d on day one then 250 mg/d for next 4 days) with HCQ. Patients (n = 16) who rejected the treatment or had relegation criteria, served as controls	Control patients were younger than HCQ-treated patients (37.3 years vs.51.2 years) At day 6 post-inclusion, 70% of HCQ-treated patients were negative compared with 12.5% in the control group (P = 0.001) At day 6 of post-inclusion, 1005 of patients treated with combination of HCQ and AZT were negative compared with 57.1% in patients cured with HCQ only and 12.5% in the control group

(continued)

Research reference	Population (n Patients)	Therapeutic methodology	Key outcomes
[117]	Confirmed COVID-19 patients. Thirty patients were randomly chosen and assigned to treatment and control groups	Oral HCQ sulfate 400 mg OD × 5 days ($n = 15$) NO HCQ was provided to patients ($n = 15$)	On day 7, the number of negative samples did not differ and was (13 (86.7%) in the HCQ group versus 14 (93.3%) in the control group; $P > 0.05$) The period from hospitalization to virus-free nucleic acid cells did not differ (4 ± 1, 9 days in HCQ versus 2 ± 1.4 days in the control group; $P > 0.05$) The time for body temperature normalization was comparable (1 ± 0.2 day one HCQ group versus 1 ± 0.3 days in the control group) Radiological progress was noted on CT images in 7 cases (46.7%) of the control group and 5 cases (33.3%) of the HCQ group, and all patients revealed amelioration in follow-up examinations Three cases (20%) of the control group and four cases (26.7%) of the HCQ group showed abnormal liver function and transient diarrhea ($P > 0.05$)
[118]	COVID-19-exposed patients (211 containing 22 healthcare workers and 189 patients) with negative PCR tests for COVID-19 in a long-term care hospital in Korea. Four patients and one coworker were not finally completed	COVID-19-exposed patients were administered HCQ at 400 mg OD×14 days during the quarantine. There were no control groups	At the ending of 2 weeks of quarantine, follow-up PCR tests were negative A sum of 32 individuals (15.6%) mentioned one or more symptoms through postexposure prophylaxis The most common symptoms were the skin rash (4.3%), loose stool or diarrhea (9%), bradycardia (0.95%), and gastrointestinal disorders (0.95%). Postexposure prophylaxis was terminated in 5 patients (2.7%) because of the requirement for fasting, bradycardia, and gastrointestinal disorder

(continued)

Research reference	Population (*n* Patients)	Therapeutic methodology	Key outcomes
[119]	Patients ($n = 95$) were 18 years of age or older and suspected of having COVID-19 disease	Loading dose of CQ was 600 mg followed by 300 mg BD (starting 12 h after the loading dose), for the next 4 days	CQ treatment in patients with COVID-19 markedly extended the QTc interval by 34–35 ms; 23% of the patients had QTc interval exceeding 500 ms. Statistically distinct effects were recorded on QRS interval (mean difference 6 ms), PR interval (mean difference 6 ms), PR interval (mean difference 8 ms), and heart rate (mean difference-10 bpm)
[120]	A retrospective investigation of 251 patients having COVID-19	HCQ was orally administered at 400 mg BD for 1 day (loading dose) then 200 mg BD for 4 days. AZT was orally administered for 5 days at a dose of 500 mg OD	QTc interval extended from a baseline of 439 ± 29 ms to a maximum value of 473 ± 36 ms ($P < 0.001$), which happen on day 4.1 ± 2 of treatment Extreme novel QTc interval extension to >500 ms revealed in 23% of patients One patient showed polymorphic ventricular tachycardia
[121]	A retrospective investigation of 1376 patients having COVID-19	HCQ ($n = 811$) was provided at 600 mg BD on day 1, followed by 400 mg/d for 4 in the succeeding days. Control group patients were less adversely affected at baseline than the HCQ-treated cohort group ($n = 565$; the ratio of the partial pressure of arterial oxygen to the fraction of inspired oxygen, 223 vs 360)	HCQ use was not accompanied with a markedly lower or higher hazard of death or intubation (hazard ratio, 1.04; 95% Cl, 082 to 1.32)

(continued)

Research reference	Population (*n* Patients)	Therapeutic methodology	Key outcomes
	A retrospective investigation of 368 patients having COVID-19	HCQ ($n = 97$) alone and HCQ+ AZT ($n = 113$) in combination. In the control group ($n = 158$), no HCQ was provided	The hazard of death from any reason was elevated in HCQ group (adjusted hazard ratio, 2.61; 95% CI, 1.10 to 6.17; $P = 0.03$) The risk of death was similar in the HCQ+AZ group (adjusted hazard ratio, 1.14; 95% CI, 0.56 to 2.32: $P = 0.72$) The hazard of ventilation was comparable in the HCQ group (adjusted hazard ratio, 1.43; 95% CI, 0.53 to 3.79; $P = 0.48$) and the HC + AZ group (adjusted hazard ratio, 0.43; 95% CI, 0.16 to 1.12; $P = 0.09$)
[122]	A retrospective investigation of 181 patients having COVID-19 and requiring oxygen ≥2 L/min.	HCQ ($n = 84$) 600 mg/d for 7 days; in the control group ($n = 97$), no HCQ was provided	20.2% of the patients in HCQ group died within 7 days or moved to the ICU vs. 22.1% in the no-HCQ group (16 vs. 21 events, the relative hazard of 0.91, 95% CI 0.47–1.80) in the HCQ group The death of 2.8% of the patients was within 7 days vs. 4.6% in the no-HCQ group (3–4 events, the relative risk of 0.61, 95% CI 0.13–2.89) 27.4% in the HCQ group versus 24.1% in the control group patients developed acute respiratory distress syndrome within 7 days (24 vs. 23 events, relative risk of 1.14, 95% CI 0.65–2.00) 8 patients receiving HCQ (9.5%) revealed electrocardiogram modifications requesting HCQ stop
[123]	A retrospective investigation of 181 patients having COVID-19	HCQ at 200–600 mg OD/BD ($n = 271$) alone; HCQ+AZT ($n = 735$) in combination; AZT 200–500 mg once/OD/BD ($n = 211$). And no drug ($n = 221$)	The death of patients treated with AZT alone, 21/211 (10.0% (95% CI, 5.9–14.0%)), the rate HCQ+ AZT was 189/735 (25.7% (95% CI, 22.3–28.9%)), HCQ alone, 54/271 (19.9% (95% CI, 15.2–24.7%)), and neither drug, 28/221 (12.7% (95% CI, 8.3–17.1%))

(continued)

Research reference	Population (*n* Patients)	Therapeutic methodology	Key outcomes
[124]	A retrospective investigation of 181 patients having COVID-19 and treated with HCQ	HCQ 400 mg/d (200 mg BD) for 7–10 days ($n = 48$) In the control group ($n = 520$), no HCQ was provided	Mortalities reduced in HCQ group (18.8% (9/48) versus 45.8% (238/520) in control group ($P < 0.05$) The level of inflammatory cytokine IL-6 markedly decreased from 22.2 (8.3–118.9) pg/ml ($P < 0.05$) in the HCQ group, but there is no alteration in the control group
[125]	An ongoing randomized controlled trial of more than 11,000 COVID-19 patients to date	HCQ (200 mg tablet containing 155 mg base equivalent) received a loading dose of four tablets (800 mg) at zero and 6 h, then two tablets (400 mg) at zero and 6 h, then two tablets (400 mg) starting at 12 h after the initial dose, and then every 12 h for the next 9 days or until discharge	28-day mortality was 26.8% and 25% in the HCQ and standard of care groups HCQ treatment was markedly accompanied with an elevated length of hospital stay and elevated hazard of developing to death
[126]	An ongoing randomized controlled trial of more than 5000 COVID-19 patients to date	HCQ standard care	Not available; details were not published
[127]	An internet-based randomized controlled trial in nonhospitalized patients in the United States and Canada	HCQ (800 mg once, followed by 600 mg in 6–8 h, then 600 mg daily for 4 more days) placebo	Symptom severity did not significantly differ over 14 days (−0.27 points (95% CI, −0.61 to 0.07 points): $P = 0.17$) At 14 days, 24% of the participants receiving placebo ($P = 0.21$) Medication adverse effects occurred in 43% of HCQ group compared to 22% in the placebo group ($P < 0,001$)

Abbreviations: *HCQ* hydroxychloroquine, *CQ* chloroquine, *OD* one a day, *BD* twice a day, *TD* thrice a day, *CI* confidence interval, *hrs*. hours, EC_{50} half maximum effective concentration, *CC50* 50% cytotoxic concentration, *SI* selectivity index, *Moi* multiplicity of infection, *AZT* azidothymidine

Nevertheless, the cellular immunity is a key element in deterring viral infection and comes to action when the virus enters our system. First, the invading virus encounters the antigen-presenting cells (APCs) whose main function is to present the virus mediated by major histocompatibility class II (MHC II), a pathway regarded as integral to the T-lymphocyte activation and key determinant in pathophysiology of COVID-19 [128].

Scientists who adopt the view that HCQ and CQ has a positive role in *malaria* treatment, and prevention of COVID-19 indicate that these drugs disrupt this pathway reducing T-cell activation, thus hindering the release of inflammatory cytokines as well as the signals that co-stimulate this process [129].

Reports that documented the possible additional mechanisms through which HCQ exerts a positive role in reducing COVID-19 indicate that the high pH of this drug causes inhibition of the lysosomes and consequently prevents antigen presentation within major histocompatibility class II (MHC II) and T-cell activation, which are considered elements essential for inflammatory cytokine release. Moreover, altered intracellular pH (pHi) will interfere with Toll-like receptors (TLR) particularly TLR7 and TLR9 [130, 131].

The latter receptors are expressed in dendritic and epithelial cells, macrophages, and fibroblasts, considered as the frontier immune mechanism that neutralize pathogens and endogenous substances liberated by necrotic cells. In order to accomplish this function, they utilize the cGAS-STING pathway that activates interferon (inflammatory) genes. HCQ produces a weakened inflammatory response by disrupting the above pathway. In view of this mechanism, massive immune response is observed in SARS-Cov-2 patients within cytokine release syndrome (CRS) [107].

Upon successful binding of the virus-ACE2 receptor, the SARS-Cov-2 virus uses endosomes to enter host target cells. The fusion process occurs through lysosomal proteases which cleave the virus spike proteins and initiate viral replication. Lysosomal proteases activity is dependent upon low pH, and therefore a high pH will hinder the process [132, 133].

Because of the characteristically low pHi of the endosomes, administered HCQ and CQ enter the cell, accessing the endosomes and eventually heightening endosomal pHi. This chemical alteration disrupts the endosomes and SARS-Cov-2 viral fusion process [105]. The HCQ and CQ also act against COVID-19 infection by impairing the ability of the virus to bind to host ACE2 receptors and to the membrane fusion process [107]. In addition to disruption of ACE2-virus binding, both these drugs impair the attachment of the virus spike proteins to the cell receptors by inducing alterations in the glycosylation of ACE2 receptors.

13 Synopsis

The above discussion points to an important fact that the of HCQ and CQ not only counter the *malaria* infection but may also play a role in limiting COVID-19 spread in the affected population. This contrast between *malaria* frequency and COVID-19 is correlated with the mechanism of action of these medications. As preventive and therapeutic drugs, HCQ and CQ halt disease progression through direct antiviral function and impediment of viral replication by altering the intracellular pH and also by suppressing cytokine release and associated storm by lessening T lymphocyte stimulation. Antimalaria medications have been revised, and artemisinins class of drugs have been proven to be efficacious against *malaria* including drug-resistant strains but not against COVID. Artemisinins, extracted from sweet wormwood (*Artemisia annua*), seem to be highly effective against phylogenetically different parasitic infections such as schistosomiasis.

14 Conclusion

Despite the potential dual role of HCQ and CQ as a therapy and prevention against SARS-CoV2, a clear-cut evidence is yet to be established. Several factors affect the progression of COVID-19 infection in *malaria*-endemic populations. Aside from healthcare delivery system and the availability or lack of associated organization, the difference in the prevalence of these diseases could be linked in part to ACEI/D and ACE2 polymorphism. The fact that *malaria* patients naturally form anti-glycosylphosphatidylinositol (GPI) antibodies that recognize the SARS-CoV-2 glycoproteins may be responsible for a protective and suppressive role preventing or lessening the severity of symptoms caused by the virus.

References

1. Zhou P, Yang XL, Wang XG, Hu B, Zhang L, Zhang W, et al. A pneumonia outbreak associated with a new coronavirus of probable bat origin. Nature. 2020;579:270–3.
2. Dowd JB, Andriano L, Brazel DM, Rotondi V, Block P, Ding X, et al. Demographic science aids in understanding the spread and fatality rates of COVID-19. Proc Natl Acad Sci U S A. 2020;117:9696–8.
3. Rice GI, Thomas DA, Grant PJ, Turner AJ, Hooper NM. Evaluation of angiotensin-converting enzyme (ACE), its homologue ACE2 and neprilysin in angiotensin peptide metabolism. Biochem J. 2004;383:45–51.
4. Ondoa P, Kebede Y, Loembe MM, Bhiman JN, Tessema SK, Sow A, et al. COVID-19 testing in Africa: lessons learnt. Lancet Microbe. 2020;1:e103–4.
5. Guan W, Ni Z, Hu Y, Liang W, Ou C, He J, et al. Clinical characteristics of coronavirus disease 2019 in China. N Engl J Med. 2020;382:1708–20.
6. Schofield L, Mueller I. Clinical immunity to *malaria*. Curr Mol Med. 2006;6(2):205–21.
7. Bustinduy AL, et al. Age stratified profiles of serum IL-6, IL-10, and TNF-alpha cytokines among Kenyan children with schistosoma haematobium, plasmodium falciparum and other chronic parasitic co-infections. Am J Trop Med Hyg. 2015;95(5):945–51.
8. O'Brien D, Tobin S, Brown GV, Torresi J. Fever in returned travelers: review of hospital admissions for a 3-year period. Clin Infect Dis. 2001;33:603–9.
9. Gostic KM, Gomez ACR, Mummah RO, Kucharski AJ, Lloyd-Smith JO. Estimated effectiveness of symptom and risk screening to prevent the spread of COVID-19. elife. 2020;9:e55570.
10. Huang C, Wang Y, Li X, Ren L, Zhao J, Hu Y, et al. Clinical features of patients infected with 2019 novel coronavirus in Wuhan. China Lancet. 2020;395:497–506.
11. Ye Q, Wang B, Mao J. The pathogenesis and treatment of the 'Cytokine Storm' in COVID-19. J Infect. 2020;80:607–13.

12. Liao Y-C, Liang W-G, Chen F-W, Hsu J-H, Yang J-J, Chang M-S. IL-19 induces production of IL-6 and TNF-α and results in cell apoptosis through TNF-α. J Immunol. 2002;169:4288–97.
13. Wu Z, McGoogan JM. Characteristics of and important lessons from the coronavirus disease 2019 (COVID-19) outbreak in China: summary of a report of 72314 cases from the Chinese Center for Disease Control and Prevention. JAMA. 2020;323:1239–42.
14. Bucşan AN, Williamson KC. Setting the stage: the initial immune response to blood-stage parasites. Virulence. 2020;11:88–103.
15. Rudragouda C, Fehr AR, Jian Z, Christine W-L, Abrahante JE, Matthias M, Ramakrishna S, McCray PB, Meyerholz DK, Stanley P. IFN-I response timing relative to virus replication determines MERS coronavirus infection outcomes. J Clin Investig. 2019;129(9):3625–39.
16. Strayer D, Dickey R, Carter W. Sensitivity of SARS/MERS CoV to interferons and other drugs based on achievable serum concentrations in humans. Infect Disord Drug Targets. 2014;14:37–43.
17. Fauci AS, Lane HC, Redfield RR. COVID-19—navigating the uncharted. N Engl J Med. 2020;382:1268–9.
18. Akanmori BD, Kurtzhals JA, Goka BQ, Adabayeri V, Ofori MF, Nkrumah FK, et al. Distinct patterns of cytokine regulation in discrete clinical forms of Plasmodium falciparum *malaria*. Eur Cytokine Netw. 2000;11:113–8.
19. Mazhar F, Haider N. Respiratory manifestation of *malaria*: an update. Int J Med Res Health Sci. 2016;5:59–65.
20. Jin Y, Yang H, Ji W, Wu W, Chen S, Zhang W, et al. Virology, epidemiology, pathogenesis, and control of COVID-19. Viruses. 2020;12:372.
21. Visseren FL, Bouwman JJ, Bouter KP, Diepersloot RJ, de Groot PH, Erkelens DW. Procoagulant activity of endothelial cells after infection with respiratory viruses. Thromb Haemost. 2000;84:319–24.
22. Klok FA, Kruip MJHA, van der Meer NJM, Arbous MS, Gommers DAMPJ, Kant KM, et al. Incidence of thrombotic complications in critically ill ICU patients with COVID-19. Thromb Res. 2020;191:145–7.
23. Oxley TJ, Mocco J, Majidi S, Kellner CP, Shoirah H, Singh IP, et al. Large-vessel stroke as a presenting feature of Covid-19 in the young. N Engl J Med. 2020;382:e60.
24. Fox SE, Akmatbekov A, Harbert JL, Li G, Quincy Brown J, Vander Heide RS. Pulmonary and cardiac pathology in African American patients with COVID-19: an autopsy series from New Orleans. Lancet Respir Med. 2020;8:681–6.
25. Tang N, Li D, Wang X, Sun Z. Abnormal coagulation parameters are associated with poor prognosis in patients with novel coronavirus pneumonia. J Thromb Haemost. 2020;18:844–7.
26. Srichaikul T. Hemostatic alterations in *malaria*. Southeast Asian J Trop Med Public Health. 1993;24:86–91.
27. Angchaisuksiri P. Coagulopathy in *malaria*. Thromb Res. 2014;133:5–9.
28. Krishnan A, Karnad DR, Limaye U, Siddharth W. Cerebral venous and dural sinus thrombosis in severe falciparum *malaria*. J Infect. 2004;48:86–90.
29. Schwartz J, Musoke C, Ssendikadiwa C, Babua C. Severe falciparum *malaria* associated with massive pulmonary embolism. Ann Afr Med. 2014;13:47.
30. Zhang H, Baker A. Recombinant human ACE2: acing out angiotensin II in ARDS therapy. Crit Care. 2017;21:305.
31. Silva LS, Silva-Filho JL, Caruso-Neves C, Pinheiro AAS. New concepts in *malaria* pathogenesis: the role of the renin-angiotensin system. Front Cell Infect Microbiol. 2016;5:103.
32. Faik I, van Tong H, Lell B, Meyer CG, Kremsner PG, Velavan TP. Pyruvate Kinase and Fcγ receptor gene copy numbers associated with *malaria* phenotypes. J Infect Dis. 2017;216:276–82.
33. Munde EO, Okeyo WA, Raballah E, Anyona SB, Were T, Ong'echa JM. Association between Fcγ receptor IIA, IIIA, and IIIB genetic polymorphisms and susceptibility to severe *malaria* anemia in children in western Kenya. BMC Infect Dis. 2017;17:289.

34. Hatami N, Ahi S, Sadeghinikoo A, Foroughian M, Javdani F, Kalani N, et al. Worldwide ACE (I/D) polymorphism may affect COVID-19 recovery rate: an ecological meta-regression. Endocrine. 2020;68:479–84.
35. Saab YB, Gard PR, Overall ADJ. The geographic distribution of the ACE II genotype: a novel finding. Genet Res. 2007;89:259–67.
36. Gelabert P, Olalde I, de-Dios T, Civit S, Lalueza-Fox C. *Malaria* was a weak selective force in ancient Europeans. Sci Rep. 2017;7:1377.
37. Sampson UKA, Edwards TL, Jahangir E, Munro H, Wariboko M, Wassef MG, et al. Factors associated with the prevalence of hypertension in the southeastern United States: insights from 69,211 blacks and whites in the Southern Community Cohort Study. Circ Cardiovasc Qual Outcomes. 2014;7:33–54.
38. Cappuccio FP. Ethnicity and cardiovascular risk: variations in people of African ancestry and South Asian origin. J Hum Hypertens. 1997;11:571–6.
39. Gallego-Delgado J, Walther T, Rodriguez A. The high blood pressure-malaria protection hypothesis. Circ Res. 2016;119:1071–5.
40. WHO. Responding to community spread of COVID-19. Interim Guide 7 March. 2020. p. 1–6.
41. King T, Lamb T. Interferon-γ: the Jekyll and Hyde of *malaria*. PLoS Pathog. 2015;11:e1005118.
42. de Mendonça VR, Barral-Netto M. Immunoregulation in human *malaria*: the challenge of understanding asymptomatic infection. Mem Inst Oswaldo Cruz. 2015;110:945–55.
43. Gomes LR, Martins YC, Ferreira-Da-Cruz MF, Daniel-Ribeiro CT. Autoimmunity, phospholipid-reacting antibodies, and *malaria* immunity. Lupus. 2014;23:1295–8.
44. Rodrigo C, Fernando SD, Rajapakse S. Clinical evidence for repurposing chloroquine and hydroxychloroquine as antiviral agents: a systematic review. Clin Microbiol Infect. 2020;26:979.
45. Al-Karmalawy AA, Khattab M. Molecular modelling of mebendazole polymorphs as a potential colchicine binding site inhibitor. New J Chem. 2020;44:13990–6.
46. Eliaa SG, et al. Empagliflozin and doxorubicin synergistically inhibit the survival of triple-negative breast cancer cells via interfering with the mTOR pathway and inhibition of calmodulin: in vitro and molecular docking studies. ACS Pharmacol Translat Sci. 2020;3:1330–8.
47. Khattab M, Al-Karmalawy AA. Revisiting activity of some nocodazole analogues as a potential anticancer drugs using molecular docking and DFT calculations. Front Chem. 2021;9:9.
48. Singh TU, et al. Drug repurposing approach to fight COVID-19. Pharmacol Rep. 2020;72(6):1479–1508.
49. Xue H, et al. Review of drug repositioning approaches and resources. Int J Biol Sci. 2018;14(10):1232.
50. Tripathy S, et al. A review on possible modes modes of action of chloroquine/hydroxychloroquine: repurposing against SAR-CoV-2 (COVID-19) pandemic. Int J Antimicrob Agents. 2020;56(2):106028.
51. Glebov OO. Understanding SARS-CoV-2 endocytosis for COVID-19 drug repurposing. FEBS J. 2020;287(17):3664–367. Al-Karmalawy AA, Eiss IHJPS. Molecular docking and dynamics simulations reveal the potential of anti-HCV drugs to inhibit COVID-19 main protease. Pharm Sci. 2021;27:S109–S121.
52. Alnajjar R, et al. Molecular docking, molecular dynamics, and in vitro studies reveal the potential of angiotensin II receptor blockers to inhibit the COVID-19 main protease. Heliyon. 2020;6:e0564112.
53. Eissa I, et al. Molecular docking and dynamics simulation revealed the potential inhibitory activity of ACEIs against SARS-CoV-2 targeting hACE2 receptor. Front Chem. 2021;9:227.
54. Ghanem A, et al. Tanshinone IIA synergistically enhances the antitumor activity of doxorubicin by interfering with the PI3K/AKT/mTOR pathway and inhibition of topoisomerase II: in vitro and molecular docking studies. New J Chem. 2020;44:17374–1738140.
55. Zaki AA, et al. Molecular docking reveals the potential of Cleome amblyocarpa isolated compounds to inhibit COVID-19 virus main protease. New J Chem. 2020;44:16752–1675839.

56. Elmaaty AA, et al. Revisiting activity of some glucocorticoids as a potential inhibitor of SARS-CoV-2 main protease: theoretical study. RSC Adv. 2021;11(17):10027–42.
57. Elmaaty AA, et al. *In* a search for potential drug candidates for combating COVID-19: computational study revealed salvianolic acid B as a potential therapeutic targeting 3CLpro and spike proteins. J Biomol Struct Dyn. 2022;40(19):8866–8893.
58. Mohanty S, et al. Application of artificial intelligence in COVID-19 drug repurposing. Diabetes Metab Syndr. 2020;14:1027.
59. Krishna S, et al. Repurposing antimalarials to tackle the COVID-19 pandemic. Trends Parasitol. 2021;37(1):8–11
60. Altay O, et al. Current status of COVID-19 therapies and drug repositioning applications. Iscience. 2020;23:101303.
61. Pollard CA, Morran MP, Nestor-Kalinoski AL. The COVID-19 pandemic: a global health crisis. Physiol Genomics. 2020;52(11):549–57.
62. Atri D, et al. COVID-19 for the cardiologist: basic virology, epidemiology, cardiac manifestations, and potential therapeutic strategies. Basic Transl Med. 2020;5(5):518–36.
63. Gao J, Tian Z, Yang X. Breakthrough: chloroquine phosphate has shown apparent efficacy in treatment of COVID-19 associated pneumonia in clinical studies. Biosci Trends. 2020;14:72.
64. Gao J, Hu S. Update on use of chloroquine/hydroxychloroquine to treat coronavirus disease 2019 (COVID-19). Biosci Trends. 2020;14:156–8.
65. Spinelli FR, et al. To consider or not antimalarials as a prophylactic intervention in the SARS-CoV-2 (Covid-19) pandemic. Ann Rheum Dis. 2020;79(5):666–7.
66. Amanat F, Krammer F. SARS-CoV-2 vaccines: status report. Immunity. 2020;52:583–9.
67. Nakamura K, Hikone M, Shimizu H, Kuwahara Y, Tanabe M, Kobayashi M, et al. A sporadic COVID-19 pneumonia treated with extracorporeal membrane oxygenation in Tokyo, Japan: a case report. J Infect Chemother. 2020;26:756–61.
68. Shukla AM, Archibald LK, Shukla AW, Mehta HJ, Cherabuddi K. Chloroquine and hydroxychloroquine in the context of COVID-19. Drugs Context. 2020;9:4–5.
69. Chu CM, Cheng VCC, Hung IFN, Wong MML, Chan KH, Chan KS, et al. Role of lopinavir/ritonavir in the treatment of SARS: initial virological and clinical findings. Thorax. 2004;59:252–6.
70. Cao B, Wang Y, Wen D, Liu W, Wang J, Fan G, et al. A trial of lopinavir-ritonavir in adults hospitalized with severe COVID-19. N Engl J Med. 2020;382:1787–99.
71. Wang M, Cao R, Zhang L, Yang X, Liu J, Xu M, et al. Remdesivir and chloroquine effectively inhibit the recently emerged novel coronavirus (2019-nCoV) in vitro. Cell Res. 2020;30:269–71.
72. Khalili JS, Zhu H, Mak NSA, Yan Y, Zhu Y. Novel coronavirus treatment with ribavirin: groundwork for an evaluation concerning COVID-19. J Med Virol. 2020;92:740–6.
73. Li G, Clercq ED. Therapeutic options for the 2019 novel coronavirus (2019-nCoV). Nat Rev Drug Discov. 2020;19:149–50.
74. Marotto D, Sarzi-Puttini P. What is the role of rheumatologists in the era of COVID-19? Autoimmun Rev. 2020;19:102539.
75. Sallard E, Lescure FX, Yazdanpanah Y, Mentre F, Peiffer-Smadja N. Type 1 interferons as a potential treatment against COVID-19. Antivir Res. 2020;178:104791.
76. Long Q, Liu B, Deng H, Wu GC, Deng K, Yao-Kai Chen YK, et al. Antibody responses to SARS-CoV-2 in patients with COVID-19. Nat Med. 2020;26:845–8.
77. Jiang S, Hillyer C, Du L. Neutralizing antibodies against SARS-CoV-2 and other human coronaviruses. Trends Immunol. 2020;41:355–9.
78. Zha L, Li S, Pan L, Tefsen B, Li Y, French N, et al. Corticosteroid treatment of patients with coronavirus disease 2019 (COVID-19). Med J Aust. 2020;212:416–20.
79. Arrow KJ, Panosian C, Gelband H. Saving lives, buying time: economics of malaria drugs in an age of resistance. Washington, DC: National Academies Press; 2004.
80. Browning DJ. Pharmacology of chloroquine and hydroxychloroquine. Hydroxychloroquine Chloroquine Retinopathy. 2014;4:35–63.

81. Shimizu Y, Yamamoto S, Homma M, Ishida N. Effect of chloroquine on the growth of animal viruses. Archiv für die gesamte Virusforschung. 1972;36:93–104. https://doi.org/10.1007/bf01250299
82. Glushakova SE, Lukashevich IS. Early events in arenavirus replication are sensitive to Lysosomotropic compounds. Arch Virol. 1989;104:157–61.
83. Vincent MJ, Bergeron E, Benjannet S, Erickson BR, Rollin PE, Ksiazek TG, et al. Chloroquine is a potent inhibitor of SARS coronavirus infection and spread. Virol J. 2005;2:69.
84. Jang CH, Choi JH, Byun MS, Jue DM. Chloroquine inhibits production of TNF-α, IL-1β and IL-6 from lipopolysaccharide-stimulated human monocytes/macrophages by different modes. Rheumatology. 2006;45:703–10.
85. Farias KJ, Machado PR, de Almeida Junior RF, de Aquino AA, da Fonseca BA. Chloroquine interferes with dengue-2 virus replication in U937 cells. Microbiol Immunol. 2014;58:318–26.
86. Keyaerts E, Li S, Vijgen L, Rysman E, Verbeeck J, Ranst MV, et al. Antiviral activity of chloroquine against human coronavirus OC43 infection in newborn mice. Antimicrob Agents Chemother. 2009;53:3416–21.
87. Tan YW, Yam WK, Sun J, Chu JJH. An evaluation of chloroquine as a broad-acting antiviral against hand, foot and mouth disease. Antivir Res. 2018;149:143–9.
88. Li C, Zhu X, Ji X, Quanquin N, Deng YQ, Tian M, et al. Chloroquine, an FDA-approved drug, prevents Zika virus infection and its associated congenital microcephaly in mice. EBioMedicine. 2017;24:189–94.
89. Yan Y, Zou Z, Sun Y, Li X, Xu K-F, Wei Y, et al. Anti-malaria drug chloroquine is highly effective in treating avian influenza a H5N1 virus infection in an animal model. Cel Res. 2013;23:300–2.
90. Roques P, Thiberville SD, Dupuis-Maguiraga L, Lum FK, Labadie K, Martinon F. Paradoxical effect of chloroquine treatment in enhancing chikungunya virus infection. Viruses. 2018;10:268.
91. Liu J, Cao R, Xu M, Wang X, Zhang H, Hu H, et al. Hydroxychloroquine, a less toxic derivative of chloroquine, is effective in inhibiting SARS-CoV-2 infection in vitro. Cell Discov. 2020;6:16.
92. Silva JC, Mariz HA, Rocha LF, Oliveira PSS, Dantas AT, Duarte ALBP, et al. Hydroxychloroquine decreases Th17-related cytokines in systemic lupus erythematosus and rheumatoid arthritis patients. Clinics (Sao Paulo). 2013;68:766–71.
93. Gevers S, Kwa MSG, Wijnans E, van Nieuwkoop C. Safety considerations for chloroquine and hydroxychloroquine in the treatment of COVID-19. Clin Microbiol Infect. 2020;26:1276–7.
94. Yao X, Ye F, Zhang M, Cui C, Huang B, Niu P, et al. *In vitro* antiviral activity and projection of optimized dosing Design of Hydroxychloroquine for the treatment of severe acute respiratory syndrome coronavirus 2 (SARS-CoV-2). Clin Infect Dis. 2020;71(15):732–9.
95. Chandler LC, Yusuf IH, McClements ME, Barnard AR, MacLaren RE, Xue K. Immunomodulatory effects of hydroxychloroquine and chloroquine in viral infections and their potential application in retinal gene therapy. Int J Mol Sci. 2020;21:4972.
96. Pereira BB. Challenges and cares to promote rational use of chloroquine and hydroxychloroquine in the management of coronavirus disease 2019 (COVID-19) pandemic: a timely review. J Toxicol Environ Health B Crit Rev. 2020;23:177–81.
97. Becker RC. Covid-19 treatment update: follow the scientific evidence. J Thromb Thrombolysis. 2020;50:43–53.
98. Lythgoe MP, Middleton P. Ongoing clinical trials for the management of the COVID-19 pandemic. Trends Pharmacol Sci. 2020;41:363–82.
99. (clinicaltrials.gov, WHO). Clinical Trial Table (2021). WHO. Available at: https://clinicaltrials.gov/ct2/who_table.
100. Derwand R, Scholz M, Zelenko V. COVID-19 outpatients: early risk-stratified treatment with zinc plus low-dose hydroxychloroquine and azithromycin: a retrospective case series study. Int J Antimicrob Agents. 2020;56:106214.

101. Chen Y, Guo Y, Yihang P, Zhao ZJ. Structure analysis of the receptor binding of 2019-nCoV. Biochem Biophys Res Commun. 2020;525:135–40.
102. Gautret P, Lagier JC, Parola P, Hoang VT, Meddeba L, Mailhe M, et al. Hydroxychloroquine and azithromycin as a treatment of COVID-19: results of an open-label non-randomized clinical trial. Int J Antimicrob Agents. 2020;56:105949.
103. Napoli PE, Nioi M. Global spread of coronavirus disease 2019 and *malaria*: an epidemiological paradox in the early stage of a pandemic. J Clin Med. 2020;9:1138.
104. Savarino A, Boelaert JR, Cassone A, Majori G, Cauda R. Effects of chloroquine on viral infections: an old drug against today's diseases? Lancet Infect Dis. 2003;3:722–7.
105. Dyall J, Gross R, Kindrachuk J, Johnson RF, Olinger GG, Hensley LE, et al. Middle east respiratory syndrome and severe acute respiratory syndrome: current therapeutic options and potential targets for novel therapies. Drugs. 2017;77:1935–66.
106. Zhou D, Dai S-M, Tong Q. COVID-19: a recommendation to examine the effect of hydroxychloroquine in preventing infection and progression. J Antimicrob Chemother. 2020;75:1667–70.
107. Mehra MR, Ruschitzka F, Patel AN. Retraction—hydroxychloroquine or chloroquine with or without a macrolide for treatment of COVID-19: a multinational registry analysis. Lancet. 2020;395:1820.
108. The Lancet Editors. Expression of concern: hydroxychloroquine or chloroquine with or without a macrolide for treatment of COVID-19: a multinational registry analysis. Lancet. 2020;395:102.
109. Ocan M, Akena D, Nsobya S, Kamya MR, Senono R, Kinengyere AA, et al. Persistence of chloroquine resistance alleles in *malaria* endemic countries: a systematic review of burden and risk factors. Malar J. 2019;18:76.
110. Lother SA, Abassi M, Agostinis A, Bangdiwala AS, Cheng MP, Drobot G, et al. Post-exposure prophylaxis or pre-emptive therapy for severe acute respiratory syndrome coronavirus 2 (SARS-CoV-2): study protocol for a pragmatic randomized-controlled trial. Can J Anesth. 2020;67:1201–11.
111. Boulware DR, Pullen MF, Bangdiwala AS, Pastick KA, Lofgren SM, Okafor EC, et al. A randomized trial of hydroxychloroquine as postexposure prophylaxis for Covid-19. N Engl J Med. 2020;383:517–25.
112. Yao X, Ye F, Zhang M, et al. In vitro antiviral activity and projection of optimized dosing design of hydroxychloroquine for the treatment of severe acute respiratory syndrome coronavirus 2 (SARS-CoV-2). Clin Infect Dis. 2020;71:723–39.
113. Liu J, Cao R, Xu M, et al. Hydroxychloroquine, a less toxic derivative of chloroquine, is effective in inhibiting SARS-CoV-2 infection in vitro. Cell Discov. 2020;6:16. https://doi.org/10.1038/s41421-020-0156-0.
114. Wang M, Cao R, Zhang L, et al. Remdesivir and chloroquine effectively inhibit the recently emerged novel coronavirus (2019-nCoV) in vitro. Cell Res. 2020;30:269–71. https://doi.org/10.1038/s41422-020-0282-0.
115. Gautret P, Lagier J-C, Parola P, et al. Hydroxychloroquine and azithromycin as a treatment of COVID-19: results of an open-label non-randomized clinical trial. Int J Antimicrob Agents. 2020;56:105949. https://doi.org/10.1016/j.ijantimicag.2020.105949.
116. Chen J, Liu D, Liu L, et al. A pilot study of hydroxychloroquine in treatment of patients with moderate COVID-19. J Zhejiang Univ (Med Sci). 2020b;49:215–9. https://doi.org/10.3785/j.issn.1008-9292.2020.03.03.
117. Lee SH, Son H, Peck KR. Can post-exposure prophylaxis for COVID-19 be considered as an outbreak response strategy in long-term care hospitals? Int J Antimicrob Agents. 2020;55:105988. https://doi.org/10.1016/j.ijantimicag.2020.105988.
118. Van-den Broek MPH, Möhlmann JE, Abeln BGS, Liebregts M, Van-dijk VF, Van-de Grade EM. Chloroquine-induced QTc prolongation in COVID-19 patients. Neth Hear J. 2020;28:406–9. https://doi.org/10.1007/s12471-020-01429-7.

119. Chorin E, Wadhwani L, Magnani S, et al. QT interval prolongation and torsade de pointes in patients with COVID-19 treated with hydroxychloroquine/azithromycin. Heart Rhythm. 2020a;17:1425–33. https://doi.org/10.1016/j.hrthm.2020.05.014.
120. Geleris J, Sun Y, Platt J, et al. Observational study of hydroxychloroquine in hospitalized patients with Covid-19. N Engl J Med. 2020;382:2411–8. https://doi.org/10.1056/NEJMoa2012410.
121. Mahevas M, Tran VT, Roumier M, et al. No evidence of clinical efficacy of hydroxychloroquine in patients hospitalized for COVID-19 infection with oxygen requirement: results of a study using routinely collected data to emulate a target trial. BJM. 2020;369:m1844. https://doi.org/10.1101/2020.04.10.20060699.
122. Rosenberg ES, Dufort EM, Udo T, et al. Association of treatment with hydroxychloroquine or azithromycin with in-hospital mortality in patients with COVID-19 in New York state. JAMA. 2020;323:2493–502. https://doi.org/10.1001/jama.2020.8630.
123. Yu B, Wang DW, Li C. Hydroxychloroquine application is associated with a decreased mortality in critically ill patients with COVID-19. MedRxiv. 2020:20073379. https://doi.org/10.1101/2020.04.27.20073379.
124. Horby P, Mafham M, Linsell L, et al. Effect of hydroxychloroquine in hospitalized patients with Covid-19. N Engl J Med. 2020;383:2030–40. https://doi.org/10.1056/NEJMoa2022926.
125. WHO. WHO discontinues hydroxychloroquine and lopinavir/ritonavir treatment arms for COVID-19. 2020. Available from: www.who.int/news-room/detail/04-07-2020-who-discontinues-hydroxychloroquine-and-lopinavir-ritonavir-treatment-arms-for-covid-19.
126. Skipper CP, Pastick KA, Engen NW, et al. Hydroxychloroquine in nonhospitalized adults with early COVID-19. Ann Intern Med. 2020;173:623–31. https://doi.org/10.7326/M20-4207.
127. Lotteau V, Teyton L, Peleraux A, Nilsson T, Karlsson L, Schmid SL, et al. Intracellular transport of class II MHC molecules directed by invariant chain. Nature. 1990;348:600–5.
128. Jang C-H, Choi J-H, Jue D-M. Chloroquine inhibits production of TNF-a, IL-1b and IL-6 from lipopolysaccharide-stimulated human monocytes/macrophages by different modes. Rheumatology. 2006;45:703–10.
129. Kužnik A, Benčina M, Švajger U, Jeras M, Rozman B, Jerala R. Mechanism of endosomal TLR inhibition by antimalarial drugs and imidazoquinolines. J Immunol. 2011;186:4794–804.
130. Ewald SE, Lee BL, Lau L, Wickliffe KE, Shi GP, Chapman HA, et al. The ectodomain of toll-like receptor 9 is cleaved to generate a functional receptor. Nature. 2008;456:658–62.
131. Millet JK, Whittaker GR. Host cell proteases: critical determinants of coronavirus tropism and pathogenesis. Virus Res. 2015;202:120–34.
132. Schrezenmeier E, Dörner T. Mechanisms of action of hydroxychloroquine and chloroquine: implications for rheumatology. Nat Rev Rheumatol. 2020;16:155–66.
133. Al-Bari MAA. Targeting endosomal acidification by chloroquine analogs as a promising strategy for the treatment of emerging viral diseases. Pharmacol Res Perspect. 2017;5:e00293.

The Use of Azithromycin and Lopinavir-Ritonavir in the Treatment of COVID-19

Andang Miatmoko, Yulistiani, Melanny Ika Sulistyowati, Dwi Setyawan, Devy Maulidya Cahyani, and Purwati

Abstract Coronavirus disease 2019 (COVID-19) is a new type of disease first identified in 2019 which is caused by the Sars-CoV-2 virus that can cause illness with mild-to-severe symptoms. A common symptom of COVID-19-infected individuals is that of acute respiratory distress, such as fever, cough, and shortness of breath. COVID-19 causes pneumonia, acute respiratory syndrome, renal failure, and even, in severe cases, death. Clinical symptoms reported as presented by patients with this condition include fever, breathing difficulties, and double pneumonia affecting both lungs as confirmed by a patient's X-ray results. Curative therapy can be administered using antiviral agents such as those drugs prescribed during the MERS and SARS pandemics, one being a combination of lopinavir/ritonavir. An antibiotic such as azithromycin is additionally utilized. The combination of these drugs has been studied through both in vitro and clinical trials; the results of which have confirmed its potential application as part of COVID-19 treatment therapy. In

A. Miatmoko
Department of Pharmaceutical Sciences, Faculty of Pharmacy, Universitas Airlangga, Nanizar Zaman Joenoes Building, Surabaya, Indonesia

Stem Cell Research and Development Center, Universitas Airlangga, Institute of Tropical Disease Building, Surabaya, Indonesia

Yulistiani
Department of Pharmacy Practice, Faculty of Pharmacy, Universitas Airlangga, Nanizar Zaman Joenoes Building, Surabaya, Indonesia

M. I. Sulistyowati · D. Setyawan
Department of Pharmaceutical Sciences, Faculty of Pharmacy, Universitas Airlangga, Nanizar Zaman Joenoes Building, Surabaya, Indonesia

D. M. Cahyani
Master Program of Pharmaceutical Sciences, Faculty of Pharmacy, Universitas Airlangga, Nanizar Zaman Joenoes Building, Surabaya, Indonesia

Purwati (✉)
Stem Cell Research and Development Center, Universitas Airlangga, Institute of Tropical Disease Building, Surabaya, Indonesia

R. Shegokar, Y. Pathak (eds.), *Viral Drug Delivery Systems*,
https://doi.org/10.1007/978-3-031-20537-8_15

vitro tests conducted on Vero cells indicated that a combination of lopinavir/ritonavir and azithromycin is highly effective in inhibiting viral replication of the SARS-CoV-2 virus previously isolated from patients at Universitas Airlangga Hospital, Indonesia.

In vitro studies of mesenchymal cells have confirmed the low cytotoxicity of this drug combination. Moreover, molecular docking studies have shown the effectiveness of the combination of lopinavir/ritonavir and azithromycin at several stages of the viral infection cycle. Clinical trials involving COVID-19 patients confirmed an accelerated improvement in their clinical symptoms and virus clearance with high levels of safety within kidney and liver function parameters. The combination of these drugs has also been shown to reduce the effect of IL-6 and TNF-α as pro-inflammatory markers, as well as an enhanced anti-inflammatory response characterized by an increase in IL-10 cytokines. The combination of lopinavir/ritonavir and azithromycin represents an effective drug combination for managing the COVID-19 virus or future viral pandemic outbreaks.

Keywords COVID-19 · Virus · Lopinavir/ritonavir · Azithromycin · Illness

1 Introduction

In December 2019, a public health crisis was caused by the SARS-CoV-2 virus whose rapid spread can cause coronavirus (COVID-19), a severe respiratory system disease. The first case appeared in Wuhan, China, before it spread exponentially to produce an international pandemic. Since the World Health Organization (WHO) classified COVID-19 a global pandemic in October 2021, 237,196,253 positive cases have been reported, culminating in 4,840,189 deaths.

Coronavirus forms part of the *Coronavirinae* subfamily within the *Coronaviridae* family of the *Nidovirales* order. Spherical in shape, the virus has a surface which is sealed by the protein spike (S), a membrane (M), and an envelope (E); the latter two of which are located between the protein S [1]. SARS-CoV-2 is a virus with single-stranded RNA that has a distinctive crown-like shape due to the presence of glycoproteins on the envelope under an electron microscope [2]. This virus enters the human body through the angiotensin-converting enzyme 2 (ACE-2) receptor [3].

Coronavirus has spherical shapes belonging to the *Coronavirinae* subfamily in the *Coronaviridae* family in the order *Nidovirales*. The surface of the virus is covered with spike (S), membrane (M), and envelope (E) proteins, which are between the S proteins [1]. SARS-CoV-2 is a virus with single-stranded RNA that has a distinctive crown-like shape due to the presence of glycoproteins on the envelope under an electron microscope [2]. This virus enters the human body through the angiotensin-converting enzyme 2 (ACE-2) receptor [3].

SARS-CoV-2 is transmitted directly via droplets produced by coughing, sneezing, or talking and contact leading to the direct infecting of others. Transmission can also be indirect through physical contact with the contaminated surfaces of objects or the use of aerosols in enclosed spaces [4].

Patients infected with SARS-CoV-2 may experience reactions ranging from being asymptomatic, through mild symptoms, to presenting severe ones. However, the virus primarily affects the respiratory system, producing fever, dry cough, and dyspnea. It is widely known that the symptoms of COVID-19 are extremely diverse, encompassing mild symptoms to hypoxia with *acute respiratory distress syndrome* (ARDS) [5]. Patients experiencing mild cases may present upper respiratory tract infection with a dry cough, low-grade fever, nasal congestion, sore throat, headache, muscle aches, and malaise [6]. In moderate cases, the patient experiences respiratory tract disorders as indicated by symptoms such as a cough, shortness of breath, and tachypnea [6]. Patients with severe symptoms may experience severe pneumonia, ARDS, sepsis, and septic shock [6].

Patients with cases of mild severity can progress rapidly to severe or critical ones [6], while the over-65 s run a greater risk of progressing to a more severe phase [7]. Elderly patients and individuals with comorbidities tend to succumb to the virus, one underlying reason being that abnormal lung tissue causes dysfunctional maturation of dendritic cells, resulting in activation of damaged T cells [8].

The severity of COVID-19 is associated with an increase in inflammatory mediators, including cytokines such as interleukin (IL)-2, IL-7, and IL-10 and tumor necrosis factor (TNF), among others [9]. Within the many elevated inflammatory mediators, blood levels of IL-6 are strongly correlated with COVID-19-induced death. This disease is characterized by cytokine storm-induced cytokine release syndrome (CRS) producing a high mortality rate [10]. *Cytokine storm* is an acute hyper inflammatory response responsible for critical illnesses such as viral infections, as in the case of SARS-CoV-2. Critically ill COVID-19 patients who develop cytokine storms are known to have a poor prognosis and an increased mortality rate [11].

The therapy management aimed at treating COVID-19 is supportive in nature, treating the symptoms experienced by the patient as a means of preventing respiratory failure. Current therapeutic management using several drugs, including lopinavir-ritonavir, remdesivir, azithromycin, and hydroxychloroquine, has been clinically trialed. However, none has proven to be a definitive therapy. Therefore, determining the appropriate drug therapy is very important in dealing with the COVID-19 pandemic. Azithromycin is one of the proposed drugs for the treatment of COVID-19 [12]. On the other hand, the drug combination lopinavir-ritonavir is also used in COVID-19 therapy due to its ability to inhibit the SARS-CoV 3C protease enzyme [13].

2 Etiology

2.1 The Entry Stage of the Virus

COVID-19 is transmitted through droplets expelled from the respiratory tract or aerosols from an infected person produced while coughing or sneezing. Droplets from infected individuals can spread across a distance of up to 2 meters and adhere to surfaces which would subsequently become potential sources of transmission of the virus to healthy individuals who touch their mouth, nose, and/or eyes with unsanitized hands. Moreover, asymptomatic carriers can also spread the COVID-19 virus [15]. A study conducted by Lirong et al. [16] reported that SARS-CoV-2 survives in aerosols for 3 h. In addition, the stability of this virus enables it to survive on plastic, stainless steel, copper, and cardboard surfaces where it can be detected for up to 72 h. Therefore, the transmission of SARS-CoV-2 via both aerosols and surfaces is possible because the virus can survive for any length of time from hours to several days.

The virus infiltrates the host's body by binding to a host receptor through endocytosis or fusion [17]. As shown in Fig. 1, coronavirus consists of four protein structures, namely, spike, membrane, envelop, and nucleocapsid [18]. Spike has two functional subunits: the S1 subunit functions to bind to host cell receptors, while the S2 subunit plays a role in viral and cellular membrane fusion, as shown in Fig. 2. The functional receptor for SARS-CoV-2, angiotensin-converting enzyme 2 (ACE-2), is found in the heart, ileum, kidney, and bladder [19, 20]. Through these receptors, viral S protein initiates host cell invasion. After SARS-CoV-2 binds to the ACE-2 receptor, the S protein is activated through protease cleavage, followed by fusion of the virus and host membrane. The virus enters pulmonary alveolar epithelial cells post-fusion, thereby releasing viral content. In the host cell, the virus undergoes replication and the formation of RNA through transcription. After the

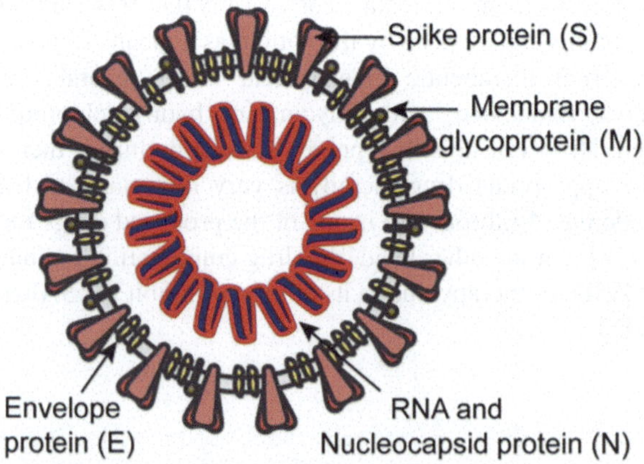

Fig. 1 Schematic structure of SARS-CoV-2 [14]

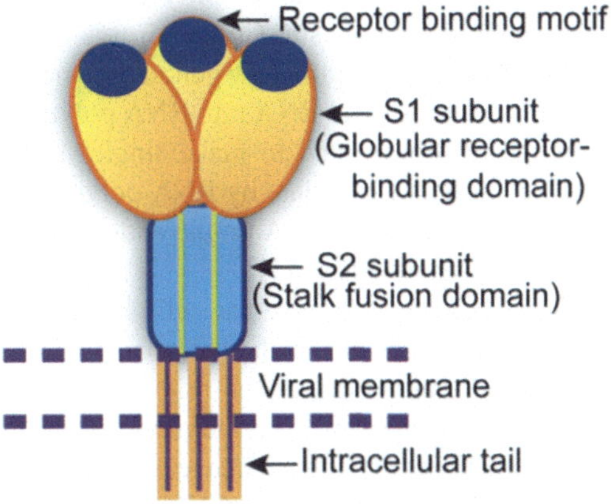

Fig. 2 The trimeric structure of the SARS-CoV-2 S protein [14]

new RNA is formed, new proteins continue to be synthesized in the cytoplasm of the cell [17].

2.2 *Incubation Stage*

It was reported that the immune response due to SARS-CoV-2 infection includes two clinical phases, incubation and infection. The World Health Organization (WHO) states that the postinfection incubation period for SARS-CoV-2 ranges from 1 to 14 days and may vary between individuals. An adaptive immune response is required during the incubation phase to eliminate the virus and prevent progression of the disease to a more severe stage [21].

It is important to increase the immune response at this stage since if it is compromised the virus can spread causing massive damage to the affected tissues, especially in organs with a high ACE-2 expression. Damaged cells induce inflammation mediated by pro-inflammatory macrophages and granulocytes. This inflammation causes pneumonia which can be life-threatening at an advanced stage [21]. Due to the targeting of SARS-CoV-2 in the respiratory system during the incubation stage, this virus attacks epithelial cells in the respiratory tract, producing symptoms in the respiratory system [22].

2.3 *Stage of Infection*

The most common symptoms experienced by patients include fever, dry cough, shortness of breath, fatigue, nausea/vomiting, diarrhea, and myalgia. Furthermore, the patient also presented other symptoms such as disorders in the digestive tract, anosmia being one of the prominent symptoms in COVID-19 patients [23].

When SARS-CoV-2 successfully infects the host cell, the genome of the uncoated virus enters its cytoplasm. Coronaviruses have an RNA genome that can directly produce proteins from the new genome in the cytoplasm. Host cell ribosomes function in translating viral RNA into RNA polymerase proteins. RNA polymerase then reads the positive strand to make single-stranded negative-sense RNA (ssRNA-) which is used as a template by RNA polymerase to make other ssRNA+. Host cell ribosomes read strand RNA on the endoplasmic reticulum to make structural components of the virus [24, 25].

Virus replication in alveolar cells induces tissue damage and an inflammatory response; the latter of which is also triggered by the entry of the virus. Damage to alveolar cells also induces the release of interferon, cytokines, and several other intracellular components. Furthermore, these inflammatory cells lead to a cytokine storm that causes the organ damage and multi-organ failure seen in severe cases [26].

It is well established that COVID-19 infection is accompanied by an aggressive inflammatory response due to the release of large numbers of pro-inflammatory cytokines known as "cytokine storms." The immune response of patients infected with SARS-CoV-2 is hyperactive, triggering an excessive inflammatory reaction. It was reported that the cytokine profile of COVID-19 patients confirmed that the onset of a cytokine storm was directly related to lung damage, multiple organ failure, and a worsening prognosis of COVID-19 [27].

Certain patients experienced more severe symptoms including hypoxia and the development of pulmonary infiltrates which progress to more severe disease. The patient's condition can rapidly deteriorate leading to organ damage and death in patients with severe cases. Some patients with dyspnea and hypoxemia may develop acute respiratory distress syndrome (ARDS), septic shock, blood clotting dysfunction, and even organ failure within a period of 1 week [28].

3 Clinical Pathology of COVID-19

3.1 Clinical Symptoms

Common clinical symptoms in adults with COVID-19 include fever, dry cough, sore throat, headache, fatigue, myalgia, and shortness of breath. However, some infected patients may not present any symptoms at all. With the global spread of the disease, resulting in an increase in the number of patients, other symptoms began to be reported [29].

A standard clinical parameter among COVID-19 patients is that of a rise in body temperature, the degree of that increase reflecting the severity of inflammation experienced. It was reported that fever was the most common symptom observed in patients with an increase in body temperature indicating a response to a foreign substance [29].

Shortness of breath also represents a typical symptom in COVID-19 patients. The virus can cause respiratory problems, including lung dysfunction with reduced oxygen saturation leading to acute respiratory distress syndrome (ARDS). There are three main mechanisms through which SARS-CoV-2 causes patients to experience shortness of breath, inflammation of the alveoli and lung tissue, thrombosis, micro cloths, and neuroinvasion [30]. Inflammation of the alveoli and lungs causes disturbances in the gas diffusion capacity with the result that the patient experiences hypoxemia [31].

The targeting of the respiratory tract by the SARS-CoV-2 infection can cause damage resulting in numerous health problems; one of which is hypoxia. A number of patients were found to have silent hypoxia, a condition involving a decrease in oxygen saturation (~50–80%), yet did not subsequently experience difficulty breathing. Normally, the respiratory rate of hypoxic patients is elevated, but this symptom was not present [32].

3.2 Clinical Laboratorium Test

COVID-19 presents clinically in a wide range of symptoms ranging from none through those of mild flu to severe life-threatening conditions. Therefore, laboratory testing is required to identify COVID-19 patients in its early phase. There are currently two types of testing for COVID-19; molecular tests to detect SARS-CoV-2 viral RNA and serological/immunological tests. Molecular testing for viral RNA involves application of the polymerase chain reaction (PCR) technique. PCR assay was used to identify individuals infected with SARS-CoV-2 in its acute phase. Serological and immunological assays are based on the detection of antibodies produced through exposure to a virus or the detection of protein antigens in an infected individual. This test is used to identify an individual's immune status, in addition to identifying those who have been able to develop antibodies against the SARS-CoV-2 virus with the result that they become convalescent plasma donors [33].

The parameters of D-dimer, a fibrin degradation product widely used as a biomarker for thrombotic disorders, can also be used in the assay. A D-dimer value less than 0.5 g/mL is considered normal. The value of D-dimer becomes greater with age and in pregnant women. This parameter also increases with the intensifying severity of pneumonia [34]. During the current pandemic, D-dimer was identified as a potential indicator of the prognosis of COVID-19 disease in patients. The D-dimer parameter has been employed in several studies to predict disease severity. This value rises as the severity of COVID-19 increases. In a study by Yao et al. [35], there was a significant correlation between the increase in the value of D-dimer and disease severity which was grouped according to lung infection, oxygen saturation, and the clinical stage of the patient.

The occurrence of lymphopenia and cytokine release syndrome (CRS) is also associated with the severity of disease. CRS is triggered by several factors characterized by increased pro-inflammatory cytokine interleukin-6 (IL-6) levels. Gorham et al.'s [36] study of 2020 reported that the IL-6 parameter increased above the normal range in COVID-19 patients. Moreover, it was also higher in patients with severe infection than those with mild-to-moderate cases.

Procalcitonin, a 116-amino acid precursor of the hormone calcitonin, can also be used as a parameter in laboratory tests. These biomarkers show increased inflammation in bacterial infections. Procalcitonin is also used to differentiate between strains of influenza involving bacterial infection and those without. In addition, it is also used to identify COVID-19 patients. It was recently reported that an increase in procalcitonin levels in such individuals was associated with the severity of the disease experienced by the patient [5, 7, 37]. In the study conducted by Hu et al. [38], the average procalcitonin level in SARS-CoV-2-infected patients presenting severe symptoms increased four times more than that in their counterparts presenting moderate symptoms. Meanwhile, critically ill patients experienced an eightfold increase compared to those with moderate symptoms.

COVID-19 not only affects the respiratory tract but reportedly also other organs such as the liver due to an increase in alanine aminotransferase (ALT) and aspartate aminotransferase (AST) values. Furthermore, COVID-19 patients present moderate microvesicular steatosis together with mild lobular and portal activity. This suggests that SARS-CoV-2 may cause liver damage [39, 40] because the coronavirus can directly affect hepatocytes. Alternatively, the liver can be injured by an intensified inflammatory response due to increased immunity and the toxicity of therapeutic drugs administered which cause an increase in liver enzymes [41].

Coronavirus can reportedly exert a direct cytopathic effect on the kidneys. This is based on detection using a polymerase chain reaction (PCR) which identifies fragments of the coronavirus in the blood and urine samples of COVID-19-infected patients. An in vitro study found that the cytopathic effect of SARS-CoV-2 on proximal tubular cells can cause acute kidney damage in COVID-19 patients, especially those with high ACE-2 receptor expression [42]. The study undertaken by Li et al. [43] reported kidney structure abnormalities in 100% of patients infected with SARS-CoV-2.

In addition to the parameters above, in the study by Wang et al. [44], the most common laboratory abnormalities found in COVID-19 patients included decreased lymphocytes, prolonged prothrombin time, and increased lactate dehydrogenase. These abnormalities suggest that SARS-CoV-2 infection is associated with cellular immune deficiency, activation of coagulation, myocardial injury, liver injury, and kidney injury. Laboratory test results of COVID-19 patients show a reduction in lymphocytes, indicating that COVID-19 infection primarily affects lymphocytes, particularly T lymphocytes. Virus particles can spread and infect other cells, induce a cytokine storm in the body, induce a series of immune responses, and cause changes in peripheral white blood cells and immune cells such as lymphocytes. In certain patients, the disease progresses rapidly, causing ARDS and septic shock which lead to organ failure [45].

4 Azithromycin for COVID-19 Infection

4.1 Physicochemical Characteristics of Azithromycin

Azithromycin is a macrolide antibiotic commonly used to treat various bacterial infections (Fig. 3). This drug constitutes a broad-spectrum macrolide that is bacteriostatically active against numerous strains of gram-positive and gram-negative bacteria, including *Mycoplasma pneumoniae*, *Treponema pallidum*, Chlamydia sp., and *Mycobacterium avium* [46]. This antibiotic is deemed appropriate for the treatment of respiratory tract, urogenital, and skin infections, in addition to specific ones caused by other bacteria that are sensitive to azithromycin. This antibiotic has a low plasma level [46] with only 37% drug bioavailability [47] reflecting its poor pharmacokinetic profile.

On the other hand, the large volume of distribution of azithromycin enables it to achieve high tissue concentrations. The blood concentration of the drug decreases polyphasically with a relatively short half-life indicating that azithromycin is rapidly distributed in tissues. One of the highest drug concentrations was found in the lungs [47]. Therefore, this drug can be delivered efficiently to the infection site. Compared with other macrolides, it is more stable in acidic media and has a longer half-life enabling its possible administration once a day [46]. Azithromycin is not metabolized by cytochrome P450. The main elimination route of this drug is through biliary excretion of the unchanged drug. Between 5% and 14% of the unchanged drug is excreted in the urine [48].

Azithromycin can be administered orally at a dose of 500 mg for 3–5 days or 500 mg as an initial dose followed by 250 mg delivered orally on Day 2 for a further

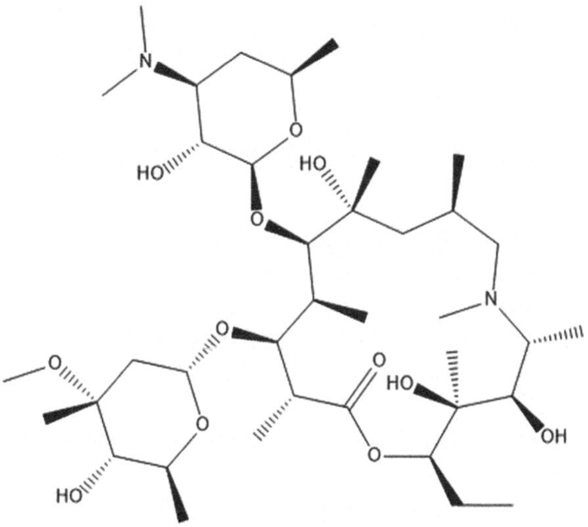

Fig. 3 Chemical structure of azithromycin

5 days [49]. However, at the time of writing, the most optimal dose in treating viral infections remains uncertain.

4.2 Action Mechanisms for COVID-19

Azithromycin has recently been shown to act on the binding of SARS-CoV-2 to respiratory cells [50] while also reported to possess anti-inflammatory properties. This drug is cationic, can accumulate in acidic cellular compartments, and binds to phospholipids, thereby increasing lysosomal pH and inducing phospholipidosis [51]. The increase in lysosomal pH due to azithromycin can also change the process of endocytosis and the function of lysosomal proteases, including cathepsin or furin, which complicates the fusion process. Reduction of furin activation by azithromycin can prevent the entry of SARS-CoV-2 into human epithelial cells [51].

In addition to antibacterial activity, azithromycin exhibits antiviral and immunomodulatory activity in treating viral infections, including COVID-19 therapy. Azithromycin is also thought to possess antiviral properties that work synergistically with antiviral drugs. Preclinical studies have found that these macrolide antibiotics can produce antiviral effects against Zika virus and induce an antiviral response to bronchial epithelial cells [52, 53]. Azithromycin has an anti-inflammatory effect that, particularly if administered early to patients, can reduce cytokine levels, in turn, preventing tissue damage and COVID-19 symptoms from becoming severe [54].

In addition, azithromycin can resemble ganglioside which has the same geometric volume and chemical properties. Spike protein in SARS-CoV-2 can bind to ganglioside, while azithromycin can inhibit SARS-CoV-2 infection by binding to the site. This prevents the viral spike protein from binding to gangliosides on the host plasma membrane [55].

4.3 The Use of Azithromycin in COVID-19 Patients

The use of azithromycin in the treatment of COVID-19 is based on the drug's mechanism and previous evidence of its use in treating pneumonia due to other virus infection, chronic lung disease, and inflammatory disorders [56]. Azithromycin has been proposed as a potential therapy in the treatment of COVID-19 because it has antiviral and immunomodulatory activities and a strong safety profile [46]. As previously explained, this drug has interesting pharmacological and therapeutic properties with the potential to treat COVID-19. Azithromycin can be widely distributed throughout tissues, especially those in the lungs, where its level in both extracellular and intracellular fluids is higher than that in plasma [57].

The desirable characteristics of azithromycin, together with its high safety profile, promote the therapeutic prescribing of this drug to COVID-19 patients.

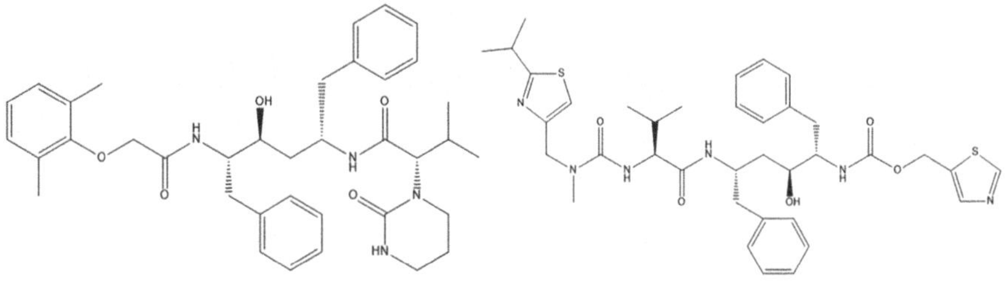

(A) (B)

Fig. 4 Chemical structure of (**a**) lopinavir and (**b**) ritonavir

Azithromycin's anti-inflammatory effect can reduce levels of cytokines, thereby preventing tissue damage and the dramatic progression of COVID-19, especially early in the course of the disease [58].

5 Lopinavir-Ritonavir for COVID-19 Infection

5.1 Physicochemical Characteristics of Lopinavir-Ritonavir

Lopinavir, an antiviral belonging to the protease inhibitor class with molecular structures shown in Fig. 4, is generally used in fixed-dose combinations with other protease class inhibitors such as ritonavir (lopinavir-ritonavir) in the treatment of human immunodeficiency virus [59].

Twice-daily administration of lopinavir-ritonavir 400/100 mg produces peak plasma concentrations at 12 h. Lopinavir given as a capsule or as a liquid dose is highly bound (98–99%) to plasma proteins [60]. The administration of lopinavir in combination with ritonavir at low booster doses can improve the pharmacokinetics of lopinavir by decelerating hepatic metabolism through inhibition of the cytotrope P450 3A4 enzyme [61]. Combination with ritonavir also decreases the elimination of lopinavir by inhibiting CYP3A4 in the liver [62]. The recommended daily dose of lopinavir in adults is 800 mg in combination with 200 mg of ritonavir, generally administered separately twice a day [61].

5.2 Mechanism of Action Lopinavir-Ritonavir for COVID-19 Therapy

Lopinavir inhibits the peptidomimetic HIV type 1 aspartate protease by binding to its catalytic site, thereby preventing the cleavage of the viral precursor polyprotein into mature functional proteins for viral replication [61]. Lopinavir-ritonavir has been shown to have antiviral effects on SARS-CoV and MERS-CoV by inhibiting protease activity of the coronavirus [63]. The drug combination demonstrated significant anti-SARS-CoV-2 activity both in terms of preventing cytotoxicity and reducing the viral load [64].

The SARS-CoV-2 virus is a single RNA virus similar to SARS-CoV and MERS-CoV which enters the body by replicating numerous copies of its genetic material. On the other hand, an enzyme 3-chymotrypsin-like protease (3CLpro) plays an important role in processing viral RNA. Lopinavir, a member of the protease inhibitor class, can inhibit the action of 3CLpro, interfering with the replication and release of the virus from host cells [65, 66].

5.3 The Use of Lopinavir-Ritonavir in COVID-19 Patients

The lopinavir-ritonavir combination is one of the drugs used in the treatment of COVID-19. Several studies have shown that the combination of lopinavir-ritonavir as an initial therapy in COVID-19 patients produces clinical improvement and treats acute respiratory disease effectively [67, 68]. The study conducted by Choy et al. [69] reported that the use of lopinavir could reach a working concentration of 26.63 M against SARS-CoV-2 in Vero E-6 cells. Meanwhile, single ritonavir did not produce in vitro activity against SARS-CoV-2. However, combining lopinavir and ritonavir produces significant antiviral activity. The current use of lopinavir-ritonavir is based on previous experience of its application in the treatment of SARS and MERS [64].

Lopinavir demonstrated inhibitory activity against the main protease SARS-CoV-2. The role of the nonstructural protein of the coronavirus, the main protease, or 3C-like protease (3CLpro), affects the proteolytic process of polyprotein replication and is crucial to viral maturation. CL-pro is a target drug in the treatment of coronavirus infections. Both Lopinavir and Ritonavir effectively interact with residues at the SARS-CoV-2 3CLpro active site [13, 70].

6 Antibiotic/Antiviral Combination as a Therapy for COVID-19 Infection: In Vitro Study

6.1 Evaluation of Antivirus Efficacy

Combining the two drugs azithromycin and lopinavir-ritonavir enhanced effectiveness with regard to reducing the viral copy number. In this study [71], an evaluation of the drug combination was carried out by testing it on Vero cells previously co-cultured with COVID-19 virus isolated from patients and testing them for IC50 at points 24, 48, and 72 h after the virus inoculation. It was reported that a combination of the two drugs proved more effective in reducing viral load, with the IC50 value decreasing in the infected cells after incubation for 24, 48, and 72 h.

6.2 Cell Cytotoxicity Test for Safety Evaluation

The combination of azithromycin with lopinavir-ritonavir at a ratio of 1:1 and 1:2 was reported as reducing the cytotoxicity of each drug in mesenchymal cells. The CC50 value for the combination of azithromycin with lopinavir-ritonavir (1:1) was 1.8×10^3 μg/mL, while for lopinavir-ritonavir (1:2) it was 1.15×10^{10} μg/mL, as analyzed using ComboSync. Combining these two drugs can reduce the risk of cytotoxicity of each on mesenchymal cells [71].

6.2.1 Analysis of Pro- and Anti-Inflammatory Marker Levels on Cells Co-incubated with Virus

The combination of azithromycin and lopinavir-ritonavir has also been reported as increasing levels of IL-10 and reducing those of IL-6 and TNF-α [71]. In the study undertaken by Sugiyama et al. [72], azithromycin can significantly increase IL-10 in dendritic cells. The anti-inflammatory effect of azithromycin is demonstrated by an increase in levels of IL-10 which is known to have the potential to successfully treat disease.

The evaluation of a combination of azithromycin with lopinavir-ritonavir indicated that it produced effective antiviral therapy in COVID-19 patients by reducing the copy number of the SARS-CoV-2 virus. Based on the cytotoxicity test results, a reduction in the toxicity of each single drug, either azithromycin or lopinavir-ritonavir, was also reported. Evaluation of pro- and anti-inflammatory markers also showed that the combination of these drugs resulted in a decrease in IL-6 and TNF-α levels and an increase in IL-10. This indicates a clinical improvement resulting from the use of azithromycin combined with lopinavir-ritonavir.

7 Antibiotic/Antiviral Combination as Therapy for COVID-19 Infection: Clinical Study

7.1 Evaluation of Antiviral Study

Evaluation of the clinical efficacy study was carried out by monitoring the improvement in clinical symptoms, including fever, cough, runny nose, shortness of breath, and sore throat. Patients treated with a combination of azithromycin and lopinavir-ritonavir reported an improvement in their clinical symptoms, and it was found that their medical conditions were no longer present on Day 4 [73].

There is a high expectation of rapid negative viral conversion in patients when managing COVID-19 cases. This is intended to inhibit an excessive physical reaction in response to the development of the virus within the body which can cause a dangerous cytokine storm. During a qualitative PCR examination of patients treated with a combination of azithromycin and lopinavir-ritonavir, 92.2% were negative on Day 3 [73].

Quantitative analysis was undertaken to find a significant reduction in the number of virus copy numbers from Day 1 to Day 3 and Day 7. Compared with the control group which only received azithromycin therapy, combination therapy involving lopinavir- ritonavir significantly reduced the viral load [73].

7.2 *Safe Use Evaluation by Monitoring Adverse Events*

Patients treated with a combination of azithromycin and lopinavir-ritonavir complained of dizziness, palpitations, impaired hearing function, and abdominal pains [73]. Regarding the evaluation of the combined use of azithromycin and lopinavir-ritonavir on renal function, an increase in serum creatinine was confirmed during clinical evaluation of COVID-19 patients. However, application of a combination of azithromycin and lopinavir-ritonavir in the treatment of COVID-19 produced no significant difference in patients' serum creatinine [73].

Liver function evaluation through analysis of SGOT and SGPT levels showed that the combination of azithromycin with lopinavir-ritonavir decreased these levels in 7 days, indicating an improvement in liver function [73]. This shows that the combination regimen of both drugs is capable of improving liver function by reducing the amount of virus circulating systemically in patients.

7.3 *Evaluation of Pro- and Anti-inflammatory Marker Levels in Patients*

Further analysis of the blood levels of cytokines including IL-6, IL-10, and TNF-α was evaluated in patients. There was a reduction in IL-6 values from Day 1 to Day 7 in patients treated with a combination of azithromycin and lopinavir-ritonavir. This combination of drugs was also reported as significantly increasing the IL-10 levels on Day 7. The TNF-α level was also found to decrease on Day 7 compared to patients who received a single dose of azithromycin, while the TNF-α value rose significantly [73].

IL-10 is a type 2 cytokine that inhibits the production of pro-inflammatory cytokines, including IFNγ, TNFα, IL-1β, and IL-6 [74]. A significant reduction in cytokine storm was characterized by a decrease in IL-6, an increase in IL-10, and a reduction in TNF-.

7.4 *Molecular Docking Study*

During the current COVID outbreak, computational drug screening has emerged as a rapid and efficient tool that is already available with the result that healthcare systems around the globe can deal with a highly contagious outbreak. In silico approaches and docking studies have become promising drug discovery and development tools. Molecular docking is an in silico approach to identifying virtual interactions between proteins and ligand molecules with low conformational energies. It involves identification of the target compound, optimization of the main compound, and virtual screening.

The SARS-CoV-2 virus that causes COVID-19 has important proteins, proteases, and spike glycoproteins, for infection and replication. The RBD (receptor-binding domain) of the spike glycoprotein (RBD-S) can bind to the ACE-2 (receptor on the protease (PD) domain (PD-ACE2) of the host cell, subsequently causing a viral infection.

The study was conducted by molecular docking using the MOE 2010 program. The protein targets selected were RBDS (PDB ID: 6LXT), PD-ACE2 (PDB ID: 6VW1), and the SARS-CoV-2 protease (PDB ID: 6LU7). The bond affinity formed is represented as a docking score. Molecular docking studies were chosen as the tool to screen the binding affinity of several natural compounds of the product to the SARS-CoV-2, RBD-S, PD-ACE2, and SARS-cov-2 marker proteins. Molecular docking studies comprising docking simulations, RMSD calculations, and visualization of binding interactions were performed using MOE 2010. The RBD-S model employs a crystal structure reported with PDB ID 6VSB due to information from the spike structure of a refused glycoprotein containing a single receptor binding domain. PDB ID 6VW1 was used as a PD-ACE2 model in a complex with SARS-CoV-2 RBD. For the crystal structure of the SARS-CoV-2 protease, PDB ID 6LU7 which informs the structure of the protease domain in complex with the protease inhibitor was employed. The molecular docking results describe the affinity represented by the docking score and the binding interaction of each compound on the target protein [75].

Ligand interactions visualized protein-ligand interactions of stably tethered lopinavir with major protease complexes. Lopinavir demonstrated hydrogen bonding interactions with Glu 166, GLN 189, and GLY 143 and pi-pi stacking with His 41 (Fig. 2a, b) and successfully docked with the highest docking score of 9.918, glide energy of 8.023, and glide of e-model 76,898 (Table 1). Therefore, from this in silico approach, the US FDA-approved drug, lopinavir, can be tested in vivo as a potent drug against SARS-CoV-2 [76].

The main protease SARS-CoV-2 (PDB ID: ALU6) is a ~306 amino acid principal protease whose crystal structure with a resolution of 1.93 A has been elucidated.

Table 1 Docking score of three drug molecules, i.e., azithromycin, lopinavir, and ritonavir to potential binding domains of SARS-CoV-2

No	Ligand molecule	Software (method)	Energy (kcal/mol)										Ref.
			Main protein (6 LU7)	Spike glycoprotein (6LXT)	Main protease (ALU6)	Spike protein (6LZG)	Main protease (7BRP)	Nsp9 replicase (6W4B)	Main protease (6Y84)	NSP15 endoribonuclease (6VWW)	S-protein ACE2 (2AJF)	Others	
1	Lopinavir	MOE 2010	−11.62	–	–	–	–	–	–	–	–	–	75
2	Lopinavir	GLIDE module	−9.918	–	–	–	–	–	–	–	–	–	76
3	Lopinavir	Schrodinger Maestro 2019-2	–	–	−28.56	–	–	–	–	–	–	–	71
4	Ritonavir	Schrodinger Maestro 2019-2	–	–	−30.47	–	–	–	–	–	–	–	71
5	Azithromycin	Schrodinger Maestro 2019-2	–	–	−22.01	–	–	–	–	–	–	–	71
6	Lopinavir	CDOCKER	−50.95	–	–	–	–	–	–	–	–	–	77
7	Ritonavir	CDOCKER	−61.21	–	–	–	–	–	–	–	–	–	77
8	Azithromycin	AutoDock Vina	−5.00	–	–	−5	–	–	–	–	–	> − 3	78
9	Lopinavir	AutoDock Vina	NT	–	–	–	−6.0	–	–	–	–	–	79
10	Ritonavir	AutoDock Vina	NT	–	–	–	−5.3	–	–	–	–	–	79

No	Ligand molecule	Software (method)	Energy (kcal/mol)										Ref.
			Main protein (6 LU7)	Spike glycoprotein (6LXT)	Main protease (ALU6)	Spike protein (6LZG)	Main protease (7BRP)	Nsp9 replicase (6W4B)	Main protease (6Y84)	NSP15 endoribonuclease (6VWW)	S-protein ACE2 (2AJF)	Others	
11	Azithromycin	AutoDock Vina	NT	–	–	–	–	–4.9	–5.8	–5.6	–6.7	–	80
12	Lopinavir	AutoDock Vina	NT	–	–	–	–	–6.0	–6.9	–8.0	–7.6	–	80
13	Ritonavir	AutoDock Vina	NT	–	–	–	–	–6.8	–6.8	–7.2	–7.5	–	80

The main protease enzyme is the optimum target for inhibiting the SARS-CoV-2 virus. This protease breaks the spike and is further formed by penetration. The results showed that all ligands, lopinavir, ritonavir, and azithromycin, can interact with the main viral proteases. It is possible that these compounds inhibit the process of viral replication and translation and may have a highly significant impact on controlling the viral load in infected individuals [71].

Since the main protease (Mpro) is a critical protein in the viral life cycle, Mpro is presented as one of the essential targets of antiviral treatment. Mpro inhibitors bind to Mpro and prevent the replication of this virus and block the proteolytic cleavage of protein precursors essential for early infection. For comparative purposes, the docking of protease inhibitor drugs such as lopinavir and ritonavir has been carried out. The 2D interactions of these protease inhibitors indicate that most of the interactions are electrostatic and van der Walls forces [77].

Azithromycin was evaluated in silico against several viral target proteins, namely, ADP ribose phosphatase (ADPRP, PDB:6WO2), main protease (Mpro, PDB:6LU7), endoribonuclease (NSP15, PDB: 6WLC), papain-like protease (PLpro, PDB: 6WXR), RNA-dependent RNA polymerase (RdRp, PDB: 7BV2), spike protein (S-protein, PDB:6LZG), adapter-associated kinase 1 (AAK1 PDB:4WSQ), cathepsin L (GDP: 2YJC), furin (GDP: 6EQX), and cyclin-G-associated kinase (GAK, PDB: 4Y8D). Of the various human and virus-based target proteins obtained by the protein data bank, only two target proteins have lower docking values: Mpro and S-protein [78].

Rachakulla and Rachakulla [79] have studied virtual interactions and docking involving the major COVID-19 protease (PDB ID 7BRP) with lopinavir and ritonavir. The docking test was carried out by measuring ten observation modes. After successfully docking these two drug compounds into the COVID-19 protease main pathway in the complex, the results showed different modes of drug-protein interaction generated by docking scores. The binding mode with the least binding energy is considered to be the most effective because it is the most stable for the ligand, i.e., mode 1. The presence of F-hydrogen bonds and F-π interactions and also the presence of β-unsaturated ketones are thought to be responsible for showing the affinity binding to lopinavir which also exhibits greater binding affinity than other drug candidates docked in the study [79].

Barros et al. [80] carried out molecular docking tests to determine the interaction of three selected ligands with four SARS-CoV-2 receptors, namely, the Nsp9 replication protein, the main protease, NSP15 endoribonuclease, and also one of the SARS-CoV-2 receptors with homology modeling of spike protein (S-protein) and human ACE2 receptors. A total of 96 molecular docking calculations were completed, and criteria for interaction efficiency were defined as the complex formation with a binding energy value of better or equal to −7 kcal/mol. Azithromycin did not interact with the four selected receptors when the binding energy value was less than −7 kcal/mol. This may indicate that the drug acts at a different stage in the viral cycle than the one involving the protein in this study. Lopinavir and ritonavir showed interactions in two of the four receptors, viz., NSP15 endonuclease and S-protein-ACE2, with binding energy values greater than −7 kcal/mol at both receptors [80].

8 Conclusion

Since the outbreak of the COVID-19 pandemic, urgent efforts have been made and various studies undertaken to identify the most effective means of eradicating the virus. A considerable body of research related to various tests on the use of combination drugs in COVID-19 patients exists.

The in vitro evaluation of the use of the drug combination, azithromycin with lopinavir-ritonavir, can increase antiviral effectiveness by reducing the copy number of the SARS-CoV-2 virus. The cytotoxicity test results of the combination of these drugs are reported as capable of reducing the toxicity of every single drug, thereby increasing the safety profile in their use. In addition, there was also a decrease in IL-6 and TNF-α levels and an increase in IL-10 in the use of azithromycin with lopinavir-ritonavir which reflected the clinical improvement in patients.

In this clinical study, a combination of the two drugs was shown to be capable of producing high clinical effectiveness by accelerating clinical improvement and eradicating patient illnesses. Moreover, with regard to the PCR examination, in patients with mild severity treated with a combination of azithromycin with lopinavir-ritonavir, 92.2% were found to be negative on Day 3. In addition, clinical improvement was also achieved with a significant reduction in cytokine storm characterized by a decrease in IL-6, an increase in IL-10, and a decrease in TNF-α.

References

1. Yang Y, et al. SARS-CoV-2: characteristics and current advances in research. Virol J. 2020;17:1–17.
2. Abebe EC, Dejenie TA, Shiferaw MY, Malik T. The newly emerged COVID-19 disease: a systemic review. Virol J. 2020;17:1–8.
3. Hoffmann M, et al. SARS-CoV-2 cell entry depends on ACE2 and TMPRSS2 and is blocked by a clinically proven protease inhibitor. Cell. 2020;181:271–280.e8.
4. Lotfi M, Hamblin MR, Rezaei N. COVID-19: transmission, prevention, and potential therapeutic opportunities. Clin Chim Acta. 2020;508:254–66.
5. Huang C, et al. Clinical features of patients infected with 2019 novel coronavirus in Wuhan, China. Lancet. 2020;395:497–506.
6. Hassan SA, Sheikh FN, Jamal S, Ezeh JK, Akhtar A. Coronavirus (COVID-19): a review of clinical features, diagnosis, and treatment. Cureus. 2020;12:e7355.
7. Guan W, et al. Clinical characteristics of coronavirus disease 2019 in China. N Engl J Med. 2020;382:1708–20.
8. Qin W, et al. Clinical course and risk factors of disease deterioration in critically ill patients with COVID-19. Hum Gene Ther. 2021;32:310–5.
9. Hojyo S, et al. How COVID-19 induces cytokine storm with high mortality. Inflamm Regen. 2020;40:1–7.
10. Hirano T, Murakami M. COVID-19: a new virus, but a familiar receptor and cytokine release syndrome. Immunity. 2020;52:731–3.
11. Bhaskar S, et al. Cytokine storm in COVID-19—Immunopathological mechanisms, clinical considerations, and therapeutic approaches: the REPROGRAM consortium position paper. Front Immunol. 2020;11:1648.

12. Butler CC, et al. Azithromycin for community treatment of suspected COVID-19 in people at increased risk of an adverse clinical course in the UK (PRINCIPLE): a randomised, controlled, open-label, adaptive platform trial. Lancet. 2021;397:1063–74.
13. Nutho B, et al. Why are lopinavir and ritonavir effective against the newly emerged coronavirus 2019? Atomistic insights into the inhibitory mechanisms. Biochemistry. 2020;59:1769–79.
14. Mittal A, et al. COVID-19 pandemic: insights into structure, function, and hACE2 receptor recognition by SARS-CoV-2. PLoS Pathog. 2020;16:e1008762.
15. Chakraborty R, Parvez S. COVID-19: an overview of the current pharmacological interventions, vaccines, and clinical trials. Biochem Pharmacol. 2020;180:114184.
16. Lirong Z, et al. Aerosol and surface stability of SARS-CoV-2 as compared with SARS-CoV-1. N Engl J Med. 2020;382:1177–9.
17. Parasher A. COVID-19: current understanding of its pathophysiology, clinical presentation and treatment. Postgrad Med J. 2021;97:312–20.
18. Yuki K, Fujiogi M, Koutsogiannaki S. COVID-19 pathophysiology: a review. Clin Immunol. 2020;215:108427.
19. Li W, et al. Angiotensin-converting enzyme 2 is a functional receptor for the SARS coronavirus. Nat Publ. 2003:450–545. https://doi.org/10.1254/fpj.147.120.
20. Zou X, et al. Single-cell RNA-seq data analysis on the receptor ACE2 expression reveals the potential risk of different human organs vulnerable to 2019-nCoV infection. Front Med. 2020;14:185–92.
21. Shi Y, et al. COVID-19 infection: the perspectives on immune responses. Cell Death Differ. 2020;27:1451–4.
22. Yi Y, Lagniton PNP, Ye S, Li E, Xu RH. COVID-19: what has been learned and to be learned about the novel coronavirus disease. Int J Biol Sci. 2020;16:1753–66.
23. Meng X, Deng Y, Dai Z, Meng Z. COVID-19 and anosmia: a review based on up-to-date knowledge. Am J Otolaryngol. 2020;41:102581.
24. Astuti I, Ysrafil. Severe Acute Respiratory Syndrome Coronavirus 2 (SARS-CoV-2): An overview of viral structure and host response. Diabetes Metab Syndr. 2020;14:407–12.
25. Rahman S, Montero MTV, Rowe K, Kirton R, Kunik F. Epidemiology, pathogenesis, clinical presentations, diagnosis and treatment of COVID-19: a review of current evidence. Expert Rev Clin Pharmacol. 2021;14:601–21.
26. Liu J, et al. Overlapping and discrete aspects of the pathology and pathogenesis of the emerging human pathogenic coronaviruses SARS-CoV, MERS-CoV, and 2019-nCoV. J Med Virol. 2020;92:491–4.
27. Ragab D, Salah Eldin H, Taeimah M, Khattab R, Salem R. The COVID-19 cytokine storm; What we know so far. Front Immunol. 2020;11:1–4.
28. Tsai PH, et al. Clinical manifestation and disease progression in COVID-19 infection. J Chinese Med Assoc. 2021;84:3–8.
29. da Rosa Mesquita R, et al. Clinical manifestations of COVID-19 in the general population: systematic review. Wien Klin Wochenschr. 2021;133:377–82.
30. Hentsch L, et al. Breathlessness and COVID-19: a call for research. Respiration. 2021;100:1016. https://doi.org/10.1159/000517400.
31. Gattinoni L, et al. COVID-19 pneumonia: different respiratory treatments for different phenotypes? Intensive Care Med. 2020;46:1099–102.
32. Rahman A, et al. Silent hypoxia in COVID-19: pathomechanism and possible management strategy. Mol Biol Rep. 2021;48:3863–9.
33. Carter LJ, et al. Assay techniques and test development for COVID-19 diagnosis. ACS Cent Sci. 2020;6:591–605.
34. Querol-Ribelles JM, et al. Plasma d-dimer levels correlate with outcomes in patients with community-acquired pneumonia. Chest. 2004;126:1087–92.
35. Yao Y, et al. D-dimer as a biomarker for disease severity and mortality in COVID-19 patients: a case control study. J Intensive Care. 2020;8:1–11.

36. Gorham J, et al. Interleukine-6 in critically ill COVID-19 patients: a retrospective analysis. PLoS One. 2020;15:1–11.
37. Zhang J, et al. Clinical characteristics of 140 patients infected with SARS-CoV-2 in Wuhan, China. Allergy. 2020;75:1730–41.
38. Hu R, Han C, Pei S, Yin M, Chen X. Procalcitonin levels in COVID-19 patients. Int J Antimicrob Agents. 2020;56:106051.
39. Cai Q, et al. COVID-19: abnormal liver function tests. J Hepatol. 2020;73:566–74.
40. Xu Z, et al. Pathological findings of COVID-19 associated with acute respiratory distress syndrome. Lancet Respir Med. 2020;8:420–2.
41. Leulseged TW, et al. Laboratory biomarkers of COVID-19 disease severity and outcome: Findings from a developing country. medRxiv. 2021;182:1–15.
42. Pan X, et al. Identification of a potential mechanism of acute kidney injury during the COVID-19 outbreak: a study based on single-cell transcriptome analysis. Intensive Care Med. 2020;46:1114–6.
43. Li Z, et al. Caution on kidney dysfunctions of COVID-19 patients. SSRN Electron J. 2020:1–25. https://doi.org/10.2139/ssrn.3559601.
44. Wang D, et al. Clinical characteristics of 138 hospitalized patients with 2019 novel coronavirus-infected pneumonia in Wuhan, China. JAMA. 2020;323:1061–9.
45. Chen N, et al. Epidemiological and clinical characteristics of 99 cases of 2019 novel coronavirus pneumonia in Wuhan, China: a descriptive study. Lancet. 2020;395:507–13.
46. Parnham MJ, et al. Azithromycin: mechanisms of action and their relevance for clinical applications. Pharmacol Ther. 2014;143:225–45.
47. Lode H. The pharmacokinetics of azithromycin and their clinical significance. Eur J Clin Microbiol Infect Dis. 1991;10:807–12.
48. Rapp. Azithromycin iv oral. 1998;32:785–793.
49. Zheng S, Matzneller P, Zeitlinger M, Schmidt S. Development of a population pharmacokinetic model characterizing the tissue distribution of azithromycin in healthy subjects. Antimicrob Agents Chemother. 2014;58:6675–84.
50. Echeverría-Esnal D, et al. Azithromycin in the treatment of COVID-19: a review. Expert Rev Anti-Infect Ther. 2021;19:147–63.
51. Nujić K, Banjanac M, Munić V, Polančec D, Eraković Haber V. Impairment of lysosomal functions by azithromycin and chloroquine contributes to anti-inflammatory phenotype. Cell Immunol. 2012;279:78–86.
52. Gielen V, Johnston SL, Edwards MR. Azithromycin induces anti-viral responses in bronchial epithelial cells. Eur Respir J. 2010;36:646–54.
53. Bosseboeuf E, et al. Azithromycin inhibits the replication of Zika virus. J Antivir Antiretrovir. 2018;10:6–11.
54. Oliver ME, Hinks TSC. Azithromycin in viral infections. Rev Med Virol. 2021;31:1–13.
55. Gautret P, et al. Hydroxychloroquine and azithromycin as a treatment of COVID-19: results of an open-label non-randomized clinical trial. Int J Antimicrob Agents. 2020;56:105949.
56. Gyselinck I, Janssens W, Verhamme P, Vos R. Rationale for azithromycin in COVID-19: an overview of existing evidence. BMJ Open Respir Res. 2021;8:1–10.
57. Rodvold KA, Gotfried MH, Danziger LH, Servi RJ. Intrapulmonary steady-state concentrations of clarithromycin and azithromycin in healthy adult volunteers. Antimicrob Agents Chemother. 1997;41:1399–402.
58. Oldenburg CE, et al. Effect of Oral azithromycin vs placebo on COVID-19 symptoms in outpatients with SARS-CoV-2 infection: a randomized clinical trial. JAMA. 2021;326:490–8.
59. Su B, et al. Efficacy and tolerability of lopinavir/ritonavir- And efavirenz-based initial antiretroviral therapy in HIV-1-infected patients in a tertiary care hospital in Beijing, China. Front Pharmacol. 2019;10:1–8.
60. Cvetkovic RS, Goa KL. Lopinavir/ritonavir a review of its use in the management of HIV infection. Adis Drug Eval. 2003;63:769–802.

61. Meini S, et al. Role of lopinavir/ritonavir in the treatment of covid-19: a review of current evidence, guideline recommendations, and perspectives. J Clin Med. 2020;9:1–15.
62. Sham HL, et al. ABT-378, a highly potent inhibitor of the human immunodeficiency virus protease. Antimicrob Agents Chemother. 1998;42:3218–24.
63. Nukoolkarn V, Lee VS, Malaisree M, Aruksakulwong O, Hannongbua S. Molecular dynamic simulations analysis of ritronavir and lopinavir as SARS-CoV 3CLpro inhibitors. J Theor Biol. 2008;254:861–7.
64. Kang CK, et al. In vitro activity of lopinavir/ritonavir and hydroxychloroquine against severe acute respiratory syndrome coronavirus 2 at concentrations achievable by usual doses. Korean J Intern Med. 2020;35:782–7.
65. Anand K, Ziebuhr J, Wadhwani P, Mesters JR, Hilgenfeld R. Coronavirus main proteinase (3CLpro) Structure: Basis for design of anti-SARS drugs. Science (80-.). 2003;300:1763–7.
66. Zhang L, et al. Crystal structure of SARS-CoV-2 main protease provides a basis for design of improved a-ketoamide inhibitors. Science (80-.). 2020;368:409–12.
67. Zhang Z, et al. A comparative study on the time to achieve negative nucleic acid testing and hospital stays between danoprevir and lopinavir/ritonavir in the treatment of patients with COVID-19. J Med Virol. 2020;92:2631–6.
68. Chan KS, et al. Treatment of severe acute respiratory syndrome with lopinavir/ritonavir: a multicentre retrospective matched cohort study. Hong Kong Med J. 2003;9:399–406.
69. Choy KT, et al. Remdesivir, lopinavir, emetine, and homoharringtonine inhibit SARS-CoV-2 replication in vitro. Antivir Res. 2020;178:104786.
70. Wang J. Fast identification of possible drug treatment of coronavirus disease-19 (COVID-19) through computational drug repurposing study. J Chem Inf Model. 2020;60:3277–86.
71. Purwati, et al. An in vitro study of dual drug combinations of anti-viral agents, antibiotics, and/or hydroxychloroquine against the SARS-CoV-2 virus isolated from hospitalized patients in Surabaya, Indonesia. PLoS One. 2021;16:1–27.
72. Sugiyama K, et al. Differing effects of clarithromycin and azithromycin on cytokine production by murine dendritic cells. Clin Exp Immunol. 2007;147:540–6.
73. Purwati, et al. A randomized, double-blind, multicenter clinical study comparing the efficacy and safety of a drug combination of lopinavir/ritonavir-azithromycin, lopinavir/ritonavir-doxycycline, and azithromycin-hydroxychloroquine for patients diagnosed with Mild to Mo. Biochem Res Int. 2021; 2021.
74. Costela-Ruiz VJ, Illescas-Montes R, Puerta-Puerta JM, Ruiz C, Melguizo-Rodríguez L. SARS-CoV-2 infection: the role of cytokines in COVID-19 disease. Cytokine Growth Factor Rev. 2020;54:62–75.
75. Peele KA, Potla C, Srihansa T, Krupanidhi S. Since January 2020 Elsevier has created a COVID-19 resource centre with free information in English and Mandarin on the novel coronavirus COVID- 19 . The COVID-19 resource centre is hosted on Elsevier Connect , the company's public news and information. 2020.
76. Yudi Utomo R, Meiyanto E. Revealing the potency of citrus and galangal constituents to halt SARS-CoV-2 infection. 2020; 2:1–8.
77. Aghaee E, Ghodrati M, Ghasemi JB. In silico exploration of novel protease inhibitors against coronavirus 2019 (COVID-19). Informatics Med Unlocked J. 2020;23:100516.
78. Sayed AM, Khalaf AM, Abdelrahim MEA, Elgendy MO. Repurposing of some anti-infective drugs for COVID-19 treatment: a surveillance study supported by an in silico investigation. Int J Clin Pract. 2021;75:1–9.
79. Rachakulla VSR, Rachakulla HD. Potential docking affinity of three approved drugs against SARS-CoV-2 for COVID-19 treatment. ChemRxiv. 2020; https://doi.org/10.26434/chemrxiv.12548063.v1.
80. Barros RO, Junior FLCC, Pereira WS, Oliveira NMN, Ramos RM. Interaction of drug candidates with various SARS-CoV-2 receptors: an in silico study to combat COVID-19. J Proteome Res. 2020;19:4567–75.

Current Strategies to Combat COVID-19

Vidhi Shah and Tejal Mehta

Abstract The outbreak of the novel coronavirus disease (COVID-19) has created global crisis. It originated from Wuhan, China, and has outspread globally. The higher infectivity and asymptomatic transmission is the major concern for this disease. The primary mode of transmission is by droplet and fomite. The symptoms are fever, dry cough, fatigue, shortness of breath, sore throat which advances to respiratory failure, septic shock, and multiple organ failure. As the current treatment for COVID-19 is not available, the preventive measures are the only available options at an individual level to combat COVID-19. In this review, we have discussed traditional prophylactic measures, diagnosis, and therapeutics for management of COVID-19. The traditional medicines, herbs, vitamins, and essential minerals having antiviral and immunobooster activity known for the prevention of COVID-19 are summarized in the article. Currently used therapeutics for COVID-19 like remdesivir, chloroquine, and hydroxychloroquine, lopinavir/ritonavir, ribavirin, favipiravir, Arbidol, convalescent plasma therapy are discussed.

Keywords Coronavirus · SARS-CoV-2 · COVID-19 · Prevention · Diagnosis · Treatment

V. Shah (✉)
L. M. College of Pharmacy, Ahmedabad, Gujarat, India

T. Mehta
Institute of Pharmacy, Nirma University, Ahmedabad, Gujarat, India

R. Shegokar, Y. Pathak (eds.), *Viral Drug Delivery Systems*,
https://doi.org/10.1007/978-3-031-20537-8_16

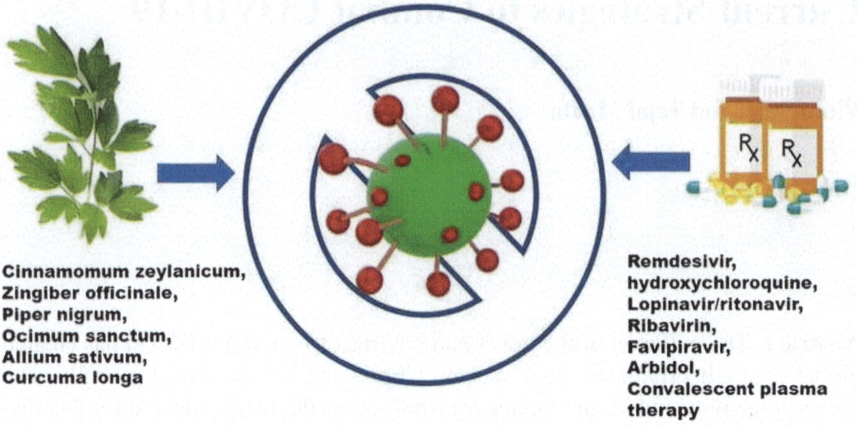

1 Introduction

The global health emergency has been created due to novel coronavirus 2019 (COVID-19) caused by severe acute respiratory syndrome coronavirus-2 (SARS-CoV-2) [1]. It originated in Wuhan, Hubei Province, Central China, in November 2019. In about 1 year, it has migrated across the world affecting billions of people and mortality to greater extent.

Coronavirus affects the individual by entering through respiratory system by droplets formed during coughing and sneezing. The spike protein on the surface of SARS-CoV-2 binds to the catalytic domain of angiotensin-converting enzyme-2 (ACE2) receptor on type II alveolar cells of the host. The virus multiplies in host cell through RNA replication and is released to infect nearby cells. The immune system of the body gets activated to attack the SARS-CoV-2 virus. Thus, the initial symptoms of sore throat, cough, fever, headache, body ache, and chills begin to appear. The fluid begins to accumulate in alveoli causing pneumonia and shortness of breath. Further progression of the virus leads to severe stage where patient may experience dyspnea and/or hypoxemia. The virus advances from the respiratory tract to the extrapulmonary organs by exocytosis [2]. The ACE2 receptors are also located on the cells of other organs such as the heart, kidney, liver, gastrointestinal tract, and bladder, increasing its susceptibility to virus. Due to the spread of infection to other organs and aggravated immune response of the body, it reaches to the critical stage of multiple organ failure and respiratory failure, and septic shock occurs. The rapid viral replication causes increase in pro-inflammatory cytokines, chemokine response, and inflammatory cell infiltration. Cytokine release syndrome is known to aggravate disease progression. Severe COVID-19 patients show increase in IL-6 and IL-10 level while lower CD4 + T and CD8+ level [3]. Thus, immunocompromised patient or elderly patients are more prone to the infection and has higher mortality rate.

The contact transmission is also reported by direct contact with infected person or indirect contact with virus-contaminated surfaces or object. The person gets infected when he touches the eyes, nose, or mouth with these contaminated hands. SARS-CoV-2 remains in air for 3 hours and on plastic and stainless steel for 24 to 72 hours. Thus, frequent washing of the hands, sanitization, and avoidance of touching eyes, nose, and mouth are recommended [4].

The mean incubation period of SARS-CoV-2 is 5 days (ranging from 1 to 14 days), so symptoms may start to develop after 14 days of infection which complicates it by causing difficulty in early diagnosis and treatment. Thus, active monitoring or quarantine of suspected patient for 14 days is recommended by the WHO [4].

This small invisible virus has caused massive destruction in terms of healthcare, employment, economy, and many more. The entire world is looking for the effective treatment and vaccine for the earliest rescue against COVID-19. In this chapter, we have summarized the preventive approach using traditional medicines, diagnosis, and currently used therapeutics against COVID-19.

2 Diagnosis

Real-Time Reverse Transcription-Polymerase Chain-Reaction (RT-PCR)
Currently, RT-PCR is the method of choice for the diagnosis of COVID-19. Nasal swab or oropharyngeal swab is used in this test to detect COVID-19. This assay is based on different primer and probe sets which are specific for SARS-CoV-2 genome. The common target genes for assay are RNA-dependent RNA polymerase (RdRp)/helicase (Hel), ORF1a, ORF1ab, envelope (E), spike (S), and nucleocapsid (N) genes of SARS-CoV-2. Chan et al. compared the performance RdRp/helicase (Hel), spike (S), and nucleocapsid (N) genes in RT-PCR assay. The COVID-19-RdRp/Hel assay was found to be effective, and no cross-reactivity with other human pathogenic coronaviruses and respiratory pathogens were observed [5]. Despite its high sensitivity, it is complex, expensive, and time-consuming.

Reverse Transcription Loop-Mediated Isothermal Amplification (RT-LAMP)
In this method, reverse transcription of RNA is followed by isothermal amplification reaction. The amplification of a target sequence is carried out with four to six primers under isothermal conditions (63–65 °C), using a polymerase with high strand displacement activity. The advantage is that the result of the test is determined through naked eyes by observing the change in color of the sample. This method has high sensitivity and specificity with the ability to provide positive results in as little as 5 minutes and negative results in 13 minutes [6].

Serological Methods
The infection with coronavirus develops the antibodies in the body. The IgG and IgM antibodies are developed against immunogenic N protein and S protein of SARS-CoV-2. This can be analyzed by different serological tests, including ELISA

(enzyme- linked immunosorbent assay), lateral flow immunoassays, IIFT (indirect immunofluorescence test), and neutralization tests. However, the limitation of serological test is that the antibodies against COVID-19 are generated 5–6 days or more after the onset of illness. Thus, during this lag time, even if the infected person is tested, the negative result will be obtained. Moreover, the cross-reactivity with other virus can lead to positive result. As a result, though it is quick, the reliability of the test is less [6].

CT Scan
The chest CT scan of COVID-19-infected patients shows ground-glass opacities in the lung and bilateral pulmonary infiltrates. The other common CT features of COVID-19 patients are consolidation with or without vascular enlargement, surrounding halo sign, air bronchogram, and interlobular septal thickening [7]. The CT scan of other viral infection of the same family may result into similar features of CT, and thus the expertise of the radiologist is an important parameter.

3 Therapeutics

3.1 Chloroquine and Hydroxychloroquine

Chloroquine and hydroxychloroquine are widely known for their antimalarial activity. Currently, few reports have shown their effectiveness against COVID-19 due to its immunomodulatory activity. Various mechanisms are reported for its efficacy against SARS- CoV-2. Chloroquine lowers the expression of phosphatidylinositol binding clathrin assembly protein (PICALM), which is responsible for the endocytosis of SARS-CoV-2. It also hinders viral binding by impairing terminal glycosylation of the ACE2 receptor. Additionally, it increases the endosomal pH, which prevents fusion of virus with host cell [8].

Gautret et al. reported that COVID-19 patients show efficient clearing of viral nasopharyngeal carriage of SARS-CoV-2 by hydroxychloroquine in 3–6 days, in most of the patients. The combination therapy of hydroxychloroquine and azithromycin is effective to cure the infection and to minimize the transmission of the virus to other people [9]. Gao et al. stated that treatment with chloroquine phosphate in more than 100 patients have demonstrated improved lung imaging with negative virus results indicating the shortening of diseases [10]. On contrary, Boulware et al. studied the postexposure prophylaxis with hydroxychloroquine after 4 days of exposure with COVID-19 patients. They found that hydroxychloroquine did not prevent illness compatible with COVID-19 or confirmed infection [11]. Geleris et al. performed similar study on hospitalized COVID-19 patients and observed no significant improvement in the condition of COVID-19 patient on treatment with hydroxychloroquine [12].

However, these drugs exhibit narrow margin between the therapeutic and toxic dose. Chloroquine poisoning may lead to life-threatening cardiovascular disorders of QTc prolongation with ventricular dysrhythmia. The other adverse effects on the hematologic, hepatic, and renal systems are also observed. Thus, these drugs should be used carefully, and self-treatment is not advisable; hydroxychloroquine decreases the severe progression of COVID-19 by suppressing T-cell activation and inhibiting cytokine storm. It also has lesser side effects compared to chloroquine with same antiviral efficacy [13].

3.2 *Lopinavir/Ritonavir*

Lopinavir/ritonavir combination (Kaletra) was previously found to be effective in SARS-CoV and MERS-CoV, so its use for treatment of COVID-19 is attempted clinically [1].

Lopinavir/ritonavir combination is used as HIV-1 protease inhibitors which might inhibit the protein synthesis of the SARS-CoV-2. The combinational of lopinavir with other effective drugs against COVID-19 virus might increase synergy and reduce the inhibitory concentration of lopinavir [14].

In vitro studies show that lopinavir is a potent inhibitor of HIV-1 than ritonavir. However, lopinavir has poor bioavailability in vivo due to rapid catabolism by the cytochrome P450 3A4 enzyme system. Ritonavir inhibits HIV-1 protease as well as cytochrome P450 3A4 enzyme and thereby extends the bioavailability of lopinavir in vivo when given in combination.

According to Cao et al., the randomized control trial was conducted of 199 severe COVID-19 patients of which 99 were treated with lopinavir/ritonavir combination and 100 as standard control. They found no significant benefit of lopinavir/ritonavir in hospitalized COVID-19 patients compared to standard care. The time for clinical improvement, mortality at 28 days, and viral loads at different time were similar between both the groups [15]. Cheng et al. showed that lopinavir/ritonavir combination did not shorten the duration of SARS-CoV-2 viral shedding in patients with mild pneumonia [16].

3.3 *Remdesivir*

Gilead Sciences has developed an antiviral drug, remdesivir (GS-5734), which is a ray of hope for COVID-19 treatment. Remdesivir is a nucleoside analog drug which was previously found to be effective against SARS-CoV, MERS-CoV, and Ebola virus infection. Remdesivir is RNA-dependent RNA polymerase (RdRp) inhibitor, which inhibits the replication of SARS- CoV-2 in respiratory epithelial cells. It evades proofreading of viral exoribonuclease leading to premature termination of viral RNA transcription. It has been reported that remdesivir competes with natural counterpart adenosine triphosphate, confuses viral RdRp, and acts as a delayed RNA chain terminator [17].

Remdesivir has shown good safety and pharmacokinetics in both phases I and II clinical trials [2]. However, it has not been approved in any country as its safety and effectiveness must be confirmed by further clinical trials.

In the United States, the first COVID-19 patient treated with remdesivir showed significant improvement in clinical symptoms within 24 hours of treatment. Grein et al. reported the compassionate use of remdesivir for severe COVID-19 patient. The clinical improvement was observed in 36 of 53 patients indicating 68% successful result [18]. Wang and colleagues found reduction in time to clinical improvement but no statistically significant clinical benefits in severe COVID-19 patient [19]. Antinoti et al. stated better clinical outcome with fewer adverse effect by remdesivir for COVID-19 patients hospitalized outside the ICU [20]. Beigel et al. reported the randomized clinical trial with 1059 patients enrolled where 538 received remdesivir and 521 had placebo. The mean recovery time with remdesivir was found to be 11 days while with placebo it was 15 days. Thus, remdesivir reduced the duration of illness in hospitalized COVID-19 patients [21]. The duration of treatment for 5 days and 10 days with remdesivir to severe COVID-19 patients did not show significant difference [22].

3.4 Ribavirin

Ribavirin is a guanosine analogue that prevents replication of RNA viruses by suppressing inosine monophosphate dehydrogenase activity which is responsible for the synthesis of guanosine triphosphate. In vitro studies have shown that ribavirin inhibits replication of MERS-CoV and HCoV-OC43 [23]. However, the significant effects are produced at high dose leading to severe adverse effects like hemolysis, increased bilirubin, hepatotoxicity, etc. Thus, it is used in combination with other drugs in order to minimize the therapeutic dose of the drug. Hung et al. reported phase 2 trial for the treatment of 127 COVID-19 patients with triple combination of ribavirin, interferon beta-1b, and lopinavir–ritonavir. The combination was found to improve the symptoms and shorten the duration of viral shedding and hospital stay of mild-to-moderate COVID-19 patients [24].

3.5 Favipiravir

Favipiravir is a purine nucleosides analogue which competes with nucleoside and interferes with viral replication by incorporation into the virus RNA and thereby inhibiting the RdRp of RNA viruses [25].

It is a substitute for the compassionate use in COVID-19 patients. The drug is under clinical trials. It has been used in Japan to treat influenza, and now it is approved as an experimental treatment for COVID-19 infections. Cai et al. reported the clinical trials of favipiravir against lopinavir/ritonavir as control for the treatment of COVID-19 patients. A shorter viral clearance time and significant improvement in chest imaging was observed for the favipiravir along with fewer adverse effects compared to control [26]. Chen et al. conducted randomized clinical trial to compare the efficacy of favipiravir and Arbidol for 7 days. They observed favipiravir did not improve clinical recovery rate at day 7 (61.21%) compared to Arbidol group (51.67%). However, it remarkably improved the latency to cough relief and decreased duration of pyrexia [27].

3.6 Arbidol

Arbidol is an indole derivative which blocks viral fusion with host cell and also regulates the immune system by inducing the production of interferon and the activation of macrophages. Previous studies have shown its efficacy against, Zika, Ebola, influenza A and B viruses, and hepatitis C viruses. In vitro studies have also shown its efficacy against SARS-CoV [16].

Li et al. studied the efficacy of Arbidol in COVID-19 patients. The results indicated the efficacy of Arbidol depends on the severity of COVID-19 patients. For patients with mild illness, Arbidol was found to enhance the viral clearance, improve focal absorption on radiologic images, and reduce the need for oxygen therapy in hospitalization. However, for patients with severe illness at admission, Arbidol presented no apparent advantages on virus clearance or clinical and radiologic recovery [28]. Zu et al. reported the superiority of Arbidol compared to lopinavir/ritonavir therapy as no viral load was detected in Arbidol group after 14 days compared to 44.1% in lopinavir/ritonavir group. However, no difference in fever duration or apparent side effects were observed in both group [29]. Deng et al. investigated the combination therapy of Arbidol and lopinavir/ritonavir compared to lopinavir/ritonavir alone for 33 COVID-19 patients. After 7 days, SARS-CoV-2 was undetectable in the nasopharyngeal specimens of 75% patient with combination therapy compared to 35% in monotherapy. However, after 14 days, it improved to 94% patient with combination therapy compared to 53% in monotherapy. The chest CT scans showed improvement after 7 days in the combination group [30]. The combination of Arbidol along with IFN-a2b was also studied. However, no significant difference in COVID-19 RNA clearance and hospitalization was observed with Arbidol and IFN-a2b combination compared to IFN-a2b monotherapy [31].

Few other drugs reported for COVID-19 treatment are listed in Table 1.

Table 1 Drugs effective against COVID-19 based on in vitro studies

Therapeutic drugs	Mechanism of action	Remarks	References
Teicoplanin	It inhibits the low pH cleavage of the viral S protein by cathepsin L in endosomes. Therefore, preventing the release of viral RNA into the cytoplasm of the host cell and inhibiting replication of virus	Reported to be effective against Ebola virus and able to block the MERS and SARS virus	[32]
Ivermectin	It inhibits IMPα/β1-mediated nuclear import of viral proteins	The single addition to Vero-hSLAM cells with SARS-CoV-2 was found to effect ~5000-fold reduction in viral RNA at 48 hours	[33]
Interferon-α (IFN-α)	IFN-α suppresses virus infection by directly interfering with viral replication and thereby promoting both innate and adaptive immune responses to infection	The clinical trials are under progress for combination of lopinavir/ritonavir and IFNα2b (ChiCTR2000029387) or a combination of IFNβ1b with lopinavir/ritonavir and ribavirin (NCT04276688) for the treatment of COVID-19	[34]

3.7 Angiotensin II Receptor Blockers

Few reports have also suggested the use of angiotensin-converting enzyme inhibitors and angiotensin II receptor blockers (ACEIs/ARBs) for the treatment of COVID-19 patients. The SARS-CoV-2 virus uses angiotensin-converting enzyme 2 (ACE2) as a receptor to bind the virus to the bronchial cell membrane. Neprilysin inhibitor and ACE2 receptor blocker (sacubitril and valsartan) reduces the concentration of pro-inflammatory cytokines and neutrophils with simultaneous increase in lymphocyte count [35]. Gurwitz et al. suggests the use of angiotensin receptor 1 (AT1R) blockers, such as losartan, might reduce the aggressiveness and mortality from SARS-CoV-2 virus infections [36]. However, no clinical evidences on the effectives of these drugs are reported till date. Table 2 shows the few marketed products of the above drugs for COVID-19.

4 Convalescent Plasma

Earlier convalescent plasma (CP) therapy was found to be effective against SARS, Ebola, H1N1, and Spanish flu. In convalescent plasma therapy, the plasma containing antiviral antibodies produced from COVID-19-recovered patients can be used to treat severely ill COVID-19 patients. The immunoglobulin antibodies present in the

Table 2 Marketed products of the drugs used for COVID-19 [37]

Drug	Brand name	Dosage form	Company
Chloroquine, hydroxychloroquine	Aralen, Plaquenil	Tablet	Sanofi-Aventis
Lopinavir and ritonavir combination	Kaletra	Tablet, capsule, and oral solution	AbbVie Corporation
Remdesivir	Veklury	Injection	Gilead Sciences
Ribavirin	Virazole	Inhalation solution	Bausch Health
	Moderiba	Tablet	AbbVie Corporation
Favipiravir	Fabiflu	Tablet	Glenmark Pharmaceuticals
	Avigan	Tablet	FUJIFILM Toyama Chemical Co., Ltd.
Umifenovir	Arbidol	Tablet	Pharmstandard

plasma of patients recovering from COVID-19 infection might suppress viremia in critical patients. The administration of CP has shown to shorten the hospital stay and mortality in severely ill COVID-19 patients. Duan et al. reported the effectiveness of CP in 10 severe COVID-19 patients. The clinical symptoms were improved within 3 days along with increase in lymphocyte count and decrease in C-reactive protein. The radiological examination presented varying degrees of absorption of lung lesions within 7 days. The viral load was undetectable after 7 days of transfusion without any severe adverse effect [38]. Sui et al. reported the successful treatment of a severe COVID-19 patient by plasma exchange immunoglobulin therapy [39]. Shen et al. investigated effectivity of CP therapy on five severely ill COVID-19 patients. After plasma transfusion, the body temperature was reduced in 3 days, viral load was decreased in 12 days, and neutralizing antibody titers increased in 7 days [40].

5 Prevention

Considering the current scenario of lack of specific treatment against SARS-CoV-2 and emergence of second wave of COVID-19, the only available option is prevention. The preventive approach is mask, social distancing, frequent sanitization, and immunity booster. Various approaches to boost the immunity of an individual are by different traditional medicines. Few commonly used natural products in China for COVID-19 management are hesperidin, glycyrrhizin, baicalin, and quercetin. Hesperidin inhibits the entry of the virus through ACE2 receptors, minimizes the release of inflammatory mediators, improves the cellular immunity, and prevents venous thromboembolism [41]. Glycyrrhizin is also known to inhibit viral absorption, penetration, and replication of SARS-CoV and so might be useful against SARS-CoV-2 also [42]. In India, the Ministry of Ayush has recommended few

Indian traditional medicines to fight against COVID-19 infection. According to Ayurveda, the decoction of *Cinnamomum zeylanicum*, *Zingiber officinale*, *Piper nigrum*, and *Ocimum sanctum* is suggested in order to improve the immunity of the individual. The consumption of turmeric, onion, ginger, and garlic is also advised considering its antioxidant, anti-inflammatory, antiviral, and immunomodulatory activity. Garlic represses the production and secretion of pro-inflammatory cytokines, suppresses TNF-α and C reactive protein, stimulates NK cells, and increases cytotoxic and helper T-cell concentration [43].

The decoction of *Andrographis paniculata* can be useful for COVID-19; which is widely used against Chikungunya and dengue. *Andrographis paniculata* suppresses the increased NOD-like receptor protein 3, interleukin-1β, and caspase-1 which are involved in the pathogenesis of coronavirus [44].

According to homoeopathic medicine, Arsenicum album 30 can be used as preventive measure against flu-like illness including coronavirus. The formulation containing Arsenicum album has shown to decrease NF-κB hyperactivity in HT29 cells and decrease TNF-α release in macrophages [45].

The vitamin C, vitamin D, and vitamin E supplements are reported for protection against COVID-19. Vitamin C reduces the susceptibility of lower respiratory tract infection. Vitamin E increases NK cells activity, enforces B cells as well as antibody responses, and initiates T-cell activation signals [46]. Vitamin D modulates adaptive immunity and increases cellular immunity thereby reducing the cytokine storm induced by the innate immune system [47].

Moreover, the application of sesame oil or coconut oil in both the nostrils aids to prevent the spread of SARS-CoV-2. Sesame oil has lower interfacial tension and thus coats the nasal mucosa and might disrupt the structure of virus. Additionally, it has high boiling point, so it remains on the surface for longer time [48].

The micronutrients like selenium and zinc is known to be beneficial for COVID-19 infection. Selenium inhibits the entry of viruses into the host cell and abolishes their infectivity. Organic form of selenium like sodium selenite oxidizes the thiol groups in the virus protein disulfide isomerase; consequently, it prevents the exchange reaction with disulfide groups of cell membrane proteins which inhibits its entry in host cell. It also increases the proliferation of natural killer (NK) cells. However, the toxicity should be considered as mild toxicity may result in reversible form of hair loss and nail [49]. Zinc-deficient populations showed higher susceptibility to infectious diseases, including viral infections. Zinc exhibits antiviral activity through the generation of both innate and acquired (humoral) immune responses, inhibition of viral entry, and replication through interference with the viral genome transcription, protein translation, polyprotein processing, viral attachment, and uncoating [50]. Here, we have outlined that a few traditional prophylactic approaches might be helpful against COVID-19. However, further evidence-based research is necessary to derive a final conclusion.

6 Conclusion

The COVID-19 pandemic has created public health emergency at global level. The transmission and treatment of COVID-19 has increased strain for healthcare system. The vigorous research for vaccine and specific drug against SARS-CoV-2 is under progress. Remdesivir is a hope for people, but still complete clinical trial needs to be done in order to get a final conclusion. Chloroquine and hydroxychloroquine along with azithromycin, favipiravir, and ivermectin are the current treatment strategy for the treatment of COVID-19. For convalescent plasma therapy, risk to benefit ratio needs to be considered before implementing.

Due to high infectivity and no specific treatment, prevention is the only possible option for the individual. Social distancing, frequent hand wash, and wearing of mask are the few precautionary measures to prevent infection. The decoction of *Cinnamomum zeylanicum*, *Zingiber officinale*, *Piper nigrum*, and *Ocimum sanctum* is suggested in order to improve the immunity of the individual. Additionally, immunity boosting agents like Arsenicum album, sesame oil, garlic, and *Andrographis paniculata* are recommended to fight against COVID-19. It is expected that the discovery of new medicine and vaccine could only nullify the corona virus totally.

References

1. Han Q, Lin Q, Jin S, You L. Coronavirus 2019-nCoV: a brief perspective from the front line. J Infect. 2020;80:373–7. https://doi.org/10.1016/j.jinf.2020.02.010.
2. Cao Y, Deng Q, Dai S. Remdesivir for severe acute respiratory syndrome coronavirus 2 causing COVID-19: an evaluation of the evidence. Travel Med Infect Dis. 2020;35:101647. https://doi.org/10.1016/j.tmaid.2020.101647.
3. Li H, Liu S-M, Yu X-H, Tang S-L, Tang C-K. Coronavirus disease 2019 (COVID-19): current status and future perspective. Int J Antimicrob Agents. 2020;2019:105951. https://doi.org/10.1016/j.ijantimicag.2020.105951.
4. van Doremalen N, Bushmaker T, Morris DH, Holbrook MG, Gamble A, Williamson BN, Tamin A, Harcourt JL, Thornburg NJ, Gerber SI, Lloyd-Smith JO, de Wit E, Munster VJ. Aerosol and surface stability of SARS-CoV-2 as compared with SARS-CoV-1. N Engl J Med. 2020;382:1564–7. https://doi.org/10.1056/NEJMc2004973.
5. Chan JF-W, Yip CC-Y, To KK-W, Tang TH-C, Wong SC-Y, Leung K-H, Fung AY-F, Ng AC-K, Zou Z, Tsoi H-W, Choi GK-Y, Tam AR, Cheng VC-C, Chan K-H, Tsang OT-Y, Yuen K-Y. Improved molecular diagnosis of COVID-19 by the novel, highly sensitive and specific COVID-19-RdRp/Hel real-time reverse transcription-PCR assay validated in vitro and with clinical specimens. J Clin Microbiol. 2020;58 https://doi.org/10.1128/JCM.00310-20.
6. Ortiz-Prado E, Simbaña-Rivera K, Gómez-Barreno L, Rubio-Neira M, Guaman LP, Kyriakidis NC, Muslin C, Jaramillo AMG, Barba-Ostria C, Cevallos-Robalino D, Sanches-SanMiguel H, Unigarro L, Zalakeviciute R, Gadian N, López-Cortés A. Clinical, molecular, and epidemiological characterization of the SARS-CoV-2 virus and the Coronavirus Disease 2019 (COVID-19), a comprehensive literature review. Diagn Microbiol Infect Dis. 2020;98:115094. https://doi.org/10.1016/j.diagmicrobio.2020.115094.

7. Li Y, Xia L. Coronavirus Disease 2019 (COVID-19): role of chest CT in diagnosis and management. AJR Am J Roentgenol. 2020;214:1280–6. https://doi.org/10.2214/AJR.20.22954.
8. Hu TY, Frieman M, Wolfram J. Insights from nanomedicine into chloroquine efficacy against COVID-19. Nat Nanotechnol. 2020;15:247–9. https://doi.org/10.1038/s41565-020-0674-9.
9. Gautret P, Lagier J, Parola P, Doudier B, Courjon J, La Scola B, Rolain J, Brouqui P, Raoult D, Mailhe M, Doudier B, Courjon J. Department of Virology, Biological and Pathological Center, Centre Hospitalier. Int J Antimicrob Agents. 2020;56:105949. https://doi.org/10.1016/j.ijantimicag.2020.105949.
10. Gao J, Tian Z, Yang X. Breakthrough: chloroquine phosphate has shown apparent efficacy in treatment of COVID-19 associated pneumonia in clinical studies. Biosci Trends. 2020;14:72–3. https://doi.org/10.5582/BST.2020.01047.
11. Boulware DR, Pullen MF, Bangdiwala AS, Pastick KA, Lofgren SM, Okafor EC, Skipper CP, Nascene AA, Nicol MR, Abassi M, Engen NW, Cheng MP, LaBar D, Lother SA, MacKenzie LJ, Drobot G, Marten N, Zarychanski R, Kelly LE, Schwartz IS, McDonald EG, Rajasingham R, Lee TC, Hullsiek KH. A randomized trial of hydroxychloroquine as postexposure prophylaxis for Covid-19. N Engl J Med. 2020;383:517–25. https://doi.org/10.1056/NEJMoa2016638.
12. Geleris J, Sun Y, Platt J, Zucker J, Baldwin M, Hripcsak G, Labella A, Manson DK, Kubin C, Barr RG, Sobieszczyk ME, Schluger NW. Observational study of hydroxychloroquine in hospitalized patients with Covid-19. N Engl J Med. 2020;382:2411–8. https://doi.org/10.1056/NEJMoa2012410.
13. Zhou D, Dai SM, Tong Q. COVID-19: a recommendation to examine the effect of hydroxychloroquine in preventing infection and progression. J Antimicrob Chemother. 2020;75:1667–70. https://doi.org/10.1093/jac/dkaa114.
14. Choy K-T, Wong AY-L, Kaewpreedee P, Sia SF, Chen D, Hui KPY, Chu DKW, Chan MCW, Cheung PP-H, Huang X, Peiris M, Yen H-L. Remdesivir, lopinavir, emetine, and homoharringtonine inhibit SARS-CoV-2 replication in vitro. Antivir Res. 2020;178:104786. https://doi.org/10.1016/j.antiviral.2020.104786.
15. Cao B, Wang Y, Wen D, et al. A trial of lopinavir–ritonavir in adults hospitalized with severe Covid-19. N Engl J Med. 2020;382:1787–99. https://doi.org/10.1056/nejmoa2001282.
16. Cheng C-Y, Lee Y-L, Chen C-P, Lin Y-C, Liu C-E, Liao C-H, Cheng S-H. Lopinavir/ritonavir did not shorten the duration of SARS CoV-2 shedding in patients with mild pneumonia in Taiwan. J Microbiol Immunol Infect. 2020;53:488–92. https://doi.org/10.1016/j.jmii.2020.03.032.
17. Martinez MA. Compounds with therapeutic potential against novel respiratory 2019 coronavirus. Antimicrob Agents Chemother. 2020;64 https://doi.org/10.1128/AAC.00399-20.
18. Grein J, Ohmagari N, Shin D, Diaz G, Asperges E, et al. Compassionate use of remdesivir for patients with severe Covid-19. N Engl J Med. 2020;382:2327–36. https://doi.org/10.1056/NEJMoa2007016.
19. Wang Y, Zhang D, Du G, et al. Remdesivir in adults with severe COVID-19: a randomised, double-blind, placebo-controlled, multicentre trial. Lancet. 2020;395:1569–78. https://doi.org/10.1016/S0140-6736(20)31022-9.
20. Antinori S, Cossu MV, Ridolfo AL, et al. Compassionate remdesivir treatment of severe Covid-19 pneumonia in intensive care unit (ICU) and Non-ICU patients: clinical outcome and differences in post-treatment hospitalisation status. Pharmacol Res. 2020;158:104899. https://doi.org/10.1016/j.phrs.2020.104899.
21. Beigel JH, Tomashek KM, Dodd LE, et al. Remdesivir for the treatment of Covid-19 – final report. N Engl J Med. 2020;383:1813–26. https://doi.org/10.1056/NEJMoa2007764.
22. Goldman JD, Lye DCB, Hui DS, Marks KM, et al. Remdesivir for 5 or 10 days in patients with severe Covid-19. N Engl J Med. 2020;383:1827–37. https://doi.org/10.1056/NEJMoa2015301.
23. Khalili JS, Zhu H, Mak A, Yan Y, Zhu Y. Novel coronavirus treatment with ribavirin: groundwork for evaluation concerning COVID-19. J Med Virol. 2020;92:740–6. https://doi.org/10.1002/jmv.25798.

24. Hung IF, Lung K, Tso EY, et al. Triple combination of interferon beta-1b, lopinavir – ritonavir, and ribavirin in the treatment of patients admitted to hospital with COVID-19: an open-label, randomised, phase 2 trial. Lancet. 2020;6736:1–10. https://doi.org/10.1016/S0140-6736(20)31042-4.
25. Du Y-X, Chen X-P. Favipiravir: pharmacokinetics and concerns about clinical trials for 2019-nCoV infection. Clin Pharmacol Ther. 2020;108:242–7. https://doi.org/10.1002/cpt.1844.
26. Cai Q, Yang M, Liu D, Chen J, Shu D, Xia J, Liao X, Gu Y, Cai Q, Yang Y, Shen C, Li X, Peng L, Huang D, Zhang J, Zhang S, Wang F, Liu J, Chen L, Chen S, Wang Z, Zhang Z, Cao R, Zhong W, Liu Y, Liu L. Experimental treatment with favipiravir for COVID-19: an open-label control study. Engineering. 2020;6:1192–8. https://doi.org/10.1016/j.eng.2020.03.007.
27. Chen C, Huang J, Cheng Z, Wu J, Chen S, Zhang Y, Chen B, Lu M, Luo Y, Zhang J, Yin P, Wang X. Favipiravir versus arbidol for COVID-19: a randomized clinical trial. Med Rxiv. 2020:2020.03.17.20037432. https://doi.org/10.1101/2020.03.17.20037432.
28. Xu KL, Chen Y, Yuan J, Yi P, Ding C, Wu W, Li Y, Ni Q, Zhou R, Li X, Xu M, Zhang Y, Zhao H, Zhang X, Yu L, Su J, Lang G. Clinical efficacy of arbidol in patients with 2019 novel coronavirus-infected pneumonia: a retrospective cohort study. SSRN Electron J. 2020; https://doi.org/10.2139/ssrn.3542148.
29. Zhu Z, Lu Z, Xu T, Chen C, Yang G, Zha T, Jianchun YX. Arbidol monotherapy is superior to lopinavir/ritonavir in treating COVID-19. J Infect. 2020;81:19–21. https://doi.org/10.1016/j.jinf.2020.03.060.
30. Deng L, Li C, Zeng Q, Liu X, Li X, Zhang H, Hong Z, Xia J. Arbidol combined with LPV/r versus LPV/r alone against Corona Virus Disease 2019: a retrospective cohort study. J Infect. 2020;81:1–5. https://doi.org/10.1016/j.jinf.2020.03.002.
31. Xu P, Huang J, Fan Z, Huang W, Qi M, Lin X, Song W, Yi L. Arbidol/IFN-α2b therapy for patients with corona virus disease 2019: a retrospective multicenter cohort study. Microbes Infect. 2020;22:200–5. https://doi.org/10.1016/j.micinf.2020.05.012.
32. Baron SA, Devaux C, Colson P, Raoult D, Rolain JM. Teicoplanin: an alternative drug for the treatment of COVID-19? Int J Antimicrob Agents. 2020;2:18–9. https://doi.org/10.1016/j.ijantimicag.2020.105944.
33. Caly L, Druce JD, Catton MG, Jans DA, Wagstaff KM. The FDA-approved drug ivermectin inhibits the replication of SARS-CoV-2 in vitro. Antivir Res. 2020;178:104787. https://doi.org/10.1016/j.antiviral.2020.104787.
34. Sallard E, Lescure F-X, Yazdanpanah Y, et al. Type 1 interferons as a potential treatment against COVID-19. Antivir Res. 2020;178:104791. https://doi.org/10.1016/j.antiviral.2020.104791.
35. Acanfora D, Ciccone MM, Scicchitano P, Acanfora C, Casucci G. Neprilysin inhibitor-angiotensin II receptor blocker combination (sacubitril/valsartan): rationale for adoption in SARS-CoV-2 patients. Eur Hear Journal Cardiovasc Pharmacother. 2020;6:135–6. https://doi.org/10.1093/ehjcvp/pvaa028.
36. Gurwitz D. Angiotensin receptor blockers as tentative SARS-CoV-2 therapeutics. Drug Dev Res. 2020;81:537–40. https://doi.org/10.1002/ddr.21656.
37. https://www.rxlist.com/.
38. Duan K, Liu B, Li C, et al. Effectiveness of convalescent plasma therapy in severe COVID-19 patients. Proc Natl Acad Sci U S A. 2020;117:9490–6. https://doi.org/10.1073/pnas.2004168117.
39. Shi H, Zhou C, He P, Huang S, Duan Y, Wang X, Lin K, Zhou C, Zhang X, Zha Y. Successful treatment of plasma exchange followed by intravenous immunogloblin in a critically ill patient with 2019 novel coronavirus infection. Int J Antimicrob Agents. 2020;56:105974. https://doi.org/10.1016/j.ijantimicag.2020.105974.
40. Shen C, Wang Z, Zhao F, Yang Y, Li J, Yuan J, Wang F, Li D, Yang M, Xing L, Wei J, Xiao H, Yang Y, Qu J, Qing L, Chen L, Xu Z, Peng L, Li Y, Zheng H, Chen F, Huang K, Jiang Y, Liu D, Zhang Z, Liu Y, Liu L. Treatment of 5 critically ill patients with COVID-19 with convalescent plasma. JAMA J Am Med Assoc. 2020;323:1582–9. https://doi.org/10.1001/jama.2020.4783.

41. Haggag YA, El-Ashmawy NE, Okasha KM. Is hesperidin essential for prophylaxis and treatment of COVID-19 infection? Med Hypotheses. 2020;144:109957. https://doi.org/10.1016/j.mehy.2020.109957.
42. Cinatl J, Morgenstern B, Bauer G, Chandra P, Rabenau H, Doerr HW. Glycyrrhizin, an active component of liquorice roots, and replication of SARS-associated coronavirus. Lancet (London, England). 2003;361:2045–6. https://doi.org/10.1016/s0140-6736(03)13615-x.
43. Deng J-G, Hou X-T, Zhang T-J, Bai G, Hao E-W, Chu JJH, Wattanathorn J, Sirisa-Ard P, Soo Ee C, Low J, Liu C-X. Carry forward advantages of traditional medicines in prevention and control of outbreak of COVID-19 pandemic. Chinese Herb Med. 2020;12:207–13. https://doi.org/10.1016/j.chmed.2020.05.003.
44. Vellingiri B, Jayaramayya K, Iyer M, et al. COVID-19: a promising cure for the global panic. Sci Total Environ. 2020;725:138277. https://doi.org/10.1016/j.scitotenv.2020.138277.
45. Bellavite P, Signorini A, Marzotto M, Moratti E, Bonafini C, Olioso D. Cell sensitivity, non-linearity and inverse effects. Homeopathy. 2015;104:139–60. https://doi.org/10.1016/j.homp.2015.02.002.
46. Jayawardena R, Sooriyaarachchi P, Chourdakis M, Jeewandara C, Ranasinghe P. Enhancing immunity in viral infections, with special emphasis on COVID-19: a review. Diabetes Metab Syndr Clin Res Rev. 2020;14:367–82. https://doi.org/10.1016/j.dsx.2020.04.015.
47. Ilie PC, Stefanescu S, Smith L. The role of vitamin D in the prevention of coronavirus disease 2019 infection and mortality. Aging Clin Exp Res. 2020;32:1195–8. https://doi.org/10.1007/s40520-020-01570-8.
48. Meng X, Shi L, Yao L, Zhang Y, Cui L. Calcination induced PEG-Ni-ZnO nanorod composite and its biomedical applications. Colloids Surfaces A Physicochem Eng Asp. 2020;2019:124658. https://doi.org/10.1016/j.colsurfa.2020.124658.
49. Kieliszek M, Lipinski B. Selenium supplementation in the prevention of coronavirus infections (COVID-19). Med Hypotheses. 2020;143:109878. https://doi.org/10.1016/j.mehy.2020.109878.
50. Kumar A, Kubota Y, Chernov M, Kasuya H. Potential role of zinc supplementation in prophylaxis and treatment of COVID-19. Med Hypotheses. 2020;144:109848. https://doi.org/10.1016/j.mehy.2020.109848.

Phytomolecules and Novel Drug Delivery Approach for COVID-19

Mittal Maheshwari, Bharat Patel, and Niyati Acharya

Abstract Coronavirus disease 2019 (COVID-19) has been a recent pandemic where very rarely few therapeutic modalities are effective in the treatment and management of the complications. Some of the pharmacological treatment approaches still in use include hydroxychloroquine/chloroquine, repurposed antiviral medications, monoclonal antibodies or IL-6 pathway inhibitors, corticosteroids, convalescent plasma, and cell and biological therapy. It has been shown that type I and type II interferons have antiviral action, and it is thought that the human immune system contributes significantly to the viral eradication process.

Numerous herbal drugs such as tulsi, ginger, clove, dalchini, garlic, ashwagandha, giloy, black pepper, black cumin, amla, turmeric, and garlic have been cited in Ayurvedic texts for their diversified use as anti-inflammatory, antioxidant, and immunomodulatory agents. Many herbs and natural phytoconstituents are efficient home remedies for COVID-19 therapy for quick recovery and immune modulators as a preventive measure. In this chapter, we've discussed the value of natural substances in the treatment of COVID-19 as well as contemporary integrated pharmaceutical therapies that use herbal formulations and their clinical results. So many studies have been done in recent years to demonstrate the antiviral efficacies of medicinal plant extracts and secondary metabolites. We discussed the safety and effectiveness of phytomolecules, conventional medication's quality problems, and analytical challenges with a new delivery approach combined with polymer science, pharmaceutics, bioconjugate chemistry, and molecular biology. It also emphasizes

M. Maheshwari
A-One Pharmacy College, Anasan, Ahmedabad, Gujarat, India

B. Patel
Institute of Pharmacy, Nirma University, Ahmedabad, Gujarat, India

N. Acharya (✉)
Institute of Pharmacy, Nirma University, Ahmedabad, Gujarat, India

Department of Pharmacognosy, Institute of Pharmacy, Nirma University, Ahmedabad, Gujarat, India

R. Shegokar, Y. Pathak (eds.), *Viral Drug Delivery Systems*,
https://doi.org/10.1007/978-3-031-20537-8_17

the use of newly developed drug delivery systems of already developed therapeutic agents to address the gaps left by latent targeted administration.

Keywords COVID-19 · Phytoconstituents · Antiviral · Herbal drugs · Giloy · Ashwagandha

1 Introduction

Coronavirus disease 2019 (COVID-19) caused by the severe acute respiratory syndrome coronavirus 2 (SARS-CoV-2) was initially identified in December 2019 at Wuhan, China [1]. In India, 43 million confirmed cases and 516.3 lakhs deaths have been reported till March 2022 [2].

The structural contribution of SARS-CoV-2 is an enveloped β-coronavirus with a large positive-sense, single-stranded RNA genome ranging from 26 to 32 kb in size. The four secondary components present in the virus are the spike (S) protein, envelope (E), membrane (M) protein, and nucleocapsid (N) protein. The spike protein interacts with the host cell membrane, allowing the virus to enter despite the infections [3, 4]. Angiotensin-converting enzyme 2 (ACE2) is among the potential target receptors for SARS-CoV-2 in the human body. It is distributed in various tissues such as the heart, gastrointestinal tract, vessel, intestine, liver, kidney, spleen, and skin [5, 6]. This could contribute to the possible involvement of numerous human organs in the short and long term. It was also a challenge to battle the newly evolving viruses and its variants. Viral RNA polymerase has a high risk of mutation due to the lack of capacity to proofread [7], and this function helps RNA-genome viruses establish resistance to existing antiviral drugs [8]. COVID-19 has a high replication rate because the RNA-dependent polymerase RNA (RdRP) jumps and continuously generates transcription errors [9]. Because of its high levels of mutation, COVID-19 is zoonotic pathogens that can infect different humans and animals, resulting in a broad variety of clinical characteristics ranging from acute to multiorgan failure [10, 11]. Currently, there are no appropriate therapeutic approaches available for treating COVID-19 infection, and there is an inadequate amount of research reported in this field. Non-corona viral therapies attack either the human cells or the virus themselves. Human immune system is considered to play a major role in destroying the virus and research has shown that the type I and type II interferon antiviral efficacy. Interferon-beta (IFN-β) has been proven to decrease Middle East respiratory syndrome coronavirus (MERS-CoV) replication in vitro [12]. The other targets in human cells are blocking the cell surface receptors for coronavirus binding and the cell signaling pathways that aid in viral replication. ACE2 is among the targets suggested to promote drug target therapy to avoid virus infection. Anti-ACE2 monoclonal antibodies, anti-SARS-CoV-2 neutralizing monoclonal antibodies, peptide fusion inhibitors, and anti-proteases therapy will invade cell entry via ACE2 receptors [13, 14].

The COVID-19 study has shown in recovering patients that it influences quality of life, but this deadly respiratory virus may have long-lasting adverse effects. Viral load may have been reduced for many, but some common symptoms such as cough, sore throat, and weakness may continue for weeks after fighting the disease, and even the recovered patients may return to the hospital with complaints of cardiac attacks, emotional illness, and comorbid condition. It is therefore much more important that researchers not only wait for a vaccine to conquer the COVID-19 tide but also take rigorous steps to ensure the safety for the community. In Ayurveda, Acharya Charaka clarified that Panchakarma (five purification procedures), Sadvritta (good behavior), and Rasayana dravyas (immunomodulators) of Ayurveda could be used for immune modulation, prevention, and reduction in burden of various diseases [15–17]. We have explained the importance of natural ingredients in the management of COVID and current integrated pharmacological interventions using herbal formulations with their clinical outcomes.

2 Methodology

Literature review was performed on PubMed, Google Scholar, and ScienceDirect by Boolean, and specific searches to identify terms included "coronavirus," "severe acute respiratory syndrome", "COVID-19," "novel formulations drug delivery system," and "phytoproducts" in combination with "treatment" and "pharmacology action." The data was collected from different reviews, case reports, case series, and research articles published on some of the mentioned topics. Ongoing clinical trials were cited from the official website of Clinical Trial Registry of India (CTRI).

3 Pathology and Current Pharmacological Intervention for COVID-19

Various pharmacological treatment options still going on are hydroxychloroquine/chloroquine, repurposed antiviral medications, monoclonal antibodies or IL-6 pathway inhibitors, and corticosteroids along with the convalescent plasma and cell and biological therapy.

To date, no specific medicines have been developed to prevent COVID-19, and the drugs being used are only to help in reducing the viral load and to offer symptomatic treatment. The US Food and Drug Administration (US FDA) has granted an antiviral drug remdesivir emergency use authorization, while the UK government has approved the use of low-cost steroidal medication [18] to manage respiratory conditions. India has also approved the use of immunosuppressive tocilizumab and convalescent plasma therapy off label on specific patient groups apart from oxygen or mechanical ventilation. In addition, India is also using favipiravir and

Table 1 Pharmacological target for the COVID-19

Synthetic drugs	Targets
Hydroxychloroquine sulfate/chloroquine phosphate	Inhibit terminal glycosylation
Lopinavir/ritonavir	3 cl (chymotrypsin-like protease)
Umifenovir	S-protein/ACE2, membrane fusion inhibitor
Remdesivir	RNA polymerase inhibitor
Favipiravir	RNA polymerase inhibitor
Tocilizumab	IL-6 inhibitor in cytokine storm

hydroxychloroquine as prophylaxis drugs. Hydroxychloroquine also has role with azithromycin which collapsed with time in treating COVID-19 patients [19].

It is useful to know what drug should be given at what time and stage of the disease for, e.g., tocilizumab when the levels of IL-6 (IL-6 is among the main cytokine storm mediators) are significantly high. The drug tocilizumab is still under investigation as a treatment for COVID-19, but the outcomes of phase-three trials have reportedly shown no clinical or mortality advantage [20]. The use of antiviral drugs (lopinavir and ritonavir) for further treatment was investigated in the beginning phases against COVID-19; however, both have wasted favors [21]. Patients only have supporting treatment like IV fluids, ventilation systems, and steroids. The convalescent plasma therapy showed mild improvement but did not reveal better results in overall condition. The function of the anticoagulants has become increasingly important in patients with mild and severe disease or with low risk of internal bleeding are also being prescribed [22] (Table 1).

4 Overview of the Phytoproducts

Ayurveda is a 5000-year-old medicine system that originated in India. Some theories stated that it is one of the oldest healing sciences. In Ayurveda, integrated therapy includes phytoproducts (mostly obtained from nature that involves metal and mineral products), diet, and exercise, meditation, and lifestyle modification. Because of its potential usefulness and safety, the use of traditional compounds was also found for this perspective in view of the medicinal benefits. Some of the potent bioactive compounds including curcumin, *Withania* alkaloids, cannabidiol, etc. are a perfect example of the bioactive compound's multi-mechanistic mode of action for favorable therapy. It acts against infectious diseases including inflammatory diseases, neurological diseases, cardiovascular diseases, pulmonary diseases, metabolic diseases, liver diseases, and cancers. There has been increasing evidence on the antiviral efficacy of the traditional compounds, and also Ayurvedic therapy and home remedies support to increase the immune system preventing the viral disease from spreading.

4.1 Ashwagandha

Withania somnifera (WS), widely known as ashwagandha, is an Ayurvedic herb that has recently gained recognition as a treatment for anxiety and stress in India. WS is classified as an anti-inflammatory supplement with antioxidant and immunomodulatory bioactives. The WS is also classified as an adaptogen, indicating its capacity to control physiological processes and stabilize the response of the body to stress [23]. Generally, traditional medical practitioners were using natural constituents of the WS plant called withanolides to treat various diseases. Withaferin shows real potential as a treatment modality to treat or prevent COVID-19 spread due to the recorded interaction with the host receptor binding of the viral S-protein; however, there is a lack of effect on ACE2 expression in the respiratory system [24]. Research investigated that WS extract-treated rats responded better to symptoms of induced chronic stress, similar to those of standard benzodiazepines [25]. The results of this specific study show that herbal supplementation is equally effective in anxiety management as are standard prescription drugs in a rodent model, without the harmful adverse effects. In people with mild cognitive impairment, WS may be effective in improving both immediate and general memory as well as improving cognitive function, recognition, and speed for processing of information. WS root extract is also being used as a rejuvenating agent for enhanced muscle strength, fatigue resistance, exercise recovery, and as an energy booster. In healthy athletic adults, the oral administration (high concentration) of WS root extract improves cardiovascular dynamics by increasing the maximum levels of oxygen and thus improving cardiorespiratory endurance as well as improving quality of life (QOL) in healthy adults [26].

4.2 Tulsi

Ocimum sanctum is not only an immunity boosting herb; it is also the most commonly found plant in Indian households. There are different variants of tulsi like Rama tulsi, Krishna tulsi, and Vana tulsi that can be consumed for different health benefits. Active phytoconstituents of tulsi have been proven to significantly inhibit the key protease COVID-19 [27]. Compounds such as ursolic acid, carnosol, rosmarinic acid, cirsilineol, apigenin, eugenol, and cirsimaritin present in *Ocimum sanctum* increase the concentration of hemoglobin and increase the sheep red blood cell agglutinin titer; decrease the activity of cyclo-oxygenase-2, lipoxygenase-5, and NF-3B pathway; and increase the IL-2, IFN-γ, and TNF-alpha regulations [28]. Tulsi improves the defense mechanism against infection by boosting immune reaction. Several studies stated that tulsi extracts have antimicrobial (including antibacterial, antiviral), antioxidant, anti-inflammatory, cardio-protective, analgesic, antipyretic, and immune-modulatory properties [29].

Tulsi is known as the Elixir of Life in Ayurveda for its healing capacity and promising potential to cure various conditions, i.e., bronchitis, pyrexia, rheumatism, asthma and microbial infections, gastric and hepatic disorders, etc. The role of tulsi in post COVID-19 control to treat pain, diarrhea, cough, and fever which are common symptoms associated with COVID-19 is still not clear. Tulsi has significant importance and role in boosting the immune system that assists the human body to fight against unwanted microbial strangers including bacteria and viruses. This study confirmed that the use of tulsi extract against COVID-19 is due to inhibition of replication of SAR-CoV with ACE2 blocking properties [30].

4.3 Curcumin

Curcumin is said to be one of the most effective Ayurvedic herbs as it has bioactive compounds with medicinal properties. Curcumin may even have promising impacts toward COVID-19 infection through its ability to alter the diverse molecular components that lead to SARS-CoV attachment and internalization to organs [31]. Curcumin may also regulate pathways for regulating cellular activities such as inflammation, apoptosis, and transcription of RNA. In infection with COVID-19, curcumin may also suppress pulmonary edema and associated pathways to fibrosis. The anti-inflammatory and immunomodulatory activity of curcumin together with proof of this phytochemical's anti-fibrotic and pulmonary protective activity on the lung tissue renders it a suitable candidate for the management of post COVID-19. Curcumin has reported strong inhibitory effects on NF-kB and other pro-inflammatory cytokines that can be particularly useful as an adjunct in preventing the destructive cytokine storm [32]. Curcumin as an antiviral and anti-inflammatory agent can be helpful in both the prevention and treatment of new emerging COVID-19 symptoms. Nonetheless, excellently designed clinical trials are needed to demonstrate curcumin's possible effectiveness against post COVID-19 complication.

4.4 Cannabidiol

Cannabidiol (CBD), a non-psychotropic phytocannabinoid, has the potential for so many rationales to limit the severity and disease progression. High cannabidiol levels can reduce the expression of the two main SARS-CoV2 receptors in several human epithelial models [33–35]. Cannabidiol has a wide range of immunomodulatory and anti-inflammatory effects and can reduce the excessive production of cytokines responsible for acute lung injury. It can demonstrate direct antiviral activity as a PPARγ agonist which is a regulator for activation of fibroblast/myofibroblast and can prevent the development of pulmonary fibrosis [34]. Significant protective function for CBD during ARDS can expand CBD by reducing cytokine storm, protecting lung tissues, and re-establishing inflammatory homeostasis as part of COVID-19 therapy [34, 35].

Recent results suggesting the ability of CBD in the treatment of acute respiratory distress syndrome (ARDS) in a follow-up study indicate that CBD can improve the symptoms of ARDS through upregulation of apelin, a peptide with an important role in the central and peripheral regulation of immunity, CNS, metabolic, and cardiovascular system [36].

Anxiety and PTSD linked to COVID-19 are likely to be a major long-term problem arising from the current pandemic. Research assumes that CBD, a product with reported anxiolytic properties extracted from *Cannabis sativa* may be a therapeutic alternative for the treatment of anxiety disorders linked to COVID-19. Anxiety, stress, depression, and sleep disorders are consistently the primary reasons individuals use CBD in the regional over-the-counter. Observational and preclinical evidence also support the therapeutic value of CBD in enhancing sleep (increased duration of sleep/quality and decrease in nightmares) and depression, frequently followed by anxiety. Together with these CBD characteristics, it is an attractive new therapeutic choice for COVID-19-related PTSS that merits study and testing through carefully regulated randomized controlled trials.

Immune support or boosting is one of the most common claims for CBD. Immunomodulation can be mediated by cannabinoid receptors or by various other pharmacological targets. These net effects can be summarized as anti-inflammatory and for a number of immune-mediated disorders, including autoimmune conditions and neurodegeneration, are the sought-after therapeutic effects of CBD [33].

4.5 Giloy

Tinospora cordifolia, a very important medicinal plant, is also known as Guduchi in Sanskrit and Galo in Gujarati. Giloy stems are widely used in folk and Ayurvedic systems of medicine. The chemical constituents of *T. cordifolia* are classified into different classes that include alkaloids, steroids, glycosides, diterpenoid lactones, polysaccharides, aliphatic compounds, phenolics, and sesquiterpenoids. It appears that *T. cordifolia* improves the phagocytic function. It will also occur without affecting the humoral or cell-mediated immune system. It is categorized as Rasayana and used for its anti-inflammatory, immunomodulatory, anti-allergic, and anti-diabetic properties, etc. [37]. The whole plant is utilized medicinally; however, the stem is approved and useful in medicine as listed by the Ayurvedic Pharmacopeia of India. This is due to higher alkaloid content in the stems than in the leaves. *Guduchi Ghana* (concentrated form of decoction) is the secondary *Kalpana* (formulation) derived from the primary *Kalpana*, i.e., *Kwatha* (decoction). Several research works have been carried out regarding the anti-inflammatory activity of the decoction, alcohol extract, and water extract of the stem of Giloy. The water extract of the plant is found to be more potent than the other extract. Hence, it has been planned to study the comparative anti-inflammatory activity of classically prepared and market samples of *Guduchi Ghana.*

Until now, three countries, including India, China, and South Korea, have issued guidelines on traditional regimens for the prevention and management of COVID-19.

The Indian Traditional System of Medicine is one of the oldest systems of medical practice in the world. India has the exclusive distinction of its own recognized traditional medicine: Ayurveda, Yoga, Unani, Siddha, and Homeopathy (AYUSH). Approaches of these systems are holistic, and the pharmacological modalities are based on natural products of plants, animals, and/or mineral origin. After the successful treatment of plague, cholera, and Spanish flu with the help of AYUSH, to combat the current deadly pandemic of COVID-19, repurposing the traditional uses of Indian medicinal plants and formulation is a need [21, 38].

AYUSH suggested a few known traditional formulations of immunity modulators, which were in use for centuries in some allergic conditions and respiratory disorders. The Government of India has listed a few of them as prophylactic measures in red zones and containment zones as well as for corona warriors. A few of them are under clinical trials now for COVID-19 patients. Some antiviral medicinal plants are listed (Table 2).

Some homeopathic formulations like arsenic album, *Bryonia alba*, and *Rhus toxicodendron* were also used for management of COVID-19.

Ashwagandha, Giloy, ginger, cinnamon, tulsi, black pepper, black cumin, amla, turmeric, garlic, flax seeds, etc. are routinely used common Indian medicinal plants for management of COVID-19 (Table 3).

5 Prospective Phyto-ingredients for COVID-19 and Their Possible Mode of Action

The emergence of infectious diseases caused by novel viral strains that are resistant to common antiviral drugs is a major worldwide issue. Interestingly, herbal medicines, also known as phytomedicines derived from traditional Chinese, Japanese, Indian, and European herbal medicine systems, are promising candidates for the discovery and development of novel antiviral drugs. Therefore, in recent years, a huge number of experiments confirming the antiviral efficacies of medicinal plant extracts and secondary metabolites (i.e., such as flavonoids, naphthodianthrones, and anthraquinones) have been conducted. Particularly, in the last two decades, a number of medicinal plant extracts and/or related physiologically active ingredients have been reported to exhibit antiviral activities. Some of the phytochemicals proven effective against the viruses or symptoms related to what is shown by COVID-19 along with mode of action are discussed in this section.

Chinese traditional medicine (TCM) is highly recommended by the government of China for the eradication of SARS-CoV-2. It was reported that the following medicinal plants and their derived formulations have been used in 23 provinces of

Table 2 Potential traditional Indian/AYUSH formulations for the management of COVID-19 [38]

Sr no.	Type and name of formulations	Constituents	Sources	Activity
	Ayurvedic approaches			
1	AYUSH KWATH ("AYUSH Kudineer" or "AYUSH Joshanda") [39–50]	*Ocimum sanctum* L. leaves, *Cinnamomum verum* J. Presl. stem barks, *Zingiber officinale* Roscoe rhizomes *Piper nigrum* L. fruits	Powder and tablet	Boost immunity; antiviral remedies
2	Samshamani Vati (Guduchi Ghan Vati) [51–53]	*Tinospora cordifolia* (Willd.) Miers (family Menispermaceae)	Powder extract	Antipyretic and anti-inflammatory remedy
3	AYUSH-64 [54–56]	*Alstonia scholaris* (L.) R. Br. bark, *Picrorhiza kurroa* Royle ex Benth. rhizomes, *Swertia chirayita* (Roxb.) H. Karst. whole plant, *Caesalpinia crista* L. seed pulp	Tablet	Antiviral, anti-asthmatic, and immune boosting
4	Agastya Haritaki (Avaleha kalpana)	More than 15 herbal ingredients including Chitrak, Apamarga, Haritaki, Shankhpushpi, Kachur, Dashamoola, and Pushkarmool	Avaleha	Antiviral, anti-asthmatic, anti-inflammatory, and immunomodulatory activities
5	Anuthaila	*Leptadenia reticulata* *Ocimum sanctum* L. *Sesamum indicum* L. oil *S. indicum* seeds with *Trachyspermum ammi* (L.) Sprague seeds	Oil	Anti-allergic Anti-fever, cough, malaria; migraine and respiratory infections; lung disease
	Unani approaches			
6	Tiryaq-e-Arba [57–59]	*Laurus nobilis* L. berries, *Bergenia ciliata* Sternb. stem, *Aristolochia indica* L. roots, *Commiphora myrrha* (Nees) Engl.	Powder form	Detoxifying agent; potent antiviral agent against SARS-CoV, hepatitis C, HIV virus
7	Roghan-e-Baboona [60]	Flowers of *Matricaria chamomilla* L	Liquid preparation	Anti-asthmatic and inflammatory and against acute viral nasopharyngitis, as well as for sore throat
8	Arq-e-Ajeeb [61, 62]	Thymol, menthol, and camphor	Liquid preparation	Topical antiviral agent; anti-inflammatory; proven against swine flu

(continued)

Table 2 (continued)

Sr no.	Type and name of formulations	Constituents	Sources	Activity
9	Khamira-e-Banafsha [63, 64]	Decoction of flowers of *Viola odorata* L. to a sugar or sugar with honey base	Semisolid	Treatment of ailments of respiratory system and chest diseases, bronchitis, whooping cough; decrease viral load
10	Laooq-e-Sapistan [65, 66]	Ripe fruit of *Cordia myxa L.* and Ziziphus *mauritiana* fruit and *Viola odorata L.*	Semisolid sugar-based polyherbal Unani formulation	Treatment of cold and cough, whooping cough, and phlegm; antiviral and antitussive immunity booster
11	Sharbat-e-Sadar [67, 68]	*Trachyspermum ammi* Sprague, *Adhatoda vasica* *Bombyx mori*, etc.	Unani syrup	Common cold, cough and respiratory diseases; immunomodulator
12	Khamira Marwareed [69]	Compound sugar-based semisolid	Compound sugar-based semisolid	Immunomodulator; antiviral
13	Asgandh Safoof [68]	Asgand (*Withania somnifera*) root extract	Suspension	Immunomodulation and antiviral
14	Habb-e-Bukhar [70, 71]	Cinchona bark, *Tinospora cordifolia*	Poly herbal tablet	Elephantiasis and malarial fever; antiviral
15	Sharbat-e-Toot Siyah [72]	*Morus nigra L.* juice	Liquid/juice with sugar base	Treat tonsillitis and sore throat; analgesics; immunomodulatory activity
16	Laooq-e-Katan [73]	*Linum usitatissimum L.* seed	Sugar-based semisolid	Antiviral, anti-inflammatory, and immunomodulatory activities
	Siddha approaches			
17	Nilavembu Kudineer [74]	Nilavembu Kudineer polyherbal formulation	Polyherbal formulation	Immunomodulator activity against the dengue fever and chikungunya, malaria, typhoid and viral fever
18	Ahatodai Manapagu [75]	*Adhatoda vasica* Nees. leaves	Juice, syrup	To treat respiratory disorders and immunomodulator activity
19	Kabasura Kudineer [40, 76, 77]	Kabasura Kudineer herbs	More than herbs liquid formulation	Common respiratory complaints; for severe phlegm, dry cough, and fever; COVID-19 symptomatic management

China and proved effective for the treatment of COVID-19. These include *Agastache rugosa, Astragalus membranaceus, Radix platycodonis, Atractylodis rhizoma, Cyrtomium fortunei, Lonicerae japonicae, Glycyrrhiza uralensis, Fructus*

Table 3 Routinely used common Indian medicinal plants and traditional Indian formulations for the management of COVID-19

Sr. no.	Type and name of formulations	Part used	Sources/forms	Activity
1	*Allium sativum L.* (garlic) [78]	*A. sativum* peel	Oil	Immunomodulatory activity; antiviral; anti-inflammatory
2	*Cinnamomum verum J. Presl.* (cinnamon) or *Cinnamomum zeylanicum Blume* [39]	*C. verum* bark *C. zeylanicum* bark	Essential oil and powder; bark extract	Antioxidant, immunostimulant, and antiviral activity; immunomodulatory activity
3	*Curcuma longa L.* (turmeric) [79, 80]	*C. longa* oleo resins	Extract	Immunomodulatory activity; allergic disorders; antiviral; anti-inflammatory
4	*Linum usitatissimum L.* (flaxseed) [81]	Heteropolysaccharide and phenolic compound from flaxseed	Extract	Immunomodulatory; immunostimulant and vaccine adjuvant
5	*Nigella sativa L.* (black cumin) [82]	Seed of *Nigella sativa*	Seed extract	Inhibitors of COVID-19-associated disorder like obstructive lung diseases; immunosuppressive activity and immunomodulator
6	*Ocimum sanctum L.* (tulsi) [41]	*Ocimum sanctum* leaves	Oil/leaves	Antioxidant; immunomodulator; anti-allergic and anti-asthma
7	*Phyllanthus emblica L.* [83] (amla)	Fruits of *Phyllanthus emblica*	Fruit extract	Antioxidant; immunomodulator
8	*Piper nigrum L.* (black pepper) [84]	Piperamides isolated from *P. nigrum* fruits	Fruit extract	Antioxidant; immunomodulator; anti-allergic and anti-asthma
9	*Tinospora cordifolia* (Willd.) Miers (giloe) [85, 86]	*T. cordifolia* silver nanoparticles	Stem; leaves extract	Proven against chikungunya virus; immunity booster; anti-inflammatory
10	*Withania somnifera (L.) Dunal* (ashwagandha) [87]	*W. somnifera* formulation (supplemented with minerals)	Leaves extract	Antiviral and immunomodulatory
11	*Zingiber officinale Roscoe* (ginger) [40]	*Z. officinale* extract	Soft gel capsule; oil	Antiviral and immunomodulatory

(continued)

Table 3 (continued)

Sr. no.	Type and name of formulations	Part used	Sources/forms	Activity
12	Chyawanprash [88, 89]	Polyherbal health supplement with amla fruit base and nutrient rich herbs and minerals	Semisolid *paka* preparation	Antioxidant, free radical scavenging, antibacterial, antiviral, anti-inflammatory, anti-allergic, and antithrombotic effects; pulmonary tuberculosis; immunomodulatory
13	Triphala [90, 91]	Polyherbal Ayurvedic medicine consisting of equal proportions of fruits of *Phyllanthus emblica L.*, *Terminalia bellirica* (Gaertn.) *Roxb.* and *Terminalia chebula Retz.*	Powder churna; tablet; capsules	Digestive disorders; antioxidants, anti-inflammatory, antineoplastic, antimicrobial, antidiabetic, etc.
14	Rooh Afza Sharbat [92, 93]	Seeds of khurfa (*Portulaca oleracea L.*), kasni (*Cichorium intybus L.*), angoor (*Vitis vinifera L.*), nilofar (*Nymphaea alba L.*), Neel Kamal (*Nymphaea nouchali Burm. f.*), kamal (*Nelumbo nucifera Gaertn.*), Gaozaban (*Borago officinalis L.*), badiyan (*Coriandrum sativum L.*), fruits/juices of santara (*Citrus sinensis (L.) Osbeck*), ananas (*Ananas comosus (L.) Merr.*), seb (*Malus domestica* (Suckow) Borkh., berries (*Rubus fruticosus L.*), vegetables like palak (*Spinacia oleracea L.*), gazar (*Daucus carota L.*), and pudina (*Mentha arvensis L.*)	Concentrated squash prepared as sugar syrup	Refreshing; antiviral, immunomodulatory, and anti-allergic against respiratory disorders

forsythiae, Saposhnikoviae divaricata, and *Rhizoma atractylodis*. Although most of the treatments were found to lack proper statistical designs, effectiveness of these trials could be questioned. However, some TCM formulations and their possible mode of actions reported against novel coronavirus are listed in Table 4.

In addition to these formulations, many herbal extracts have been proposed as supplements to treat symptoms of COVID-19. For example, *Tinospora cordifolia* extract having an immunomodulatory effect against human immunodeficiency virus is effective to treat related symptoms. Similarly, herbal extracts of *Anthemis hyalina, Nigella sativa*, and *Citrus sinensis* decreased the coronavirus replication and downregulated TRP genes that may be involved in the survival of coronavirus in epithelial cells in a study conducted by Ulasli et al. [94]. Likewise, medicinal plants

Table 4 Phytochemicals from some medicinal herbs and their mode of action against treatment of COVID-19

Herbs	Phytochemicals	Mode of action	Reference
Isatis indigotica	Phenolic compounds like aloe emodin, hesperidin, quercetin, and naringenin in plant extract	Inhibition of the cleavage activity of SARS-3CLpro enzyme	[95]
Houttuynia cordata	Phenolic quercetin 7-rhamnoside in plant extract	Inhibition of viral RNA-dependent RNA polymerase activity (RdRp)	[96]
Multiple herbs	Isobavachalcone, herbacetin, helichrysetin, quercetin, 3-β-glucoside	Inhibition of cleavage activity of MERS-3CLpro enzyme	[97]
Glycyrrhizae Radix	Glycyrrhizin	Inhibition of viral attachment and penetration	[98]
Litchi chinensis and *Rheum palmatum* (Chinese rhubarb)	Flavonoids such as rhoifolin, pectolinarin, epigallocatechin gallate, gallocatechin gallate, quercetin, and herbacetin	Inhibition of SARS-3CLpro activity	[99, 100]
Scutellaria baicalensis	Baicalin	Inhibition of angiotensin-converting enzyme (ACE)	[101]

such as *Heteromorpha* spp. and *Scrophularia scorodonia* possess various phytochemicals, for example, saikosaponins, a derivative of triterpene-oleanane found abundantly across many angiosperm families. It is reported to possess medicinal functions such as modulation of immune function, anti-inflammation, anti-hepatoma, and antimicrobial effects; therefore, it has been shown to be active against measles, herpes simplex, influenza, varicella zoster, and human immunodeficiency viruses and related symptoms. A study conducted by Cheng et al. [102] indicated that saikosaponin B2 has potent antiviral property against infection caused by human coronavirus 229E, and possible mode of action includes inhibitory effect on attachment, penetration, and replication of the novel coronavirus. Similarly, *Zingiber officinale*-derived phytochemical, 6-gingerol, showed promising anti-coronavirus properties due to its high binding affinity against multiple SARS-CoV-2 targets, namely, RNA-binding protein, proteases, and spike proteins.

6 Herbal Formulations Used in Market for COVID-19

In this pandemic condition of COVID-19, it's a requirement of human being for using herbal remedies to boost the innate and acquired immunity to fight against viruses. To boost the "immune system," ways like active lifestyle, healthy diet, physical exercise, relaxation, and sound sleep are needed. Home remedies also played a vital role as immunity modulator agents. Ayurveda treatises have described several herbal drugs like tulsi, ginger, clove, dalchini, turmeric, garlic, and Marich as effective home remedies for viral infections, as COVID-19 therapy for speedy recovery, and as immunity modulators as a preventive solution [103]. Herbal medicine includes herbs, herbal materials (like plant parts) or preparations, processed and finished herbal products, and active ingredients [104].

Many herbal extracts are proposed as supplements to treat COVID-19 symptoms by different modes of action. For example, *Tinospora cordifolia* extract having an immunomodulatory effect against human immunodeficiency virus is used to treat related symptoms. Herbal extracts of *Anthemis hyalina*, *Nigella sativa*, and *Citrus sinensis* decreased the coronavirus replication and downregulated TRP genes. Likewise, plants such as *Heteromorpha* spp. and *Scrophularia scorodonia* have phytoconstituents, for example, saikosaponins, which has been proven to be active against measles, herpes simplex, influenza, varicella zoster, and human immunodeficiency viruses and like symptoms.

COVID-19 is emerging as a very severe danger to global health. Unfortunately, no agents (even currently approved remdesivir) are showing clinical efficacy against SARC-CoV-2 and its complications, thus needing a drug repurposing strategy or newer drug delivery systems. Pharmaceutical research shows a key role by implementation of novel formulations for old phytocompounds that are able to enhance the delivery of bioactives to the site of infection and also to improve safety by minimizing the side effects for phytomolecules like dietary supplements, herbal compounds, medical gas mixtures, etc.

6.1 Dietary Supplements, Micronutrients, and Herbal Medicines

Several dietary supplements, micronutrients, nutraceuticals, probiotics, and herbal medicine formulations were suggested as repurposed compounds for the management of COVID-19. Although evidence of their benefit in viral infections (e.g., influenza, common cold, or SARS) is limited, different ongoing trials are currently investigating the efficacy of several dietary supplements and herbal compounds in COVID-19 when used alone or in combination with "traditional" repurposed agents. Notably, medicinal products are well defined for specific indications, and they must follow specific legislation to demonstrate quality, efficacy, and safety to obtain marketing authorization. Dietary supplements, nutraceuticals, and herbal products do not follow a similar procedure, given that marketing authorization may be required. Thus, their efficacy remains poorly supported, and open questions sometimes remain on safety. As regards Chinese herbal medicine, different meta-analyses of randomized controlled trials reported a better outcome in patients treated with Chinese herbal medicine in association with traditional western medicine compared to traditional Western medicine alone, although significant biases in included studies existed. Herbal formulations exhibit a wide range of pharmacological functions, including anti-inflammatory, antiviral, antipyretic, expectorant, anti-asthmatic, and antitussive effects. Licorice root (Gancao, Glycyrrhizae Radix) was the most commonly administered compound. Different formulations of Chinese herbal products exist, including decoction, granule, capsule, oral liquid, pill, and injection, with decoction being the most commonly used. However, it is noteworthy that more evidence in terms of not only efficacy but also, above all, safety are required for these compounds. We focused on novel pharmaceutical formulations of dietary supplements and herbal medicines (namely, zinc, essential oils, and glycyrrhizin) developed and implemented in COVID-19 setting.

6.2 Zinc Supplementation

Zinc exhibits a wide variety of direct and indirect antiviral activities against different species, including rhinovirus and influenza virus, enhancing both immune and adaptive immunity, as well as affecting virus attachment and replication. Although the efficacy of zinc supplementation in treating the common cold caused by rhinoviruses is debated, it is important to underline that intranasal zinc gluconate gel formulations exist, potentially providing for direct micronutrient delivery at the site of infection. The administration of intranasal zinc formulations could also be repurposed as adjuvant treatment in patients affected by COVID-19, particularly concerning the prevention of disease transmission and the treatment of nasal symptoms. However, zinc toxicity involving the olfactory system was found in preclinical models, and several cases of zinc-induced anosmia syndrome were reported.

6.3 Essential Oils

Essential oils (EOs) include a complex mixture of volatile phytochemicals from diverse classes, including monoterpenes, sesquiterpenes, and phenylpropanoids, showing anti-inflammatory, immunomodulatory, bronchodilatory, and antiviral properties. EOs usually contain about 20–60 components showing widely different concentrations, of which two or three are present at higher concentrations (20–70%; major components) compared to the others (retrieved in trace amounts), thus determining the biological properties of the compound. The chemical profiles of the EOs differ not only in the number and type of molecules but also in their stereochemical structures and can be very different according to the selected method of extraction. Notably, EOs exhibited in vitro activity against several viruses, including influenza and other respiratory viral infections. Different EOs have been investigated through different repurposing approaches (including the in silico approach, in vitro assays, molecular docking) against COVID-19, being eucalyptus oil from *Eucalyptus globulus*, jensenone, and eucalyptol as major components and garlic oil, coupled with several single major components (viz., farnesol, anethole, cinnamaldehyde, carvacrol, geraniol, cinnamyl acetate, L-4-terpineol, thymol, pulegone, eugenol, menthol, and carvacrol) the most promising compounds. EOs are usually used by external application (gargles or inhalation), with the respiratory tract exhibiting the most rapid route of administration, followed by the dermal pathway. However, unfavorable chemical properties of EOs—namely, poor solubility, solvent toxicity, high volatility, low bioavailability, and physicochemical instability (responsible for degradation of EOs components)—limit their use as active compounds in several formulations. Consequently, the search for different novel formulations arose as an urgent need for pharmaceutical research in this area, also in order to potentially improve and implement the use of EOs in viral infections, including COVID-19, leading to the development of many nanotechnology-based carriers, namely, liposomes, dendrimers, nanoparticles, nanoemulsion, and microemulsion. Encapsulation of bioactive and major compounds of EOs represents a feasible approach to modulate drug release, increase the physical stability of the active substances, protect them from interactions with the environment, decrease their volatility, enhance their bioactivity, reduce toxicity, and improve patient compliance and convenience. Currently, there are different ongoing trials investigating EOs in the management of COVID-19, including their efficacy for anosmia recovery in post-COVID infection.

6.4 Glycyrrhizin

Glycyrrhizin or glycyrrhizic acid is a natural product isolated from the roots (Glycyrrhizae Radix) of the plants *Glycyrrhiza glabra* (typically cultivated in Europe, henceforth called European licorice) and *G. uralensis Fisch* and *G. inflata Bat* (used in the Chinese pharmacopeia). It has been demonstrated to exhibit antiviral (based on cytoplasmic and membrane effects) and anti-inflammatory/immunomodulatory properties (through the activation of multiple pathways involving Toll-like receptors and inhibition of pro-inflammatory cytokines), inhibiting in vitro isolates of SARS-associated coronavirus and other respiratory viruses.

However, the extensive first-pass metabolism strongly reduces plasmatic exposure of glycyrrhizic acid, with the consequent achievement of inadequate serum concentrations, well below the IC50 for SARS-CoV. To overcome this issue, different pharmaceutical approaches have been explored: (a) the modification of chemical structure of glycyrrhizic acid, in order to develop amide derivatives and amino-acid conjugates that may considerably enhance the activity against SARS-CoV but with increased cytotoxicity, and (b) the development of drug delivery systems consisting of the encapsulation of glycyrrhizic acid into nanoliposomes, hyalurosomes, or niosomes. The latter formulations not only may improve plasmatic bioavailability and exposure of glycyrrhizic acid but also facilitate the transportation and delivery of co-transported drugs, given the amphiphilic properties of vesicles, allowing for the enhancement of poorly soluble drugs and increasing the passive diffusion through cellular membranes of co-transported agents.

As previously mentioned, hydroxychloroquine, a repurposed drug widely used in the first phase of the COVID-19 pandemic, is one of the compounds combined with glycyrrhizic acid in novel nano-formulation by virtue of the potential antiviral and anti-inflammatory synergism coupled with an improvement in delivery (Table 5).

7 Intention for Newer Novel Drug Delivery Approach

7.1 Gap or Loopholes for Phytomolecule in Terms of Safety and Effectiveness; Quality Issues, and Analytical Challenges for the Traditional Medicines [105]

- The current case study of Villena-Tejada M et al. [106] reported an association between 17 medicinal plants' uses and prevention and treatment of the respiratory symptoms related to COVID-19; the major plants were eucalyptus, ginger, spiked pepper, chamomile, garlic, etc. It was observed that the population in the present study had used a greater number of plants for disease prevention when respondents were older and if one of the friends or family members came out from COVID-19. It was also found that respondents who achieved technical or higher education were using less plants for its treatment. The latent use of medicinal plants for respiratory conditions was approved but more and more research is essential to get solid confirmation of their effectiveness and standardized process to isolate compounds to achieve their potential pharmacological use. A few studies are necessary to determine proper effective forms and their doses and potential combination of these plants [106].
- Phytoformulations, especially homeopathic formulations, are prepared by dilutions in such a way that no single detectable molecule is present in the final formulation, which results in controversy.
- Non-evidential rationale to determine the biological effects of solutions containing unmeasurable starting material led to criticism.
- Challenge is for a pharmacologist to validate the therapeutic claims of homeopathic-typed phytomedicines through experiments. Low acceptance of these formulations is due to the absence of standardized protocols to justify their pharmacological potential.
- The time and processes required to develop an herbal medicine of prime quality and consistency for therapeutic use with sufficient safety data is extremely protracted. This is due to the nature of medicinal plants having multiple phytochemicals, getting easily affected by agronomic factors [107].
- Even identifying, isolating, and producing reference standards required for the standardization of medicinal plants is difficult, compared to straight synthetic chemical entities. Standardization of herbal products based on bioactive markers remains vital to ensure consistency and efficacy for different batches.

Table 5 Clinical trial on photoproducts on COVID-19 (ctri.nic.in)

Sr no.	CTRI No.	Intervention	Clinical outcome of the phytoproducts	Sample size
1	CTRI/2020/04/024882	Kashaya of Tinospora cordifolia added with 2 gm dried powder of *Piper longum* fruit	Assess the effect of the Ayurveda drug combination of *Tinospora cordifolia* and *Piper longum* in the progression of the disease, its severity, and clinical outcome	60
2	CTRI/2020/05/024981	Dabur Chyawanprash	Comparative assessment of the severity of COVID-19, changes in quality of life, assessment of incidence, and severity of disease	600
3	CTRI/2020/05/025171	1. *Guduchi Ghana Vati* 2. Anu Taila 3. Rock salt and turmeric	Improvement in *Bala* of an individual resulting in immune-stimulation, improvement in quality of life	50,000
4	CTRI/2020/05/025069	Sudarshana Ghana Vati or ashwagandha	Effectiveness of such preventive measures and Ayurveda advocacies	1324
5	CTRI/2020/05/025093	Yashtimadhu tablet/ ashwagandha tablet/ Guduchi tablet	Comparative assessment of occurrence of COVID-19 infection, assessment of severity, assessment of subjects not requiring hospitalization, severity of symptoms	1200
6	CTRI/2020/05/025178	Tab Samshamani Vati, Herbal tea, Anu Taila (into nostrils), Haridra Khanda	Improvement in *Bala* of an individual resulting in immune-stimulation leading to nondevelopment of symptoms of COVID-19, improvement in quality of life	140
7	CTRI/2020/05/025156	AYUSH-64 as add-on to standard treatment	Clinical cure rate, duration of fever and each of the respiratory symptoms. Improvement in hematological and laboratory parameters, required invasive or noninvasive oxygen, progressed to multi-organ failure	60
8	CTRI/2020/05/025222	AOIM-Z Tablets	Prevention of incidence of COVID-19 infection, comparative assessment of occurrence of COVID-19 infection, assessment of severity, assessment of subjects not requiring hospitalization	275

(continued)

Table 5 (continued)

Sr no.	CTRI No.	Intervention	Clinical outcome of the phytoproducts	Sample size
9	CTRI/2020/05/025341	Kwath (Kiratiktadi Kwath) and ashwagandha Churna along with yoga exercises (pranayama, Surya Namaskar	Management of mild and asymptomatic cases of COVID-19, rate of recovery	30

- Development of evidence-based validated methods and advancements in the traditional system to justify its measurable dilutions, which may help to understand the mechanism of action and these formulations.
- Due to the variation in the formulations existing for medicinal plants, there are ample safety and toxicity studies related to the formulation of importance. Due to so many challenges, it is highly unlikely to build up new products from scratch in time for emergency use during COVID-19 pandemic type of crises. In emergency time, generally accelerated approvals for therapeutic molecules with proven safety with lowest risk of toxicity and having potential for benefits are considered [108].

7.2 *Potentials of Novel Herbal Drug Delivery System for Traditional Medicines*

- The novel drug delivery technology in herbal medicine may help to increase the efficacy of targeted active phyto-ingredient and even to reduce its side effects; untoward effects are likely to happen in phytomedicine, as people don't know their limited dose and dosing frequency, and, of course, herbs may contain polyphytocompounds which may be less targeted to provide effectiveness and more to give side effects.
- Earlier times, herbal medicines were not being taken risk for development as novel formulations due to their lack of scientific justifications and their processing problems, like standardization, extraction, and even identification of key components in polyherbal complex systems. Modern phytopharmaceutical research may be able to solve the scientific needs like pharmacokinetics determination, mechanism and mode of action, site of action, accurate dosing and dosage regimen, etc. of herbal medicines within novel drug delivery system, examples nanoparticles, microemulsions, matrix systems, solid dispersions, liposomes, solid lipid nanoparticles, etc.
- New drug delivery is an interdisciplinary approach that combines polymer science, pharmaceutics, bioconjugate chemistry, and molecular biology. Novel drug systems can deliver a herbal drug constituent with significant efficacy. Some constituents have an optimal concentration range, which can bring its maximum

benefit, and above or below its concentration range may be toxic or unsafe or may have less or slow effect in severe disease. Henceforth, a multidisciplinary way of delivering key phytoconstituents to targeted site tissues, for required amount for required time by controlling the pharmacokinetics, pharmacodynamics, nonspecific safety and toxicity, immunogenicity, biorecognition and efficacy of drugs came in existence.

A few case studies may be presented for reference [109].

Phytopharmaceuticals are pharmaceuticals using traditional compounds derived from botanicals instead of chemicals. Natural ingredients are more easily and more readily metabolized by the body. Therefore, they produce fewer, if any, side effects and provide increased absorption in the bloodstream resulting in more thorough and effective treatments. Pharmaceuticals made from chemical compounds are prone to adverse side effects. The human body will have a tendency to reject certain chemical compounds which do not occur naturally. These rejections occur in the form of side effects; some may be as mild as minor headaches and others as severe as to be potentially lethal. It is important to note that, while phytopharmaceuticals produce fewer to no side effects, chemical interactions with other prescription drugs can occur.

Furthermore, as they are single and purified compounds, they can be easily standardized making it easier to incorporate them in modern drug delivery systems compared to herbs.

8 Lipid-Based Drug Delivery Systems Have Proven Their Potential in Controlled and Targeted Drug Delivery

Pharmacosomes are amphiphilic phospholipid complexes of drugs bearing active hydrogen that bind to phospholipids. They impart better biopharmaceutical properties to the drug, resulting in improved bioavailability.

Phytosomes are novel compounds comprising of lipophilic complexes of components of plant origin like *Silybum marianum*, *Ginkgo biloba*, ginseng, etc. with phospholipid [110]. They are also called as phytolipid's delivery system. It has high lipophilicity and improved bioavailability and therapeutic effects. These are advanced forms of herbal extract having improved pharmacokinetic and pharmacological parameters, advantageous in treatment of acute liver diseases, either metabolic or infective origin. Phytosomes have wide potential in cosmetology in Indian Ayurvedic medicines to combat serious ailments using both benefits like novel drug delivery and phytosomes.

Novel drug delivery systems like mouth dissolving tablets; mucoadhesive systems; sustained, controlled, prolonged, and extended release formulations; transdermal dosage forms; micro/nanoparticles; microcapsules; implants; etc. are exhaustively researched. A few have reached the market as well (Table 6).

Table 6 Case studies of successful herbal novel drug delivery systems

Form of medicine	Solution of problem by mechanism of improvement	Results
Asoka Lifescience Limited Res-Q, the world's first poly-herbal mouth dissolving tablet	Dissolves in the mouth by mixing with the saliva and gets absorbed; drug reaches the blood directly and the first pass metabolism is bypassed; imparts increased efficacy for lung problems and other respiratory ailments	Relief from respiratory distress within 15 min; resembles the efficacy of Sorbitrate, a revolutionary mouth dissolving drug used in cardiac distress
Patented orally administrable matrix tablet/microcapsule in 2-piece capsule for the controlled release or stable storage of a granulated herb	A granulated herb and a carrier, the formulation release of 75% of the active in between 4 and 18 h	Steady supply of the active components for a sustained period So convenient oral dosage form for user compliance
Process and product US patented new stable *Gingko biloba* extract formulation in the form of sustained-release microgranules	Poor flowability and compressibility properties can be improved by excipients (pellets can be prepared by extrusion-spheronization, fluid air bed process, or a coating-pan method)	Product and process is improved without costly equipment
Palatal mucoadhesive tablet of sage, *Echinacea, Lavender,* and *Mastic gum*	Sustained release abilities of buccal tablets having mucoadhesive polymers	Effective in reducing oral malodor and VSC levels
Transdermal films of boswellic acid (*Boswellia serrata)* and curcumin (*Curcuma longa)*	Targeted action, avoids the first pass metabolism, no pain associated with injection, may be sustained drug delivery	Targeted action with patient compliance and effectiveness; new version of Ayurvedic turmeric *poultice* or *lepa*
Effect and mechanism of Shuanghua aerosol (SHA) checked on upper respiratory tract infections in children from 3 to 14 years	SHA has obvious anti-inflammatory and antiviral effects and has proven good curative effect in treating infantile upper respiratory tract infections	SHA has obvious anti-inflammatory and antiviral effects and has proven good curative effect in treating infantile upper respiratory tract infections
Microparticles of Gugulipid—oleo-gum resin of *Commiphora wightii* using chitosan, egg albumin, sodium alginate, ethyl cellulose, cellulose acetate, gelatin, and beeswax	Optimum physicochemical characteristics and HPLC showed distinct separation of E- and Z-guggulsterone; confirming entrapment of Gugulipid in microparticle	Good release can be achieved

(continued)

Table 6 (continued)

Form of medicine	Solution of problem by mechanism of improvement	Results
Sustained release implant of danshen (*radix Salvia miltiorrhiza*) extract using chitosan	In the CS/gelatin (1:2) matrix, drug release was effectively controlled by the drug amount loaded in the matrix Film degradation was being seen in rats	Implant for the promotion of anastomosing and healing of muscles and tissues at the organic incision site in abdominal cavities
Nanoparticles of TCH (traditional Chinese herbs) *peach seed, safflower, angelica root, Szechuan lovage rhizome, Rehmannia* root, red peony root, leech, gadfly, earthworm, and ground beetle	The thrombolytic effects of nanoparticles of TCHs are much more intense than their non-nanoparticle form	Improve their absorption and distribution in the body, and so enhance their efficacies
ArthriBlend—SR is a proprietary clinically validated blend of glucosamine sulfate, Boswellin (*Boswellia serrata* extract), and Curcumin C3 Complex (curcuminoids from *Curcuma longa*) natural actives for joint care applications	Slow release profile of 80–90% active ingredient release, in an 8-h period Sustained release technology benefits the continuous management of symptoms of arthritis	Support healthy joints and connective tissues in the body Relevant to the bioavailability of glucosamine

9 Novel Technologies for Drug Delivery Approach [111]

Novel drug delivery systems (NDDS) will be a boon for the better and target-specific delivery of repurposed drugs. Currently ongoing research exemplify the execution of NDDS in prevention and treatment of COVID-19 as well as in the vaccine development process. Here, emphasis is on how NDDS of already developed therapeutic agents in the form of improved convenient delivery systems (dosage forms) fill the void spaces for latent targeted delivery. Therefore, NDDS and nanotechnology intervention in product development may play a vital safeguard to humanity in this difficult era.

- *Novel drug delivery system (NDDS) for the delivery of vaccines*
  - Transdermal vaccine
  - Oral vaccine
  - Intranasal vaccine
  - Biosynthetic nanoparticle-based vaccines
- *NDDS in antiviral therapy*
  - Microemulsions
  - Liposomes and ethosomes

  - Microspheres
  - Iontophoresis

- *Repurposing of drugs*
  - Hydroxychloroquine
  - Remdesivir
  - Tocilizumab
  - Azithromycin
  - Convalescent plasma

- *NDDS for drugs in treatment of COVID-19*
  - Nanoparticles of drugs
  - Intranasal delivery nanoparticles

Plant-based supplements which are used in different countries other than India:

1. *Echinacea purpurea*: Purple coneflower is one of the most popular herbal medicines in form of extracts, tinctures, teas, and sprays used in Europe and North America due to several bioactive compounds present like chicoric acid and caffeic acids, alkylamides, and polysaccharides. A preliminary in vitro study found that Echinaforce®, an *E. purpurea* preparation, inactivated SARS-CoV-2 in a clinical trial in Iran on 100 adults based on chest computed tomography (CT) scan or x-ray analysis.
2. *Curcuma longa*: Curcumin is an alkaloid from rhizomes of turmeric and is used for treatment for hypertension, which is the most common comorbidity in COVID-19 patients (23.7%) as per studies used across the globe.
3. Cinchona bark (Cinchona species): Quinine alkaloids isolated from cinchona bark found in the Andean mountain forests, which were the most wanted drugs in the society for COVID-19 treatment.
4. Xanthorrhizol: Java turmeric *or Curcuma xanthorrhiza* Roxb. is a herbal plant widely used as an immunosuppressant in the treatment for COVID-19. This plant is commonly found growing in Southeast Asian countries like Indonesia, Thailand, Philippines, Sri Lanka, and Malaysia; however, using xanthorrhizol for treatment and prevention in COVID-19 still requires more evaluation, especially in the clinical trial setting [112, 113].
5. *Andrographis paniculata*: A few in vitro studies suggest that andrographolide isolated from andrographis might bind the main protease of SARS-CoV-2, thereby inhibiting its replication, transcription, and host cell recognition as observed in small clinical trial in Thailand; when given 3 times per day in 12 people with mild-to-moderate COVID-19 symptoms, marked improvement was observed after patients started taking the low dose (60 mg) andrographis, and all patients recovered after 3 weeks.
6. *Sambucus nigra:* Elder berry fruits growing in North America, Europe, and parts of Africa and Asia and having rich chemical profile including anthocyanins, fla-

vonols, and phenolic acids. Owing to its antioxidant, anti-inflammatory, antiviral, antimicrobial, and immune-stimulating effects, sales of supplements containing elder berry has been doubled shortly after the COVID-19 pandemic began in the United States.

10 Conclusion and Future Aspects

The active constituents of Ayurvedic/AYUSH origin for COVID-19 patients can be utilized in a better form with enhanced efficacy by involvement of modern dosage delivery forms. Phytotherapeutics require a scientific policy to deliver the key components in a novel active manner to boost patient compliance and reduce repeated administration. Novel drug delivery systems utilized for formulations of COVID-19 may not only reduce the frequent administration to conquer noncompliance but may also help to improve the therapeutic value by reducing side effects and toxicity and by increasing the bioavailability at required sites.

The medicinal plant species mentioned and categorized for preclinical and clinical investigation may be further taken up by research organizations on priority basis, which may result in the development of key marker molecules against SARS-CoV-2 and COVID-19.

By seeing the potential of AYUSH medicines and medicinal plants' diversity in India, the herbal drug manufacturers and various government and nongovernment research organizations have developed essential strategies for proceeding with preclinical and clinical research on these promising therapeutic remedies.

The potential use of various medicinal plants for respiratory and other symptomatic conditions were successfully acknowledged to be positive as a prevention and remedy for COVID-19, but more and more research is needed to have solid proof of their effectiveness for specific isolated compounds with budding pharmacological use. Furthermore, studies are needed to determine effective and safe doses, forms of formulation, and their effective combinations of medicinal plants. This may open up a new market for Ayurvedic pharmaceuticals.

References

1. Zhu H, Wei L, Niu P. The novel coronavirus outbreak in Wuhan, China. Glob Health Res Policy. 2020;5:6. https://doi.org/10.1186/s41256-020-00135-6.
2. WHO Coronavirus Disease (COVID-19) Dashboard. WHO coronavirus disease (COVID-19) dashboard | WHO coronavirus disease (COVID-19) dashboard. 2020. https://covid19.who.int/?gclid=CjwKCAiA4o79BRBvEiwAjteoYEGw-zpU8s5e-r22Sfy5ibF2prHJjEoUFJtVTa-D7Wtmk3SBRAxUhRoC0zgQAvD_BwE. https://covid19.who.int/?gclid=CjwKCAiA4o79BRBvEiwAjteoYEGw-zpU8s5e-r22Sfy5ibF2prHJjEoUFJtVTa-D7Wtmk3SBRAxUhRoC0zgQAvD_BwE. Accessed 6 Nov 1980.

3. Li X, Geng M, Peng Y, Meng L, Lu S. Molecular immune pathogenesis and diagnosis of COVID-19. J Pharm Anal. 2020;10:102–8.
4. Shafik NS. An overview of immunopathogenesis towards COVID-19 infection. Sohag Med J. 2020;24:69–75. https://doi.org/10.21608/smj.2020.28056.1144.
5. Datta PK, Liu F, Fischer T, Rappaport J, Qin X. SARS-CoV-2 pandemic and research gaps: understanding SARS-CoV-2 interaction with the ACE2 receptor and implications for therapy. Theranostics. 2020;10:7448–64.
6. Xiao L, Sakagami H, Miwa N. ACE2: the key molecule for understanding the pathophysiology of severe and critical conditions of COVID-19: demon or angel? Viruses. 2020;12:s.
7. Elena SF, Sanjuán R. Virus evolution: insights from an experimental approach. Annu Rev Ecol Evol Syst. 2007;38:27–52.
8. Sun Z, Ostrikov KK. Future antiviral surfaces: lessons from COVID-19 pandemic. Sustain Mater Technol. 2020;25:e00203.
9. Drexler JF, Gloza-Rausch F, Glende J, Corman VM, Muth D, Goettsche M, Seebens A, Niedrig M, Pfefferle S, Yordanov S, Zhelyazkov L, Hermanns U, Vallo P, Lukashev A, Müller MA, Deng H, Herrler G, Drosten C. Genomic characterization of severe acute respiratory syndrome-related coronavirus in European bats and classification of coronaviruses based on partial RNA-dependent RNA polymerase gene sequences. J Virol. 2010;84:11336–49. https://doi.org/10.1128/jvi.00650-10.
10. Grubaugh ND, Hanage WP, Rasmussen AL. Making sense of mutation: what D614G means for the COVID-19 pandemic remains unclear. Cell. 2020;182:794–5. https://doi.org/10.1016/j.cell.2020.06.040.
11. Wang Y, Wang Y, Chen Y, Qin Q. Unique epidemiological and clinical features of the emerging 2019 novel coronavirus pneumonia (COVID-19) implicate special control measures. J Med Virol. 2020b;92:568–76.
12. Haji Abdolvahab M, Moradi-kalbolandi S, Zarei M, Bose D, Majidzadeh-A K, Farahmand L. Potential role of interferons in treating COVID-19 patients. Int Immunopharmacol. 2021;90:107171.
13. Jahanshahlu L, Rezaei N. Monoclonal antibody as a potential anti-COVID-19. Biomed Pharmacother. 2020;129:110337.
14. Wang C, Li W, Drabek D, Okba NMA, van Haperen R, Osterhaus ADME, van Kuppeveld FJM, Haagmans BL, Grosveld F, Bosch BJ. A human monoclonal antibody blocking SARS-CoV-2 infection. Nat Commun. 2020a;11:2251. https://doi.org/10.1038/s41467-020-16256-y.
15. Mishra A, Bentur SA, Thakral S, Garg R, Duggal B. The use of integrative therapy based on Yoga and Ayurveda in the treatment of a high-risk case of COVID-19/SARS-CoV-2 with multiple comorbidities: a case report. J Med Case Rep. 2021;15:1–12. https://doi.org/10.1186/s13256-020-02624-1.
16. Niraj S, Varsha S. A review on scope of immuno-modulatory drugs in Ayurveda for prevention and treatment of Covid-19. Plant Sci Today. 2020;7:417–23.
17. Umesh C, Ramakrishna KK, Jasti N, Bhargav H, Varambally S. Role of Ayurveda and yoga-based lifestyle in the COVID-19 pandemic – a narrative review, vol. 13. Med: J. Ayurveda Integr; 2022. p. 100493.
18. Nabil A, Uto K, Elshemy MM, Soliman R, Hassan AA, Ebara M, Shiha G. Current coronavirus (Sars-cov-2) epidemiological, diagnostic and therapeutic approaches: an updated review until June 2020. EXCLI J. 2020;19:992–1016.
19. Nina PB, Dash AP. Hydroxychloroquine as prophylaxis or treatment for COVID-19: what does the evidence say? Indian J Public Health. 2020;64:S125–7.
20. Quartuccio L, Sonaglia A, McGonagle D, Fabris M, Peghin M, Pecori D, De Monte A, Bove T, Curcio F, Bassi F, De Vita S, Tascini C. Profiling COVID-19 pneumonia progressing into the cytokine storm syndrome: results from a single Italian Centre study on tocilizumab versus standard of care. J Clin Virol. 2020;129:104444. https://doi.org/10.1016/j.jcv.2020.104444.
21. Alvi MM, Sivasankaran S, Singh M. Pharmacological and non-pharmacological efforts at prevention, mitigation, and treatment for COVID-19. J Drug Target. 2020;28:1–13.

22. Gavriatopoulou M, Ntanasis-Stathopoulos I, Korompoki E, Fotiou D, Migkou M, Tzanninis IG, Psaltopoulou T, Kastritis E, Terpos E, Dimopoulos MA. Emerging treatment strategies for COVID-19 infection. Clin Exp Med. 2020:1–13.
23. Priyanka G, Anil Kumar B, Lakshman M, Manvitha V, Kala Kumar B. Adaptogenic and immunomodulatory activity of Ashwagandha root extract: an experimental study in an equine model. Front Vet Sci. 2020;7:700. https://doi.org/10.3389/fvets.2020.541112.
24. Straughn AR, Kakar SS. Withaferin A: a potential therapeutic agent against COVID-19 infection. J Ovarian Res. 2020;13:79.
25. Pratte MA, Nanavati KB, Young V, Morley CP. An alternative treatment for anxiety: a systematic review of human trial results reported for the Ayurvedic herb Ashwagandha (Withania somnifera). J Altern Complement Med. 2014;20:901–8.
26. Choudhary B, Shetty A, Langade D. Efficacy of Ashwagandha (Withania somnifera [L.] Dunal) in improving cardiorespiratory endurance in healthy athletic adults. Ayu. 2015;36:63. https://doi.org/10.4103/0974-8520.169002.
27. Shree P, Mishra P, Selvaraj C, Singh SK, Chaube R, Garg N, Tripathi YB. Targeting COVID-19 (SARS-CoV-2) main protease through active phytochemicals of ayurvedic medicinal plants–Withania somnifera (Ashwagandha), Tinospora cordifolia (Giloy) and Ocimum sanctum (Tulsi)–a molecular docking study. J Biomol Struct Dyn. 2022;40:190–203. https://doi.org/10.1080/07391102.2020.1810778.
28. Kousik DM, Baldev K. A review on therapeutic uses of Ocimum sanctum Linn (TULSI) with its pharmacological actions. Int J Res Ayurveda Pharm. 2012;3:645–7.
29. Gautam S, Gautam A, Chhetri S, Bhattarai U. Immunity against COVID-19: potential role of Ayush Kwath. J Ayurveda Integr Med. 2022;13:100350.
30. Srivastava AK, Chaurasia JP, Khan R, Dhand C, Verma S. Role of medicinal plants of traditional use in recuperating devastating role of medicinal plants of traditional use in recuperating devastating COVID-19 situation. Med Aromat Plants. 2020;9:1–16.
31. Nugraha RV, Ridwansyah H, Ghozali M, Khairani AF, Atik N. Traditional herbal medicine candidates as complementary treatments for COVID-19: a review of their mechanisms, pros and cons. Evid Based Complement Alternat Med. 2020;2020:2560645.
32. Peter AE, Sandeep BV, Rao BG, Kalpana VL. Calming the storm: natural immunosuppressants as adjuvants to target the cytokine storm in COVID-19. Front Pharmacol. 2021;11:583777.
33. Brown JD. Cannabidiol as prophylaxis for SARS-CoV-2 and COVID-19? Unfounded claims versus potential risks of medications during the pandemic. Res Soc Adm Pharm. 2020;17:2053.
34. Esposito G, Pesce M, Seguella L, Sanseverino W, Lu J, Corpetti C, Sarnelli G. The potential of cannabidiol in the COVID-19 pandemic. Br J Pharmacol. 2020;177:4967–70. https://doi.org/10.1111/bph.15157.
35. Khodadadi H, Salles ÉL, Jarrahi A, Chibane F, Costigliola V, Yu JC, Vaibhav K, Hess DC, Dhandapani KM, Baban B. Cannabidiol modulates cytokine storm in acute respiratory distress syndrome induced by simulated viral infection using synthetic RNA. Cannabis Cannabinoid Res. 2020;5:197–201. https://doi.org/10.1089/can.2020.0043.
36. Salles ÉL, Khodadadi H, Jarrahi A, Ahluwalia M, Paffaro VA, Costigliola V, Yu JC, Hess DC, Dhandapani KM, Baban B. Cannabidiol (CBD) modulation of apelin in acute respiratory distress syndrome. J Cell Mol Med. 2020;24:12869–72. https://doi.org/10.1111/jcmm.15883.
37. Devasagayam TPA, Sainis KB. Immune system and antioxidants, especially those derived from Indian medicinal plants. Indian J Exp Biol. 2002;40:639–55.
38. Ahmad S, Zahiruddin S, Parveen B, Basist P, Parveen A, Gaurav PR, Ahmad M. Indian medicinal plants and formulations and their potential against COVID-19–preclinical and clinical research. Front Pharmacol. 2021;11:2470.
39. Brochot A, Guilbot A, Haddioui L, Roques C. Antibacterial, antifungal, and antiviral effects of three essential oil blends. Microbiology. 2017;6:e00459. https://doi.org/10.1002/MBO3.459.

40. Chang JS, Wang KC, Yeh CF, Shieh DE, Chiang LC. Fresh ginger (Zingiber officinale) has anti-viral activity against human respiratory syncytial virus in human respiratory tract cell lines. J Ethnopharmacol. 2013;145:146–51. https://doi.org/10.1016/J.JEP.2012.10.043.
41. Ghoke SS, Sood R, Kumar N, Pateriya AK, Bhatia S, Mishra A, Dixit R, Singh VK, Desai DN, Kulkarni DD, Dimri U, Singh VP. Evaluation of antiviral activity of Ocimum sanctum and Acacia arabica leaves extracts against H9N2 virus using embryonated chicken egg model. BMC Complement Altern Med. 2018;18:174. https://doi.org/10.1186/S12906-018-2238-1.
42. Goel A, Singh DK, Kumar S, Bhatia AK. Immunomodulating property of Ocimum sanctum by regulating the IL-2 production and its mRNA expression using rat's splenocytes. Asian Pac J Trop Med. 2010;3:8–12. https://doi.org/10.1016/S1995-7645(10)60021-1.
43. Kandhare AD, Bodhankar SL, Singh V, Mohan V, Thakurdesai PA. Anti-asthmatic effects of type-A procyanidine polyphenols from cinnamon bark in ovalbumin-induced airway hyperresponsiveness in laboratory animals. Biomed Aging Pathol. 2013;3:23–30. https://doi.org/10.1016/J.BIOMAG.2013.01.003.
44. Khan AM, Shahzad M, Raza Asim MB, Imran M, Shabbir A. Zingiber officinale ameliorates allergic asthma via suppression of Th2-mediated immune response. Pharm Biol. 2015;53:359–67. https://doi.org/10.3109/13880209.2014.920396.
45. Mair C, Liu R, Atanasov A, Schmidtke M, Dirsch V, Rollinger J. Antiviral and antiproliferative in vitro activities of piperamides from black pepper. Planta Med. 2016;82:P807. https://doi.org/10.1055/S-0036-1596830.
46. Majdalawieh AF, Carr RI. In vitro investigation of the potential immunomodulatory and anticancer activities of black pepper (Piper nigrum) and cardamom (Elettaria cardamomum). J Med Food. 2010;13:371–81. https://doi.org/10.1089/JMF.2009.1131.
47. Niphade SR, Asad M, Chandrakala GK, Toppo E, Deshmukh P. Immunomodulatory activity of Cinnamomum zeylanicum bark. Pharm Biol. 2009;47:1168–73. https://doi.org/10.3109/13880200903019234.
48. Soni K, Lawal T, Wicks S, Patel U, Mahady G. Boswellia serrata and Ocimum sanctum extracts reduce inflammation in an ova-induced asthma model of BALB/c mice. Planta Med. 2015;81:PB4. https://doi.org/10.1055/S-0035-1556201.
49. Tasleem F, Azhar I, Ali SN, Perveen S, Mahmood ZA. Analgesic and anti-inflammatory activities of Piper nigrum L. Asian Pac J Trop Med. 2014;7S1:S461–8. https://doi.org/10.1016/S1995-7645(14)60275-3.
50. Zhou HL, Deng YM, Xie QM. The modulatory effects of the volatile oil of ginger on the cellular immune response in vitro and in vivo in mice. J Ethnopharmacol. 2006;105:301–5. https://doi.org/10.1016/j.jep.2005.10.022.
51. Alsuhaibani S, Khan MA. Immune-stimulatory and therapeutic activity of Tinospora cordifolia: double-edged sword against salmonellosis. J Immunol Res. 2017;2017:1. https://doi.org/10.1155/2017/1787803.
52. Pruthvish R, Gopinatha SM. Antiviral prospective of Tinospora cordifolia on HSV-1. Int J Curr Microbiol Appl Sci. 2018;7:3617–24. https://doi.org/10.20546/ijcmas.2018.701.425.
53. Tiwari M, Dwivedi UN, Kakkar P. Tinospora cordifolia extract modulates COX-2, iNOS, ICAM-1, pro-inflammatory cytokines and redox status in murine model of asthma. J Ethnopharmacol. 2014;153:326–37. https://doi.org/10.1016/J.JEP.2014.01.031.
54. Ramesh BN, Girish TK, Raghavendra RH, Naidu KA, Prasada Rao UJS, Rao KS. Comparative study on anti-oxidant and anti-inflammatory activities of Caesalpinia crista and Centella asiatica leaf extracts. J Pharm Bioallied Sci. 2014;6:86–91. https://doi.org/10.4103/0975-7406.129172.
55. Sharma ML, Rao CS, Duda PL. Immunostimulatory activity of Picrorhiza kurroa leaf extract. J Ethnopharmacol. 1994;41:185–92. https://doi.org/10.1016/0378-8741(94)90031-0.
56. Woo SY, Win NN, Noe Oo WM, Ngwe H, Ito T, Abe I, Morita H. Viral protein R inhibitors from Swertia chirata of Myanmar. J Biosci Bioeng. 2019;128:445–9. https://doi.org/10.1016/J.JBIOSC.2019.04.006.

57. Ahmad M, Butt MA, Zhang G, Sultana S, Tariq A, Zafar M. Bergenia ciliata: a comprehensive review of its traditional uses, phytochemistry, pharmacology and safety. Biomed Pharmacother. 2018;97:708–21. https://doi.org/10.1016/J.BIOPHA.2017.10.141.
58. Aurori AC, Bobiş O, Dezmirean DS, Mărghitaş LA, Erler S. Bay laurel (Laurus nobilis) as potential antiviral treatment in naturally BQCV infected honeybees. Virus Res. 2016;222:29–33. https://doi.org/10.1016/J.VIRUSRES.2016.05.024.
59. Loizzo MR, Saab AM, Tundis R, Statti GA, Menichimi F, Lampronti D, Gambari R, Cinatl J, Doerr HW. Phytochemical analysis and in vitro antiviral activities of the essential oils of seven Lebanon species. Chem Biodivers. 2008;5:461–70. https://doi.org/10.1002/cbdv.200890045.
60. Srivastava JK, Shankar E, Gupta S. Chamomile: a herbal medicine of the past with bright future. Mol Med Rep. 2010;3:895. https://doi.org/10.3892/MMR.2010.377.
61. Sharifi-Rad J, Salehi B, Schnitzler P, Ayatollahi SA, Kobarfard F, Fathi M, Eisazadeh M, Sharifi-Rad M. Susceptibility of herpes simplex virus type 1 to monoterpenes thymol, carvacrol, p-cymene and essential oils of Sinapis arvensis L., Lallemantia royleana Benth. and Pulicaria vulgaris Gaertn. Cell Mol Biol (Noisy-le-grand). 2017;63:42–7. https://doi.org/10.14715/CMB/2017.63.8.10.
62. Zaia MG, Cagnazzo TDO, Feitosa KA, Soares EG, Faccioli LH, Allegretti SM, Afonso A, Anibal FDF. Anti-inflammatory properties of menthol and menthone in Schistosoma mansoni infection. Front Pharmacol. 2016;7:170. https://doi.org/10.3389/FPHAR.2016.00170.
63. Gerlach S, Chandra P, Roy U, Gunasekera S, Göransson U, Wimley W, Braun S, Mondal D. The membrane-active phytopeptide cycloviolacin O2 simultaneously targets HIV-1-infected cells and infectious viral particles to potentiate the efficacy of antiretroviral drugs. Medicines (Basel, Switzerland). 2019;6:33. https://doi.org/10.3390/MEDICINES6010033.
64. Koochek MH, Pipelzadeh MH, Mardani H. The effectiveness of Viola odorata in the prevention and treatment of formalin-induced lung damage in the rat. J Herbs Spices Med Plants. 2008;10:95–103. https://doi.org/10.1300/J044V10N02_11.
65. Hong EH, Song JH, Bin KK, Sung SH, Ko HJ, Yang H. Anti-influenza activity of betulinic acid from Zizyphus jujuba on influenza A/PR/8 virus. Biomol Ther (Seoul). 2015;23:345. https://doi.org/10.4062/BIOMOLTHER.2015.019.
66. Jamkhande PG, Barde SR, Patwekar SL, Tidke PS. Plant profile, phytochemistry and pharmacology of Cordia dichotoma (Indian cherry): a review. Asian Pac J Trop Biomed. 2013;3:1009–12.
67. Lü P, Pan Y, Yang Y, Zhu F, Li C, Yao Q, Chen K. Discovery of anti-viral molecules and their vital functions in Bombyx mori. J Invertebr Pathol. 2018;154:12–8.
68. Roy S, Chaurvedi P, Chowdhary A. Evaluation of antiviral activity of essential oil of Trachyspermum Ammi against Japanese encephalitis virus. Pharm Res. 2015;7:263–7. https://doi.org/10.4103/0974-8490.157977.
69. Khan F, Ali S, Ganie BA, Rubab I. Immunopotentiating effect of Khamira Marwarid, an herbo-mineral preparation. Methods Find Exp Clin Pharmacol. 2009;31:513–22. https://doi.org/10.1358/mf.2009.31.8.1419719.
70. Devaux CA, Rolain JM, Colson P, Raoult D. New insights on the antiviral effects of chloroquine against coronavirus: what to expect for COVID-19? Int J Antimicrob Agents. 2020;55:105938. https://doi.org/10.1016/J.IJANTIMICAG.2020.105938.
71. Rastogi S, Pandey DN, Singh RH. COVID-19 pandemic: a pragmatic plan for ayurveda intervention. J Ayurveda Integr Med. 2022;13:100312. https://doi.org/10.1016/j.jaim.2020.04.002.
72. Chen H, Pu J, Liu D, Yu W, Shao Y, Yang G, Xiang Z, He N. Anti-inflammatory and antinociceptive properties of flavonoids from the fruits of black mulberry (Morus nigra L.). PLoS One. 2016;11:e0153080. https://doi.org/10.1371/JOURNAL.PONE.0153080.
73. Miccadei S, Masella R, Mileo AM, Gessani S. ω3 polyunsaturated fatty acids as immunomodulators in colorectal cancer: new potential role in adjuvant therapies. Front Immunol. 2016;7:486. https://doi.org/10.3389/FIMMU.2016.00486.

74. Mahadevan H, Palraj V. Literature review on siddha herbal formulations (Kudineer) available for the management of Dengue. Int J Pharmacol Clin Sci. 2016;5:90–6. https://doi.org/10.5530/IJPCS.5.3.5.
75. Sampath Kumar KP, Bhowmik D, Tiwari P, Kharel R. Indian traditional herbs Adhatoda vasica and its medicinal application. J Chem Pharm Res. 2010;2:240–5.
76. Jin JH, Lee DU, Kim YS, Kim HP. Anti-allergic activity of sesquiterpenes from the rhizomes of Cyperus rotundus. Arch Pharm Res. 2011;34:223–8. https://doi.org/10.1007/S12272-011-0207-Z.
77. Vinothapooshan G, Sundar K. Immunomodulatory activity of various extracts of Adhatoda vasica Linn. in experimental rats. Afr J Pharm Pharmacol. 2011;5:306–10. https://doi.org/10.5897/AJPP10.126.
78. Jeong YY, Ryu JH, Shin JH, Kang MJ, Kang JR, Han J, Kang D. Comparison of anti-oxidant and anti-inflammatory effects between fresh and aged black garlic extracts. Molecules. 2016;21:430. https://doi.org/10.3390/MOLECULES21040430.
79. Ichsyani M, Ridhanya A, Risanti M, Desti H, Ceria R, Putri DH, Sudiro TM, Dewi BE. Antiviral effects of Curcuma longa L. against dengue virus in vitro and in vivo. IOP Conf Ser Earth Environ Sci. 2017;101:012005. https://doi.org/10.1088/1755-1315/101/1/012005.
80. Sornpet B, Potha T, Tragoolpua Y, Pringproa K. Antiviral activity of five Asian medicinal pant crude extracts against highly pathogenic H5N1 avian influenza virus. Asian Pac J Trop Med. 2017;10:871–6. https://doi.org/10.1016/j.apjtm.2017.08.010.
81. Kasote DM, Zanwar AA, Devkar ST, Hegde MV, Deshmukh KK, Kasote DM. Immunomodulatory activity of ether insoluble phenolic components of n-butanol fraction (EPC-BF) of flaxseed in rat. Asian Pac J Trop Biomed. 2012;2:S623.
82. Bouchentouf S, Missoum N. Identification of compounds from nigella sativa as new potential inhibitors of 2019 novel coronasvirus (Covid-19): molecular docking study. 2020. https://doi.org/10.26434/CHEMRXIV.12055716.V1.
83. Xiang Y, Pei Y, Qu C, Lai Z, Ren Z, Yang K, Xiong S, Zhang Y, Yang C, Wang D, Liu Q, Kitazato K, Wang Y. In vitro anti-herpes simplex virus activity of 1,2,4,6-tetra-O-galloyl-β--D-glucose from Phyllanthus emblica L. (Euphorbiaceae). Phytother Res. 2011;25:975–82. https://doi.org/10.1002/PTR.3368.
84. Bui TT, Fan Y, Piao CH, Van Nguyen T, Shin D, Jung SY, Hyeon E, Song CH, Lee SY, Shin HS, Chai OH. Piper Nigrum extract improves OVA-induced nasal epithelial barrier dysfunction via activating Nrf2/HO-1 signaling. Cell Immunol. 2020;351:104035. https://doi.org/10.1016/j.cellimm.2019.104035.
85. Philip S, Tom G, Vasumathi AV. Evaluation of the anti-inflammatory activity of Tinospora cordifolia (Willd.) Miers chloroform extract – a preclinical study. J Pharm Pharmacol. 2018;70:1113–25. https://doi.org/10.1111/jphp.12932.
86. Sharma V, Kaushik S, Pandit P, Dhull D, Yadav JP, Kaushik S. Green synthesis of silver nanoparticles from medicinal plants and evaluation of their antiviral potential against chikungunya virus. Appl Microbiol Biotechnol. 2019;103:881–91. https://doi.org/10.1007/s00253-018-9488-1.
87. Chandran U, Patwardhan B. Network ethnopharmacological evaluation of the immunomodulatory activity of Withania somnifera. J Ethnopharmacol. 2017;197:250–6. https://doi.org/10.1016/J.JEP.2016.07.080.
88. Sastry JLN, Gupta DA, Brindavanam NB, Kanjilal DS, Kumar S, Setia DM, Vedula DS, Srivastava DR. Quantification of immunity status of Dabur Chyawanprash – A review part-2 (clinical studies). Indian J Appl Res. 2011;4:205–11. https://doi.org/10.15373/2249555X/MAR2014/61.
89. Gupta A, Kumar S, Dole S, Deshpande S, Deshpande V, Singh S, Sasibhushan V. Evaluation of Cyavanaprāśa on health and immunity related parameters in healthy children: a two arm, randomized, open labeled, prospective, multicenter, clinical study. Anc Sci Life. 2017;36:141. https://doi.org/10.4103/ASL.ASL_8_17.

90. Peterson CT, Denniston K, Chopra D. Therapeutic uses of Triphala in Ayurvedic medicine. J Altern Complement Med. 2017;23:607–14. https://doi.org/10.1089/ACM.2017.0083.
91. Prasad S, Srivastava SK. Oxidative stress and cancer: chemopreventive and therapeutic role of Triphala. Antioxidants (Basel). 2020;9:72. https://doi.org/10.3390/ANTIOX9010072.
92. Arora P, Ansari SH, Najmi AK, Anjum V, Ahmad S. Investigation of anti-asthmatic potential of dried fruits of Vitis vinifera L. in animal model of bronchial asthma. Allergy, Asthma Clin Immunol. 2016;12:1–12. https://doi.org/10.1186/S13223-016-0145-X/FIGURES/6.
93. Jiang Y, Tzi Bun NG, Zhaokun LIU, Wang C, Ning LI, Qiao W, Liua F. Immunoregulatory and anti-HIV-1 enzyme activities of antioxidant components from lotus (Nelumbo nucifera Gaertn.) rhizome. Biosci Rep. 2011;31:381–90. https://doi.org/10.1042/BSR20100062.
94. Ulasli M, Gurses SA, Bayraktar R, Yumrutas O, Oztuzcu S, Igci M, Igci YZ, Cakmak EA, Arslan A. The effects of Nigella sativa (Ns), Anthemis hyalina (Ah) and Citrus sinensis (Cs) extracts on the replication of coronavirus and the expression of TRP genes family. Mol Biol Rep. 2014;41(3):1703–1711. https://doi.org/10.1007/s11033-014-3019-7.
95. Lin CW, Tsai FJ, Tsai CH, Lai CC, Wan L, Ho TY, Hsieh CC, Chao PDL. Anti-SARS coronavirus 3C-like protease effects of Isatis indigotica root and plant-derived phenolic compounds. Antivir Res. 2005;68:36. https://doi.org/10.1016/J.ANTIVIRAL.2005.07.002.
96. Lau KM, Lee KM, Koon CM, Cheung CSF, Lau CP, Ho HM, Lee MYH, Au SWN, Cheng CHK, Lau CBS, Tsui SKW, Wan DCC, Waye MMY, Wong KB, Wong CK, Lam CWK, Leung PC, Fung KP. Immunomodulatory and anti-SARS activities of Houttuynia cordata. J Ethnopharmacol. 2008;118:79–85. https://doi.org/10.1016/J.JEP.2008.03.018.
97. Jo S, Kim H, Kim S, Shin DH, Kim MS. Characteristics of flavonoids as potent MERS-CoV 3C-like protease inhibitors. Chem Biol Drug Des. 2019;94:2023–30. https://doi.org/10.1111/CBDD.13604.
98. Wolkerstorfer A, Kurz H, Bachhofner N, Szolar OHJ. Glycyrrhizin inhibits influenza A virus uptake into the cell. Antivir Res. 2009;83:171–8. https://doi.org/10.1016/J.ANTIVIRAL.2009.04.012.
99. Jo S, Kim S, Shin DH, Kim MS. Inhibition of SARS-CoV 3CL protease by flavonoids. J Enzyme Inhib Med Chem. 2020;35:145–51. https://doi.org/10.1080/14756366.2019.1690480.
100. Xu X, Xie H, Hao J, Jiang Y, Wei X. Flavonoid glycosides from the seeds of Litchi chinensis. J Agric Food Chem. 2011;59:1205–9. https://doi.org/10.1021/JF104387Y.
101. Deng YF, Aluko RE, Jin Q, Zhang Y, Yuan LJ. Inhibitory activities of baicalin against renin and angiotensin-converting enzyme. Pharm Biol. 2012;50:401–6. https://doi.org/10.3109/13880209.2011.608076.
102. Cheng, Pei-Win, et al. Antiviral effects of saikosaponins on human coronavirus 229E in vitro. Clinical and Experimental Pharmacology and Physiology 2006;33(7):612–16.
103. Aware D, Rohane S. A role of herbal drug as an immunity booster during Covid-19 pandemic. Asian J Pharm Res. 2021;11:206–11. https://doi.org/10.52711/2231-5691.2021.00037.
104. Sen S, Chakraborty R. Revival, modernization and integration of Indian traditional herbal medicine in clinical practice: importance, challenges and future. J Tradit Complement Med. 2016;7:234–44. https://doi.org/10.1016/J.JTCME.2016.05.006.
105. Lim XY, Teh BP, Tan TYC. Medicinal plants in COVID-19: potential and limitations. Front Pharmacol. 2021;12:1–8. https://doi.org/10.3389/fphar.2021.611408.
106. Villena-Tejada M, Vera-Ferchau I, Cardona-Rivero A, Zamalloa-Cornejo R, Quispe-Florez M, Frisancho-Triveño Z, Abarca-Meléndez RC, Alvarez-Sucari SG, Mejia CR, Yañez JA. Use of medicinal plants for COVID-19 prevention and respiratory symptom treatment during the pandemic in Cusco, Peru: a cross-sectional survey. PLoS One. 2021;16:e0257165. https://doi.org/10.1371/JOURNAL.PONE.0257165.
107. Süntar I. Importance of ethnopharmacological studies in drug discovery: role of medicinal plants. Phytochem Rev. 2019;195(19):1199–209. https://doi.org/10.1007/S11101-019-09629-9.

108. Van Norman GA. Update to drugs, devices, and the FDA: how recent legislative changes have impacted approval of new therapies. JACC Basic Transl Sci. 2020;5:831–9. https://doi.org/10.1016/J.JACBTS.2020.06.010.
109. Devi VK, Jain N, Valli KS. Importance of novel drug delivery systems in herbal medicines. Pharmacogn Rev. 2010;4:27. https://doi.org/10.4103/0973-7847.65322.
110. Semalty A, Semalty M, Rawat BS, Singh D, Rawat MSM. Pharmacosomes: the lipid-based new drug delivery system. Expert Opin Drug Deliv. 2009;6:599–612. https://doi.org/10.1517/17425240902967607.
111. Mittal N, Garg V, Bhadada SK, Katare OP. Role of novel drug delivery systems in coronavirus disease-2019 (COVID-19): time to act now. Curr Drug Deliv. 2021;18:289–96. https://doi.org/10.2174/1567201817666200916090710.
112. Hudson JB. Applications of the phytomedicine Echinacea purpurea (purple coneflower) in infectious diseases. J Biomed Biotechnol. 2012;2012:769896.
113. Manayi A, Vazirian M, Saeidnia S. Echinacea purpurea: pharmacology, phytochemistry and analysis methods. Pharmacogn Rev. 2015;9:63–72.

Index

R. Shegokar, Y. Pathak (eds.), *Viral Drug Delivery Systems*, https://doi.org/10.1007/978-3-031-20537-8

SPRINGER NATURE

GPSR Compliance

The European Union's (EU) General Product Safety Regulation (GPSR) is a set of rules that requires consumer products to be safe and our obligations to ensure this.

If you have any concerns about our products, you can contact us on ProductSafety@springernature.com

In case Publisher is established outside the EU, the EU authorized representative is:

Springer Nature Customer Service Center GmbH
Europaplatz 3
69115 Heidelberg, Germany

The manufacturer's authorised representative in the EU is Springer Nature Customer Service Centre GmbH, Europaplatz 3, 69115 Heidelberg, Germany. If you have any concerns regarding our products, please contact ProductSafety@springernature.com

Printed and bound by CPI Group (UK) Ltd, Croydon, CR0 4YY
01/10/2026
02225893-0012